POLLUTION CONTROL TECHNOLOGY FOR INDUSTRIAL WASTEWATER

POLLUTION CONTROL TECHNOLOGY FOR INDUSTRIAL WASTEWATER

Edited by D.J. De Renzo

NOYES DATA CORPORATION

Park Ridge, New Jersey, U.S.A.

1981

Sole distribution by:
Gothard House Publications
Gothard House
Henley-on-Thames, Oxon
RG9 1AJ Tel. 049 12 3602

Library of Congress Catalog Card Number: 81-38394
ISBN: 0-8155-0855-7
Printed in the United States

Published in the United States of America by
Noyes Data Corporation
Noyes Building, Park Ridge, New Jersey 07656

Library of Congress Cataloging in Publication Data
Main entry under title:

Pollution control technology for industrial
wastewater.

(Pollution technology review ; no. 80)
Bibliography: p.
Includes index.
1. Factory and trade waste. 2. Sewage--Purification. I. De Renzo, D. J. II. Series.
TD897.P62 628.1'683 81-38394
ISBN 0-8155-0855-7 AACR2

Foreword

This book provides an extensive survey of the reliability and effectiveness of 56 unit operations in industrial water pollution control. These operations include 32 generic wastewater treatment technologies, classified as preliminary, primary, secondary or tertiary, and 24 sludge treatment and disposal technologies.

Each process is briefly described by the type of control equipment required, the major variations of design, flow diagrams, and information on the following: design criteria, common modifications, typical performance, applications and limitations of the process, reliability, chemicals required for operation, residuals generated, and environmental impacts.

A summary table for most technologies is provided showing the concentrations of various pollutants in the effluents, the minimum, maximum, median, and mean removal efficiencies, and the number of data points used to generate this information. Conventional pollutants as well as EPA-categorized "priority pollutants" and "hazardous substances" are covered. Data sheets summarizing the results of tests at specific installations are also included.

The data in this book is from *Treatability Manual, Volume III, Technologies for Control/Removal of Pollutants* (EPA 600/8-80-042c), issued by the Office of Research and Development of the U.S. Environmental Protection Agency, July 1980.

The information reviewed here should be extremely useful to engineers, management personnel, and others involved with regulatory requirements, guidelines, and decisions.

The table of contents will serve as a subject index and provide easy access to the information contained in the book. A complete set of references has been included.

In order to keep the price of this large and extensive book to a reasonable level, it has been reproduced by photo-offset directly from the original report and the cost savings passed on to the reader. Due to this method of publishing, certain portions of the report may be less legible than desired.

Acknowledgements

The sheer size and comprehensiveness of this document should make it obvious that this had to be the effort of a large number of people. It is the collection of contributions from throughout the Environmental Protection Agency, particularly from the Office of Enforcement, Office of Water and Hazardous Materials and the Office of Research and Development. Equally important to its success were the efforts of the employees of the Aerospace Corporation, Mathtech, Inc., and the Monsanto Research Corporation who participated in this operation.

No list of the names of everyone who took part in the effort would in any way adequately acknowledge the effort which those involved in preparing this volume made toward its development. Equally difficult would be an attempt to name the people who have made the most significant contributions both because there have been too many and because it would be impossible to adequately define the term "significant." This document exists because of major contributions by the contractor's staff and by members of the following:

Effluent Guidelines Division
Office of Water and Waste Management

Permits Division
Office of Water Enforcement

National Enforcement Investigation Center
Office of Enforcement

Center for Environmental Research Information

Municipal Environmental Research Laboratory

Robert S. Kerr Environmental Research Laboratory

Industrial Environmental Research Laboratory
Research Triangle Park, NC

Industrial Environmental Research Laboratory
Cincinnati, OH
Office of Research and Development

Table of Contents

1. Introduction

This volume presents performance data and related technical information for 56 unit operations used in industrial water pollution control. These 56 unit operations include 24 sludge treatment and disposal technologies and 32 generic wastewater treatment technologies classified as preliminary, primary, secondary, or tertiary treatment. Section 2 discusses the rationale used to segregate the 32 wastewater treatment technologies into four classifications.

In Sections 3 through 8, each wastewater or sludge treatment/disposal technology is briefly described and generalized performance characteristics are given for the preliminary wastewater treatment (conditioning) and sludge processing technologies. However, emphasis is placed on the pollutant removal capabilities of the 28 primary, secondary, and tertiary wastewater treatment technologies. Both concentration and removal efficiency data are given for the following group of pollutants:

(1) Conventional pollutants[a] such as biochemical oxygen demand (BOD_5), chemical oxygen demand (COD), total organic carbon (TOC), total suspended solids (TSS), oil and grease, total phenol, total phosphorus, total Kjeldahl nitrogen (TKN), and total organic chlorine (TOCl),

(2) 129 toxic pollutants derived by EPA from the 65 "priority pollutants" listed in a Consent Agreement, Natural Resources Defense Council vs Train, 8 ERC 2120 (D.D.C. 1976),

(3) Compounds selected from the list of substances designated by EPA as hazardous under authority of Section 311 of the CWA, based on the availability of either a consensus analytical methods or one promulgated under authority of Section 204(h) of the CWA, and

(4) Other nonconventional pollutants of concern in specific industrial wastewaters.

---

[a]Section 301 of the Clean Water Act defines conventional pollutants to be oil and grease, BOD, TSS, fecal coliform, and pH. The Treatability Manual lists other pollutants under the general heading "conventional pollutants" while recognizing that they are not so defined by the Clean Water Act.

The technology descriptions presented in Sections 3 through 8 discuss the primary functions and basic operating principles of each treatment process. They also discuss major design criteria, common modifications and applications, reliability and inherent technical limitations, technological status and extent of industry utilization, chemical requirements, and environmental impacts of each treatment process. However, the technology descriptions do not provide detailed information on process design or operation. They are intended for overview purposes only. Similarly, the performance characteristics given for the preliminary wastewater treatment and sludge treatment/disposal technologies are intended only as general guidelines.

Pollutant removal data for the primary, secondary, and tertiary treatment technologies are presented in two forms: plant specific data sheets and statistical summary tables. Each plant-specific data sheet lists the concentrations of various pollutants in the influent and effluent to the treatment operation and the corresponding removal efficiencies for these pollutants. When available, the following types of information are also provided.

- Point source category, subcategory and identification code of the plant discharging the waste
- Scale of the treatment operation (e.g., full scale, pilot scale, bench scale)
- Location of the treatment operation in the overall waste treatment system for the plant (e.g., primary, secondary, tertiary treatment)
- Design and operating parameters
- Reference from which the information was taken

References for the plant-specific data include Effluent Guidelines development documents and contractor reports, other EPA reports, journal articles, and conference papers. The data are reported as they appear in the original references, except that certain concentration and removal efficiency values are rounded to fewer significant figures. Conventional pollutant concentrations are reported to a maximum of three significant figures, while removal efficiencies and concentration data for the other groups of pollutants are limited to two significant figures. This convention has been adopted for formating purposes only and does not necessarily reflect the accuracy and reproducibility of the data. The confidence limits associated with individual concentration values and removal efficiencies are unknown unless otherwise noted on the data sheets.

In many cases, the concentrations of toxic organic pollutants in treatment system effluents are reported as "not detected" or "below detectable limits" in the original references and no detection limits are specified. These concentrations are also reported as "not detected" or "below detectable limits" on the plant specific data sheets.

For removal efficiency calculations, however, "nondetectable" organic pollutant concentrations are assumed to be either (a) <10 μg/L if the influent concentration exceeds 10 μg/L, or (b) less than the corresponding influent concentration if a finite influent concentration <10 μg/L is reported. These assumptions reflect EPA's experience with a draft analytical screening protocol (Sampling and Analysis Procedures for Screening of Industrial Effluents for Priority Pollutants, U.S. EPA, Environmental Monitoring and Support Laboratory, Cincinnati, Ohio 45268, March 1977, Revised April 1977) over the last 18 months.

In other cases, treatment system effluents have been reported to contain higher concentrations of certain pollutants than the untreated wastewaters. However, "negative removals" are not reported on the plant-specific data sheets. Where the effluent concentration for a given pollutant exceeds the corresponding influent concentration, the removal efficiency is reported as zero and treated as such in the data summarization.

The statistical summary table for each primary, secondary, and tertiary wastewater treatment technology incorporates all effluent concentration and removal efficiency data contained in the plant-specific data base for that technology. Minimum, maximum, median, and mean effluent concentrations and removal efficiencies are given for each pollutant listed on one or more of the data sheets. These statistics are intended only as general performance indicators for the treatment technologies since they do not account for differences in system design and operation, influent pollutant loadings, or the types of industrial wastewaters being treated. Median/mean effluent concentrations and removal efficiencies reported for a given treatment technology are not necessarily indicative of the technology's pollutant removal capabilities when applied to a specific industrial wastewater.

2. Technology Overview

The 56 wastewater and sludge treatment/disposal technologies addressed in this volume are divided into six groups, based on their primary functions. These are (1) wastewater conditioning, or preliminary treatment, (2) primary wastewater treatment, (3) secondary wastewater treatment, (4) tertiary wastewater treatment, (5) sludge treatment, and (6) sludge disposal. Figure 1 identifies the technologies included in each of these groups.

The four wastewater conditioning technologies are designed to prepare wastewater streams for further treatment. Screening and grit removal separate coarse materials from the waste stream to prevent damage to downstream pumps, sedimentation tank sludge collectors, and other process equipment. Equalization damps out fluctuations in hydraulic flow and pollutant loading from the plant production process, and neutralization renders acidic or basic waste streams suitable for pH sensitive treatment processes (e.g., biological treatment). Neutralization may also be used as the final step in a treatment process to meet pH standards. None of these wastewater conditioning technologies are designed to remove specific pollutants from wastewater, however.

The remaining 28 wastewater treatment technologies are arbitrarily classified as primary, secondary, or tertiary treatment based on the types of pollutants they are designed to remove. This classification procedure is adapted only for organizational purposes in this volume; it is not meant to imply that technologies classed as primary, secondary, or tertiary are always used in these treatment applications. The seven generic technologies classified as primary treatment are designed to remove suspended or colloidal materials from wastewater. Gravity oil separation, sedimentation, and gas flotation (e.g., dissolved air flotation) remove free oil and grease and suspended solids, as well as specific compounds locked in these matrices. When chemical addition (coagulants or settling aids) is used in conjunction with sedimentation or gas flotation, dispersed oil and grease and colloidal solids can also be removed. Ultrafiltration performs a similar function. Filtration is primarily used for effluent polishing, in terms of suspended solids, or as a pretreatment step for other processes that are adversely affected by suspended solids. Although these technologies are classified as primary treatment, they are not always used as the initial treatment step. For example, filtration is frequently used as a tertiary operation following secondary clarification. Ultrafiltration and sedimentation or gas flotation with chemical addition are often used as "secondary" treatment processes, following gravity oil separation for free oil removal. In some cases, these processes may also be applied for tertiary treatment. Lime treatment of secondary effluents for phosphorus removal is an example of this type of application.

The technologies classified as secondary treatment include two physical/chemical processes and four generic biological processes. For performance data summary purposes, lagooning is subdivided according to the types of biological activity involved and other basic operating principles (e.g., mechanical vs. natural aeration).[a] These technologies are classified as secondary treatment because their primary function is to remove dissolved organic materials from wastewater. These processes are normally preceded by primary treatment for suspended solids removal, particularly steam stripping and solvent extraction where contactor fouling can be a major problem.

Steam stripping and solvent extraction are frequently used in the chemical industry, but usually in the production process itself rather than for wastewater treatment. These processes are most applicable for treatment of concentrated waste streams containing organic materials that are refractory to biological oxidation. Steam stripping may also be used as a pretreatment step for activated sludge or other biological treatment processes to remove volatile organics that evaporate before biological oxidation occurs.

Activated sludge processes, trickling filters, and lagoons are by far the most common treatment processes for dissolved organic materials, primarily because they are less expensive and easier to operate than physical/chemical treatment alternatives. Rotating bilogical contactors, relatively new innovations in the wastewater treatment field, are also being used in some applications.

The 15 technologies classifed as tertiary wastewater treatment processes are primarily designed to remove dissolved organics or inorganics that are refractory to primary and secondary treatment. Processes such as activated carbon adsorption, chemical oxidation, and ozonation may be used in secondary treatment applications, but they tend to be more expensive than biological treatment. However, the use of powdered activated carbon in conjunction with the activated sludge process is gaining favor

---

[a]Sedimentation with chemical addition and gas flotation with chemical addition are also subdivided for data summarization according to the type(s) of coagulants or settling aids used.

as a method to improve refractory organic removal and secondary settling.

For wastewaters containing little or no suspended or biodegradable organic material, tertiary technologies may be used to remove selected materials from the raw waste stream without recourse to standard primary or secondary treatment processes. Examples of this include chromate removal from cooling tower blowdown via ion exchange or reverse osmosis. In most wastewater treatment applications, however, primary and secondary treatment processes are used upstream from the tertiary technologies listed in Figure 1. Most of these tertiary technologies are rendered uneffective or more expensive to operate by high suspended solids or organic loadings.

The 15 sludge treatment technologies include various thickening, digestion, dewatering, disinfection, and other conditioning alternatives. Many of these processes are used consecutively in wastewater treatment plants; thickening, digestion, and dewatering for example. In general, they are designed to render sludge suitable for a particular disposal alternative and/or to facilitate handling and transportation.

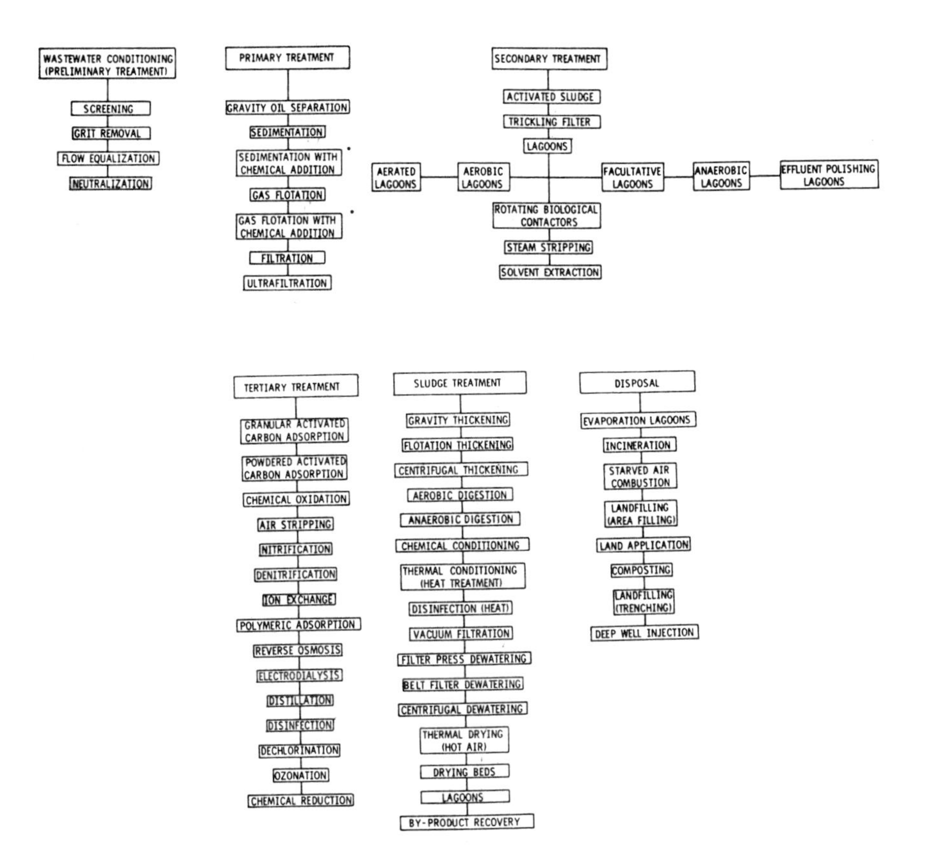

Figure 1. Treatment technology overview.

3. Wastewater Conditioning (Preliminary Treatment)

3.1 SCREENING [1]

3.1.1 Function

Screening is used to remove coarse and/or gross solids from untreated wastewater before subsequent treatment.

3.1.2 Description

There are two major types of screening processes. These are termed wedge wire screening and rotating horizontal shaft screening.

Wedge Wire Screen. A wedge wire screen is a device onto which wastewater is directed across an inclined stationary screen or a drum screen of uniform sized openings. Solids are trapped on the screen surface while the wastwater flows through the openings. The solids are moved either by gravity (stationary) or by mechanical means (rotating drum) to a collecting area for discharge. Stationary screens introduce the wastewater as a thin film flowing downward with a minimum of turbulence across the wedge wire screens, which is generally in three sections of progressively flatter slope. The drum screen employs the same type of wedge wire wound around its periphery. Wastewater is introduced as a thin film near the top of the drum and flows through the hollow drum and out the bottom. The solids retained by the peripheral screen follow the drum rotation until removed by a doctor blade located at about 120° from the introduction point. Wedge wire spacing can be varied to best suit the application. For municipal wastewater applications spacings are generally between 0.01 and 0.06 inches (0.25 to 1.5 mm). Inclined screens can be housed in stainless steel or fiberglass; wedge wires may be curved or straight; the screen face may be a single multi-angle unit, three separate multi-angle pieces, or a single curved unit. Rotary screens can have a single rotation speed drive or a variable speed drive.

Rotating Horizontal Shaft Screen. A rotating horizontal shaft screen is an intermittently or continously rotating drum covered with a plastic or stainless steel screen of uniform sized openings, installed and partially submerged in a chamber. The chamber is designed to permit the entry of wastewater to the interior of the drum and collection of filtered (or screened) wastewater from the exterior side of the drum. With each revolution, the solids are flushed by sprays from the exposed screen surface into a collecting trough. Coarse screens have openings of less than 1/4 inch. Screen with openings of 20 to 70 microns are called microscreens or microstrainers. Drum diameters are 3 to 5 feet with 4- to 12-feet lengths.

3.1.3 Technology Status

Wedge Wire Screen. Wedge wire screens have been used in industry since 1965 and in municipal wastewater treatment since 1967. There are over 100 installations to date.

Rotating Horizontal Shaft Screen. Rotating horizontal shaft screens are in widespread use for roughing pretreatment and for secondary biological plant effluent polishing.

3.1.4 Applications

Wedge Wire Screen. Stationary and rotary drum screens are ideally suited and usually employed after bar screens and prior to grit chambers. They have also been employed for primary treatment, scum dewatering, sludge screening, digester cleaning, and storm water overflow treatment.

Rotating Horizontal Shaft Screen. Used for removal of coarse wastewater solids from the wastewater treatment plant influent after bar screen treatment with screen openings 150 microns to 0.4 inches; also used for polishing activated sludge effluent with screen openings 20 to 70 microns.

3.1.5 Limitations

Wedge Wire Screen. Require regular cleaning and prompt residuals disposal.

Rotating Horizontal Shaft Screen. Dependence on pretreatment and inability to handle solids fluctuations in tertiary applications; reducing speed of rotation of drum and less frequent flushing of screen has resulted in increased removal efficiencies, but reduced capacities.

3.1.6 Residuals Generated

Wedge Wire Screen. Solids trapped on the screen surface (1 to 2 yd^3/Mgal for domestic waste).

Rotating Horizontal Shaft Screen. Sidestream of solids accumulations backwashed from screen.

3.1.7 Reliability

Wedge Wire Screen. Very high reliability for process and mechanical areas when maintained.

Rotating Horizontal Shaft Screen. High degree of reliability for both the process and mechanical areas; process is simple to operate; mechanical equipment is generally simple and straightforward; occasional problems may arise because of incomplete solids removal by flushing (hand cleaning may be required with acid solution for stainless steel cloths); blinding by grease can be a problem in pretreatment applications.

3.1.8 Environmental Impact

Wedge Wire Screen. Can create odors if screenings are not disposed of properly; impact on land is practically nil; screenings are generally disposed of in a landfill or by incineration, no impact on water.

Rotating Horizontal Shaft Screen. Odor problems around equipment may be created if solids are not flushed frequently enough from the screen (pretreatment); disposal of solid by incineration can affect air quality; disposal of solids in landfill has neglible impact; no impact on water.

3.1.9 Design Criteria

Wedge Wire Screen. In screening of raw wastewater (0.05 to 36 Mgal/d):

Stationary	Parameter	Rotary Drum
0.01 to 0.06 in.	Screen opening	0.01 to 0.06 in.
4 to 7 ft	Head required	2.5 to 4.5 ft
10 to 750 ft	Space required	10 to 100 ft^2
-	Motor size	0.5 to 3 hp

Rotating Horizontal Shaft Screen.

Screen submergence: 70 to 80 percent.
Loading rate: 2 to 10 gal/min/ft^2 of submerged area, depending on pretreatment and mesh size.
Screen openings: 150 microns to 0.4 inches for pretreatment; 20 to 70 microns for tertiary treatment.
Drum rotation: 0 to 7 revolutions/min
Screen materials: Stainless steel or plastic cloth
Washwater = 2 to 5 percent of flow being treated.
Performance of fine screen device varies considerably on influent solids type, concentration and loading patterns; mesh size; hydraulic head; and degree of biological conditioning of solids.

3.1.10 Flow Diagrams

Wedge Wire Screen.

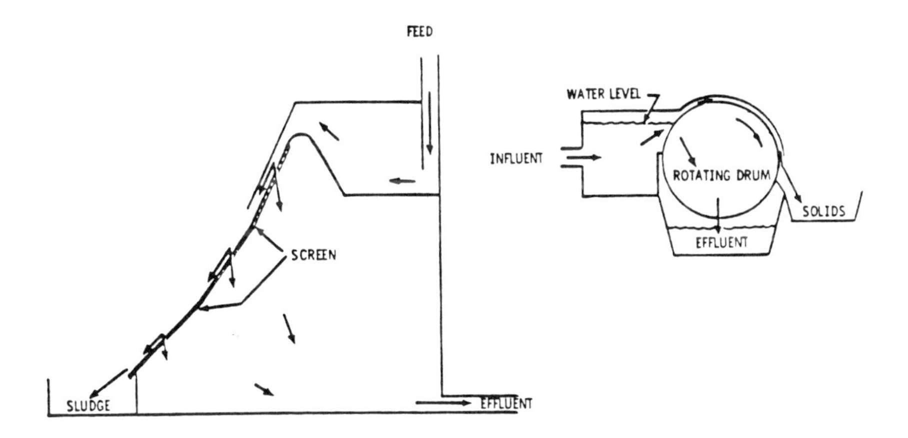

Rotating Horizontal Shaft Screen.

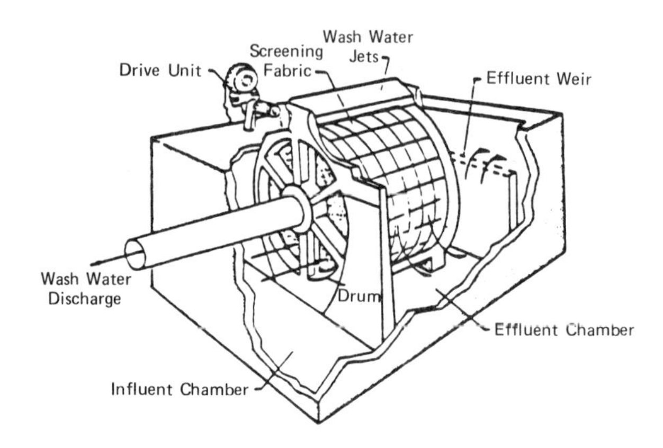

3.1.11 Performance

The following data is characteristic of typical waste:

Wedge Wire Screen.

Pollutant	Typical percent removal
BOD	5 to 20
SS	5 to 25

Rotating Horizontal Shaft Screen. (Tertiary applications)

Pollutant	Typical percent removals
BOD_5	40 to 60
SS	50 t0 70

3.1.12 References

1. Innovative and Alternative Technology Assessment Manual. EPA-430/9-78-009 (draft), U.S. Environmental Protection Agency, Cincinnati, Ohio, 1978. 252 pp.

3.2 GRIT REMOVAL (PRELIMINARY TREATMENT) [1]

3.2.1 Function

The purpose of preliminary treatment is to remove large objects, such as rocks, logs, and cans, as well as grit, in order to prevent damage to subsequent treatment and process equipment. Objects normally removed by preliminary treatment steps can be extremely harmful to pumps, and can increase downtime due to pipe clogging and clarifier scraper mechanism failures.

3.2.2 Description

Preliminary treatment usually consists of two separate and distinct unit operations: bar screening and grit removal. There are two types of bar screens (or racks). The most commonly used, and oldest technology, consists of hand-cleaned bar racks, which are generally used in smaller treatment plants. The second type of bar screen is the type that is mechanically cleaned, which is commonly used in larger facilities.

Grit is most commonly removed in chambers that are capable of settling out high density solid materials, such as sand, gravel, and cinders. There are two types of grit chambers: horizontal flow, and aerated; in both types the settleables collect at the bottom of the unit. Horizontal units are designed to maintain a relatively constant velocity by use of proportional weirs or flumes in order to prevent settling of organic solids, while simultaneously obtaining relatively complete removal of inorganic particles (grit). Aerated grit chambers produce spiral action whereby the heavier particles remain at the bottom of the tank to be removed, while organic particles are maintained in suspension by rising air bubbles. One main advantage of aerated units is that the amount of air can be regulated to control the grit/organic solids separation, and less offensive odors are generated. The aeration process also facilitates cleaning of the grit. Grit removed from horizontal flow units usually needs additional cleaning steps prior to disposal.

3.2.3 Common Modifications

Many plants also use comminutors, which are mechanical devices that cut up the material normally removed in the screening process. Therefore, these solids remain in the wastewater to be removed in downstream unit operations, rather than being removed immediately from the wastewater. In recent years, the use of static or rotating wedge-wire screens has increased to remove large organic particulates just prior to degritting. These units have been found to be superior to comminutors in that they remove the material immediately from the waste instead of creating additional loads downstream. Other grit chamber designs are available including swirl concentrators and square tanks.

3.2.4 Technology Status

Preliminary treatment has been widely used since the early days of wastewater treatment. Wedge-wire screens are newer technology (approximately 13 years old).

3.2.5 Applications

Bar screens are also normally used prior to wastewater pumping stations.

3.2.6 Limitations

High fiber and grease content in waste will require frequent changing of screens. Corrosive wastewaters require special materials of construction.

3.2.7 Performance

Bar screens are designed to remove all large debris, such as stones, wood, cans, etc.; grit chambers are designed to remove virtually all inorganic particles, such as sand and gravel; wedge-wire screens remove up to 25 percent suspended solids and associated BOD_5 and possibly reduce digester scum.

3.2.8 Chemicals Required

None.

3.2.9 Residuals Generated

All unit operations, except comminutors, will generate solids that need disposal; screenings removed by fine wedge-wire screens have amounted to approximately 1 to 2 yd^3/Mgal of wastewater treated; grit and other solids are often landfilled.

3.2.10 Design Criteria

In bar screens, bar size is 1/4 to 5/8 in. width by 1 to 3 in. depth; spacing is 0.75 to 3 in.; slope from vertical is 0 to 45°; velocity is 1.5 to 3 ft/s; criteria for wedge-wire screens is shown in Section III.3.1; in grit chambers, horizontal velocities are 0.5 to 1.25 ft/s; units are sufficiently long to settle lightest and smallest (usually 0.2 mm) grit particles with an additional factor of safety (up to 50 percent); weir crests are generally set 4 to 12 in. above the bottom.

3.2.11 Reliability

Preliminary treatment systems are extremely reliable and, in fact, are designed to improve the reliability of downstream treatment systems.

3.2.12 Environmental Impact

Requires relatively little use of land; requires minimal amounts of energy; solids will be generated, requiring disposal; odors are common when removed grit contains excess organic solids and is not disposed of within a short time after removal.

3.2.13 Flow Diagram

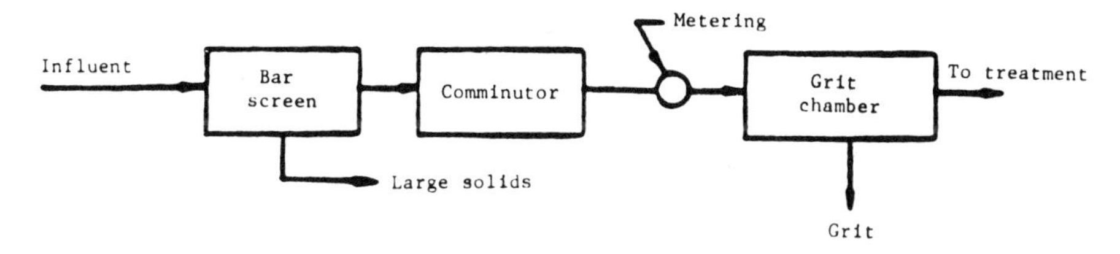

3.2.14 References

1. Innovative and Alternative Technology Assessment Manual. EPA-430/9-78-009 (draft), U.S. Environmental Protection Agency, Cincinnati, Ohio, 1978. 252 pp.

3.3 FLOW EQUALIZATION [1]

3.3.1 Function

Flow equalization is used to balance the quantity and quality of wastewater before subsequent downstream treatment.

3.3.2 Description

Wastewater flows into treatment facilities are subjected to diurnal and seasonal fluctuation in quality and in quantity. Most waste treatment processes are sensitive to such changes. An equalization basin serves to balance the extreme quality and quantity of these fluctuations to allow normal contact time in the treatment facility. This section of the manual addresses only equalization basins that are used to equalize flow; however, it should be noted that the quality of wastewater will also equalize to a degree.

Equalization basins may be designed as either in-line or side-line units. In the in-line design, the basin receives the wastewater directly from the collection system, and the discharge from the basin through the treatment plant is kept essentially at constant rate. In the side-line design, flows in excess of the average are diverted to the equalization basin and, when the plant flow falls below the average, wastewater from the basin is discharged to the plant to bring the flow to the average level. The basins are sufficiently sized to hold the peak flows and discharge at constant rate.

Pump stations may or may not be required to discharge into or out of the equalization basin, depending upon the available head. Where pumping is found necessary, the energy requirements will be based on total flow for in-line basins and on excess flow for side-line basins.

Aeration of the wastewater in the equalization basin is normally required for mixing and maintaining aerobic conditions.

3.3.3 Common Modifications

There are various methods of aeration, pumping and flow control. Tanks or basins can be manufactured from steel or concrete, or excavated and of the lined or unlined earthen variety.

3.3.4 Technology Status

Flow equalization has been used in the municipal and industrial sectors for many years.

3.3.5. Applications

Can be used to equalize the extremes of diurnal and wet weather flow fluctuations; secondary benefits are equalization of quality and the potential for protection from toxic upsets.

3.3.6. Limitations

Application to equalize diurnal fluctuation is rather limited because the cost is unjustifiable when compared to the benefits; may require substantial land area.

3.3.7 Residuals Generated

Due to the settling characteristics of influent wastewater solids, some materials will collect at bottom of basin, and will need to be periodically discarded; provisions must be made to accommodate this need.

3.3.8 Reliability

Units are reliable from both a unit and process standpoint and used to increase the reliability of the flow-sensitive treatment processes that follow.

3.3.9 Environmental Impact

Can consume large land areas; impact upon air quality and noise levels are minimal; some sludge may be generated that will require disposal. Aeration is also required to prevent odor problems.

3.3.10 Design Criteria

Design of an equalization basin is highly site-specific and dependent upon the type and magnitude of the input flow variations and facility configuration; pumping/flow control mode, aeration, mixing and flushing methods are dependent upon the size and site conditions; grit removal should be provided upstream of the basin; mechanical mixing at 20 to 40 hp/Mgal of storage; aeration is a function of wastewater strength.

3.3.11 Flow Diagram

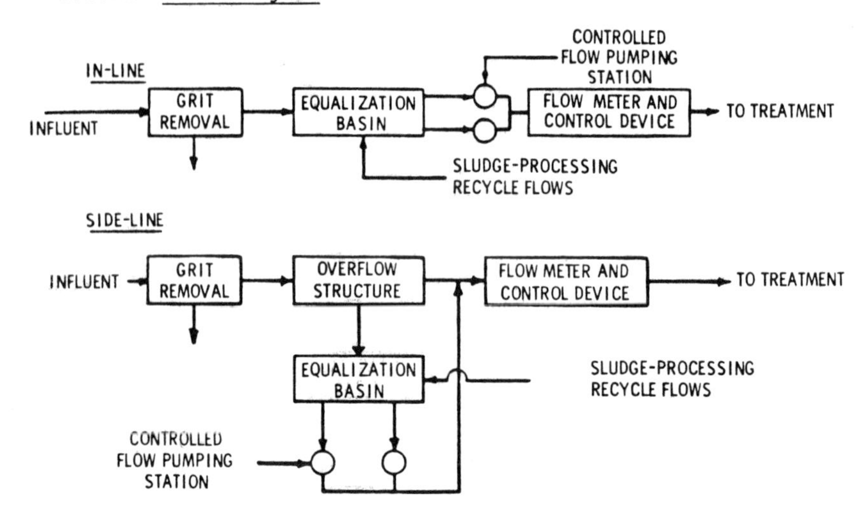

3.3.12 Performance

Flow equalization basins are easily designed to achieve the objective; use of aeration, in combination with the relatively long detention times afforded can produce BOD_5 reductions of 10 to 20 percent.

3.3.13 References

1. Innovative and Alternative Technology Assessment Manual. EPA-430/9-78-009 (draft), U.S. Environmental Protection Agency, Cincinnati, Ohio, 1978. 252 pp.

3.4 NEUTRALIZATION [1]

3.4.1 Function

Neutralization is the process of adjusting either an acidic or a basic wastestream to a pH in the range of seven.

3.4.2 Description

Many manufacturing and processing operations produce effluents that are acidic or alkaline in nature. Neutralization of an excessively acidic or basic waste stream is necessary in a variety of situations, for example: (1) to prevent metal corrosion and/or damage to other construction materials; (2) to protect aquatic life and human welfare; (3) as a preliminary treatment, allowing effective operation of biological treatment processes, and (4) to provide neutral pH water for recycle, either as process water or as boiler feed. Treatment to adjust pH may also be desirable to break emulsions, to insolubilize certain chemical species, or to control chemical reaction rates, e.g., chlorination. Although natural waters may differ widely in pH, changes in a particular pH level could produce detrimental effects on the environment. To minimize any undesirable consequences, the effluent limitations guidelines for industrial sources set the pH limits for most industries between 6.0 and 9.0 for 1977 and 1983.

Simply, the process of neutralization is the interaction of an acid with a base. The typical properties exhibited by acids in solution are due to the hydrogen ion, (H^+). Similarly, alkaline (or basic) properties are a result of the hydroxyl ion, (OH^-). In aqueous solutions, acidity and alkalinity are defined with respect to pH, where pH = log (H^+) and, at room temperature, pH = 14 + log (OH^-). In the strict sense, neutralization is the adjustment of pH to 7, at which level the concentrations of hydrogen and hydroxyl ions are equal. Solutions with excess hydroxyl ion concentration (pH >7) are said to be basic; solutions with excess hydrogen ions (pH <7) are acidic. Since adjustment of the pH to 7 is not often practical or even desirable in waste treatment, the term "neutralization" is sometimes used to describe adjustment of pH to values other than 7.

The actual process of neutralization is accomplished by the addition of an alkaline to an acidic material or by adding an acidic to an alkaline material, as determined by the required final pH. The primary products of the reaction are a salt and water. A simple example of acid-base neutralization is the reaction between hydrochloric acid and sodium hydroxide:

$$HCl + NaOH \rightarrow H_2O + NaCl$$

The product, sodium chloride in aqueous solution, is neutral with pH = 7.0.

3.4.3 Technology Status

Neutralization is considered to be demonstrated technology and is widely used in industrial waste treatment.

3.4.4 Applications

Finds widest application in the treatment of aqueous wastes containing strong acids such as sulfuric and hydrochloric, or bases such as caustic soda and ammonium hydroxide; however, process can be used with nonaqueous materials (for example, acidic phenols, which are insoluble in water); although neutralization is a liquid phase phenomenon, it can also treat both gaseous and solid waste streams; gases can be handled by absorption in a suitable liquid phase, as in the alkali scrubbing of acid vapors; slurries can be neutralized, with due consideration for the nature of the suspended solid and its dissolution properties; sludges are also amenable to pH adjustment, but the viscosity of the material complicates the process of physical mixing and contact between acid and alkali that is essential to the treatment; in principle, even tars can be neutralized, although the problems of reagent mixing and contact are usually severe, making the process impractical in most instances; solids and powders that are acidic or basic salts could also be neutralized if dissolved; can be used to treat both inorganic and organic waste streams that are either excessively acidic or alkaline; often used to precipitate heavy metal ions, e.g., Zn^{++}, Pb^{++}, or Cu^{++} by the addition of an alkali (usually lime) to a waste stream; organic compounds that can be treated include carboxylic acids, sulfonic acids, phenols, and many other materials.

INDUSTRIES USING NEUTRALIZATION

Industry	Wastewater pH range
Pulp and paper	Acidic and basic
Dairy products	Acidic and basic
Textiles	Basic
Pharmaceuticals	Acidic and basic
Leather tanning and finishing	Acidic and basic
Petroleum refining	Acidic and basic
Grain milling	Acidic and basic
Fruits and vegetables	Acidic and basic
Beverages	Acidic and basic
Plastic and synthetic materials	Acidic and basic
Steel pickling	Acidic
Byproduct coke	Basic
Metal finishing	Acidic
Organic chemicals	Acidic and basic
Inorganic chemicals	Acidic and basic
Fertilizer	Acidic and basic
Industrial gas products	Acidic and basic
Cement, lime, and concrete products	Basic
Electric and steam generation	Acidic and basic
Nonferrous metals - aluminum	Acidic

3.4.5 Limitations

Subject to influence of temperature and resulting heat effects common to most chemical reactions; generally, in water-based reactions, increasing the temperature of reactants increases the rate of reaction; in neutralization, the interaction of acid and alkali is frequently exothermic (evolves heat), with an accompanying rise in temperature; an average value for the heat released during the neutralization of dilute solutions of strong acids and bases is 13,360 cal/g mole of water formed; by controlling the rate of addition of neutralizing reagent, the heat produced may be dissipated and temperature increase minimized. For each reaction, the final temperature depends on initial reactant temperatures, chemical species participating in the reaction (and their heats of solution and reaction), concentrations of the reactants, and relative quantities of the reactants; in general, concentrated solutions can produce large temperature increases as relative quantities of reactants approach stoichiometric proportions; this can result in boiling and splashing of the solution, and accelerated chemical attack on materials; in most cases, proper planning of the neutralization scheme with respect to concentration of neutralizing agent, rate of addition, reaction time, and equipment design can alleviate the heating problem.

3.4.6 Residuals Generated/Environmental Impact

After neutralization a waste stream will usually show an increased total dissolved solids content due to addition of chemical agent, but there may also be an accompanying reduction in the concentration of heavy metals if the treatment proceeds to alkaline pH's; conversely, in neutralizations involving the addition of acid to alkali, there is the possibility of solids dissolution, which may, on occasion, be disadvantageous, particularly if the suspended matter is slated for removal, e.g., by filtration; anions resulting from neutralization of sulfuric and hydrochloric acids are sulfate and chloride, respectively, which are not considered hazardous, but recommended limits exist for discharge, based primarily on problems in drinking water; common cations present after neutralization involving caustic soda and lime (or limestone) are sodium and calcium (possibly magnesium), respectively, which are not toxic and have no recommended limits; however, calcium and magnesium are responsible for water hardness and accompanying scaling problem; carbonate produced during limestone neutralization is also harmless both in solution and as carbon dioxide gas.

With regard to atmospheric emissions, one must be cautious not to indiscriminately neutralize wastewater streams; acidification of streams containing certain salts, such as sulfide, will produce toxic gases; if there is no satisfactory alternative, the gas must be removed through scrubbing or some other treatment; where solid products are formed (as in precipitation of calcium sulfate, or heavy metal hydroxides), clarifier/thickeners and filters must be provided; if precipitate is of sufficient purity, it would be a salable product; otherwise, a disposal scheme must be devised.

3.4.7 Reliability

Process is highly reliable if properly monitored.

3.4.8 Flow Diagram

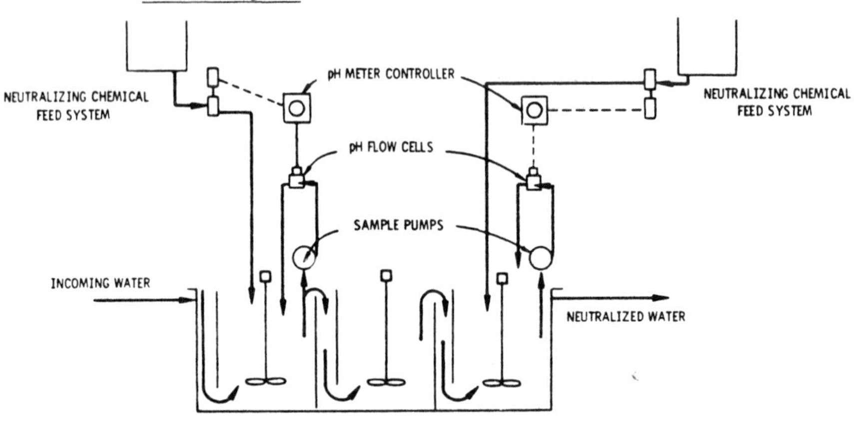

3.4.9 References

1. Physical, Chemical, and Biological Treatment Techniques for Industrial Wastes, PB 275 287, U.S. Environmental Protection Agency, Washington, D.C., November 1976. pp. 33-1 to 33-18.

4. Primary Wastewater Treatment

4.1 GRAVITY OIL SEPARATION [1]

4.1.1 Function

Gravity oil separation is used for the removal of floatable oil and grease.

4.1.2 Description

A gravity oil separator (skimming tank) is a chamber so arranged that floating matter rises and remains on the surface of the wastewater until removed, while the liquid flows out continuously through deep outlets or under partitions, curtain walls, or deep scum boards. This may be accomplished in a separate tank or combined with primary sedimentation, depending on the process and nature of the wastewater.

The objective of skimming tanks is the separation from the wastewater of the lighter floating substances. The material collected on the surface of skimming tanks, whence it can be removed, includes oil, grease, soap, pieces of cork and wood, and vegetable debris. The outlet, which is submerged, is opposite the inlet and at a lower elevation to assist in flotation and to remove any solids that may settle.

Gravity-type separators are the most common devices employed in oily waste treatment. The effectiveness of a gravity separator depends upon proper hydraulic design, and the design period of wastewater retention. Longer retention times allow better separation of the floatable oils from the water. Short detention times of less than 20 minutes result in less than 50% oil-water separation, while more extended holding periods improve oil separation from the waste stream.

Gravity separators are equally effective in removing both greases and nonemulsified oils. The standard unit in refinery waste treatment is the API separator, based upon design standards published by the American Petroleum Institute. Separators used for metal and food processing oily wastes operate upon the same principle of floating the oil, and many are designed in a similar fashion to the API process insofar as skimming, retention time, etc. Separators may be operated as batch vats, or as continuous flow-through basins, depending upon the volume of waste to be treated.

4.1.3 Technology Status

Gravity oil separation is well-developed for many industrial waste treatment applications, especially refinery wastes.

4.1.4 Applications

Used in refinery, steel rolling, metal processing, food processing, and most other industrial waste treatment where oil is present; in addition, recovery of skimmed oil or grease from all major types of oily waste is increasingly common, as the value of the recoverable oil is realized; frequently a substantial savings is possible through recovery or recycle of oily material.

4.1.5 Limitations

To meet increasingly more stringent regulations, gravity oil separation will usually require subsequent coagulant addition or other treatment in order to increase oil removal efficiencies to required levels.

4.1.6 Residuals Generated/Environmental Impact

If skimmings cannot be reused, they are typically disposed of by burial, lagooning, or incineration; odor and nuisance-free oil sludge incineration has been reported.

4.1.7 Reliability

Highly dependable, if regularly maintained. Variable wastewater characteristics such as flow, temperature, and pH can adversely affect performance.

4.1.8 Flow Diagram

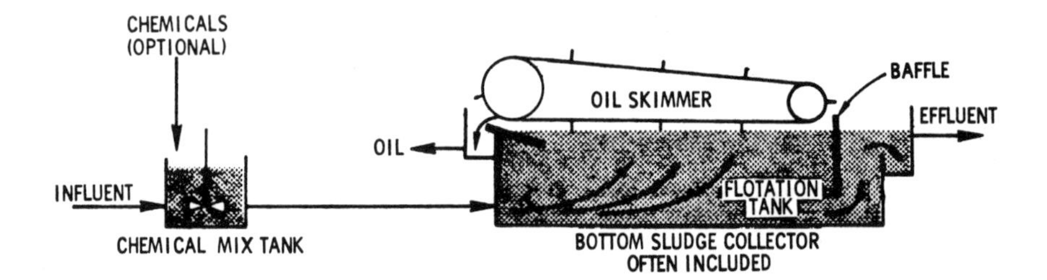

4.1.9 Performance

Subsequent data sheets provide performance data from studies on the following industries and/or wastestreams:

Iron and steel industry
  Cold rolling

Petroleum refining
  Cracking

Timber products processing
  Wood preserving/steaming

4.1.10 References

1. Patterson, J. W. Wastewater Treatment Technology. Ann Arbor Science Publishers, Inc. Ann Arbor, Mighigan, 1975. pp. 179-185.

CONTROL TECHNOLOGY SUMMARY FOR GRAVITY OIL SEPARATION

Pollutant	Number of data points	Effluent concentration			
		Minimum	Maximum	Median	Mean
Conventional pollutants, mg/L:					
BOD_5	17	24	1,650	190	376
COD	27	83	6,450	425	848
TOC	25	25	401	81	130
TSS	16	17	380	58	101
Oil and grease	23	11.9	1,350	66.1	157
Total phenol	29	0.063	189	5.9	25
Toxic pollutants, μg/L:					
Antimony	5	1	840	290	290
Arsenic	12	3	440	7	45
Beryllium	1				2
Cadmium	3	1	200	6	69
Chromium	24	1	25,000	430	1,700
Copper	15	6	450	44	110
Cyanide	15	10	1,300	40	170
Lead	13	4	920	36	150
Mercury	11	0.5	3	1.3	1.4
Nickel	11	4	500	26	69
Selenium	11	1	76	12	17
Silver	2	1	250		130
Thallium	4	1	3	2	2
Zinc	16	56	870	340	380
Bis(2-ethylhexyl) phthalate	9	9.5	700	300	310
Di-n-butyl phthalate	1				1.3
Diethyl phthalate	1				12
Acrylonitrile	1				30
2,4-Dimethylphenol	4	71	650	>100	230
Pentachlorophenol	2	120	850		490
Phenol	12	13	16,000	>130	2,000
Benzene	3	>100	>>100	>>100	>100
1,2-Dichlorobenzene	1				<10[a]
1,3-Dichlorobenzene	1				<10[a]
Ethylbenzene	1				>100
Toluene	4	23	>>100	>100	>81
Acenaphthene	4	37	3,000	300	910
Acenaphthylene	3	4	87	67	53
Anthracene/phenanthrene	8	4.6	~230	120	120
Benz(a)anthracene	1				55
Benzo(a)pyrene	2	5.5	15		10
Benzo(ghi)perylene	2	2	~1,100		550
Benzo(k)fluoranthene	2	37	270		150
Chrysene	2	1.7	20		11
Chrysene/benz(a)anthracene	4	30	~50	~35	~38
Fluoranthene	3	7.5	170	8.5	62
Fluoranthene/pyrene	2	30	220		130
Fluorene	3	34	300	80	140
Indeno(1,2,3-cd)pyrene	1				40
Naphthalene	10	50	1,100	~300	400
Pyrene	3	7	99	16	41
Aroclor 1016	3	0.2	1.9	1.8	1.3
Aroclor 1221	1				0.1
Aroclor 1232	3	0.5	0.9	0.5	0.63
Aroclor 1242	3	0.5	5.2	5.0	3.6
Carbon tetrachloride	1				43
Chloroform	5	<10	67	~15	~25
1,1-Dichloroethane	1				93
1,2-*Trans*-dichloroethylene	1				~20
Methylene chloride	5	~10	660	>>50	~170
Tetrachloroethylene	2	>50	71		>60
1,1,1-Trichloroethane	1				190
Trichloroethylene	1				>>100
Other pollutants, μg/L:					
NH_3	15	5.7	470	30	150

Note: Blanks indicate data not applicable.

[a]Not detected, assumed to be <10 μg/L.

TREATMENT TECHNOLOGY: Gravity Oil Separation

Data source: Effluent Guidelines		Data source status:	
Point source category: Iron and Steel		Engineering estimate	___
Subcategory: Cold rolling		Bench scale	___
Plant: 105 (also coded VV-2)		Pilot scale	___
References: A41, p. 261-262		Full scale	x

Usē in system: Primary
Pretreatment of influent:

DESIGN OR OPERATING PARAMETERS

Unit configuration:
Wastewater flow:
Hydraulic detention time:
Hydraulic loading:
Sludge overflow:

REMOVAL DATA

Sampling period:

Pollutant	Influent	Effluent	Percent Removal
Conventional pollutants, mg/L:			
TSS	290	295	0[a]
Oil and grease	1,860	1,350	27
Total phenol	53	54	0[a]
Toxic pollutants, μg/L:			
Antimony	16	290	0[a]
Arsenic	30	31	0[a]
Cadmium	140	200	0[a]
Chromium	170	240	0[a]
Copper	240	450	0[a]
Cyanide	15	13	13
Lead	420	600	0[a]
Nickel	350	500	0[a]
Selenium	26	76	0[a]
Silver	175	250	0[a]
Zinc	200	680	0[a]
1,2-Dichlorobenzene	11	ND[b]	>9
1,3-Dichlorobenzene	17	ND	>41
Carbon tetrachloride	33	43	0[a]
Chloroform	39	67	0[a]
1,1-Dichloroethane	14	93	0[a]
Tetrachloroethylene	82	71	13
1,1,1-Trichloroethane	140	190	0[a]
Other pollutants, μg/L:			
Xylene	12	<10	

[a]Acutal data indicate negative removal.
[b]Not detected, assumed to be <10 μg/L.

Note: Blanks indicate information was not specified.

TREATMENT TECHNOLOGY: Gravity Oil Separation

Data source: Effluent Guidelines
Point source category: Petroleum refining
Subcategory: Cracking
Plant: 2
References: A9, p. 30

Data source status:
Engineering estimate ___
Bench scale ___
Pilot scale ___
Full scale x

Use in system: Primary
Pretreatment of influent:

DESIGN OR OPERATING PARAMETERS

Unit configuration: API design
Wastewater flow:
Hydraulic detention time:
Hydraulic loading:
Sludge overflow:

REMOVAL DATA

Sampling period:

Pollutant/parameter	Effluent concentration
Conventional pollutants, mg/L:	
Total phenol	31.0
Other pollutants, mg/L:	
NH_3	30

Note: Blanks indicate information was not specified.

TREATMENT TECHNOLOGY: Gravity Oil Separation

Data source: Effluent Guidelines
Point source category: Petroleum refining
Subcategory: Cracking
Plant: 3
References: A9, p. 30

Data source status:
- Engineering estimate ___
- Bench scale ___
- Pilot scale ___
- Full scale _x_

Use in system: Primary
Pretreatment of influent:

DESIGN OR OPERATING PARAMETERS

Unit configuration: API design
Wastewater flow:
Hydraulic detention time:
Hydraulic loading:
Sludge overflow:

REMOVAL DATA

Sampling period:

Pollutant/parameter	Effluent concentration
Conventional pollutants, mg/L:	
BOD_5	641
COD	1,500
TOC	352
Oil and grease	96.1
Total phenol	31.7
Toxic pollutants, μg/L:	
Chromium	450
Other pollutants, mg/L:	
NH_3	320

Note: Blanks indicate information was not specified.

TREATMENT TECHNOLOGY: Gravity Oil Separation

Data source: Effluent Guidelines
Point source category: Petroleum refining
Subcategory: Cracking
Plant: 4
References: A9, p. 30

Data source status:
- Engineering estimate ___
- Bench scale ___
- Pilot scale ___
- Full scale x

Use in system: Primary
Pretreatment of influent:

DESIGN OR OPERATING PARAMETERS

Unit configuration: API design
Wastewater flow:
Hydraulic detention time:
Hydraulic loading:
Sludge overflow:

REMOVAL DATA

Sampling period:

Pollutant/parameter	Effluent concentration
Conventional pollutants, mg/L:	
BOD_5	354
COD	1,220
TOC	158
Oil and grease	48.8
Total phenol	92.6
Toxic pollutants, µg/L:	
Chromium	280
Other pollutants, mg/L:	
NH_3	82

Note: Blanks indicate information was not specified.

TREATMENT TECHNOLOGY: Gravity Oil Separation

Data source: Effluent Guidelines
Point source category: Petroleum refining
Subcategory: Cracking
Plant: 7
References: A9, p. 30

Data source status:
Engineering estimate ___
Bench scale ___
Pilot scale ___
Full scale x

Use in system: Primary
Pretreatment of influent:

DESIGN OR OPERATING PARAMETERS

Unit configuration: API design
Wastewater flow:
Hydraulic detention time:
Hydraulic loading:
Sludge overflow:

REMOVAL DATA

Sampling period:

Pollutant/parameter	Effluent concentration
Conventional pollutants, mg/L:	
BOD_5	1,650
COD	6,450
TOC	401
Oil and grease	915
Total phenol	2.70
Toxic pollutants, μg/L:	
Chromium	2,000
Other pollutants, mg/L:	
NH_3	470

Note: Blanks indicate information was not specified.

TREATMENT TECHNOLOGY: Gravity Oil Separation

Data source: Effluent Guidelines
Point source category: Petroleum refining
Subcategory: Cracking
Plant: 10
References: A9, p. 30

Data source status:
Engineering estimate ___
Bench scale ___
Pilot scale ___
Full scale x

Use in system: Primary
Pretreatment of influent:

DESIGN OR OPERATING PARAMETERS

Unit configuration: API design
Wastewater flow:
Hydraulic detention time:
Hydraulic loading:
Sludge overflow:

REMOVAL DATA

Sampling period:

Pollutant/parameter	Effluent concentration
Conventional pollutants, mg/L:	
BOD_5	720
COD	2,260
TOC	229
Oil and grease	147
Total phenol	189
Other pollutants, mg/L:	
NH_3	600

Note: Blanks indicate information was not specified.

TREATMENT TECHNOLOGY: Gravity Oil Separation

Data source: Effluent Guidelines
Point source category: Petroleum refining
Subcategory: Cracking
Plant: 15
References: A9, p. 30

Data source status:
Engineering estimate ___
Bench scale ___
Pilot scale ___
Full scale x

Use in system: Primary
Pretreatment of influent:

DESIGN OR OPERATING PARAMETERS

Unit configuration: API design
Wastewater flow:
Hydraulic detention time:
Hydraulic loading:
Sludge overflow:

REMOVAL DATA

Sampling period:

Pollutant/parameter	Effluent Concentration
Conventional pollutants, mg/L:	
BOD_5	37.5
COD	309
TOC	71.2
Oil and grease	66.1
Total phenol	6.40
Toxic pollutants, µg/L:	
Chromium	2,100
Other pollutants, mg/L:	
NH_3	15

Note: Blanks indicate information was not specified.

TREATMENT TECHNOLOGY: Gravity Oil Separation

Data source: Effluent Guidelines
Point source category: Petroleum refining
Subcategory: Cracking
Plant: 17
References: A9, p. 30

Data source status:
- Engineering estimate ___
- Bench scale ___
- Pilot scale ___
- Full scale x

Use in system: Primary
Pretreatment of influent:

DESIGN OR OPERATING PARAMETERS

Unit configuration: API design
Wastewater flow:
Hydraulic detention time:
Hydraulic loading:
Sludge overflow:

REMOVAL DATA

Sampling period:

Pollutant/parameter	Effluent concentration
Conventional pollutants, mg/L:	
BOD_5	1,620
COD	2,890
TOC	31.6
Oil and grease	16.2
Total phenol	17.5
Other pollutants, mg/L:	
NH_3	17

Note: Blanks indicate information was not specified.

TREATMENT TECHNOLOGY: Gravity Oil Separation

Data source: Effluent Guidelines
Point source category: Petroleum refining
Subcategory: Cracking
Plant: 18
References: A9, p. 30

Data source status:
- Engineering estimate ___
- Bench scale ___
- Pilot scale ___
- Full scale x

Use in system: Primary
Pretreatment of influent:

DESIGN OR OPERATING PARAMETERS

Unit configuration: API design
Wastewater flow:
Hydraulic detention time:
Hydraulic loading:
Sludge overflow:

REMOVAL DATA

Sampling period:

Pollutant/parameter	Effluent concentration
Conventional pollutants, mg/L:	
BOD_5	156
COD	546
TOC	171
Oil and grease	17.3
Total phenol	2.42
Toxic pollutants, µg/L:	
Chromium	180
Other pollutants, mg/L:	
NH_3	5.7

Note: Blanks indicate information was not specified.

TREATMENT TECHNOLOGY: Gravity Oil Separation

Data source: Effluent Guidelines
Point source category: Petroleum refining
Subcategory: Cracking
Plant: 19
References: A9, p. 30

Data source status:
Engineering estimate ___
Bench scale ___
Pilot scale ___
Full scale x

Use in system: Primary
Pretreatment of influent:

DESIGN OR OPERATING PARAMETERS

Unit configuration: API design
Wastewater flow:
Hydraulic detention time:
Hydraulic loading:
Sludge overflow:

REMOVAL DATA

Sampling period:

Pollutant/parameter	Effluent concentration
Conventional pollutants, mg/L:	
COD	425
TOC	286
Oil and grease	170
Total phenol	25.9
Toxic pollutants, µg/L:	
Chromium	140
Other pollutants, mg/L:	
NH_3	91

Note: Blanks indicate information was not specified.

TREATMENT TECHNOLOGY: Gravity Oil Separation

Data source: Effluent Guidelines
Point source category: Petroleum refining
Subcategory: Cracking
Plant: 25
References: A9, p. 30

Data source status:
- Engineering estimate ___
- Bench scale ___
- Pilot scale ___
- Full scale x

Use in system:
Pretreatment of influent:

DESIGN OR OPERATING PARAMETERS

Unit configuration: API design
Wastewater flow:
Hydraulic detention time:
Hydraulic loading:
Sludge overflow:

REMOVAL DATA

Sampling period:

Pollutant/parameter	Effluent concentration
Conventional pollutants, mg/L:	
BOD_5	190
COD	432
TOC	66.4
Oil and grease	9.22
Total phenol	50.8
Toxic pollutants, µg/L:	
Chromium	25,000
Other pollutants, mg/L:	
NH_3	160

Note: Blanks indicate information was not specified.

TREATMENT TECHNOLOGY: Gravity Oil Separation

Data source: Effluent Guidelines
Point source category: Petroleum refining
Subcategory: Petro chemical
Plant: 16
References: A9, p. 30

Data source status:
- Engineering estimate ___
- Bench scale ___
- Pilot scale ___
- Full scale x

Use in system: Primary
Pretreatment of influent:

DESIGN OR OPERATING PARAMETERS

Unit configuration: API design
Wastewater flow:
Hydraulic detention time:
Hydraulic loading:
Sludge overflow:

REMOVAL DATA

Sampling period:

Pollutant/parameter	Effluent concentration
Conventional pollutants, mg/L:	
BOD_5	202
COD	1,100
Oil and grease	11.9
Total phenol	29.1
Toxic pollutants, µg/L:	
Chromium	280
Other pollutants, mg/L:	
NH_3	350

Note: Blanks indicate information was not specified.

TREATMENT TECHNOLOGY: Gravity Oil Separation

Data source: Effluent Guidelines
Point source category: Petroleum refining
Subcategory:
Plant: A
References: A3, p. IV-36

Data source status:
Engineering estimate ___
Bench scale ___
Pilot scale ___
Full scale x

Use in system: Primary
Pretreatment of influent:

DESIGN OR OPERATING PARAMETERS

Unit configuration: API design
Wastewater flow:
Hydraulic detention time:
Hydraulic loading:
Sludge overflow:

REMOVAL DATA

Sampling period:

Pollutant/parameter	Effluent concentration[a]
Conventional pollutants, mg/L:	
BOD_5	24
COD	107
TOC	29
TSS	380
Total phenol	0.1
Toxic pollutants, μg/L:	
Arsenic	12
Chromium	270
Copper	26
Lead	130
Nickel	23
Zinc	260
Di-n-butyl phthalate	1.3[b]
Diethyl phthalate	12[b]
Phenol	13
Benzene	>>100
Ethylbenzene	>100
Acenaphthene	37[b]
Acenaphthylene	4[b]
Anthracene/phenanthrene	4.6[b,c]
Naphthalene	68[b]
Aroclor 1232	0.9
1,2-*Trans*-dichloroethylene	~20
Methylene chloride	>>50[d]
Tetrachloroethylene	>50
Trichloroethylene	>>100

[a]Concentrations from several days averaged.

[b]This extract was diluted 1:10 before analysis.

[c]Concentrations represent sums for these two compounds which elute simultaneously and have the same major ions for GC/MS.

[d]Possibly due to laboratory contamination.

Note: Blanks indicate information was not specified.

TREATMENT TECHNOLOGY: Gravity Oil Separation

Data source: Effluent Guidelines
Point source category: Petroleum refining
Subcategory:
Plant: C
References: A2, p. IV-36

Data source status:
- Engineering estimate ___
- Bench scale ___
- Pilot scale ___
- Full scale x

Use in system: Primary
Pretreatment of influent:

DESIGN OR OPERATING PARAMETERS

Unit configuration:
Wastewater flow:
Hydraulic detention time:
Hydraulic loading:
Sludge overflow:

REMOVAL DATA

Sampling period:

Pollutant/parameter	Effluent concentration[a]
Conventional pollutants, mg/L:	
COD	320
TOC	71
TSS	28
Oil and grease	93
Total phenol	5.6
Toxic pollutants, μg/L:	
Arsenic	8
Chromium	850
Copper	190
Cyanide	430
Lead	12
Mercury	3
Selenium	15
Zinc	640
Bis(2-ethylhexyl) phthalate	290
Phenol	2,200
Naphthalene	950
Anthracene/phenanthrene	∿190
Other pollutants, mg/L:	
NH_3	38

[a]Concentrations from several days were averaged.

Note: Blanks indicate information was not specified.

TREATMENT TECHNOLOGY: Gravity Oil Separation

Data source: Effluent Guidelines
Point source category: Petroleum refining
Subcategory:
Plant: J
References: A3, p. IV-36

Data source status:
Engineering estimate ___
Bench scale ___
Pilot scale ___
Full scale x

Use in system: Primary
Pretreatment of influent:

DESIGN OR OPERATING PARAMETERS

Unit configuration: API design
Wastewater flow:
Hydraulic detention time:
Hydraulic loading:
Sludge overflow:

REMOVAL DATA

Sampling period:

Pollutant/parameter	Effluent concentration[a]
Conventional pollutants, mg/L:	
COD	550
TOC	160
TSS	120
Oil and grease	160
Total phenol	1.8
Toxic pollutants, μg/L:	
Arsenic	5
Chromium	650
Copper	60
Cyanide	10
Lead	920
Mercury	2
Nickel	31
Selenium	12
Thallium	2
Zinc	870
Bis(2-ethylhexyl) phthalate	300
Phenol	160
Anthracene/phenanthrene	90[b]
Chrysene/benz(a)anthracene[c]	30[b]
Naphthalene	350
Aroclor 1016	0.2
Aroclor 1232	0.5
Aroclor 1242	0.5

[a]Concentrations from several days were averaged.

[b]Approximately.

[c]Assume 50% mixture.

Note: Blanks indicate information was not specified.

TREATMENT TECHNOLOGY: Gravity Oil Separation

Data source: Effluent Guidelines
Point source category: Petroleum refining
Subcategory:
Plant: J
References: A3, p. IV-36

Data source status:
- Engineering estimate ___
- Bench scale ___
- Pilot scale ___
- Full scale x

Use in system: Primary
Pretreatment of influent:

DESIGN OR OPERATING PARAMETERS

Unit configuration: API design
Wastewater flow:
Hydraulic detention time:
Hydraulic loading:
Sludge overflow:

REMOVAL DATA

Sampling period:

Pollutant/parameter	Effluent concentration[a]
Conventional pollutants, mg/L:	
COD	340
TOC	74
TSS	52
Oil and grease	83
Total phenol	4.3
Toxic pollutants, μg/L:	
Antimony	1
Arsenic	3
Chromium	1,500
Copper	38
Cyanide	60
Lead	40
Mercury	2
Selenium	16
Zinc	420
Bis(2-ethylhexyl) phthalate	600
2,4-Dimethylphenol	650
Pentachlorophenol	850
Phenol	16,000
Anthracene/phenanthrene[b]	∿230
Chrysene/benz(a)anthracene[b]	∿40
Fluoranthene/pyrene[b]	∿20
Fluorene	80
Naphthalene	50

[a]Concentrations from several days were averaged.
[b]Assume 50% mixture.

Note: Blanks indicate information was not specified.

TREATMENT TECHNOLOGY: Gravity Oil Separation

Data source: Effluent Guidelines
Point source category: Petroleum refining
Subcategory:
Plant: J
References: A3, p. IV-36

Data source status:
- Engineering estimate ___
- Bench scale ___
- Pilot scale ___
- Full scale x

Use in system: Primary
Pretreatment of influent:

DESIGN OR OPERATING PARAMETERS

Unit configuration: API design
Wastewater flow:
Hydraulic detention time:
Hydraulic loading:
Sludge overflow:

REMOVAL DATA

Sampling period:

Pollutant/parameter	Effluent concentration[a]
Conventional pollutants, mg/L:	
COD	83
TOC	25
TSS	30
Oil and grease	14
Total phenol	0.251
Toxic pollutants, μg/L:	
Arsenic	9
Beryllium	2
Cadmium	6
Chromium	2,500
Copper	75
Cyanide	20
Lead	52
Mercury	0.8
Nickel	40
Selenium	20
Thallium	3
Zinc	580

[a]Concentrations from several days were averaged.

Note: Blanks indicate information was not specified.

TREATMENT TECHNOLOGY: Gravity Oil Separation

Data source: Effluent Guidelines
Point source category: Petroleum refining
Subcategory:
Plant: J
References: A3, p. IV-36

Data source status:
Engineering estimate ___
Bench scale ___
Pilot scale ___
Full scale x

Use in system: Primary
Pretreatment of influent:

DESIGN OR OPERATING PARAMETERS

Unit configuration: API design
Wastewater flow:
Hydraulic detention time:
Hydraulic loading:
Sludge overflow:

REMOVAL DATA

Sampling period:

Pollutant/parameter	Effluent concentration[a]
Conventional pollutants, mg/L:	
COD	190
TOC	53
TSS	45
Oil and grease	34
Total phenol	0.75
Toxic pollutants, μg/L:	
Arsenic	3
Chromium	720
Copper	15
Cyanide	10
Lead	36
Mercury	0.6
Nickel	32
Selenium	17
Zinc	230
Bis(2-ethylhexyl) phthalate	50
Chrysene/benz(a)anthracene[b]	50[c]

[a]Concentrations from several days were averaged.
[b]Assume 50% mixture.
[c]Approximately.

Note: Blanks indicate information was not specified.

TREATMENT TECHNOLOGY: Gravity Oil Separation

Data source: Effluent Guidelines
Point source category: Petroleum refining
Subcategory:
Plant: J
References: A3, p. IV-36

Data source status:
Engineering estimate ___
Bench scale ___
Pilot scale ___
Full scale x

Use in system: Primary
Pretreatment of influent:

DESIGN OR OPERATING PARAMETERS

Unit configuration: API design
Wastewater flow:
Hydraulic detention time:
Hydraulic loading:
Sludge overflow:

REMOVAL DATA

Sampling period:

Pollutant/parameter	Effluent concentration[a]
Conventional pollutants, mg/L:	
COD	180
TOC	51
TSS	53
Oil and grease	77
Total phenol	0.7
Toxic pollutants, μg/L:	
Arsenic	3
Chromium	150
Copper	290
Cyanide	10
Lead	32
Mercury	1.4
Nickel	26
Selenium	8
Zinc	300
Bis(2-ethylhexyl) phthalate	180
Phenol	420[b]
Anthracene/phenanthrene[c]	30[b]
Chrysene/benz(a)anthracene[c]	30[b]
Fluoranthene/pyrene	30[b]

[a]Concentrations from several days were averaged.

[b]Approximately.

[c]Assume 50% mixture.

Note: Blanks indicate information was not specified.

TREATMENT TECHNOLOGY: Gravity Oil Separation

Data source: Effluent Guidelines
Point source category: Petroleum refining
Subcategory:
Plant: I
References: A3, p. VI-36

Data source status:
Engineering estimate ___
Bench scale ___
Pilot scale ___
Full scale x

Use in system: Primary
Pretreatment of influent:

DESIGN OR OPERATING PARAMETERS

Unit configuration: API design
Wastewater flow:
Hydraulic detention time:
Hydraulic loading:
Sludge overflow:

REMOVAL DATA

Sampling period:

Pollutant/parameter	Effluent concentration[a]
Conventional pollutants, mg/L:	
BOD_5	49
COD	260
TOC	81
TSS	39
Oil and grease	32
Total phenol	5.1
Toxic pollutants, μg/L:	
Arsenic	5
Chromium	3
Copper	6
Cyanide	10
Mercury	0.6
Nickel	4
Selenium	4
Zinc	100
Bis(2-ethylhexyl) phthalate	300
Phenol	390
Naphthalene	290

[a]Concentrations from several days were averaged.

Note: Blanks indicate information was not specified.

TREATMENT TECHNOLOGY: Gravity Oil Separation

Data source: Effluent Guidelines
Point source category: Petroleum refining
Subcategory:
Plant: H
References: A3, p. IV-36

Data source status:
- Engineering estimate ___
- Bench scale ___
- Pilot scale ___
- Full scale x

Use in system: Primary
Pretreatment of influent:

DESIGN OR OPERATING PARAMETERS

Unit configuration: API design
Wastewater flow:
Hydraulic detention time:
Hydraulic loading:
Sludge overflow:

REMOVAL DATA

Sampling period:

Pollutant/parameter	Effluent concentration[a]
Conventional pollutants, mg/L:	
BOD_5	42
COD	190
TOC	54
TSS	102
Oil and grease	52
Total phenol	2.1
Toxic pollutants, µg/L:	
Chromium	10
Copper	23
Cyanide	100
Zinc	56

[a]Concentrations from several days were averaged.

Note: Blanks indicate information was not specified.

TREATMENT TECHNOLOGY: Gravity Oil Separation

Data source: Effluent Guidelines
Point source category: Petroleum refining
Subcategory:
Plant: L
References: A3, p. IV-36

Data source status:
Engineering estimate ___
Bench scale ___
Pilot scale ___
Full scale x

Use in system: Primary
Pretreatment of influent:

DESIGN OR OPERATING PARAMETERS

Unit configuration: API design
Wastewater flow:
Hydraulic detention time:
Hydraulic loading:
Sludge overflow:

REMOVAL DATA

Sampling period:

Pollutant/parameter	Effluent concentration[a]
Conventional pollutants, mg/L:	
BOD_5	37
COD	190
TOC	50
TSS	42
Total phenol	11.8
Toxic pollutants, μg/L:	
Antimony	840
Copper	44
Cyanide	150
Lead	17
Mercury	0.5
Nickel	16
Zinc	320
2,4-Dimethylphenol	>100
Phenol	>100
Benzene	>>100
Toluene	>>100
Acenaphthene	3,000[b]
Chrysene	1.7
Fluoranthene	8.5[b]
Flourene	300[b]
Naphthalene	280[b]
Pyrene	7[b]
Chloroform	<10
Methylene chloride	~10
Other pollutants, mg/L:	
NH_3	11

[a]Concentrations from several days were averaged.
[b]This extract was diluted 1:10 before analysis.

Note: Blanks indicate information was not specified.

TREATMENT TECHNOLOGY: Gravity Oil Separation

Data source: Effluent Guidelines
Point source category: Petroleum refining
Subcategory:
Plant: L
References: A3, p. lV-36

Data source status:
Engineering estimate ___
Bench scale ___
Pilot scale ___
Full scale _x_

Use in system: Primary
Pretreatment of influent:

DESIGN OR OPERATING PARAMETERS

Unit configuration: API design
Wastewater flow:
Hydraulic detention time:
Hydraulic loading:
Sludge overflow:

REMOVAL DATA

Sampling period:

Pollutant/parameter	Effluent concentration[a]
Conventional pollutants, mg/L:	
BOD_5	130
COD	420
TOC	120
TSS	120
Total phenol	54.7
Toxic pollutants, µg/L:	
Chromium	400
Copper	120
Cyanide	380
Lead	45
Mercury	1.3
Nickel	70
Zinc	360
2,4-Dimethylphenol	>100
Phenol	>100
Benzene	>100
Toluene	>100
Anthracene/phenanthrene	∿230[b,c]
Benzo(k)fluoranthene	270[b]
Chrysene	20[b]
Naphthalene	500[b]
Aroclor 1242	5.2
Chloroform	∿10
Methylene chloride	∿30

[a]Concentrations from several days were averaged.

[b]This extract was diluted 1:10 before analysis.

[c]Concentrations represent sums for these two compounds which elute simultaneously and have the same major ions for GC/MS.

Note: Blanks indicate information was not specified.

TREATMENT TECHNOLOGY: Gravity Oil Separation

Data source: Effluent Guidelines
Point source category: Petroleum refining
Subcategory:
Plant: N
References: A3, p. lV-36

Data source status:
- Engineering estimate ___
- Bench scale ___
- Pilot scale ___
- Full scale x

Use in system: Primary
Pretreatment of influent:

DESIGN OR OPERATING PARAMETERS

Unit configuration: API design
Wastewater flow:
Hydraulic detention time:
Hydraulic loading:
Sludge overflow:

REMOVAL DATA

Sampling period:

Pollutant/parameter	Effluent concentration[a]
Conventional pollutants, mg/L:	
COD	410
TOC	100
TSS	85
Total phenol	5.9
Toxic pollutants, μg/L:	
Chromium	1,300
Copper	38
Cyanide	40
Lead	18
Nickel	16
Zinc	600
2,4-Dimethylphenol	71
Phenol	>100
Toluene	>100
Acenaphthene	520[b]
Acenaphthylene	87[b]
Anthracene/phenanthrene	140[b,c]
Benzo(a)pyrene	5.5[b]
Fluoranthene	7.5
Naphthalene	~300
Pyrene	16[b]
Aroclor 1016	1.9
Aroclor 1221	0.1
Aroclor 1232	0.5
Chloroform	~15
Methylene chloride	>90[d]

[a]Concentration from several days were averaged.

[b]This extract was limited 1:10 before analysis.

[c]Concentration represent sums for those two compounds which elute simltaneously and have the same ions for GC/MS.

[d]Possibly due to laboratory contamination.

Note: Blanks indicate information was not specified.

TREATMENT TECHNOLOGY: Gravity Oil Separation

Data source: Effluent Guidelines
Point source category: Petroleum refining
Subcategory:
Plant: P
References: A3, p. IV-36

Data source status:
- Engineering estimate ___
- Bench scale ___
- Pilot scale ___
- Full scale x

Use in system: Primary
Pretreatment of influent:

DESIGN OR OPERATING PARAMETERS

Unit configuration: API design
Wastewater flow:
Hydraulic detention time:
Hydraulic loading:
Sludge overflow:

REMOVAL DATA

Sampling period:

Pollutant/parameter	Effluent concentration[a]
Conventional pollutants, mg/L:	
BOD_5	190
COD	540
TOC	150
TSS	63
Total phenol	68
Toxic pollutants, μg/L:	
Antimony	300
Arsenic	6
Chromium	500
Cyanide	60
Selenium	8
Thallium	1
Zinc	61

[a]Concentrations from several days were averaged.

Note: Blanks indicate information was not specified.

TREATMENT TECHNOLOGY: Gravity Oil Separation

Data source: Effluent Guidelines
Point source category: Petroleum refining
Subcategory:
Plant: Q
References: A3, p. IV-36

Data source status:
- Engineering estimate ___
- Bench scale ___
- Pilot scale ___
- Full scale x

Use in system: Primary
Pretreatment of influent:

DESIGN OR OPERATING PARAMETERS

Unit configuration: API design
Wastewater flow:
Hydraulic detention time:
Hydraulic loading:
Sludge overflow:

REMOVAL DATA

Sampling period:

Pollutant/parameter	Effluent concentration[a]
Conventional pollutants, mg/L:	
COD	320
TOC	80
TSS	17
Oil and grease	38
Total phenol	0.112
Toxic pollutants, μg/L:	
Arsenic	440
Chromium	1
Copper	160
Cyanide	20
Lead	10
Mercury	2
Selenium	8
Zinc	430
Bis(2-ethylhexyl) phthalate	320
Phenol	60

[a]Concentrations from several days were averaged.

Note: Blanks indicate information was not specified.

TREATMENT TECHNOLOGY: Gravity Oil Separation

Data source: Effluent Guidelines
Point source category: Petroleum refining
Subcategory:
Plant: 6
References: A3, p. IV-36

Data source status:
- Engineering estimate ___
- Bench scale ___
- Pilot scale ___
- Full scale x

Use in system: Primary
Pretreatment of influent:

DESIGN OR OPERATING PARAMETERS

Unit configuration: API design
Wastewater flow:
Hydraulic detention time:
Hydraulic loading:
Sludge overflow:

REMOVAL DATA

Sampling period:

Pollutant/parameter	Effluent concentration[a]
Conventional pollutants, mg/L:	
BOD_5	260
COD	840
TOC	230
TSS	140
Oil and grease	93
Total phenol	24
Toxic pollutants, μg/L:	
Cyanide	1,300
Bis(2-ethylhexyl) phthalate	700
Phenol	4,900
Benzo(ghi)perylene	∿1,100
Naphthalene	1,100
Aroclor 1016	1.8
Aroclor 1242	5
Other pollutants, mg/L:	
NH_3	9

[a]Concentrations from several days were averaged.

Note: Blanks indicate information was not specified.

TREATMENT TECHNOLOGY: Gravity Oil Separation

Data source: Effluent Guidelines
Point source category: Petroleum refining
Subcategory:
Plant: 26
References: A9, p. 30

Data source status:
Engineering estimate ___
Bench scale ___
Pilot scale ___
Full scale x

Use in system: Primary
Pretreatment of influent:

DESIGN OR OPERATING PARAMETERS

Unit configuration: API design
Wastewater flow:
Hydraulic detention time:
Hydraulic loading:
Sludge overflow:

REMOVAL DATA

Sampling period:

Pollutant/parameter	Effluent concentration
Conventional pollutants, mg/L:	
BOD_5	94.4
COD	442
TOC	167
Oil and grease	57.0
Total phenol	0.063
Toxic pollutants, μg/L:	
Chromium	1,200
Other pollutants, mg/L:	
NH_3	15

Note: Blanks indicate information was not specified.

TREATMENT TECHNOLOGY: Gravity Oil Separation

Data source: Effluent Guidelines
Point source category: Timber products processing
Subcategory: Wood preserving/steaming, no discharger
Plant: 495
References: A1, p. 7-30

Data source status:
Engineering estimate ___
Bench scale ___
Pilot scale ___
Full scale x

Use in system: Primary
Pretreatment of influent:

DESIGN OR OPERATING PARAMETERS

Unit configuration:
Wastewater flow:
Hydraulic detention time:
Hydraulic loading:
Sludge overflow:

REMOVAL DATA

Sampling period:

Pollutant/parameter	Effluent concentration
Conventional pollutants, mg/L:	
COD	374[a]
Oil and grease	70[a]
Total phenol	0.154[a]
Toxic pollutants, μg/L:	
Antimony	1[a]
Arsenic	15[a]
Cadmium	1
Chromium	4[a]
Copper	37[a]
Lead	4
Mercury	1.3[a]
Nickel	5.5[a]
Selenium	1
Silver	1
Thallium	2
Zinc	110[a]
Bis(2-ethylhexyl) phthalate	9.5[a]
Acrylonitrile	30
Pentachlorophenol	120[a]
Phenol	15
Toluene	23
Acenaphthene	78[a]
Acenaphthylene	67
Anthracene/phenanthrene	45[a]
Benz(a)anthracene	55
Benzo(a)pyrene	15
Benzo(ghi)perylene	2
Benzo(k)fluoranthene	37
Fluoranthene	170[a]
Fluorene	34[a]
Indeno(2,2,3-cd)pyrene	40
Naphthalene	86[a]
Pyrene	99[a]
Chloroform	23
Methylene chloride	660

[a]Average of two studies conducted one year apart.

Note: Blanks indicate information was not specified.

4.2 CLARIFICATION/SEDIMENTATION [1]

4.2.1 Function

Clarification/sedimentation is used to remove suspended solids by settling.

4.2.2 Description

Primary Rectangular Clarification. Primary rectangular clarification involves a relatively long period of quiescence in a basin (depths of 10 to 15 feet) where most of the settleable solids fall out of suspension by gravity. The solids are mechanically transported along the bottom of the tank by a scraper mechanism and pumped as a sludge underflow.

The maximum length of rectangular tanks has been approximately 300 feet. When widths greater than 20 feet are required, multiple bays with individual cleaning equipment may be employed. Influent channels and effluent channels are normally located at opposite ends of the tank.

Sludge removal equipment usually consists of a pair of endless conveyor chains. Attached to the chains at about 10 foot intervals are wooden crosspieces or flights, extending the full width of the tank or bay. Linear conveyor speeds of 2 to 4 ft/min are common. The settled solids are scraped to sludge hoppers in small tanks and to transverse troughs in large tanks. The troughs, in turn, are equipped with cross collectors, usually of the same type as the longitudinal collectors, which convey solids to one or more sludge hoppers. Screw conveyors are also used for the cross collectors. Tanks also may be cleaned by a bridge-type mechanism that travels up and down the tank on rails supported on the sidewalls. Scraper blades are suspended from the bridge and are lifted clear of the sludge on the return travel. For very long tanks, it is desirable to use two sets of chains and flights in tandem with a central hopper to receive the sludge. Tanks in which mechanisms that move the sludge toward the effluent end in the same direction as the density current have shown superior performance in some instances.

Floatables are usually skimmed to the effluent end of rectangular tanks by the flights returning at the liquid surface. The floatables are then scraped manually up an inclined apron, or they can be removed hydraulically or mechanically. Types of removal processes include rotating slotted pipes, transverse rotating helical wiper, chain and flight collectors, and scum rakes.

Primary Circular Clarification. Primary circular clarification also involves a relatively long period of quiescence at depths of 10 to 15 feet. The settled solids are mechanically collected on the bottom to a central sludge hopper and pumped as a sludge underflow.

The clarifier bottom is typically sloped one inch per foot. Floating scum is trapped inside a peripheral scum baffle and squeegeed into a scum discharge box. The clarifier contains a center motor-driven turntable drive supported by a bridge spanning the top of the tank, or supported by a vertical steel center pier. The turntable gear rotates a vertical cage or torque tube, which in turn rotates the truss arms (preferably two long arms). The truss arms carry multiple flights (plows) on the bottom chord that are set at a 30° angle of attack and literally "plow" heavy fractions of sludge and grit along the bottom slope toward a center blow-down hopper. A peripheral weir overflow system carries the effluent. An inner diffusion chamber receives influent flow and distributes it (by means of about a four-inch water head loss) inside of the large diameter feedwell skirt. Approximately three percent of the clarifer surface area is used for the feed-well. The depth of the feed-wells are generally about one-half of the tank depth. The center sludge hopper should be less than two feet deep and less than four square feet in cross section.

Secondary Rectangluar Clarification. Secondary rectangular clarifiers when used in conjunction with the activated sludge process are similar to primary rectangular clarifiers except that the large volume of flocculent solids in the mixed liquor are considered during the design and in sizing the sludge pumps. Since mixed liquor, on entering the tank, has a tendency to flow as a density current, interfering with the separation of the solids and the thickening of the sludge. The following factors are considered in the design of these tanks: (1) type of tank to be used, (2) surface loading rate, (3) solids loading rate, (4) flowthrough velocities, (5) weir placement and loading rates, and (6) scum removal.

Secondary Circular Clarification. Secondary circular clarifiers have been constructed with diameters ranging from 12 to 200 feet with depths of 12 to 15 feet. There are two basic types: the center-feed and the rim-feed. Both utilize a revolving mechanism to transport and remove the sludge from the bottom of the clarifier. Mechanisms are of two types: those that scrape or plow the sludge to a center hopper similar to the types used in primary sedimentation tanks, and those that remove the sludge directly from the tank bottom through suction orifices that serve the entire bottom of the tank in each revolution. In one system of the latter type, the suction is maintained by reduced static head on the individual drawoff pipes. In another patented suction system, sludge is removed through a manifold either hydrostatically or by pumping.

Secondary circular clarifiers are made with effluent overflow weirs located near the center or near the perimeter of the tank. Skimming facilities are generally included on all secondary clarifiers.

The design is similar to primary clarifiers. When used in conjunction with the activated sludge process, the large volume of flocculent solids in the mixed liquor to the secondary circular clarifier requires that special consideration be given to the design. Design factors to be considered are similar to those of the secondary rectangular clarifier.

4.2.3 Common Modifications in Rectangular and Circular Clarification

Secondary Clarification, High Rate Trickling Filter. The design of clarifiers that follow high rate trickling filters is similar to that of primary clarifiers, except that the surface loading rate is based on the plant flow plus the effluent recycle flow minus the underflow (often neglected). These clarifiers differ from secondary clarifiers following activated sludge processes in that the sludge recirculation is not used. Also, solids loading limits are not involved in the sizing. Recirculation of the supernatant from the clarifier to the trickling filter can range from one to four times the plant influent flow rate. Under suitable trickling filter operating conditions, it is more economical to recirculate the clarifier influent to reduce the flow sizing requirements in the clarifier.

Primary Rectangular Clarification. Tanks may be cleaned by a bridge-type mechanism which travels up and down the tank on rails supported on the sidewalls. Scraper blades are suspended from the bridge and are lifted clear of the sludge on the return travel. Chemical coagulants may be added to improve BOD_5 and suspended solid removals and to remove phosphorous ion.

Primary Circular Clarification. Two short auxiliary scraper arms are added perpendicular to the two long arms on medium-to-large tanks. This makes the use of deep spiral flights practicable, which aid in center region plowing where ordinary shallow straight plows (30° angle of attack) are nearly useless. Peripheral feed systems are sometimes used in lieu of central feed. Also, central effluent weirs are sometimes used. Flocculating feed wells also may be provided if coagulants are to be added to assist sedimentation.

Secondary Circular Clarification. Multiple inlets are used with balanced flow at various spacings with target baffles to reduce velocity of streams. Hydraulic balancing is used between parallel clarifier units. Wind effects on water surface are controlled. Sludge hopper collection systems and flocculation inlet structures are used. Traveling bridge sludge collectors and skimmers are used as an alternate to chair and flight systems. Steeply inclined tube settlers are used to enhance suspended solids removal in either new or rehabilitated clarifiers. Wedge wire settler panels are used at peak hydraulic loading of less than 800 gpd/ft^2 for improved suspended solids removal.

4.2.4 Technology Status

Rectangular Clarification. Rectangular clarification is in widespread use. Methodology to obtain design parameters is completely developed.

Circular Clarification. Circular clarification is in widespread use.

4.2.5 Applications

Primary Rectangular Clarification. Used for removal of readily settleable solids and floating material to reduce total suspended solids and BOD_5; can accept high solids loading; primary clarifiers are generally employed as preliminary step in further processing; rectangular tanks also lend themselves to nesting with preaeration tanks and aeration tanks in activated sludge plants.

Primary Circular Clarification. Used for removal of readily settleable solids and floating material to reduce suspended solids content and BOD_5. Can accept high solids loading. Primary clarifiers are generally employed as a preliminary step in further processing.

Secondary Rectangular Clarification. Used for solids separation and for production of a concentrated return sludge flow to sustain biological treatment; multiple rectangular tanks require less area than multiple circular tanks and are used where ground area is at premium; rectangular tanks lend themselves more readily to nesting with primary tanks and aeration tanks in activated sludge plants, and are also used generally where tank roofs or covers are required.

Secondary Circular Clarification. Used to separate the activated sludge solids from the mixed liquor, to produce the concentrated solids for the return flow required to sustain biological treatment, and to permit settling of solids resulting from low-rate trickling filter treatment.

4.2.6 Limitations

Primary Rectangular Clarification. Maximum length of tank is about 300 feet; horizontal velocities in clarifier must be limited to prevent "scouring" of settled solids from the sludge bed and their eventual escape in the effluent. Depth of the tank is limited by particle settling rates.

Primary Circular Clarification. Maximum diameter is 200 feet; larger tanks are subject to unbalanced radial diffusion and wind action, both of which can reduce efficiency; horizontal velocities in the clarifier must be limited to prevent "scouring" of settled solids from the sludge bed and their eventual escape in the effluent.

Secondary Rectangular Clarification. Must operate at relatively low hydraulic loadings (large space requirements); maximum length of tank has been about 300 feet; horizontal velocities in clarifier must be limited to prevent "scouring" of settled solids from the sludge bed and their eventual escape to the effluent.

Secondary Circular Clarification. Must operate at relatively low hydraulic loadings (large space requirements); maximum diameter is 200 feet; larger tanks are subject to unbalanced radial diffusion and wind action, both of which can reduce efficiency; horizontal velocities in clarifier must be limited to prevent "scouring" of settled solids from the sludge bed and eventual escape to the effluent.

.4.2.7 Chemicals Required

Use of chemical addition to rectangular and circular clarifiers is discussed in another section of this manual entitled "Clarification/Sedimentation with Chemical Addition," Section III.4.3.

4.2.8 Reliability

Primary Rectangular Clarification. In general, reliability is very high; however, broken links in collector drive chain can cause outages; pluggage of sludge hoppers also has been a problem when cross collectors are not provided.

Primary Circular Clarification. In general, reliability is high; however, clarification of solids into a packed central mass may cause collector arm stoppages; attention to design of center area bottom slope, number of arms, and center area scraper blade design is required to prevent such problems.

Secondary Rectangular Clarification. Mechanical reliability can be considered high provided suitable preventive maintenance and inspection procedures are observed; pluggage of sludge hoppers has been a problem when cross collectors are not provided; process reliability is highly dependent upon the upstream performance of the aerator for the production of good settling sludge with acceptable compactability; rising sludge by denitrification of the sludge is a problem in certain cases.

Secondary Circular Clarification. In general, reliability is very high; however, rising sludge due to denitrification and sludge bulking may cause problems, which may be overcome by proper operational techniques.

4.2.9 Environmental Impact

Primary Rectangular Clarification. Multiple rectangular tanks require less land than multiple circular tanks and are used where ground area is at a premium; however, they require relatively large space for the level of treatment imparted.

Primary Circular Clarification. Scum on surface is a source of odors that can be controlled by masking with chemical additives and aerosols; large land use requirement for degree of treatment imparted, even more so than rectangular units.

Secondary Rectangular Clarification. Offers a higher space efficiency than circular clarifiers, although it requires large land areas.

Secondary Circular Clarification. Circular units require greater land area than rectangular units.

4.2.10 Design Criteria

Design criteria based on: Settling rate of suspended solids; Degree of clarification required; Concentration/loading of solids; Hydraulic loading; Sludge underflow concentration and solids flux rates; Sludge storage; and removal of floatables.

Primary Clarification

Criteria	Units	Primary clarification
Surface loading	gpd/ft²	600 - 1,200
Detention time	hr	1.5 - 3.0
Wier loading	gpd/linear ft	10,000 - 30,000
Rectangular basin, length/width ratio	-	At least 4
Circular basin, sludge collector tip speed	ft/min	10 - 15

Secondary Clarification

Criteria	Units	Secondary Rectangular Clarification Following Activated Sludge	Secondary Clarification Following: Trickling Filtration	Activated Sludge	Extended Aeration	Activated Sludge Using Pure Oxygen
Hydraulic loading (average)	gpd/ft²	400-800	400-600	400-800	200-400	400-800
Hydraulic loading (peak)	gpd/ft²	700-1,200	1,000-1,200	1,000-2,000	800	1,000-1,200
Solids loading (average)	lb/day/ft²	14-28	–	20-30	20-30	25-35
Solids loading (peak)	lb/day/ft²	30-48	–	50	50	50
Depth	ft	12-15	10-12	12-15	12-15	12-15
Inlet baffle diameter	%*	–	15-20	15-20	15-20	15-20
Weir loading rate	gpd/linear ft	10,000-30,000	10,000-30,000	10,000-30,000	10,000-30,000	10,000-30,000
Maximum upflow velocity	ft/hr	12-24	12-24	12-24	12-24	12-24

*Of tank diameter.

4.2.11 Flow Diagrams

Primary Rectangular Clarification.

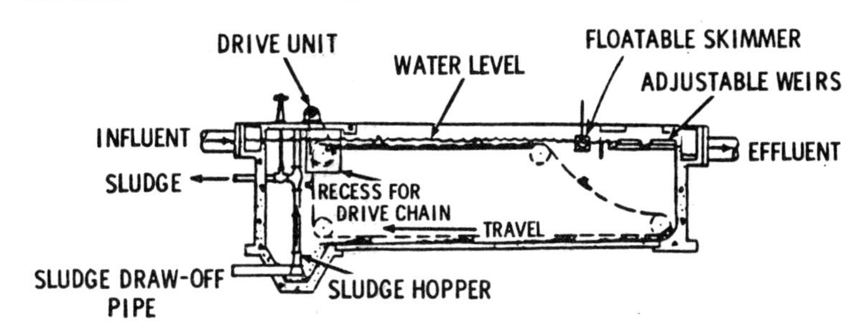

Primary Circular Clarification.

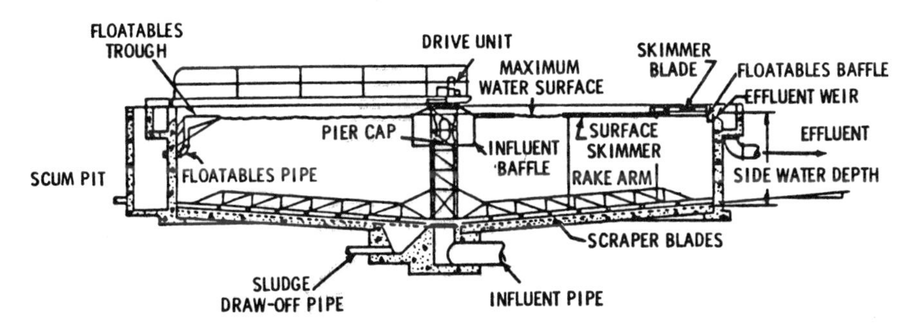

Secondary Circular and Rectangular Clarification.

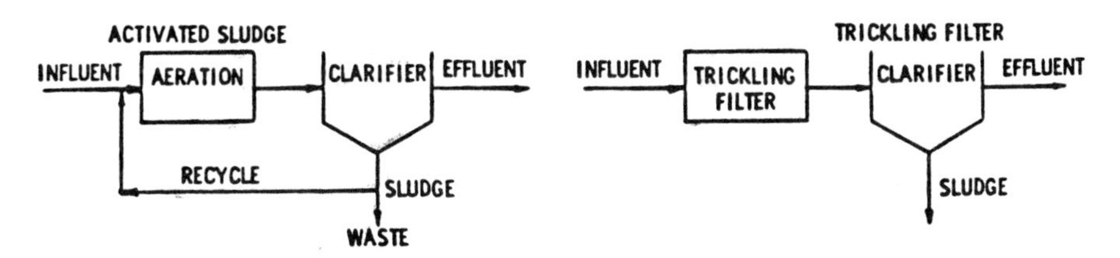

4.2.12 Performance

Subsequent data sheets provide performance data from studies on the following industries and/or wastestreams:

Adhesives and sealants production

Coal mining
- Alkaline mines
- Coal preparation plants and associated areas

Electroplating

Foundry industry
- Copper and copper alloy foundries, mold cooling and casting quench
- Ferrous foundry dust collection
- Ferrous foundry melting furnace scrubber
- Ferrous foundry sand washing
- Steel foundry mold cooling and casting quench
- Steel foundry sand washing and reclaiming

Ink manufacturing
- Water and/or caustic wash

Inorganic chemicals production
- Hydrofluoric acid
- Titanium dioxide

Iron and steel industry
- Bee-hive coke manufacturing
- Cold rolling
- Continuous casting
- Hot forming - section
- Wet open combustion, basic oxygen furnace
- Wet suppressed basic oxygen furnace

Leather tanning and finishing

Mineral mining and processing
- Construction sand and gravel
- Crushed stone
- Dimension stone
- Industrial sand

Ore mining and dressing
- Bauxite mining
- Copper mining/milling/smelting/refining
- Ferroalloy mining/milling
- Iron ore mining/milling
- Lead/zinc mining/milling/smelting/refining

Placer mining
Silver mining/milling
Titanium mining/milling

Paint manufacturing

Steam electric power generation

Textile milling
Wool scouring

4.2.13 References

1. Innovative and Alternative Technology Assessment Manual, EPA-430/9-78-009, (draft), U.S. Environmental Protection Agency, Cincinnati, Ohio, 1978. 252 pp.

CONTROL TECHNOLOGY SUMMARY FOR CLARIFICATION/SEDIMENTATION

Pollutant	Number of data points	Effluent concentration				Removal efficiency, %			
		Minimum	Maximum	Median	Mean	Minimum	Maximum	Median	Mean
Conventional pollutants, mg/L:									
BOD_5	5	980	6,670	1,150	2,500	0[a]	69	25	33
COD	23	4	25,300	18	1,620	0[a]	>99	91	71
TOC	20	1	940	12	63	0[a]	>99	29	40
TSS	73	<1	5,700	29	212	0[a]	>99	97	82
Oil and grease	19	2.7	522	9	70	0[a]	99	40	47
Total phenol	19	0.006	84	0.027	6.3	0[a]	96	45	43
Total phosphorus	1				13.9				3
Toxic pollutants, µg/L:									
Antimony	12	2	1,000	40	<130	0[a]	86	45	35
Arsenic	24	<2	1,200	<5	<78	0[a]	>99	>87	65
Asbestos (total), fibers/L	16	4.6×10^6	3.3×10^{10}	7.0×10^8	3.4×10^9	0[a]	>99	>99	71
Asbestos (chrysotile), fibers/L	12	$<3.3 \times 10^5$	5×10^{11}	1.9×10^6	5.0×10^{10}	0[a]	>99	>99	81
Beryllium	7	<1	20	<10	<11	70	>98	>85	>85
Cadmium	15	2	200	<10	<28	0	>99	77	>73
Chromium	22	6	30,000	15	<2,900	0[a]	>99	>96	76
Chromium (+6)	1				5				0
Copper	32	<4	380	70	<72	0[a]	>99	85	66
Cyanide	12	2	4,500	<20	<410	0[a]	>90	29	34
Lead	26	<5	6,800	45	<450	0[a]	>99	>96	73
Mercury	20	0.1	84	<0.5	<9.4	0[a]	>99	47	44
Nickel	24	<5	2,000	40	<210	0[a]	>9	>93	61
Selenium	16	3	32	<5	<11	0[a]	98	83	63
Silver	12	3	<100	<10	<21	0[a]	>96	>86	71
Thallium	4	<5	80	<5	<24	>16	>83	52	>51
Zinc	32	10	49,000	120	2,100	0[a]	>99	80	65
Bis(2-ethylhexyl) phthalate	9	0.02	81	12	20	0[a]	80	17	30
Butyl benzyl phthalate	1				4				96
Di-n-butyl phthalate	5	10	36	30	25	0[a]	83	0[a]	17
Diethyl phthalate	1				22				0[a]
Dimethyl phthalate	2	55	93		74	0[a]	98		49
Di-n-octyl phthalate	1				<10[b]				>50
N-nitrosodiphenylamine	1				<10[b]				>77
2-Chlorophenol	1				<10[b]				>88
2,4-Dichlorophenol	1				48				98
2,4-Dimethylphenol	1				<10[b]				>55
2-Nitrophenol	1				<10[b]				>47
Pentachlorophenol	1				24				55

CONTROL TECHNOLOGY SUMMARY FOR CLARIFICATION/SEDIMENTATION (continued)

Pollutant	Number of data points	Effluent concentration				Removal efficiency, %			
		Minimum	Maximum	Median	Mean	Minimum	Maximum	Median	Mean
Toxic pollutants, µg/L (continued):									
Phenol	1				33				>99
p-Chloro-m-cresol	1				<10[b]				>91
4,6-Dinitro-o-cresol	1				<10[b]				>95
Benzene	2	<10[b]	96		<91	>33	56		>45
2,4-Dinitrotoluene/2,6-dinitrotoluene	1				10				80
Ethylbenzene	2	3.0	2,400		1,200	12	20		16
Nitrobenzene	1				<10[b]				>52
Toluene	4	10	1,100	10	280	0[a]	69	6	22
1,2,4-Trichlorobenzene	1				53				0[a]
Acenaphthene	1				<10[b]				>17
Acenaphthylene	1				<10				>17
Anthracene/phenanthrene	4	0.4	51	21	<23	0[a]	92	64	55
Benzo(a)pyrene	3	<0.02	<10[b]	6	<5.3	>33	>98	80	>70
Benzo(b)fluoranthene	1				6				83
Benzo(ghi)perylene	1				<10[b]				>17
Benzo(k)fluoranthene	2	<0.02	<10[b]		<5	>17	>97		>57
Fluoranthene	3	0.4	33	<10[b]	<15	0[a]	64	>37	34
Fluorene	1				<10[b]				>79
Naphthalene	2	<10[b]	<10		<10	>41	>98		>70
Pyrene	3	0.2	21		<10	>64	79	>64	>73
Chlorodibromomethane	1				<10[b]				>77
Chloroform	1				22				0[a]
Methyl chloride	1				3				84
Methylene chloride	5	1.5	29	5.6	12	0[a]	>99	77	25
Tetrachloroethylene	3	1.1	93	<10[b]	<35	0[a]	76	>55	44
1,1,1-Trichloroethane	2	<10[b]	44		<27	0[a]	>57		29
Trichloroethylene	1				56				0[a]
Isophorone	2				<46				10[b]
Other pollutants, µg/L:									
Fluoride	1				320				51

Note: Blanks indicate data not applicable.

[a]Actual data indicate negative removal.

[b]Not detected; assumed to be <10 µg/L.

TREATMENT TECHNOLOGY: Sedimentation

Data source: Government report
Point source category: Adhesives and sealants
Subcategory:
Plant: San Leandro
References: B10, p. 66

Data source status:
Engineering estimate ___
Bench scale ___
Pilot scale _x_
Full scale ___

Use in system: Primary
Pretreatment of influent: Equalization

DESIGN OR OPERATING PARAMETERS

Unit configuration: Four section settling/flotation tank, the first and third sections are settling areas and the second and fourth sections act as flotation tanks
Wastewater flow: 0.076 m^3/min (20 gpm)
Hydraulic detention time: ~2 hr
Hydraulic loading:
Weir loading:
Sludge underflow:
Percent solids in sludge:
Scum overflow:

REMOVAL DATA

Sampling period:

Pollutant/parameter	Concentration Influent	Concentration Effluent	Percent removal
Conventional pollutants, mg/L:			
BOD_5	8,740	6,670	24
COD	27,100	25,300	7
TSS	10,600	2,260	79
Oil and grease[a]	2,220	522	76
Total phenol[a]	154	84	45
Toxic pollutants, µg/L:			
Cyanide[a]	1,900	4,500	0[b]
Zinc	99,000	49,000	51

[a]Positive interference in assay suspected.
[b]Actual data indicate negative removal.

Note: Blanks indicate information was not specified.

TREATMENT TECHNOLOGY: Sedimentation

Data source: Effluent Guidelines
Point source category: Coal mining
Subcategory: Alkaline mines
Plant: PN-11
References: All, p. IV-28[a]

Data source status:
- Engineering estimate ___
- Bench scale ___
- Pilot scale ___
- Full scale x

Use in system: Primary
Pretreatment of influent:

[a]Also, see (Treated Wastewater Analyses)

DESIGN OR OPERATING PARAMETERS

Unit configuration: Settling pond
Wastewater flow: 15.2 m^3/d (4,000 gpd)
Hydraulic detention time:
Hydraulic loading:
Weir loading:
Sludge underflow:
Percent solids in sludge:
Scum overflow:

REMOVAL DATA

Sampling period: 24-hr composite

Pollutant/parameter	Concentration		Percent removal
	Influent	Effluent	
Conventional pollutants, mg/L:			
COD	9.7	<2.0	>79
TSS	16.4	4.4	73
Toxic pollutants, μg/L:			
Antimony	2	2	0
Arsenic	3	3	0
Mercury	2.2	5.6	0[a]
Selenium	4	3	25
Zinc	160	140	8

[a]Actual data indicate negative removal.

Note: Blanks indicate information was not specified.

TREATMENT TECHNOLOGY: Sedimentation

Data source: Effluent Guidelines
Point source category: Coal mining
Subcategory: Alkaline mines
Plant: V-8
References: All, p. IV-34[a]

Data source status:
- Engineering estimate ___
- Bench scale ___
- Pilot scale ___
- Full scale x

Use in system: Primary
Pretreatment of influent:

[a]Also, see (Treated Wastewater Analyses).

DESIGN OR OPERATING PARAMETERS

Unit configuration: Settling pond #4
Wastewater flow: 200 m^3/d (53,000 gpd)
Hydraulic detention time:
Hydraulic loading:
Weir loading:
Sludge underflow:
Percent solids in sludge:
Scum overflow:

REMOVAL DATA

Sampling period: Average of three 24-hr composite samples

Pollutant/parameter	Concentration		Percent
	Influent	Effluent	removal
Conventional pollutants, mg/L:			
COD	91	76	16
TOC	57	48	16
TSS	103	29	72
Toxic pollutants, μg/L:			
Antimony	6	11	0[a]
Arsenic	4	2	50

[a]Actual data indicate negative removal.

Note: Blanks indicate information was not specified.

TREATMENT TECHNOLOGY: Sedimentation

Data source: Effluent Guidelines
Point source category: Coal mining
Subcategory: Alkaline mines
Plant: V-8
References: All, p. IV-35[a]

Data source status:
Engineering estimate ___
Bench scale ___
Pilot scale ___
Full scale x

Use in system: Primary
Pretreatment of influent:

[a]Also, see (Treated Wastewater Analyses)

DESIGN OR OPERATING PARAMETERS

Unit configuration: Settling pond #6
Wastewater flow: 10.9 m^3/d (2,880 gpd)
Hydraulic detention time:
Hydraulic loading:
Weir loading:
Sludge underflow:
Percent solids in sludge:
Scum overflow:

REMOVAL DATA

Sampling period: 24-hr composite

Pollutant/parameter	Concentration Influent	Concentration Effluent[a]	Percent removal
Conventional pollutants, mg/L:			
COD	80.0	38.7	52
TOC	54.3	21.7	60
TSS	44.8	28.9	35
Toxic pollutants, μg/L:			
Antimony	6	15	0[b]
Arsenic	ND[c]	5	0[b]
Selenium	2	<2	>0

[a]Average of 3 samples.

[b]Actual data indicate negative removal.

[c]Not detected.

Note: Blanks indicate information was not specified.

TREATMENT TECHNOLOGY: Sedimentation

Data source: Effluent Guidelines
Point source category: Coal mining
Subcategory: Alkaline mines
Plant: V-9
References: All, p. IV-35, 36[a]

Data source status:
Engineering estimate ___
Bench scale ___
Pilot scale ___
Full scale x

Use in system: Primary
Pretreatment of influent:

[a]Also, see (Treated Wastewater Analyses)

DESIGN OR OPERATING PARAMETERS

Unit configuration: Settling ponds: Pond A - dugout, Pond B - pollack
Wastewater flow: Pond A - 152 m^3/d (40,000 gpd),
Pond B - 2,690 m^3/d (710,000 gpd)
Hydraulic detention time:
Hydraulic loading:
Weir loading:
Sludge underflow:
Percent solids in sludge:
Scum overflow:

REMOVAL DATA

Sampling period: 24-hr composite

Pollutant/ parameter	Pond A			Pond B		
	Concentration, mg/L		Percent	Concentration, mg/L		Percent
	Influent	Effluent	removal	Influent	Effluent	removal
Conventional pollutants:						
COD	14	18	0[a]	16.3	13.7	16
TOC	7	14.6	0[a]	10.8	9.6	11
TSS	111	46	59	59.6	78.6	0[a]

[a]Actual data indicate negative removal.

Note: Blanks indicate information was not specified.

TREATMENT TECHNOLOGY: Sedimentation

Data source: Effluent Guidelines
Point source category: Coal mining
Subcategory: Coal preparation plants and associated areas
Plant: NC-3
References: All, p. IV-41[a]

Data source status:
Engineering estimate ___
Bench scale ___
Pilot scale ___
Full scale x

Use in system: Primary
Pretreatment of influent:

[a]Also, see (Treated Wastewater Analyses)

DESIGN OR OPERATING PARAMETERS

Unit configuration: Slurry pond
Wastewater flow: 9,470 m^3/d (2.5 MGD)
Hydraulic detention time:
Hydraulic loading:
Weir loading:
Sludge underflow:
Percent solids in sludge:
Scum overflow:

REMOVAL DATA

Sampling period: 24-hr composite

Pollutant/parameter	Concentration, µg/L Influent	Concentration, µg/L Effluent	Percent removal
Toxic pollutants:			
Copper	270	<4	>98
Selenium	50	<5	>90
Zinc	1,000	49	95

Note: Blanks indicate information was not specified.

TREATMENT TECHNOLOGY: Sedimentation

Data source: Effluent Guidelines
Point source category: Coal mining
Subcategory: Coal preparation plants and associated areas
Plant: NC-8
References: All, pp. IV-43, 47

Data source status:
Engineering estimate ___
Bench scale ___
Pilot scale ___
Full scale x

Use in system: Primary
Pretreatment of influent:

DESIGN OR OPERATING PARAMETERS

Unit configuration: Slurry pond
Wastewater flow: 47,100 m^3/d (12,400,000 gpd)
Hydraulic detention time:
Hydraulic loading:
Weir loading:
Sludge underflow:
Percent solids in sludge:
Scum overflow:

REMOVAL DATA

Sampling period: Average of three 24-hr composite samples

Pollutant/parameter	Concentration		Percent removal
	Influent	Effluent	
Conventional pollutants, mg/L:			
COD	36,000	19	>99
TOC	1,490	96.8	94
TSS	34,400	8.9	>99
Toxic pollutants, μg/L:			
Antimony	<2	<6	0[a]
Arsenic	250	6	98
Beryllium	60	<1	>98
Chromium	530	13	98
Copper	1,300	<6	>99
Lead	970	<20	>97
Nickel	1,200	<5	>99
Selenium	<5	<6	0[a]
Thallium	6	<5	>16
Zinc	5,300	<60	>98
Benzene[b]	15	ND[c]	>33

[a]Actual data indicate negative removal.

[b]Only one sample.

[c]Not detected, assume to be <10 μg/L.

Note: Blanks indicate information was not specified.

TREATMENT TECHNOLOGY: Sedimentation

Data source: Effluent Guidelines
Point source category: Coal mining
Subcategory: Coal preparation plants and associated areas
Plant: NC-22
References: All, pp. IV-44, 47-48[a]

Data source status:
Engineering estimate ___
Bench scale ___
Pitot scale ___
Full scale x

Use in system: Primary
Pretreatment of influent:

[a]Also, see (Treated Wastewater Analyses)

DESIGN OR OPERATING PARAMETERS

Unit configuration: Slurry pond
Wastewater flow: 1,040 m^3/d (274,000 gpd)
Hydraulic detention time:
Hydraulic loading:
Weir loading:
Sludge underflow:
Percent solids in sludge:
Scum overflow:

REMOVAL DATA

Sampling period: 24-hr composite samples

Pollutant/parameter	Concentration Influent	Effluent	Percent removal
Conventional pollutants, mg/L:			
COD[a]	48,800	20.3	>99
TOC[a]	8,450	13.5	>99
TSS[a]	13,900	18.7	>99
Toxic pollutants, µg/L:			
Arsenic[a]	180	<5	>97
Chromium[a]	230	40	83
Copper[a]	230	8	97
Lead[a]	470	50	89
Mercury[a]	2.5	<1	>60
Nickel[a]	300	10	97
Selenium[a]	34	3	91
Thallium[a]	15	<5	>66
N-nitrosodiphenylamine[b]	44	ND[c]	>77
2-Chlorophenol[a]	86	ND	>88
2,4-Dimethylphenol[a]	22	ND	>55
2-Nitrophenol[a]	19	ND	>47
4,6-Dinitro-o-cresol[a]	190	ND	>95
Nitrobenzene[c]	21	ND	>52
Toluene[b]	12	10	17
Acenaphthylene[b]	12	ND	>17
Anthracene/phenanthrene[d]	23	<10	>56
Benzo(a)pyrene[a]	15	ND	>33
Benzo(b)fluoranthene, Benzo(k)fluoranthene	12	ND	>17
Benzo(ghi)perylene[a]	12	ND	>17
Fluoranthene[a]	16	ND	>37
Fluorene[a]	47	ND	>79
Naphthalene[a]	410	ND	>98
Pyrene[b]	28	ND	>64
Methylene chloride[a]	82	19	77
1,1,1-Trichloroethane[c]	23	ND	>57
Isophorone[a]	310	ND	>97

[a]Average of 3 samples
[b]Average of 2 samples.
[c]Not detected, assume less than 10 µg/L
[d]Only 1 sample.

Note: Blanks indicate information was not specified.

TREATMENT TECHNOLOGY: Sedimentation

Data source: Effluent Guidelines
Point source category: Electroplating
Subcategory:
Plant: 23061
References: A14, p. 149

Data source status:
Engineering estimate ___
Bench scale ___
Pilot scale ___
Full scale x

Use in system: Primary
Pretreatment of influent:

DESIGN OR OPERATING PARAMETERS

Unit configuration:
Wastewater flow:
Hydraulic detention time:
Hydraulic loading:
Weir loading:
Sludge underflow:
Percent solids in sludge:
Scum overflow:

REMOVAL DATA

Sampling period:

Pollutant/parameter	Concentration Influent	Concentration Effluent	Percent removal
Conventional pollutants, mg/L:			
TSS	67	4	94
Total phosphorus	14.3	13.9	3
Toxic pollutants, μg/L:			
Cadmium	5	2	60
Chromium (+6)	5	5	0
Chromium (total)	10	6	40
Copper	130	34	73
Cyanide (total)	5	5	0
Lead	7	14	0[a]
Nickel	380	310	17
Silver	2	3	0[a]
Zinc	40	34	15

[a]Actual data indicate negative removal.

Note: Blanks indicate information was not specified.

TREATMENT TECHNOLOGY: Sedimentation

Data source: Effluent Guidelines
Point source category: Foundry industry
Subcategory: Copper and copper alloys foundaries, mold cooling and casting quench
Plant: 6809
References: A27, pp. V-16, VI-73-80, VII-29

Data source status:
Engineering estimate ___
Bench scale ___
Pilot scale ___
Full scale x

Use in system: Primary
Pretreatment of influent:

DESIGN OR OPERATING PARAMETERS

Unit configuration: Lagoon
Wastewater flow:
Hydraulic detention time:
Hydraulic loading:
Weir loading:
Sludge underflow:
Percent solids in sludge:
Scum overflow:

REMOVAL DATA

Sampling period:

Pollutant/parameter	Concentration		Percent
	Influent	Effluent	removal
Conventional pollutants, mg/L:			
TSS	52	20	70
Oil and grease	30	6.2	76
Toxic pollutants, μg/L:			
Cadmium	100	40	60
Copper	350	110	69
Mercury	0.3	0.9	0[a]
Nickel	ND[b]	60	0[a]
Zinc	2,000	1,400	30
Dimethyl phthalate	15	93	0[a]
Tetrachloroethylene	80	93	0[a]
1,1,1,-Trichloroethane	37	44	0[a]
Trichloroethylene	50	56	0[a]

[a]Actual data indicate negative removal.
[b]Not detected.

Note: Blanks indicate information was not specified.

TREATMENT TECHNOLOGY: Sedimentation

Data source: Effluent Guidelines
Point source category: Foundry industry
Subcategory: Ferrous foundry dust collection
Plant: AAA-2A
References: A27, pp. VI-96, VII-17, 31, 57

Data source status:
Engineering estimate ___
Bench scale ___
Pilot scale ___
Full scale x

Use in system: Primary
Pretreatment of influent:

DESIGN OR OPERATING PARAMETERS

Unit configuration: Settling lagoon
Wastewater flow:
Hydraulic detention time:
Hydraulic loading:
Weir loading:
Sludge underflow:
Percent solids in sludge:
Scum overflow:

REMOVAL DATA

Sampling period:

Pollutant/parameter	Concentration		Percent
	Influent	Effluent	removal
Conventional pollutants, mg/L:			
TSS	4,200	4.6	>99
Oil and grease	15	12	20
Total phenol	1.1	0.04	96
Toxic pollutants, µg/L:			
Cyanide	37	19	49

Note: Blanks indicate information was not specified.

TREATMENT TECHNOLOGY: Sedimentation

Data source: Effluent Guidelines
Point source category: Foundry industry
Subcategory: Ferrous foundry dust collection
Plant: HHH-2B
References: A27, pp. VI-96, VII-20, 31, 67

Data source status:
Engineering estimate ___
Bench scale ___
Pilot scale ___
Full scale _x_

Use in system: Primary
Pretreatment of influent:

DESIGN OR OPERATING PARAMETERS

Unit configuration: Settling lagoon[a]
Wastewater flow:
Hydraulic detention time:
Hydraulic loading:
Weir loading:
Sludge underflow:
Percent solids in sludge:
Scum overflow:

---

[a]Treated effluent 100% recycled.

REMOVAL DATA

Sampling period:

Pollutant/parameter	Concentration		Percent removal
	Influent	Effluent	
Conventional pollutants, mg/L:			
TSS	1,500	64	96
Oil and grease	14	2.7	81
Toxic pollutants, μg/L:			
Copper	130	21	84
Zinc	1,900	1,800	5

Note: Blanks indicate information was not specified.

TREATMENT TECHNOLOGY: Sedimentation

Data source: Effluent Guidelines
Point source category: Foundry industry
Subcategory: Ferrous foundry dust collection
Plant: 291C
References: A27, pp. V-22, VI-89-96, VII-32, 70

Data source status:
Engineering estimate ___
Bench scale ___
Pilot scale ___
Full scale x

Use in system: Primary
Pretreatment of influent:

DESIGN OR OPERATING PARAMETERS

Unit configuration: Settling tank[a]
Wastewater flow:
Hydraulic detention time:
Hydraulic loading:
Weir loading:
Sludge underflow:
Percent solids in sludge:
Scum overflow:

[a]Treated effluent 100% recycled.

REMOVAL DATA

Sampling period:

Pollutant/parameter	Concentration		Percent removal
	Influent	Effluent	
Conventional pollutants, mg/L:			
TSS	410	41	90
Oil and grease	3	2.7	10
Toxic pollutants, µg/L:			
Cyanide	7	74	0[a]
Lead	30	10	67
Bis(2-ethylhexyl) phthalate	9	2	78
Anthracene/phenanthrene	3	51	0[a]

[a]Actual data indicate negative removal.

Note: Blanks indicate information was not specified.

TREATMENT TECHNOLOGY: Sedimentation

Data source: Effluent Guidelines
Point source category: Foundry industry
Subcategory: Ferrous foundry dust collection
Plant: 7929
References: A27, pp. V-23, VI-89-95, 97, VII-21, 32

Data source status:
Engineering estimate ___
Bench scale ___
Pilot scale ___
Full scale x

Use in system: Primary
Pretreatment of influent:

DESIGN OR OPERATING PARAMETERS

Unit configuration: Settling basin
Wastewater flow:
Hydraulic detention time:
Hydraulic loading:
Weir loading:
Sludge underflow:
Percent solids in sludge:
Scum overflow:

REMOVAL DATA

Sampling period:

Pollutant/parameter	Concentration Influent	Concentration Effluent	Percent removal
Conventional pollutants, mg/L:			
TSS	880	600	32
Oil and grease	3	15	0[a]
Total phenol	9.1	0.76	92
Toxic pollutants, μg/L:			
Copper	30	140	0[a]
Cyanide	47	14	70
Lead	37	200	0[a]
Nickel	10	40	0[a]
Bis (2-ethylhexyl) phthalate	ND[b]	81	0[a]
Butyl benzyl phthalate	100	4	96
Di-n-butyl phthalate	200	34	83
Diethyl phthalate	9	22	0[a]
Dimethyl phthalate	2,200	55	98
2,4-Dichlorophenol	2,200	48	98
Pentachlorophenol	53	24	55
Phenol	20,000	33	>99
Anthracene/phenanthrene	410	32	92
Benzo(a)pyrene	30	6	80
Benzo(b)fluoranthene	36	6	83
Fluoranthene	20	33	0[a]
Pyrene	98	21	79

[a]Actual data indicate negative removal.

[b]Not detected, assumed to be less than 10 μg/L.

Note: Blanks indicate information was not specified.

TREATMENT TECHNOLOGY: Sedimentation

Data source: Effluent Guidelines
Point source category: Foundry industry
Subcategory: Ferrous Foundry melting furnace scrubber
Plant: HHH-2B
References: A27, pp. VI-105, VII-20, 33, 67

Data source status:
- Engineering estimate ___
- Bench scale ___
- Pilot scale ___
- Full scale x

Use in system: Primary
Pretreatment of influent:

DESIGN OR OPERATING PARAMETERS

Unit configuration: Settling lagoon[a]
Wastewater flow:
Hydraulic detention time:
Hydraulic loading:
Weir loading:
Sludge underflow:
Percent solids in sludge:
Scum overflow:

[a]Treated effluent 100% recycled.

REMOVAL DATA

Sampling period:

Pollutant/parameter	Concentration Influent	Concentration Effluent	Percent removal
Conventional pollutants, mg/L:			
TSS	4,200	40	99
Toxic pollutants, μg/L:			
Copper	4,400	90	98
Lead	29,000	1,400	95
Mercury	6	3	50
Zinc	87,000	4,400	95

Note: Blanks indicate information was not specified.

TREATMENT TECHNOLOGY: Sedimentation

Data source: Effluent Guidelines
Point source category: Foundry industry
Subcategory: Ferrous foundry sand washing
Plant: AAA-2A
References: A27, pp. VI-130, VII-17, 37, 57

Data source status:
- Engineering estimate ___
- Bench scale ___
- Pilot scale ___
- Full scale x

Use in system: Primary
Pretreatment of influent:

DESIGN OR OPERATING PARAMETERS

Unit configuration: Settling lagoon
Wastewater flow:
Hydraulic detention time:
Hydraulic loading:
Weir loading:
Sludge underflow:
Percent solids in sludge:
Scum overflow:

REMOVAL DATA

Sampling period:

Pollutant/parameter	Concentration Influent	Concentration Effluent	Percent removal
Conventional pollutants, mg/L:			
TSS	5,900	6.6	>99
Oil and grease	8	7.8	3
Total phenol	0.59	0.021	96
Toxic pollutants, µg/L:			
Cyanide	26	14	46
Mercury	0.01	0.3	0[a]

[a]Actual data indicate negative removal.

Note: Blanks indicate information was not specified.

TREATMENT TECHNOLOGY: Sedimentation

Data source: Effluent Guidelines
Point source category: Foundry industry
Subcategory: Steel foundries - casting quench and mold cooling operations
Plant: 417A
References: A27, pp. V-41, VI-115-122, VII-36

Data source status:
Engineering estimate ___
Bench scale ___
Pilot scale ___
Full scale x

Use in system: Primary
Pretreatment of influent:

DESIGN OR OPERATING PARAMETERS

Unit configuration:
Wastewater flow:
Hydraulic detention time:
Hydraulic loading:
Weir loading:
Sludge underflow:
Percent solids in sludge:
Scum overflow:
Raw waste flow rate:
Effluent flow rate:

REMOVAL DATA

Sampling period:

Pollutant/parameter	Concentration		Percent removal
	Influent	Effluent	
Conventional pollutants, mg/L:			
TSS	90	62	35
Oil and grease	ND[a]	9	0[b]
Toxic pollutants, μg/L:			
Copper	20	50	0[b]
Cyanide	3	2	33
Lead	ND	60	0[b]
Mercury	ND	0.8	0[b]
Zinc	ND	140	0[b]
Bis(2-ethylhexyl) phthalate	ND	27	0[b]

[a]Not detected.

[b]Actual data indicate negative removal.

Note: Blanks indicate information was not specified.

TREATMENT TECHNOLOGY: Sedimentation

Data source: Effluent Guidelines
Point source category: Foundry industry
Subcategory: Steel foundries, sand washing and reclaiming
Plant: 694K
References: A27, pp. V-43, VI-123-130

Data source status:
Engineering estimate ___
Bench scale ___
Pilot scale ___
Full scale x

Use in system: Primary
Pretreatment of influent:

DESIGN OR OPERATING PARAMETERS

Unit configuration: Settling lagoon
Wastewater flow:
Hydraulic detention time:
Hydraulic loading:
Weir loading:
Sludge underflow:
Percent solids in sludge:
Scum overflow:

REMOVAL DATA

Sampling period:

Pollutant/parameter	Concentration, μg/L Influent	Concentration, μg/L Effluent	Percent removal
Toxic pollutants:			
2,4-Dinitrotoluene/2,6-Dinitrotoluene	50	10	80
1,2,4-Trichlorobenzene	30	53	0[a]

[a]Actual data indicate negative removal.

Note: Blanks indicate information was not specified.

TREATMENT TECHNOLOGY: Sedimentation

Data source: Effluent Guidelines
Point source category: Ink manufacturing
Subcategory: Water and/or caustic wash
Plant: 22
References: A10, p. VII-2 and Appendix H

Data source status:
Engineering estimate ___
Bench scale ___
Pilot scale ___
Full scale x

Use in system: Primary
Pretreatment of influent: Neutralization

DESIGN OR OPERATING PARAMETERS

Unit configuration: Oil skimming provided.
Wastewater flow:
Hydraulic detention time:
Hydraulic loading:
Weir loading:
Sludge underflow:
Percent solids in sludge:
Scum overflow:

REMOVAL DATA

Sampling period:

Pollutant/parameter	Concentration		Percent removal
	Influent	Effluent	
Conventional pollutants, mg/L:			
BOD_5	2,100	2,600	0[a]
COD	32,000	4,800	85
TOC	4,000	940	76
TSS	1,600	110	93
Oil and grease	2,400	260	89
Total phenol	330	30	91
Toxic pollutants, μg/L:			
Cadmium	90	20	78
Chromium	10,000	<50	>99
Copper	10,000	<60	>99
Lead	90,000	<200	>99
Zinc	1,000	<600	>40
Benzene	220	96	56
Ethylbenzene	6,700	2,400	64
Toluene	3,600	1,100	69
Naphthalene	17	<10	>41
Chlorodibromomethane	43	ND[b]	>77
Methylene chloride	45	29	36
Tetrachloroethylene	22	ND	>55
Isophorone	ND	46	0[a]

[a]Actual data indicate negative removal.
[b]Not detected, assumed to be <10 μg/L.

Note: Blanks indicate information was not specified.

TREATMENT TECHNOLOGY: Sedimentation

Data source: Effluent Guidelines
Point source category: Inorganic chemicals
Subcategory: Hydrofluoric acid
Plant: 251
References: A29, pp. 210-211

Data source status:
Engineering estimate ___
Bench scale ___
Pilot scale ___
Full scale x

Use in system: Primary
Pretreatment of influent:

DESIGN OR OPERATING PARAMETERS

Unit configuration: Gypsum pond
Wastewater flow:
Hydraulic detention time:
Hydraulic loading:
Weir loading:
Sludge underflow:
Percent solids in sludge:
Scum overflow:

REMOVAL DATA

Sampling period: Three 24 hr composite samples

Pollutant/parameter	Concentration		Percent
	Influent	Effluent	removal
Conventional pollutants, mg/L:			
TSS	18,600	9.7	>99
Other pollutants, μg/L			
Fluoride	660	320	51

Note: Blanks indicate information was not specified.

TREATMENT TECHNOLOGY: Sedimentation

Data source: Effluent Guidelines
Point source category: Inorganic chemicals
Subcategory: Titanium dioxide (chloride process) manufacture
Plant: 172
References: A29, pp. 270-271

Data source status:
- Engineering estimate ___
- Bench scale ___
- Pilot scale ___
- Full scale x

Use in system: Primary
Pretreatment of influent:

DESIGN OR OPERATING PARAMETERS

Unit configuration: Two retention basins in series, pH adjustment to basin effluent
Wastewater flow:
Hydraulic detention time:
Hydraulic loading:
Weir loading:
Sludge underflow:
Percent solids in sludge:
Scum overflow:
pH: 7.6-7.9

REMOVAL DATA

Sampling period: composite sample

Pollutant/parameter	Concentration		Percent
	Influent	Effluent	removal
Conventional pollutants, mg/L:			
TSS	223	6.6	97
Toxic pollutants, μg/L:			
Chromium	620	17	97
Zinc	270	84	69

Note: Blanks indicate information was not specified.

TREATMENT TECHNOLOGY: Sedimentation

Data source: Effluent Guidelines
Point source category: Iron and steel
Subcategory: Bee-hive coke manufacturing
Plant: E
References: A38, p. 252

Data source status:
- Engineering estimate ___
- Bench scale ___
- Pilot scale ___
- Full Scale x

Use in system: Primary
Pretreatment of influent:

DESIGN OR OPERATING PARAMETERS

Unit configuration: Two settling ponds in parallel
Wastewater flow: 0.022 m^3/s (340 gpm)
Hydraulic detention time:
Hydraulic loading:
Weir loading:
Sludge underflow:
Percent solids in sludge:
Scum overflow:

REMOVAL DATA

Sampling period:

Pollutant/parameter	Concentration		Percent removal
	Influent	Effluent	
Conventional pollutants, mg/L:			
TSS	165	36	78
Oil and grease	<5	<5	0
Total phenol	0.011	0.014	0[a]
Toxic pollutants, µg/L:			
Cyanide	2	4	0[a]

[a]Actual data indicate negative removal.

Note: Blanks indicate information was not specified.

TREATMENT TECHNOLOGY: Sedimentation

Data source: Effluent Guidelines
Point source category: Iron and steel
Subcategory: Continuous casting
Plant: B
References: A33, pp. 347-348

Data source status:
Engineering estimate ___
Bench scale ___
Pilot scale ___
Full scale x

Use in system: Primary
Pretreatment of influent:

DESIGN OR OPERATING PARAMETERS

Unit configuration: Two lagoons in parallel
Wastewater flow: 3.8 L/s
Hydraulic detention time:
Hydraulic loading:
Weir loading:
Sludge underflow:
Percent solids in sludge:
Scum overflow:

REMOVAL DATA

Sampling period:

Pollutant/parameter	Concentration Influent	Concentration Effluent	Percent removal
Conventional pollutants, mg/L:			
TSS	6	8	0[a]
Oil and Grease	10	6	40
Toxic pollutants, μg/L:			
Copper	87	70	24
Nickel	100	90	10
Selenium	220	5	98
Zinc	200	200	0
Di-n-butyl phthalate	7	10	0[a]
Di-n-octyl phthalate	20	ND[b]	>50
p-chloro-*m*-cresol	110	ND[b]	>91
Toluene	7	10	0[a]

[a] Actual data indicate negative removal.
[b] Not detected, assumed to be <10 μg/L.

TREATMENT TECHNOLOGY: Sedimentation

Data source: Effluent Guidelines
Point source category: Iron and steel
Subcategory: Cold rolling
Plant: XX-2
References: A41, pp. 254, 259

Data source status:
Engineering estimate ___
Bench scale ___
Pilot scale ___
Full scale x

Use in system: Primary
Pretreatment of influent:

DESIGN OR OPERATING PARAMETERS

Unit configuration: 72,800 m^2 (18 acre) lagoon divided into two segments, oil is skimmed from the top of the lagoon
Wastewater flow: 3,680 L/s (58,300 gpm)
Hydraulic detention time:
Hydraulic loading:
Weir loading:
Sludge underflow:
Percent solids in sludge:
Scum overflow:

REMOVAL DATA

Sampling period:

Pollutant/parameter	Concentration, mg/L		Percent removal
	Influent	Effluent[a]	
Conventional pollutants:			
TSS	260	30	88
Oil and grease	619	7	99

[a]Calculated from influent and percent removal.

Note: Blanks indicate information was not specified.

TREATMENT TECHNOLOGY: Sedimentation

Data source: Effluent Guidelines
Point source category: Iron and steel
Subcategory: Hot forming-section
Plant: I-2
References: A35, p. 165

Data source status:
Engineering estimate ___
Bench scale ___
Pilot scale ___
Full scale x

Use in system: Primary
Pretreatment of influent:

DESIGN OR OPERATING PARAMETERS

Unit configuration: 46,200 m^3 (12.2 Mgal) Terminal settling lagoon
Wastewater flow: 350 L/s (5,560 gpm)
Hydraulic detention time:
Hydraulic loading:
Weir loading:
Sludge underflow:
Percent solids in sludge:
Scum overflow:

REMOVAL DATA

Sampling period:

Pollutant/parameter	Concentration, mg/L Influent	Concentration, mg/L Effluent	Percent removal
Conventional pollutants:			
TSS	185	39	79
Oil and grease	120	14	89

Note: Blanks indicate information was not specified.

TREATMENT TECHNOLOGY: Sedimentation

Data source: Effluent Guidelines
Point source category: Iron and steel
Subcategory: Hot forming-section
Plant: O
References: A35, p. 164

Data source status:
- Engineering estimate ___
- Bench scale ___
- Pilot scale ___
- Full scale x

Use in system: Primary
Pretreatment of influent:

DESIGN OR OPERATING PARAMETERS

Unit configuration: Clarifier
Wastewater flow:
Hydraulic detention time:
Hydraulic loading:
Weir loading:
Sludge underflow:
Percent solids in sludge:
Scum overflow:

REMOVAL DATA

Sampling period:

Pollutant/parameter	Concentration, mg/L		Percent
	Influent	Effluent	removal
Conventional pollutants:			
TSS	15	57	0[a]
Oil and grease	4.9	12.3	0[a]

[a]Actual data indicate negative removal.

Note: Blanks indicate information was not specified.

TREATMENT TECHNOLOGY: Sedimentation

Data source: Effluent Guidelines
Point source category: Iron and steel
Subcategory: Hot forming-section
Plant: R
References: A35, p 164

Data source status:
- Engineering estimate ___
- Bench scale ___
- Pilot scale ___
- Full scale x

Use in system: Primary
Pretreatment of influent: Scale pit

DESIGN OR OPERATING PARAMETERS

Unit configuration: Settling lagoon
Wastewater flow: 90 L/s (15,700 gpm)
Hydraulic detention time:
Hydraulic loading:
Weir loading:
Sludge underflow:
Percent solids in sludge:
Scum overflow:

REMOVAL DATA

Sampling period:

Pollutant/parameter	Concentration Influent	Concentration Effluent	Percent removal
Conventional pollutants, mg/L:			
TSS	32	45	0[a]
Oil and grease	3.8	5.3	0[a]
Toxic pollutants, µg/L:			
Chromium	<10	<10	0
Copper	<10	10	0[a]
Lead	30	30	0[a]
Nickel	60	40	33
Zinc	20	20	0

[a]Actual data indicate negative removal.

Note: Blanks indicate information was not specified.

TREATMENT TECHNOLOGY: Sedimentation

Data source: Effluent Guidelines
Point source category: Iron and steel
Subcategory: Wet open combustion basic oxygen furnace
Plant: Furnace 033
References: A40, pp. 69-70

Data source status:
Engineering estimate ___
Bench scale ___
Pilot scale ___
Full scale x

Use in system: Primary
Pretreatment of influent:

DESIGN OR OPERATING PARAMETERS

Unit configuration: Clarifier
Wastewater flow: 1.01 m^3/mg of production
Hydraulic detention time:
Hydraulic loading:
Weir loading:
Sludge underflow:
Percent solids in sludge:
Scum overflow:

REMOVAL DATA

Sampling period:

Pollutant/parameter	Concentration		Percent
	Influent	Effluent	removal
Conventional pollutants, mg/L:			
TSS	7,770	52	99
Toxic pollutants, µg/L:			
Arsenic	75	17	77
Chromium	2,950	30,100	0[a]
Copper	920	69	92
Lead	13,600	942	99
Mercury	0.1	0.1	0
Nickel	710	2,020	0[a]
Selenium	37	31	16
Thallium	130	80	38
Zinc	49,100	320	99
Bis(2-ethylhexyl) phthalate	18	32	0[a]
Chloroform	13	22	0[a]

[a] Actual data indicate negative removal.

Note: Blanks indicate information was not specified.

TREATMENT TECHNOLOGY: Sedimentation

Data source: Effluent Guidelines
Point source category: Iron and steel
Subcategory: Wet suppressed, basic oxygen furnace
Plant: 034
References: A40, pp. 72-73

Data source status:
Engineering estimate ___
Bench scale ___
Pilot scale ___
Full scale x

Use in system: Primary
Pretreatment of influent: Equalization

DESIGN OR OPERATING PARAMETERS

Unit configuration: Clarifier
Wastewater flow:
Hydraulic detention time:
Hydraulic loading:
Weir loading:
Sludge underflow:
Percent solids in sludge:
Scum overflow:

REMOVAL DATA

Sampling period:

Pollutant/parameter	Concentration		Percent removal
	Influent	Effluent	
Conventional pollutants, mg/L:			
TSS	380	47	88
Toxic pollutants, μg/L:			
Copper	80	100	0[a]
Lead	242,000	820	>99
Nickel	550	690	0[a]
Zinc	610	280	54

[a]Actual data indicate negative removal.

Note: Blanks indicate information was not specified.

TREATMENT TECHNOLOGY: Sedimentation

Data source: Effluent Guidelines
Point source category: Leather tanning and finishing
Subcategory:
Plant: 10
References: A15, p. 67

Data source status:
Engineering estimate ___
Bench scale ___
Pilot scale ___
Full scale x

Use in system: Primary
Pretreatment of influent:

DESIGN OR OPERATING PARAMETERS

Unit configuration: Two circular clarifiers in series
Wastewater flow: 3,030 m^3/d (0.8 MGD)
Hydraulic detention time:
Hydraulic loading: 18.8 $m^3/d/m^2$ (460 gpd/ft^2)
Weir loading:
Sludge underflow:
Percent solids in sludge:
Scum overflow:

REMOVAL DATA

Sampling period:

Pollutant/parameter	Concentration Influent	Concentration Effluent	Percent removal
Conventional pollutants, mg/L:			
BOD_5	2,110	1,150	45
TSS	3,170	945	70
Oil and grease	490	57	88
Toxic pollutants, µg/L:			
Chromium	51,000	24,000	53

Note: Blanks indicate information was not specified.

TREATMENT TECHNOLOGY: Sedimentation

Data source: Effluent Guidelines
Point source category: Mineral mining and processing
Subcategory: See below
Plant: See below
References: A18, p. 236

Data source status:
Engineering estimate ___
Bench scale ___
Pilot scale ___
Full scale x

Use in system: Primary
Pretreatment of influent: None

DESIGN OR OPERATING PARAMETERS

Unit configuration:
Wastewater flow:
Hydraulic detention time:
Hydraulic loading:
Weir loading:
Sludge underflow:
Percent solids in sludge:
Scum overflow:

REMOVAL DATA

Sampling period:

Subcategory	Plant	TSS concentration, mg/L		Percent removal
		Influent	Effluent	
Construction sand and gravel	1044	5,110	154	97
Construction sand and gravel	1083	29,500	79	>99
Construction sand and gravel	1129	4,660	44	99
Construction sand and gravel	1391	12,700	18	>99
Crushed stone	1001	1,050	8	99
Crushed stone	1003	7,680	8	>99
Crushed stone	1004	5,710	12	>99
Crushed stone	1021	7,210	28	>99
Crushed stone	1039	10,000	14	>99
Crushed stone	1053	21,800	56	>99
Dimension stone	3001	1,810	37	98
Dimension stone	3007	2,180	80	96
Industrial sand	N01	427	56	87
Industrial sand	1019	2,010	56	97

Note: Blanks indicate information was not specified.

TREATMENT TECHNOLOGY: Sedimentation

Data source: Effluent Guidelines
Point source category: Ore mining and dressing
Subcategory: Aluminum Ore (bauxite) mine
Plant: 5102
References: A2, p. V-51, 52

Data source status:
- Engineering estimate ___
- Bench scale ___
- Pilot scale ___
- Full scale x

Use in system: Primary
Pretreatment of influent: Lime neutralization

DESIGN OR OPERATING PARAMETERS

Unit configuration:
Wastewater flow:
Hydraulic detention time:
Hydraulic loading:
Weir loading:
Sludge underflow:
Percent solids in sludge:
Scum overflow:

REMOVAL DATA

Sampling period: 24-hr composite

Pollutant/parameter	Concentration Influent	Concentration Effluent	Percent removal
Conventional pollutants, mg/L:			
TOC	2	4	0[a]
TSS	2.8	6	0[a]
Toxic pollutants, µg/L:			
Chromium	30	25	17
Copper	60	50	17
Mercury	37	84	0[a]

[a]Actual data indicate negative removal.

Note: Blanks indicate information was not specified.

TREATMENT TECHNOLOGY: Sedimentation

Data source: Effluent Guidelines
Point source category: Ore mining and dressing
Subcategory: Copper mine/mill
Plant: 2120
References: A2, pp. V-23, 24

Data source status:
- Engineering estimate ___
- Bench scale ___
- Pilot scale ___
- Full scale x

Use in system: Primary
Pretreatment of influent:

DESIGN OR OPERATING PARAMETERS

Unit configuration: Tailing pond
Wastewater flow:
Hydraulic detention time:
Hydraulic loading:
Weir loading:
Sludge underflow:
Percent solids in sludge:
Scum overflow:

REMOVAL DATA

Sampling period: 24-hr composite

Pollutant/parameter	Concentration Influent	Concentration Effluent	Percent removal
Conventional pollutants, mg/L:			
COD	3,880	12	>99
TOC	8	9	0[a]
TSS	311,000	5	>99
Total phenol	<0.01	0.01	0[a]
Toxic pollutants:			
Antimony, μg/L	300	<50	>83
Arsenic, μg/L	4,000	<2	>99
Asbestos (total), fibers/L	1.2×10^{12}	1.2×10^{9}	>99
Asbestos (chrysotile), fibers/L	3.1×10^{11}	3.0×10^{8}	>99
Cadmium, μg/L	530	<5	>99
Chromium, μg/L	670	<10	>98
Copper, μg/L	330,000	110	>99
Lead, μg/L	21,000	<20	>99
Mercury, μg/L	1.0	<0.5	>50
Nickel, μg/L	910	<20	>98
Selenium, μg/L	200	<5	>97
Silver, μg/L	540	20	96
Zinc, μg/L	280,000	50	>99
Bis(2-ethylhexyl) phthalate,[b] μg/L	4	2.6	35
Di-n-butyl phthalate,[b] μg/L	17	30	0[a]
Methyl chloride,[c] μg/L	19	3	84
Tetrachloroethylene, μg/L	4.5	1.1	76

[a]Data indicate negative removal.
[b]Possibly due to tubing used in sampling apparatus.
[c]Possibly due to laboratory contamination.

Note: Blanks indicate information was not specified.

TREATMENT TECHNOLOGY: Sedimentation

Data source: Effluent Guidelines
Point source category: Ore mining and dressing
Subcategory: Copper mine/mill/smelter
Plant: 2117
References: A2, p. V-25

Data source status:
Engineering estimate ___
Bench scale ___
Pilot scale ___
Full scale x

Use in system: Primary
Pretreatment of influent:

DESIGN OR OPERATING PARAMETERS

Unit configuration: Tailing pond
Wastewater flow:
Hydraulic detention time:
Hydraulic loading:
Weir loading:
Sludge underflow:
Percent solids in sludge:
Scum overflow:

REMOVAL DATA

Sampling period: 24 hour composite (2 sets)

Pollutant/parameter	Concentration		Percent removal
	Influent	Effluent	
Conventional pollutants, mg/L:			
COD	4,850	15	>99
TOC	29.5	5	83
TSS	207,000	2	>99
Total phenol	5.1	0.255	95
Toxic pollutants:			
Arsenic, μg/L	75	2	97
Asbestos (total), fibers/L	1.9×10^{11}	4.6×10^{6}	>99
Asbestos (chrysotile), fibers/L	5.5×10^{10}	4.4×10^{5}	>99
Beryllium, μg/L	25	5	80
Cadmium, μg/L	120	5	96
Chromium, μg/L	1,900	45	98
Copper, μg/L	59,000	20	>99
Cyanide, μg/L	200	<20	>90
Lead, μg/L	2,000	40	98
Nickel, μg/L	2,000	20	99
Selenium, μg/L	320	7	98
Silver, μg/L	200	<20	>90
Zinc, μg/L	140,000	40	>99

Note: Blanks indicate information was not specified.

TREATMENT TECHNOLOGY: Sedimentation

Data source: Effluent Guidelines
Point source category: Ore mining and dressing
Subcategory: Copper mine/mill/smelter/refinery
Plant: 2122
References: A2, pp. V-7-10

Data source status:
- Engineering estimate ___
- Bench scale ___
- Pilot scale ___
- Full scale x

Use in system: Primary
Pretreatment of influent:

DESIGN OR OPERATING PARAMETERS

Unit configuration: Tailing pond
Wastewater flow:
Hydraulic detention time:
Hydraulic loading:

Weir loading:
Sludge underflow:
Percent solids in sludge:
Scum overflow:

REMOVAL DATA

Sampling period: Average of two 24-hr composite samples

Pollutant/parameter	Concentration		Percent
	Influent	Effluent	removal
Conventional pollutants, mg/L:			
COD	530	5	99
TOC	9.5	7	26
TSS	313,000	14	>99
Total phenol	0.23	0.017	93
Toxic pollutants:			
Arsenic, μg/L	1,400	4	>99
Asbestos, fibers/L	8.7×10^{12}	2.2×10^{9}	>99
Beryllium, μg/L	30	9	70
Chromium, μg/L	9,800	20	>99
Copper, μg/L	100,000	95	>99
Cyanide, μg/L	200	<20	>90
Lead, μg/L	1,800	30	98
Nickel, μg/L	3,800	<20	>99
Selenium, μg/L	220	12	94
Silver, μg/L	100	20	81
Zinc, μg/L	3,400	35	99
Bis(2-ethylhexyl) phthalate,[a] μg/L	14	12	14
Di-n-butyl phthalate,[a] μg/L	24	36	0[c]
Methylene chloride,[b] μg/L	300	1.5	>99

[a]Possibly due to plastic tubing used during sampling.

[b]Possibly due to laboratory contamination.

[c]Actual data indicate negative removal.

Note: Blanks indicate information was not specified.

TREATMENT TECHNOLOGY: Sedimentation

Data source: Effluent Guidelines
Point source category: Ore mining and dressing
Subcategory: Ferroalloy (molybdenum) mine/mill
Plant: 6101
References: A2, pp. V-53, 53

Data source status:
Engineering estimate ___
Bench scale ___
Pilot scale ___
Full scale x

Use in system: Primary
Pretreatment of influent:

DESIGN OR OPERATING PARAMETERS

Unit configuration: Tailing pond
Wastewater flow:
Hydraulic detention time:
Hydraulic loading:
Weir loading:
Sludge underflow:
Percent solids in sludge:
Scum overflow:

REMOVAL DATA

Sampling period: 24-hr composite sample

Pollutant/parameter	Concentration		Percent removal
	Influent	Effluent	
Conventional pollutants, mg/L:			
COD	1,180	20	98
TOC	19	7	63
TSS	476,000	68	>99
Total phenol	0.02	0.01	50
Toxic pollutants:			
Antimony, µg/L	10	5	50
Asbestos, fibers/L	3.8×10^{11}	3.3×10^{10}	91
Beryllium, µg/L	130	<20	>85
Cadmium, µg/L	13	< 5	>62
Chromium, µg/L	8,300	20	>99
Copper, µg/L	10,000	<20	>99
Lead, µg/L	11,000	<20	>99
Nickel, µg/L	3,500	<20	>99
Selenium, µg/L	40	< 5	>87
Silver, µg/L	50	<10	>80
Zinc, µg/L	13,000	<20	>99
Di-n-butyl phthalate,[a] µg/L	15	15	0

[a]Possibly due to tubing used in sampling apparatus.

Note: Blanks indicate information was not specified.

TREATMENT TECHNOLOGY: Sedimentation

Data source: Effluent Guidelines
Point source category: Ore mining and dressing
Subcategory: Iron ore mine
Plant: 1105
References: A2, pp. V-3, 4

Data source status:
- Engineering estimate ___
- Bench scale ___
- Pilot scale ___
- Full scale x

Use in system: Primary
Pretreatment of influent:

DESIGN OR OPERATING PARAMETERS

Unit configuration: Settling pond
Wastewater flow:
Hydraulic detention time:
Hydraulic loading:
Weir loading:
Sludge underflow:
Percent solids in sludge:
Scum overflow:

REMOVAL DATA

Sampling period: 24-hr composite

Pollutant/parameter	Concentration		Percent
	Influent	Effluent	removal
Conventional pollutants, mg/L:			
COD	10	6	40
TOC	25	19	24
TSS	5	4	20
Toxic pollutants:			
Arsenic, μg/L	<2	5	0[a]
Asbestos (total), fibers/L	1.6×10^7	4.2×10^7	0[a]
Asbestos (chrysotile), fibers/L	3.8×10^6	3.8×10^6	0
Copper, μg/L	90	120	0[a]
Zinc, μg/L	20	30	0[a]

[a]Actual data indicate negative removal.

Note: Blanks indicate information was not specified.

TREATMENT TECHNOLOGY: Sedimentation

Data source: Effluent Guidelines
Point source category: Ore mining and dressing
Subcategory: Iron ore mine/mill
Plant: 1108
References: A2, p. V-5, 6

Data source status:
- Engineering estimate ___
- Bench scale ___
- Pilot scale ___
- Full scale x

Use in system: Primary
Pretreatment of influent:

DESIGN OR OPERATING PARAMETERS

Unit configuration: Tailing pond
Wastewater flow:
Hydraulic detention time:
Hydraulic loading:
Weir loading:
Sludge underflow:
Percent solids in sludge:
Scum overflow:

REMOVAL DATA

Sampling period: 24-hr composite

Pollutant/parameter	Concentration		Percent removal
	Influent	Effluent	
Conventional pollutants, mg/L:			
COD	96	4	96
TOC	22	11	50
TSS	110,000	<1	>99
Total phenol	<0.004	0.006	0[a]
Toxic pollutants:			
Asbestos, fibers/L	2.2×10^{11}	4.3×10^{7}	>99
Chromium, μg/L	500	10	98
Copper, μg/L	130	100	23
Lead, μg/L	80	<20	>75
Nickel, μg/L	2,700	<20	>99
Selenium, μg/L	20	<5	>75
Silver, μg/L	20	<10	>50
Zinc, μg/L	500	30	94

[a]Actual data indicate negative removal

Note: Blanks indicate information was not specified.

TREATMENT TECHNOLOGY: Sedimentation

Data source: Effluent Guidelines
Point source category: Ore mining and dressing
Subcategory: Lead/zinc mine/mill
Plant: 3101[a]
References: A2, p. V-102

Data source status:
Engineering estimate ___
Bench scale ___
Pilot scale ___
Full scale x

Use in system: Primary
Pretreatment of influent:

[a]Now closed.

DESIGN OR OPERATING PARAMETERS

Unit configuration: Tailing pond
Wastewater flow:
Hydraulic detention time:
Hydraulic loading:
Weir loading:
Sludge underflow:
Percent solids in sludge:
Scum overflow:

REMOVAL DATA

Sampling period: 24-hr composite

Pollutant/parameter	Concentration		Percent
	Influent	Effluent	removal
Conventional pollutants, mg/L:			
COD	1,240	44	96
TOC	46	19	59
TSS	152,000	5	>99
Total phenol	0.072	0.027	62
Toxic pollutants:			
Arsenic, µg/L	77	<5	>93
Asbestos (total fibers), fibers/L	2.4×10^{10}	1.9×10^{7}	>99
Asbestos (chrysotile), fibers/L	3.2×10^{9}	2.7×10^{6}	>99
Beryllium, µg/L	190	<10	>95
Cadmium, µg/L	2,800	<10	>99
Chromium, µg/L	800	25	97
Copper, µg/L	63,000	<10	>99
Lead, µg/L	97,000	140	>99
Nickel, µg/L	540	<50	>91
Selenium, µg/L	140	<10	>93
Silver, µg/L	230	<10	>96
Zinc, µg/L	560,000	70	>99

Note: Blanks indicate information was not specified.

TREATMENT TECHNOLOGY: Sedimentation

Data source: Effluent Guidelines
Point source category: Ore mining and dressing
Subcategory: Lead/zinc mine/mill
Plant: 3103
References: A2, p. V-108

Data source status:
Engineering estimate ___
Bench scale ___
Pilot scale ___
Full scale x

Use in system: Primary
Pretreatment of influent:

DESIGN OR OPERATING PARAMETERS

Unit configuration: Tailing pond
Wastewater flow:
Hydraulic detention time:
Hydraulic loading:
Weir loading:
Sludge underflow:
Percent solids in sludge:
Scum overflow:

REMOVAL DATA

Sampling period: 18-hr composite

Pollutant/parameter	Concentration Influent	Concentration Effluent	Percent removal
Conventional pollutants, mg/L:			
COD	2,100	14	99
TOC	22	15	32
TSS	124,000	3	>99
Total phenol	<0.004	0.012	0[a]
Toxic pollutants:			
Arsenic, μg/L	500	<5	>99
Asbestos (total fibers), fibers/L	2.1×10^{11}	9.9×10^{6}	>99
Asbestos (chrysotile), fibers/L	8.2×10^{10}	1.1×10^{6}	>99
Beryllium, μg/L	70	<10	>86
Cadmium, μg/L	350	<10	>97
Chromium, μg/L	200	<10	>95
Copper, μg/L	21,000	10	>99
Cyanide, μg/L	40	30	25
Lead, μg/L	120,000	240	>99
Nickel, μg/L	4,400	160	96
Silver, μg/L	150	<10	>93
Zinc, μg/L	58,000	940	98

[a]Actual data indicate negative removal.

Note: Blanks indicate information was not specified.

TREATMENT TECHNOLOGY: Sedimentation

Data source: Effluent Guidelines
Point source category: Ore mining and dressing
Subcategory: Lead/zinc mine/mill
Plant: 3110
References: A2, pp. V-36, 37

Data source status:
- Engineering estimate ___
- Bench scale ___
- Pilot scale ___
- Full scale x

Use in system: Primary
Pretreatment of influent:

DESIGN OR OPERATING PARAMETERS

Unit configuration: Tailing pond
Wastewater flow:
Hydraulic detention time:
Hydraulic loading:
Weir loading:
Sludge underflow:
Percent solids in sludge:
Scum overflow:

REMOVAL DATA

Sampling period: 24-hr composite

Pollutant/parameter	Concentration		Percent removal
	Influent	Effluent	
Conventional pollutants, mg/L:			
COD	200	6	97
TOC	3	7	0[a]
TSS	229,000	3	>99
Total phenol	0.004	0.006	0[a]
Toxic pollutants:			
Arsenic, μg/L	1,100	<2	>99
Asbestos, fibers/L	8.9×10^{11}	3.4×10^{8}	>99
Cadmium, μg/L	190	<5	>97
Chromium, μg/L	200	<10	>95
Copper, μg/L	25,000	100	>99
Lead, μg/L	20,000	<20	>99
Mercury, μg/L	0.5	<0.5	>0
Nickel, μg/L	270	<20	>93
Selenium, μg/L	20	<5	>75
Silver, μg/L	250	<10	>96
Zinc, μg/L	310,000	280	>99
Bis(2-ethylhexyl) phthalate,[b] μg/L	4.8	4	17
Methylene chloride,[c] μg/L	45	5.6	88

[a]Actual data indicate negative removal.

[b]Possibly due to tubing used in sampling apparatus.

[c]Possibly due to laboratory contamination.

Note: Blanks indicate information was not specified.

TREATMENT TECHNOLOGY: Sedimentation

Data source: Effluent Guidelines
Point source category: Ore mining and dressing
Subcategory: Lead/zinc mine/mill
Plant: 3121
References: A2, pp. V-41, 42

Data source status:
Engineering estimate ___
Bench scale ___
Pilot scale ___
Full scale x

Use in system: Primary
Pretreatment of influent:

DESIGN OR OPERATING PARAMETERS

Unit configuration: Tailing pond
Wastewater flow:
Hydraulic detention time:
Hydraulic loading:
Weir loading:
Sludge underflow:
Percent solids in sludge:
Scum overflow:

REMOVAL DATA

Sampling period: 24-hr composite

Pollutant/parameter	Concentration Influent[a]	Concentration Effluent	Percent removal
Conventional pollutants, mg/L:			
COD	970	50	95
TOC	17	15	12
TSS	12,200	14	>99
Total phenol	0.02	0.03	0[b]
Toxic pollutants:			
Antimony, μg/L	100	<50	>50
Arsenic, μg/L	30,000	<2	>99
Asbestos (total), fibers/L	1.8×10^{11}	1.6×10^{9}	99
Asbestos (chrysotile), fibers/L	2.2×10^{10}	$<3.3 \times 10^{5}$	>99
Cadmium, μg/L	670	<5	>99
Chromium, μg/L	550	<10	>98
Copper, μg/L	2,500	380	85
Lead, μg/L	150,000	20	>99
Mercury, μg/L	19	<0.5	>97
Nickel, μg/L	360	30	92
Silver, μg/L	200	<10	>95
Zinc, μg/L	240,000	440	>99

[a]Influent represents combined mine/mill water wastes to tailing pond.

[b]Actual data indicate negative removal.

Note: Blanks indicate information was not specified.

TREATMENT TECHNOLOGY: Sedimentation

Data source: Effluent Guidelines
Point source category: Ore mining and dressing
Subcategory: Lead/zinc mine/mill/smelter/refinery
Plant: 3107
References: A2, pp. VI-80-83

Data source status:
Engineering estimate ___
Bench scale ___
Pilot scale x
Full scale ___

Use in system: Secondary
Pretreatment of influent: Tailing pond, lime precipitation, aeration, flocculation, clarification, and filtration

DESIGN OR OPERATING PARAMETERS

Unit configuration:
Wastewater flow:
Hydraulic detention time: 11 hr
Hydraulic loading:
Weir loading:
Sludge underflow:
Percent solids in sludge:
Scum overflow:
pH: 7.8

REMOVAL DATA

Sampling period:

Pollutant/parameter	Concentration		Percent
	Influent	Effluent	removal
Conventional pollutants, mg/L:			
TSS	16	3	81
Toxic pollutants, μg/L:			
Cadmium	120	65	46
Copper	31	20	35
Lead	130	80	38
Zinc	2,900	790	73

Note: Blanks indicate information was not specified.

TREATMENT TECHNOLOGY: Sedimentation

Data source: Effluent Guidelines
Point source category: Ore mining and dressing
Subcategory: Placer miner
Plant: See below
References: A2, p. 142

Data source status:
- Engineering estimate ___
- Bench scale ___
- Pilot scale ___
- Full scale x

Use in system: Primary
Pretreatment of influent: None unless otherwise specified

DESIGN OR OPERATING PARAMETERS

Unit configuration: Multiple or single settling pond system

Wastewater flow:
Hydraulic detention time:
Hydraulic loading:
Weir loading:
Sludge underflow:
Percent solids in sludge:
Scum overflow:

REMOVAL DATA

Sampling period:

	TSS			Arsenic			Mercury		
	Concentration, mg/L		Percent	Concentration, μg/L		Percent	Concentration, μg/L		Percent
Plant	Influent	Effluent	removal	Influent	Effluent	removal	Influent	Effluent	removal
4114	24,000	<100	>99						
4126	14,800	76	99	1,300	250	81	2	0.2	90
4127	39,900	5,700	86	5,000	1,200	76	14	0.5	96
4132	1,540	1,040	32	50	50	0			
4133[a]	2,260	170	92	1,500	60	96	0.2	0.2	0
4135	2,890	474	84	40	22	45	20	<0.2	>99
4136[a]	64,100	150	>99	3,900	<2	>99	10	<0.2	>98
4139	9,000	230	97	1,200	12	99	4	<0.2	>95

[a]Pretreatment of influent is screening.

Note: Blanks indicate information was not specified.

TREATMENT TECHNOLOGY: **Sedimentation**

Data source: **Effluent Guidelines**
Point source category: **Ore mining and dressing**
Subcategory: **Silver mine/mill**
Plant: 4401
References: A2, pp. V-46, 47

Data source status:
- **Engineering estimate** ___
- **Bench scale** ___
- **Pilot scale** ___
- **Full scale** x

Use in system: **Primary**
Pretreatment of influent:

DESIGN OR OPERATING **PARAMETERS**

Unit configuration: **Multiple pond settling**
Wastewater flow:
Hydraulic detention time:
Hydraulic loading:
Weir loading:
Sludge underflow:
Percent solids in sludge:
Scum overflow:

REMOVAL DATA

Sampling period: **24-hr composite**

Pollutant/parameter	Concentration Influent	Concentration Effluent	Percent removal
Conventional pollutants, mg/L:			
COD	19	4	80
TOC	16	1	94
TSS	23	3	87
Toxic pollutants:			
Arsenic, µg/L	20	10	50
Asbestos (total), fibers/L	3.8×10^7	5.7×10^7	0[a]
Asbestos (chrysotile), fibers/L	1.1×10^7	1.1×10^6	90
Copper, µg/L	160	100	38
Nickel, µg/L	40	40	0
Silver, µg/L	20	30	0[a]
Zinc, µg/L	50	30	40
Bis(2-ethylhexyl) phthalate,[b] µg/L	0.1	0.02	80

[a]Actual data indicate negative removal.

[b]Possibly from tubing for sampling apparatus.

Note: Blanks indicate information was not specified.

TREATMENT TECHNOLOGY: Sedimentation

Data source: Effluent Guidelines
Point source category: Ore mining and dressing
Subcategory: Titanium mine/mill
Plant: 9905
References: A2, pp. V-70, 71

Data source status:
Engineering estimate ___
Bench scale ___
Pilot scale ___
Full scale x

Use in system: Primary
Pretreatment of influent:

DESIGN OR OPERATING PARAMETERS

Unit configuration:
Wastewater flow:
Hydraulic detention time:
Hydraulic loading:
Weir loading:
Sludge underflow:
Percent solids in sludge:
Scum overflow:

REMOVAL DATA

Sampling period: 24-hr composite

Pollutant/parameter	Concentration Influent	Concentration Effluent	Percent removal
Conventional pollutants, mg/L:			
COD	47	4	91
TOC	3	5	0[a]
TSS	57,900	<1	>99
Total phenol	0.01	0.01	0
Toxic pollutants:			
Antimony, µg/L	200	100	50
Asbestos (total), fibers/L	7.1×10^9	1.5×10^8	98
Asbestos (chrysotile), fibers/L	1.1×10^9	1.3×10^6	>99
Chromium, µg/L	740	<10	>99
Copper, µg/L	880	100	89
Lead, µg/L	50	40	20
Nickel, µg/L	630	40	94
Selenium, µg/L	15	<5	>67
Zinc, µg/L	3,500	20	99

[a]Actual data indicate negative removal.

Note: Blanks indicate information was not specified.

TREATMENT TECHNOLOGY: Sedimentation

Data source: Effluent Guidelines

Data source status: See below

- Engineering estimate ___
- Bench scale ___
- Pilot scale ___
- Full scale ___

Point source category: Ore mining and dressing
Subcategory: See below
Plant: See below
References: A2, p. VI 46, 47, 41, 39

Use in system: Primary
Pretreatment of influent: None

DESIGN OR OPERATING PARAMETERS

Unit configuration: Settling ponds or tailing ponds unless otherwise specified
Wastewater flow:
Hydraulic detention time:
Hydraulic loading:
Weir loading:
Sludge underflow:
Percent solids in sludge:
Scum overflow:

REMOVAL DATA

Sampling period:

			Total asbestos			Chrysotile asbestos		
			Concentration, fibers/L			Concentration, fibers/L		
Subcategory	Plant	Scale of treatment	Influent	Effluent	Percent removal	Influent	Effluent	removal
Asbestos cement processing	-[a]	Pilot	$5x10^{9}$	$9.3x10^{9}$	0[b]			
Asbestos mining	(in Baie Verte, Newfoundland)	Pilot	$1x10^{10}$	$5x10^{9}$	50			
Mercury mine/mill	9202	Full	$1.2x10^{12}$	$7.7x10^{8}$	>99	$1.5x10^{11}$	$5.7x10^{7}$	>99
Uranium mine/mill	9405	Full	$2x10^{8}$	$6.3x10^{8}$	0[b]	$2.25x10^{6}$	$7.5x10^{7}$	0[b]
Uranium mine/mill	-[a]	Pilot			0[b]	$4x10^{12}$	$1x10^{11}$	98
Uranium mine/mill	-[c]	Pilot				$4x10^{12}$	$5x10^{11}$	88

[a]Hydraulic detention time is 24 hr.

[b]Actual data indicate negative removal.

[c]1 hr of sedimentation.

Note: Blanks indicate information was not specified.

TREATMENT TECHNOLOGY: Sedimentation

Data source: Effluent Guidelines
Point source category: Paint manufacturing
Subcategory:
Plant: 76-J
References: A4, p. V-25

Data source status:
- Engineering estimate ___
- Bench scale ___
- Pilot scale ___
- Full scale x

Use in system: Primary
Pretreatment of influent: None

DESIGN OR OPERATING PARAMETERS

Unit configuration:
Wastewater flow:
Hydraulic detention time:
Hydraulic loading:
Weir loading:
Sludge underflow:
Percent solids in sludge:
Scum overflow:

REMOVAL DATA

Sampling period:

Pollutant/parameter	Concentration[a] Influent	Effluent	Percent removal
Conventional pollutants, mg/L:			
BOD_5	3,500	1,100	69
COD	27,900	3,300	88
TSS	15,600	1,400	91
Oil and grease	2,400	160	93
Total phenol	1.1	0.1	91
Toxic pollutants, µg/L:			
Antimony	500	70	86
Cadmium	860	200	77
Chromium	140	10	93
Copper	300	100	67
Lead	420	60	86
Mercury	1.2	0.7	42
Nickel	100	100	0
Zinc	740	100	86

[a]Average of three samples.

Note: Blanks indicate information was not specified.

TREATMENT TECHNOLOGY: Sedimentation

Data source: Effluent Guidelines
Point source category: Paint manufacturing
Subcategory:
Plant: 76-A
References: A4, p. V-25

Data source status:
Engineering estimate ___
Bench scale ___
Pilot scale ___
Full scale x

Use in system: Primary
Pretreatment of influent: None

DESIGN OR OPERATING PARAMETERS

Unit configuration:
Wastewater flow:
Hydraulic detention time:
Hydraulic loading:
Weir loading:
Sludge underflow:
Percent solids in sludge:
Scum overflow:

REMOVAL DATA

Sampling period:

Pollutant/parameter	Concentration Influent	Concentration Effluent	Percent removal
Conventional pollutants, mg/L:			
BOD_5	1,300	980	25
COD	3,000	3,500	0[a]
TSS	1,600	550	66
Oil and grease	300	220	27
Total phenol	2.5	3.5	0[a]
Toxic pollutants, μg/L:			
Antimony	1,000	1,000	0
Cadmium	10	10	0
Chromium	13,000	10,000	23
Copper	150	70	53
Lead	14,000	6,800	51
Mercury	0.9	0.5	44
Nickel	250	400	0[a]
Zinc	18,000	6,000	67

[a]Actual data indicate negative removal.

TREATMENT TECHNOLOGY: Sedimentation

Data source: Effluent Guidelines
Point source category: Steam electric power generating
Subcategory:
Plant: See below
References: A31, p. 171

Data source status:
Engineering estimate ___
Bench scale ___
Pilot scale ___
Full scale x

Use in system: Primary
Pretreatment of influent: None

DESIGN OR OPERATING PARAMETERS

Unit configuration: Combined ash pond
Wastewater flow: See below
Hydraulic detention time:
Hydraulic loading:
Weir loading:
Sludge underflow:
Percent solids in sludge:
Scum overflow:

REMOVAL DATA

Sampling period:

Plant	Wastewater flow, m^3/d(gpd)	TSS Concentration, mg/L Influent	TSS Concentration, mg/L Effluent	TSS Percent removal
5143	25,000 (6.5 x 10^6)	63,900	13	>99
7298	72,000 (19 x 10^6)	6,690	19	>99
0431	98,000 (26 x 10^6)	13,400	22	>99
4504	68,000 (18 x 10^6)	15,300	7	>99
7018	55,000 (14.5 x 10^6)	20,700	18	>99
3228	6,800 (8 x 10^6)	26,800	6	>99

Note: Blanks indicate information was not specified.

TREATMENT TECHNOLOGY: Sedimentation

Data source: Effluent Guidelines
Point source category: Steam electric power generating
Subcategory:
Plant: 4222
References: T2, pp. 238-241

Data source status:
Engineering estimate ___
Bench scale ___
Pilot scale ___
Full scale x

Use in system: Primary
Pretreatment of influent:

DESIGN OR OPERATING PARAMETERS

Unit configuration: Ash pond
Wastewater flow:
Hydraulic detention time:
Hydraulic loading:
Weir loading:
Sludge underflow:
Percent solids in sludge:
Scum overflow:

REMOVAL DATA

Sampling period:

Pollutant/parameter	Concentration, μg/L Influent	Concentration, μg/L Effluent	Percent removal
Toxic pollutants:			
Antimony	48	29	40
Arsenic	123	160	0[a]
Beryllium	100	20	80
Cadmium	10	<5	>50
Chromium	196	11	94
Copper	300	6	98
Lead	240	<5	>98
Mercury	0.62	0.21	66
Nickel	250	8	97
Selenium	<5	32	0[a]
Thallium	29	<5	>83
Zinc	400	10	98

[a]Actual data indicate negative removal.

Note: Blanks indicate information was not specified.

TREATMENT TECHNOLOGY: Sedimentation

Data source: Effluent Guidelines, Government report

Point source category: Textile mills

Subcategory: Wool scouring

Plant: A, W (different references)

References: A6, p. VII-46; B3, pp. 50-54

Data source status:

- Engineering estimate ___
- Bench scale ___
- Pilot scale _x_
- Full scale ___

Use in system: Tertiary

Pretreatment of influent: Grit removal, activated sludge (oxidation ditch plus clarifier)

DESIGN OR OPERATING PARAMETERS

Unit configuration: 6.25 m^3 (1,650 gal) clarifier

Wastewater flow:

Hydraulic detention time:

Hydraulic loading:

Weir loading:

Sludge underflow:

Percent solids in sludge:

Scum overflow:

REMOVAL DATA

Sampling period: 24-hr, toxic pollutants were composite samples, volatile organics were grab samples

Pollutant/parameter	Concentration		Percent
	Influent	Effluent	removal
Conventional pollutants, mg/L:			
Total phenol	0.016	0.049	0[a]
Toxic pollutants, μg/L:			
Antimony	540	<200	>63
Arsenic	38	39	0[a]
Cadmium	130	<40	>69
Copper	320	110	66
Cyanide	200	240	0[a]
Lead	3,500	<400	>89
Nickel	2,000	<700	>65
Silver	500	<100	>80
Zinc	1,500	190	87
Bis(2-ethylhexyl) phthalate	42	23	45
Ethylbenzene	<0.2	3.0	0[a]
Toluene	1.4	9.5	0[a]
Anthracene/phenanthrene	1.5	0.4	73
Benzo(a)pyrene	1.2	<0.02	>98
Benzo(k)fluoranthene	0.8	<0.02	>97
Fluoranthene	1.1	0.4	64
Pyrene	0.8	0.2	75
Methylene chloride[b]	<0.4	2.2	0[a]

[a]Actual data indicate negative removal.

[b]Presence may be due to sample contamination.

Note: Blanks indicate information was not specified.

4.3 CLARIFICATION/SEDIMENTATION USING CHEMICAL ADDITION [1]

4.3.1 Function

Clarification/sedimentation using chemical addition is utilized to remove suspended and colloidal solids and phosphates. Chemicals generally in use are coagulants and coagulant aides, such as lime, ferric chloride, alum, and various polymers.

4.3.2 Descriptions and Common Modifications

Lime Addition (Primary). Lime clarification of raw wastewater removes suspended solids as well as phosphates. There are two basic processes: the low-lime system and the high-lime system. The low-lime process consists of the addition of lime to obtain a pH of approximately 9 to 10. Generally, a subsequent biological treatment system is capable of readjusting the pH through natural recarbonation. The high-lime process consists of the addition of lime to obtain a pH of approximately 11 or more. In this case, the pH generally requires readjusting with carbon dioxide or acid to be acceptable to the secondary treatment system.

Lime can be purchased in many forms; quicklime (CaO) and hydrated lime [$Ca(OH)_2$] are the most prevalent forms. In either case, lime is usually purchased in the dry state, in bags, or in bulk. Bulk lime can be (1) shipped by trucks that are generally equipped with pneumatic unloading equipment; or (2) shipped by rail cars that consist of covered hoppers. The rail cars are emptied by opening a discharge gate, which discharges to a screw conveyor. The bulk lime is then transferred by the screw conveyor to a bucket elevator, which empties into the elevated storage tank. Bulk storage usually consists of steel or concrete bins. Storage vessels should be water- and air-tight to prevent the lime from "slaking".

Lime is generally made into a wet suspension or slurry before introduced into the treatment system. The precise steps involved in converting from the dry to the wet stage will vary according to the size of operation and type and form of lime used. In the smallest plants, bagged hydrated lime is often charged manually into a batch mixing tank with the resulting "milk-of-lime" (or slurry) being fed by means of a so-called solution feeder to the process. Where bulk hydrate is used, some type of dry feeder charges the lime continuously to either a batch or continuous mixer, then, by means of solution feeder, to the point of application. With bulk quicklime, a dry feeder is also used to feed a slaking device, where the oxides are converted to hydroxides, producing a paste or slurry. The slurry is then further diluted to milk-of-lime before being piped by gravity or pumped to the process. Dry feeders can be of the volumetric or gravimetric type.

Lime Addition (Two-Stage Tertiary). Lime treatment of secondary effluent for the removal of phosphorus and suspended solids is essentially the same process as high-lime clarification of raw wastewater. Calcium carbonate and magnesium hydroxide precipitate at high pH along with the phosphorus containing calcium hydroxyapatite and other suspended solids. In the two-stage system, the first-stage precipitation generally is controlled around a pH of 11, which is approximately one pH unit higher than that used in the single-stage process. After precipitation and clarification in the first stage, the wastewater is recarbonated with carbon dioxide, forming a calcium carbonate precipitate, which is removed in the second clarification stage.

Lime is generally added to a separate rapid-mixing tank or to the mixing zone of a solids-contact or sludge-blanket clarifier. After mixing, the wastewater is flocculated to allow for the particles to increase in size to aid in clarification. The clarified wastewater is recarbonated in a separate tank following the first clarifier, after which it is re-clarified in a second clarifier. Final pH adjustment may be required to meet allowable discharge limits.

Treatment systems consist of (1) separate units for flash mixing, flocculation, and clarification; or (2) specially designed solids contact or sludge-blanket units, which contain flash mix, flocculation, and clarification zones in one unit. The calcium carbonate sludge formed in the second stage can be recalcined. Final effluent can be neutralized with sulfuric acid, as well as other acids.

Alum Addition. Alum or filter alum [$Al_2(SO_4)_3 \cdot 14H_2O$] is a coagulant which, when added to wastewater, reacts with available alkalinity (carbonate, bicarbonate and hydroxide) and phosphate to form insoluble aluminum salts. The combination of alum with alkalinity or phosphate are competing reactions that are pH dependent.

Alum is an off-white crystal which when dissolved in water produces acidic conditions. As a solid, alum may be supplied in lumps, or in ground, rice, or powdered form. Shipments may be in small bags (100 lb), in drums or in bulk quantities (over 40,000 lb). In liquid form, alum is commonly supplied as a 50 percent solution delivered in minimum loads of 4,000 gallons. The choice between liquid or dry alum use is dependent on factors such as availability of storage space, method of feeding, and economics. In general, purchase of liquid alum is justified only when the supplier is close enough to make differences in transportation costs negligible. Dry alum is stored in mild steel or concrete bins with appropriate dust collection equipment. Because dry alum is slightly hydroscopic, provisions are made to avoid moisture, which could cause caking and corrosive conditions. Before addition to wastewater, dry alum must be dissolved, forming a concentrated solution. Bulk-stored or hopper-filled alum is transported to a feeder mechanism by bucket elevator, screw conveyor or a pneumatic device. Three basic types of feeders are in common use: volumetric, belt gravimetric, and loss-in-weight gravimetric. The feeder supplies a controlled quantity of dry alum (accuracy ranges from about 1% to 7%) to a mixed dissolver vessel. Because alum solubility is temperature dependent, the quantity supplied depends on the concentrate strength desired and the temperature. Because alum solution is corrosive, the dissolving chamber as well as the following storage tanks, pumps, piping and surfaces that may come in contact with the solution or generated fumes must be constructed of resistant materials such as type 316 stainless steel, fiberglass reinforced plastic (FRP), or plastics. Rubber or saran-lined pipes are commonly used. Liquid alum, which crystallizes at about 30°F and freezes at about 18°F, is stored and shipped in insulated type 316 stainless steel or rubber-lined vessels. Feeding of liquid alum (purchased or made up on site) to wastewater treatment unit processes may be accomplished by gravity, pumping, or using a Rotodip feeder. Diaphragm pumps and valves are common.

Ferric Chloride Addition. Ferric chloride ($FeCl_3$) is a chemical coagulant which, when added to wastewater, reacts with alkalinity and phosphates, forming insoluble iron salts. The colloidal particle size of insoluble ferric phosphate is small, requiring excess dosages of ferric chloride to produce a well flocculated iron hydroxide precipitate, which carries the phosphate precipitate. Large excesses of ferric chloride, and corresponding quantities of alkalinity, are required to assure phosphate removal. Exact ferric chloride dosages are usually best determined using jar tests and full-scale evaluations.

Ferric chloride is available in either dry (hydrated or anhydrous) or liquid form. Liquid ferric chloride is a dark brown oil-appearing solution supplied in concentrations ranging between 35 and 45 percent ferric chloride. Because higher concentrations of ferric chloride have higher freezing points, lower concentrations are supplied during winter. Liquid ferric chloride is shipped in 3,000- to 4,000-gallon bulk truckload lots, in 4,000- to 10,000-gallon carloads, and in 5- to 13-gallon carboys. Ferric chloride solution stains surfaces which it contacts and is highly corrosive (a one percent solution has a pH of 2.0); consequently, it must be stored and handled with care. Storage tanks are equipped with vents and vacuum relief valves. Tanks are constructed of fiberglass reinforced plastic, rubber-lined steel and plastic-lined steel. Because of freezing potential, ferric chloride solutions are either stored in heated areas or in heated and insulated vessels in northern climates.

Ferric chloride solution should not be diluted because of possible unwanted hydrolysis. Consequently, feeding at the concentration of the delivered product is common. The stored solution is transferred to a day tank using graphite or rubber-lined self-priming centrifugal pumps with corrosion resistant Teflon seals. From the day tank, controlled quantities are fed to the unit process using Rotodip feeders or diaphragm metering pumps. Rotometers are not used for ferric chloride flow measurement because the material tends to deposit on and stain the glass tubes. All pipes, valves, or surfaces that come in contact with ferric chloride must be made of corrosion resistant materials such as rubber or Saran lining, Teflon, or vinyl. Similar treatment results are obtainable by substituting ferrous chloride, ferric sulfate, ferrous sulfate, or spent pickle liquor for ferric chloride. Details of storage feeding and control for these materials are similar to those for ferric chloride. Dry ferric chloride may also be dissolved on site before use in treatment.

Polymer Addition. Polymers or polyelectrolytes are high-molecular-weight compounds (usually synthetic) which, when added to wastewater, can be used as coagulants, coagulant aids, filter aids, or sludge conditioners. In solution, polymers may carry either a positive, negative, or neutral charge and, as such, they are characterized as cationic, anionic, or nonionic. As a coagulant or coagulant aid, polymers act as bridges, reducing charge repulsion between colloidal and dispersed floc particles, and increasing settling velocities. As a filter aid, polymers strengthen fragile floc particles, controlling filter penetration and reducing particle breakthrough. Filterability and dewatering characteristics of sludges may similarly be improved through the use of polyelectrolytes.

Polymers are available in predissolved liquid or dry form. Dry polymers are supplied in relatively small quantities (up to about 100-lb bags or barrels) and must be dissolved on site prior to use. A stock solution, usually about 0.2 to 2.0 percent concentration, is made up for subsequent feeding to the treatment process. Preparation involves automatic or batch wetting, mixing, and aging. Stock polymer solutions may be very viscous. Surfaces coming in contact with the polymer stock solution should be constructed of resistant materials such as type 316 stainless steel, fiberglass reinforced plastic, or other plastic lining materials.

Polymers may be supplied as a prepared stock solution ready for feeding to the treatment process. Many competing polymer formulations with differing characteristics are available, requiring somewhat differing handling procedures. Manufacturers should be consulted for optimum practices. Polymer stock solutions are generally fed to unit processes using equipment similar to that commonly in service for dissolved coagulant addition. Because of the high viscosity of stock solutions, special attention should be paid to the diameter and slopes of pipes, as well as the size of orifices used in the feed systems.

4.3.3 Technology Status

Lime Addition (Primary). Lime addition is an established practice.

Lime Addition (Two-Stage Tertiary). These systems have been used for water softening for many decades; however, their use for phosphorus removal has been prominant only since the mid-1960's. There are presently many large-scale systems in operation.

Alum Addition. Alum addition has been used for decades for coagulation and turbidity reduction in water treatment. Its application to wastewater treatment is more recent, and the technology is well demonstrated.

Ferric Chloride Addition. Ferric chloride is commonly used in water treatment as a coagulant for turbidity reduction. Its use in wastewater treatment is more recent and well demonstrated.

Polymer Addition. Polymer or polyelectrolyte usage in wastewater and water treatment has gained widespread acceptance. The technology for its use is well demonstrated and common throughout the wastewater and water treatment fields.

4.3.4 Applications

Lime Addition (Primary). When added to a primary clarifier, used for improved removal of suspended solids and the removal of phosphates (this process is primarily used to remove phosphates); with proper application, lime addition will also remove toxic metals.

Lime Addition (Two-Stage Tertiary). Used for the removal of phosphorus from wastewater; will also remove some BOD_5 and suspended solids as well as hardness present in wastewater; will also remove metals.

Alum Addition. Used in wastewater treatment (sometimes in conjunction with polymers) for suspended solids and/or phosphorus removal; alum coagulation may be incorporated into independent physical-chemical treatment, tertiary treatment schemes, or as an add-on to existing treatment processes; in independent physical-chemical treatment (or tertiary treatment), alum is added directly to wastewater, which is intensely mixed, flocculated and settled; solids contact clarifiers may be used; in existing wastewater treatment process, alum may be added directly to primary clarifiers, secondary clarifiers, or aeration vessels to improve performance; should not be dosed directly to trickling filters because of possible deposition of chemical precipitates on filter media; has also been used as a filter aid in tertiary filtration processes and has been used to upgrade stabilization pond effluent quality.

Ferric Chloride Addition. Used (sometimes with polymer addition) in wastewater treatment for suspended solids removal and/or phosphate removal; $FeCl_3$ coagulation may be incorporated into independent physical-chemical treatment and tertiary treatment schemes; in these applications, solids contact clarifiers or separate flocculation vessels are used for treatment of either raw wastewater or secondary effluent; coagulation may also be applied to existing treatment systems; addition of ferric chloride before primary and secondary clarifiers has been practiced in both activated sludge and trickling filter plants.

Polymer Addition. Utilized in various applications in wastewater treatment ranging from flocculation of suspended or colloidal materials either alone or in conjunction with other coagulants such as lime, alum, or ferric chloride, to use as filter aid or sludge conditioner; polyelectrolytes may be added alone or with other coagulants to raw wastewater prior to primary treatment to effect or aid in suspended solids and BOD_5 removal; similarly, polymers may be used to aid coagulation or as primary coagulant in treatment of secondary effluent; as filter aid, polyelectrolytes effectively strengthen fragile chemical flocs, facilitating more efficient filter operations.

4.3.5 Limitations

Lime Addition (Primary). Will generate additional amounts of sludge, over and above that generated by normal primary clarification process (approximately twice the volume for low-lime system and five to six times for high-lime system); lime feed systems can require intensive operator attention; even low-lime system could present biological problems to fixed-growth systems with no pH adjustment; increases operator safety needs.

Lime Addition (Two-Stage Tertiary). Will generate relatively large amounts of chemical sludge; high operator skill required; in some cases, polymer or coagulant is required to assist second-stage clarification.

Alum Addition. Alum solution is corrosive; appropriate dosages are not stoichoimetric and must be frequently reconfirmed; alkalinity required for proper coagulation, and, where inadequate, supplemental alkalinity must be provided (usually by lime addition); alum sludge is voluminous and difficult to dewater.

Ferric Chloride Addition. Ferric chloride is extremely corrosive material and must be stored and transported in special corrosion resistant equipment; dosages are not stoichiometric and must be frequently rechecked using jar tests; ferric chloride coagulation requires a source of alkalinity, and, in soft wastewaters, the pH of clarified effluent might be decreased to a point requiring pH adjustment by addition of supplemental base such as lime or caustic soda; iron concentrations in plant effluents may become unacceptably high.

Polymer Addition. Frequent jar tests are necessary to assure proper dosages; overdosages (1.0 to 2.0 mg/L) can sometimes work against the treatment process.

4.3.6 Chemicals Required

Lime Addition (Primary). Lime [CaO or $Ca(OH)_2$]; CO_2 or H_2SO_4 for high-lime.

Lime Addition (Two-Stage Tertiary). Lime (CaO), CO_2 or H_2SO_4, sometimes polymer or coagulant.

Alum Addition. Amount of alum required depends on multiple factors such as alkalinity and pH of wastewater, phosphate level, and point of injection; accurate dosages should be determined using jar tests and confirmed by field trials.

Ferric Chloride Addition. Amount of ferric chloride required depends on variable factors including pH and alkalinity of the wastewater, phosphate level, point of injection, and mixing modes; accurate doses should be determined using jar tests and confirmed by field evaluations; base addition may be required when treating soft wastewaters.

Polymer Addition. Accurate dosages should be determined by bench-scale evaluation.

4.3.7 Residuals Generated

Lime Addition (Primary). Sludge (containing 1 to 1.5 pounds of dry solids per pound of lime added) plus the usual amount of solids produced in the primary settling process.

Lime Addition (Two-Stage Tertiary). In first stage: sludge containing hydroxyapatite, calcium carbonate, magnesium hydroxide, and organic solids (1 to 1.5 pounds of dry solids per pound of lime added); in second stage: sludge may contain calcium carbonate, aluminum, or ferric hydroxide, depending upon the coagulant used; quantities generated are 2.27 pounds $CaCO_3$ per pound of CO_2, 4 pounds per pound of Al in alum or 2.5 pounds per pound of Fe in ferric chloride.

Alum Addition. Alum sludges are substantially different in character from biological sludges (volumes are greater and dewatering is more difficult); alum sludge also has tendency to induce undesirable stratification in anaerobic digesters.

Ferric Chloride Addition. Used in standard biological processes, ferric chloride addition will increase volume of sludge generated; iron coagulants produce sludges that are significantly different from biological sludges, especially in terms of dewatering characteristics.

Polymer Addition. Sludges generated in conjunction with polymer addition will be somewhat different from, but not necessarily more difficult to handle than biological sludges or chemical sludges generated without polymers.

4.3.8 Reliability

Lime Addition (Primary). Process highly reliable from process standpoint, however, increased operator attention and cleaning requirements are necessary to maintain mechanical reliability of lime feed system.

Lime Addition (Two-Stage Tertiary). Systems are reliable from unit and process standpoint with skilled operator attention.

Alum Addition. Reduces phosphate and suspended solids to low levels, although effluent quality may vary unless filtration follows clarification step.

Ferric Chloride Addition. Reduces phosphate and suspended solids to low levels, although effluent quality may vary unless filtration follows clarification step.

Polymer Addition. With proper control, capable of producing consistently high quality effluents.

4.3.9 Environmental Impact

Lime Addition (Primary). Will generate relatively large amounts of inorganic sludge that will need disposal.

Lime Addition (Two-Stage Tertiary). Will generate relatively large amounts of inorganic sludge that will need disposal.

Alum Addition. Will generate relatively large amounts of inorganic sludge that will need disposal.

Ferric Chloride Addition. Will generate relatively large amounts of inorganic sludge that will need disposal.

Polymer Addition. May improve sludge dewaterability; operator safety should be carefully considered.

4.3.10 Design Criteria

Lime Addition (Primary)

Feed water alkalinity, mg/L (as $CaCO_3$)	Clarifier pH	Approximate lime dose, mg/L (as CaO)
300	9.5	185
300	10.5	270
400	9.5	230
400	10.5	380

Lime Addition (Two-Stage Tertiary). Clarifier settling rate: 1,200 to 1,400 gpd/ft^2

Secondary effluent alkalinity, mg/L (as $CaCO_3$)	Clarifier pH	Approximate lime dose, mg/L (as CaO)
300	11.0	400 - 450
400	11.0	450 - 500

Carbon dioxide

Feed tank - 5 to 15 minutes
Feed rate - 1.2 mg/L per mg/L of Ca to be precipitated.

Alum Addition. Dosage determined by jar testing, generally in the range of 5-20 mg/L as Al; in mixing, t is less than or equal to 30 s; overflow rate = 500 to 600 gpd/ft^2 (average), 800 to 900 gpd/ft^2 (peak).

Ferric Chloride Addition. Dosage determined by jar testing; dosages of 20 to 100 mg $FeCl_3$/L are common; t is less than or equal to 30 s in mixing.

Polymer Addition. Dosage determined by jar testing; materials contacting polymer solutions should be Type 316 stainless steel, FRP, or plastic; storage place must be cool and dry; storage periods should be minimized; viscosity considerations must be made in feeding system design.

4.3.11 Flow Diagrams

Lime Addition (Primary Treatment).

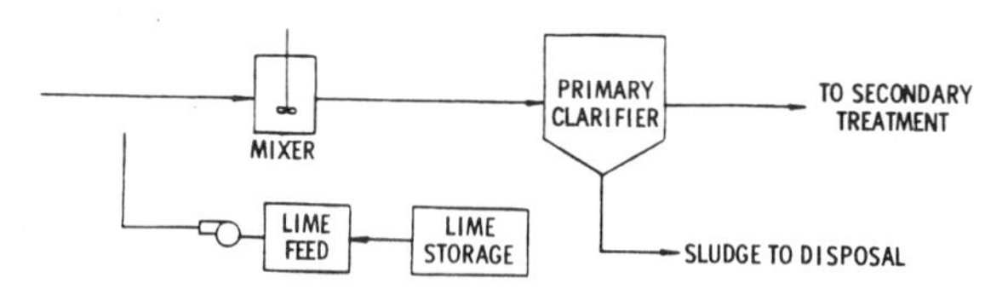

Lime Addition (Two-Stage Tertiary).

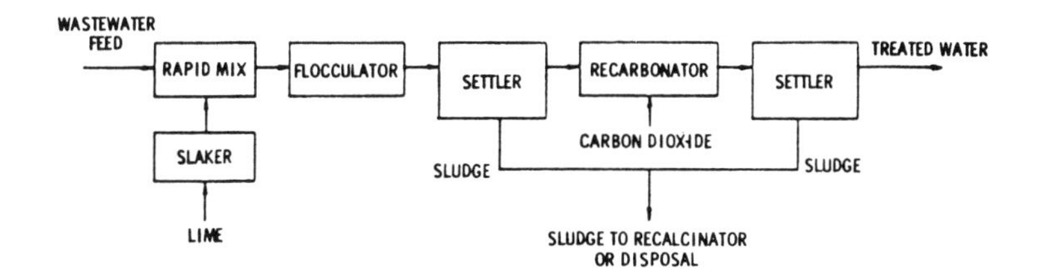

Alum Addition.

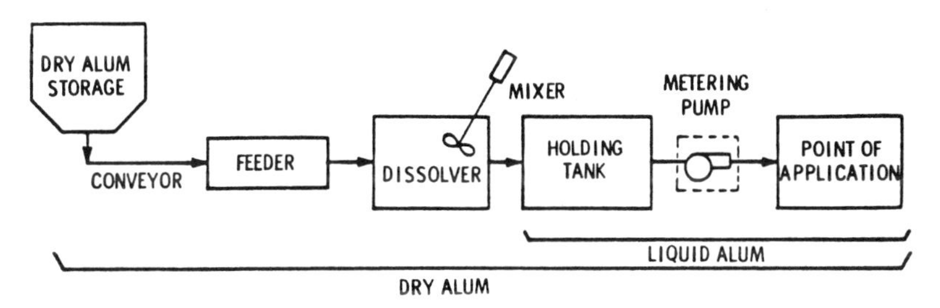

Ferric Chloride Addition.

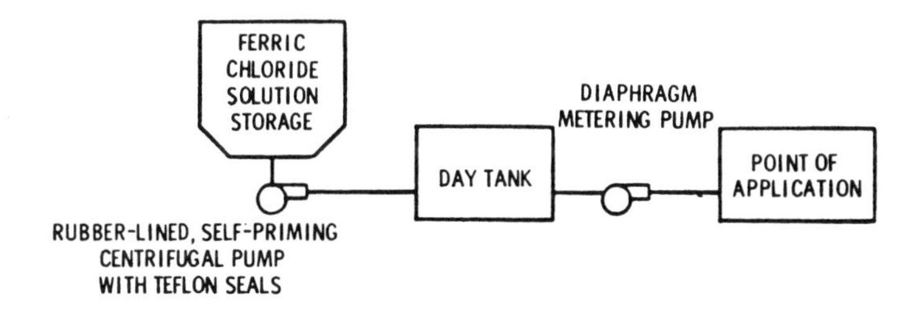

Polymer Addition.

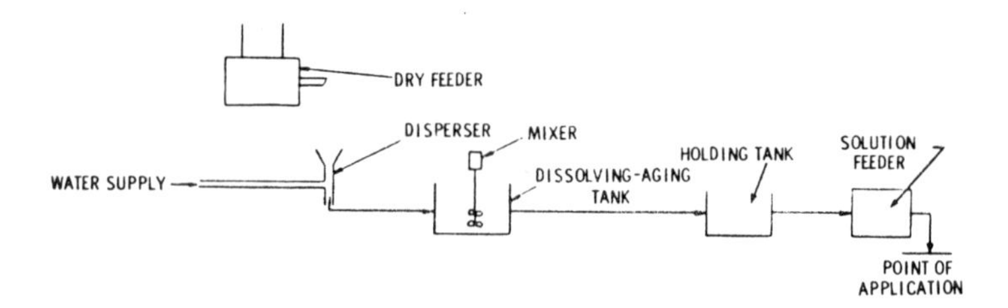

4.3.12 Performance

Subsequent data sheets provide performance data from studies on the following industries and/or wastestreams:

Auto and other laundries industry
  Power laundries

Foundry industry
  Aluminum foundries - die casting

Inorganic chemicals production
  Hydrofluoric acid

Iron and steel industry
  Combination acid pickling - batch
  Hydrochloric acid pickling
  Pipe and tube - welded

Nonferrous metals industry
  Columbium/tantalum raw waste stream
  Tungsten raw waste stream

Ore mining and dressing
  Base metal mining
  Copper mining/milling/smelting
  Lead/zinc mining/milling/smelting/refining
  Uranium mining/milling

Paint manufacturing

Steam electric power generation
  Ash sluicing

Textile milling
  Knit fabric finishing
  Wool finishing
  Woven fabric finishing

References

1. Innovative and Alternative Technology Assessment Manual, EPA-430/9-78-009 (draft), U.S. Environmental Protection Agency, Cincinnati, Ohio, 1978. 252 pp.

CONTROL TECHNOLOGY FOR SEDIMENTATION WITH CHEMICAL ADDITION (ALUM)

Pollutant	Number of data points	Effluent concentration				Removal efficiency, %			
		Minimum	Maximum	Median	Mean	Minimum	Maximum	Median	Mean
Conventional pollutants, mg/L:									
BOD_5	5	3.6	2,900	33	1,040	0[a]	82	16	<47
COD	5	212	7,600	416	2,410	4	71	61	45
TOC	4	72	1,500	105	437	5	80	63	53
TSS	5	28	122	50	55.8	0[a]	99	79	<67
Oil and grease	1				11				99
Total phenol	4	0.016	225	0.055	56.3	0[a]	31	19	<17
Total phosphorus	2	2.3	43		22.7	12	15		14
Toxic pollutants, μg/L:									
Antimony	2	23	120		72	0[a]	0[a]		0[a]
Arsenic	2	<1	62		22	0[a]	<37		19
Beryllium	1				22				0[a]
Cadmium	2	<15	29		<22	0[a]	>88		44
Chromium	4	17	280	<40	<95	0[a]	>98	44	46
Copper	4	<10	<110	13	<37	30	>78	>73	>64
Mercury	2	1.7	<150		<76	6	760		>34
Nickel	3	10	57		<54	0	>56		>41
Silver	2	72	170		120	0[a]	10		<5
Zinc	4	110	9,000	2,900	3,700	0[a]	85	30	<36
Bis(2-ethylhexyl) phthalate	2	33	44		39	0[a]	0[a]		0[a]
Di-n-butyl phthalate	2	0.6	<10		<8.3	0	>94		750
Phenol	2	<0.07	<10		<8.0	>82	>90		>86
1,2-Dichlorobenzene	2	<10	13		<12	0[a]	>50		25
Ethylbenzene	2	1.3	4,600		2,300	0[a]	0[a]		0[a]
Nitrobenzene	1				35				68
Toluene	3	1.0	2,500	14	840	0[a]	93	55	<49
1,2,4-Trichlorobenzene	1				150				90
Anthracene/Phenanthrene	1				0.1				0[a]
Chlorodibromomethane	1				<0.3				>50
Chloroform	1				22				0[a]
1,2-Dichloroethane	1				17				0[a]
Methylene chloride	2	<10	70		<40	56	>88		>72
Tetrachloroethylene	1				45				0[a]
Trichloroethylene	1				190				10

Note: Blanks indicate data not applicable.

[a]Actual data indicate negative removal.

CONTROL TECHNOLOGY SUMMARY FOR SEDIMENTATION WITH CHEMICAL ADDITION (ALUM, LIME)

Pollutant	Number of data points	Effluent concentration				Removal efficiency, %			
		Minimum	Maximum	Median	Mean	Minimum	Maximum	Median	Mean
Conventional pollutants, mg/L:									
BOD_5	1				3,900				0[a]
COD	1				7,970				95
TOC	1				2,300				82
TSS	1				480				97
Oil and grease	1				<16				>98
Total phenol	1				1.3				0[a]
Toxic pollutants, μg/L:									
Arsenic	1				62				0[a]
Chromium	1				31				72
Copper	2	13	60		37	35	88		62
Cyanide	2	<4	30		<17	>60	80		>70
Lead	1				<200				50
Mercury	1				2				71
Nickel	1				<1.0				>83
Zinc	1				1,100				>99
Bis(2-ethylhexyl) phthalate	1				44				0[a]
Di-n-butyl phthalate	1				<10				>99
Phenol	2	3	47		25	0[a]	96		<48
Benzene	1				46				50
1,2-Dichlorobenzene	1				<0.05				>99
Ethylbenzene	2	<0.2	22		<11	>96	98		>97
Toluene	2	14	72		43	55	96		76
1,2,4-Trichlorobenzene	1				150				91
Naphthalene	1				16				70
Carbon tetrachloride	1				<10				>17
Chloroform	1				74				0[a]
1,2-Dichloropropane	1				400				59
Methylene chloride	1				2,000				13
1,1,2,2-Tetrachloroethane	1				35				30
Tetrachloroethylene	1				13				95
4,4'-DDT	1				<1.0				>52
Heptachlor	1				<1.0				>29

Note: Blanks indicate data not applicable.

[a]Actual data indicate negative removal.

CONTROL TECHNOLOGY SUMMARY FOR SEDIMENTATION WITH CHEMICAL ADDITION (ALUM, POLYMER)

Pollutant	Number of data points	Effluent concentration				Removal efficiency, %			
		Minimum	Maximum	Median	Mean	Minimum	Maximum	Median	Mean
Conventional pollutants, mg/L:									
BOD_5	5	57	3,800	2,800	2,150	7	65	25	35
COD	5	125	30,000	10,000	12,100	38	80	69	62
TOC	4	40	4,800	2,850	2,640	37	71	58	56
TSS	4	46	6,000	1,370	2,200	0[a]	99	67	58
Oil and grease	4	4	880	80.5	261	48	99	80	77
Total phenol	5	0.028	0.15	0.10	0.0996	0[a]	60	26	30
Total phosphorus	1				1.6				77
Toxic pollutants, μg/L:									
Cadmium	2	12	30		21	61	76		69
Chromium	4	30	130		[illegible]	15	95	90	72
Copper	4	31	27,000		[illegible],900	0[a]	80	58	<49
Cyanide	1				74				0[a]
Lead	4	66	800	200	<320	7	>96	69	>60
Mercury	3	30	14,000	1,500	5,200	50	88	74	71
Nickel	3	<50	51,000	50	17,000	0[a]	>97	9	54
Silver	1				11				21
Zinc	4	240	1,000	700	660	60	83	70	71
Di-n-butyl phthalate	1				<10				>99
Phenol	1				2				89
Benzene	2	<10	310		<160	0[a]	>97		49
Ethylbenzene	3	<10	460	390	290	70	>94	75	>80
Toluene	4	3	2,900	540	1,000	0[a]	73	40	<57
Carbon tetrachloride	1				1,800				94
Chloroform	4	<10	550	160	<36	0[a]	>94	40	67
1,1-Dichloroethylene	1				<10				>98
1,2-Dichloroethane	2	<10	90		<50	0[a]	>60		30
1,2-*Trans*-dichloroethylene	1				190				28
Methylene chloride	4	110	13,000	7,600	7,000	0[a]	98	91	<70
Tetrachloroethylene	3	<10	700	100	400	0[a]	>44	0	15
1,1,1-Trichloroethane	2	17	120		68	0[a]	93		<47
1,1,2-Trichloroethane	1				11				0[a]
Trichloroethylene	1				12				0[a]

Note: Blanks indicate data not applicable.

[a]Actual data indicate negative removal.

CONTROL TECHNOLOGY SUMMARY FOR SEDIMENTATION WITH CHEMICAL ADDITION ($BaCl_2$)

Pollutant	Number of data points	Effluent concentration				Removal efficiency, %			
		Minimum	Maximum	Median	Mean	Minimum	Maximum	Median	Mean
Conventional pollutants, mg/L:									
COD	2	4	17		10.5	60	65		62
TOC	2	7	16		12	0[a]	98		<49
TSS	2	<1	26		14	>88	90		>89
Total phenol	1				0.01				0
Toxic pollutants, µg/L:									
Antimony	1				<50				>0
Arsenic	2	<2	15		<9	0[a]	>33		17
Asbestos, fibers/L	2	5.7×10^8	2.3×10^9		1.4×10^9	0[a]	75		38
Chromium	2	25	30		27.5	50	93		72
Copper	2	<20	30		<25	>50	73		>62
Lead	2	30	50		40	0[a]	83		<42
Mercury	1				0.5				87
Selenium	1				10				0[a]
Silver	1				20				0[a]
Zinc	2	30	30		30	50	80		65
Bis(2-ethylhexyl) phthalate	2	2.4	15		8.7	0[a]	95		<48
Other pollutants, pCi/L									
Radium (total)	5	1.1	11	<30	42	77	99	>96	>92
Radium (dissolved)	3	<0.9	<2	1.6	15	66	>99	>94	86

Note: Blanks indicate data not applicable.

[a]Actual data indicate negative removal.

CONTROL TECHNOLOGY SUMMARY FOR SEDIMENTATION WITH CHEMICAL ADDITION (Fe^{2+}, LIME)

Pollutant	Number of data points	Effluent concentration				Removal efficiency, %			
		Minimum	Maximum	Median	Mean	Minimum	Maximum	Median	Mean
Toxic pollutants, μg/L:									
Antimony	4	3.5	30	9	16	0[a]	30	0[a]	5
Arsenic	4	<1	3	<2	<2	25	>86	67	>59
Beryllium	2				<0.5				>83
Cadmium	4	<0.5	3.2	1.1	<1.5	0[a]	>50	>24	25
Chromium	4	2	4	2.5	2.7	33	>95	45	55
Copper	6	4	48	25	21	31	92	83	72
Lead	3	<3	<3	<3	<3	>0	>96	>25	>40
Mercury	2	<0.2	0.2		<0.2	0	>60		>30
Nickel	5	<0.5	6	3	3	0	>95	20	35
Selenium	2	7	32		20	12	24		18
Silver	6	0.4	10	3.1	3.0	0[a]	93	4.5	24
Thallium	2	<1.0	7.0		<4.0	22	>88		>55
Zinc	6	<2	36	<23	15	14	>97	92	79

Note: Blanks indicate data not applicable.

[a]Actual data indicate negative removal.

CONTROL TECHNOLOGY SUMMARY FOR SEDIMENTATION WITH CHEMICAL ADDITION (LIME)

Pollutant	Number of data points	Effluent concentration				Removal efficiency, %			
		Minimum	Maximum	Median	Mean	Minimum	Maximum	Median	Mean
Conventional pollutants, mg/L:									
COD	4	8	37	23.8	23.1	0[a]	50	14	32
TOC	3	9	<20	<12	13.7	>5	37	18	>20
TSS	9	4	150	12.5	44.7	0[a]	96	71	61
Oil and grease	2	1	1.5		1.25	0[a]	66		33
Total phenol	2	0.012	0.3		0.171	11	33		22
Toxic pollutants, μg/L:									
Antimony	7	1.9	180	4	30	0[a]	83	40	38
Arsenic	11	<1	110	3	<22	0[a]	>99	63	63
Asbestos, fibers/L	1				6.1×10^6				95
Beryllium	2	0.8	0.9		0.85	0	76		38
Cadmium	9	<0.5	80	3.0	17	0[a]	92	>38	43
Chromium	10	<2	1,800	40	210	0[a]	>99	88	61
Chromium (dissolved)	1				90				99
Copper	16	<10	700	54	91	29	99	79	77
Cyanide	1				45				0[a]
Lead	13	<3	200	40	67	0[a]	99	73	55
Mercury	9	<0.2	8	0.7	<1.4	0[a]	>96	>60	>35
Nickel	13	2.2	5,200	16	490	0[a]	>99	44	45
Nickel (dissolved)	1				2,500				99
Selenium	5	2.3	52	8	30	0[a]	0	0	0
Silver	6	0.4	<10	2.6	4.0	0[a]	>80	10	24
Thallium	3	<1	8	1.1	<3.4	11	>80	58	>52
Zinc	15	<2	8,200	120	680	25	>99	85	>82
Other pollutants, μg/L:									
Fluoride	3	2,500	12,000	9,100	7,900	44	98	72	71
Chloride	1				1.9×10^6				26
Aluminum	2	200	500		350	83	98		91
Iron	2				1,200	96	>99		>98
Calcium	1				230,000				58
Manganese	1				200				99
Other pollutants, μg/L									
Fluoride	1				130,000				92

Note: Blanks indicate data not applicable.

[a]Actual data indicate negative removal.

CONTROL TECHNOLOGY SUMMARY FOR SEDIMENTATION WITH CHEMICAL ADDITION (LIME, POLYMER)

Pollutant	Number of data points	Effluent concentration				Removal Efficiency, %			
		Minimum	Maximum	Median	Mean	Minimum	Maximum	Median	Mean
Conventional pollutants, mg/L:									
TSS	7	5	43	17	22.2	0[a]	>99	69	62
Oil and grease	3	4	8.5	5.5	6	68	71	70	70
Toxic pollutants, μg/L:									
Arsenic	2	<10	<10		<10	0	>0		>0
Cadmium	3	15	60	<20	<31	0	93	50	48
Chromium	3	32	360	75	160	83	90	89	87
Chromium (dissolved)	1				1,300				99
Copper	10	15	170	40	49	8.7	>99	88	>75
Cyanide	3	2	39	2.3	14	54	89	65	69
Lead	8	<20	580	130	200	0[a]	95	58	53
Nickel	3	45	330	270	220	39	96	76	70
Nickel (dissolved)	1				2,500				99
Selenium	1				<10				50
Silver	1				90				0[a]
Zinc	9	25	1,400	260	430	0[a]	>99	83	76
Bis(2-ethylhexyl) phthalate	2	32	150		91	97	99		98
Butyl benzyl phthalate	1				<10				>99
Di-n-butyl phthalate	1				1				99
Diethyl phthalate	1				<10				>99
2,4-Dimethylphenol	1				<10				>76
Phenol	1				<10				>37
p-Chloro-m-cresol	1				62				45
Anthracene	1				<10				>0
Benzo(a)pyrene	1				<10				>81
Chrysene	1				10				99
Fluoranthene	1				<10				>97
Fluorene	1				<10				>99
Naphthalene	1				3				98
Pyrene	1				<10				>87
Chloroform	2	7	42		25	0[a]	9		>5
Methylene chloride	2	13	39		26	0[a]	0[a]		0[a]
1,1,1-Trichloroethane	1				51				0[a]
Other pollutants, μg/L									
Fluoride	1				130,000				92

Note: Blanks indicate data not applicable.

[a]Actual data indicate negative removal.

CONTROL TECHNOLOGY SUMMARY FOR SEDIMENTATION WITH CHEMICAL ADDITION (POLYMER)

Pollutant	Number of data points	Effluent concentration				Removal efficiency, %			
		Minimum	Maximum	Median	Mean	Minimum	Maximum	Median	Mean
Conventional pollutants, mg/L:									
BOD_5	1				4,700				2
COD	1				8,000				71
TOC	1				1,600				82
TSS	1				39				>99
Oil and grease	1				22				98
Total phenol	2	0.082	0.3		0.191	0[a]	58		29
Toxic pollutants, μg/L:									
Antimony	1				<13				44
Cadmium	1				100				0[a]
Chromium	2	<4	25		<15	>96	97		>96
Copper	2	<4	400		200	27	>89		>58
Lead	2	<22	140		81	>12	97		>55
Mercury	2	<0.3	140		70	>25	99		>62
Nickel	1				<13				35
Zinc	2	160	6,000		3,100	89	97		93
Bis(2-ethylhexyl) phthalate	2	<10	10		<10	0	>97		>49
Di-n-butyl phthalate	2	2.8	<10		<6.4	0[a]	>99		50
Diethyl phthalate	1				<0.03				>98
Phenol	2	0.5	74		37	0[a]	29		15
Benzene	1				0.4				0[a]
Ethylbenzene	1				130				81
Toluene	2	0.4	1,900		950	0	39		20
Anthracene	1				0.9				0[a]
Chloroform	1				11				0[a]
1,2-Trans-dichloroethylene	1				21				0[a]
Methylene chloride	2	2.5	130		66	0[a]	0[a]		0[a]
Trichloroethylene	2	0.8	14		7.4	0[a]	0[a]		0[a]

Note: Blanks indicate data not applicable.

[a]Actual data indicate negative removal.

TREATMENT TECHNOLOGY: Sedimentation with Chemical Addition (Alum)

Data source: Effluent Guidelines
Point source category: Paint manufacturing
Subcategory:
Plant: 2
References: A4, Appendix G

Data source status:
- Engineering estimate ___
- Bench scale ___
- Pilot scale ___
- Full scale x

Use in system: Primary
Pretreatment of influent: None

DESIGN OR OPERATING PARAMETERS

Unit configuration:
Wastewater flow:
Chemical dosage(s):
Mix detention time:
Mixing intensity (G):
Flocculation (GCt):
pH in clarifier:
Clarifier detention time:

Hydraulic loading:
Weir loading:
Sludge underflow:
Percent solids in sludge:
Scum overflow:

REMOVAL DATA

Sampling period: Grab samples

Pollutant/parameter	Concentration[a] Influent	Effluent	Percent removal
Conventional pollutants, mg/L:			
BOD_5	2,800	2,900	0[b]
COD	26,000	7,600	71
TOC	7,500	1,500	80
TSS	9,500	50	99
Oil and grease	1,810	11	99
Total phenol	0.076	0.070	8
Toxic pollutants, μg/L:			
Cadmium	130	<15	>88
Chromium	1,700	<40	>98
Copper	470	<110	>78
Mercury	400	<150	>62
Nickel	90	<40	>56
Zinc	60,000	9,000	85
Di-n-butyl phthalate	160	ND[c]	>94
Phenol	96	ND	>90
Ethylbenzene	ND	4,600	0[b]
Nitrobenzene	110	35	68
Toluene	ND	2,500	0[b]
Chloroform	ND	22	0[b]
1,2-Dichloroethane	ND	17	0[b]
Methylene chloride	85	ND	>88
Tetrachloroethylene	ND	45	0[b]
Trichloroethylene	210	190	10

[a]Average of several samples.
[b]Actual data indicate negative removed.
[c]Not detected; assumed to be <10 μg/L.

Note: Blanks indicate information was not specified.

TREATMENT TECHNOLOGY: Sedimentation with Chemical Addition (Sodium Aluminate)

Data source: Effluent Guidelines	Data source status:		
Point source category: Paint manufacturing		Engineering estimate	___
Subcategory:		Bench scale	___
Plant: 5		Pilot scale	___
References: A4, Appendix G		Full scale	x

Use in system: Primary
Pretreatment of influent: None

DESIGN OR OPERATING PARAMETERS

Unit configuration:
Wastewater flow:
Chemical dosage(s):
Mix detention time:
Mixing intensity (G):
Flocculation (GCt):
pH in clarifier:
Clarifier detention time:
Hydraulic loading:
Weir loading:
Sludge underflow:
Percent solids in sludge:
Scum overflow:

REMOVAL DATA

Sampling period: Grab sample

Pollutant/parameter	Concentration[a] Influent	Effluent	Percent removal
Conventional pollutants, mg/L:			
BOD_5	48,000	20,400	57
COD	79,600	31,000	61
TOC	8,000	5,980	25
TSS	12,900	21	>99
Oil and grease	1,260	22	98
Total phenol	0.102	0.077	24
Toxic pollutants, µg/L:			
Antimony	55	<25	>55
Beryllium	8	<4	>50
Cadmium	40	30	25
Chromium	27,000	17,000	35
Copper	900	450	50
Cyanide	<150	<20	87
Lead	14,000	13,500	4
Mercury	540	170	69
Nickel	<50	600	0[b]
Zinc	110,000	35,000	68
Bis(2-ethylhexyl) phthalate	410	80	81
Di-n-butyl phthalate	36,000	550	98
Pentachlorophenol	2,700	200	93
Phenol	ND[c]	140	0[b]
Benzene	ND	240	0[b]
Ethylbenzene	7,800	38,000	0[b]
Nitrobenzene	1,200	ND	>99
Toluene	ND	7,200	0[b]
Naphthalene	9,000	1,300	85
Carbon tetrachloride	ND	65	0[b]
1,2-Dichloroethane	420	ND	>98
1,1-Dichlorethylene	12	22	0[b]
Methylene chloride	450	320	28
Trichloroethylene	40,500	110	>99
Isophorone	ND	220	0[b]

[a]Average of several samples.
[b]Actual data indicate negative removal.
[c]Not detected.

Note: Blanks indicate information was not specified.

TREATMENT TECHNOLOGY: Sedimentation with Chemical Addition (Alum)

Data source: Effluent Guidelines
Point source category: Textile mills
Subcategory: Knit fabric finishing
Plant:
References: A6, p. VII-38

Data source status:
Engineering estimate ___
Bench scale ___
Pilot scale ___
Full scale x

Use in system: Tertiary
Pretreatment of influent: Equalization, aerated lagoon plus clarifier

DESIGN OR OPERATING PARAMETERS

Unit configuration:
Wastewater flow:
Chemical dosage(s):
Mix detention time:
Mixing intensity ($\bar{G}$):
Flocculation ($\bar{G}$Ct):
pH in clarifier:
Clarifier detention time:
Hydraulic loading:
Weir loading:
Sludge underflow:
Percent solids in sludge:
Scum overflow:

REMOVAL DATA

Sampling period: Daily samples for one year, phenol and metals sampled once per month.

Pollutant/parameter	Concentration Influent	Concentration Effluent	Percent removal
Conventional pollutants, mg/L:			
BOD_5	122	33	73
COD	1,060	416	61
TOC	200	105	47
TSS	368	122	67
Total phenol	0.030	0.040	0[a]
Toxic pollutants, µg/L:			
Chromium	360	280	22
Copper	30	ND[b]	>67
Lead	28	23	18
Mercury	1.8	1.7	6
Nickel	10	10	0
Zinc	220	110	50

[a] Actual data indicate negative removal.
[b] Not detected; assumed to be <10 µg/L.

Note: Blanks indicate information was not specified.

TREATMENT TECHNOLOGY: Sedimentation with Chemical Addition (Alum)

Data source: Effluent Guidelines
Point source category: Textile mills
Subcategory: Woven fabric finishing
Plant: V, C (different references)
References: A6, pp. VII-43-44; B3, pp. 45-49

Data source status:
Engineering estimate ___
Bench scale ___
Pilot scale _x_
Full scale ___

Use in system: Tertiary
Pretreatment of influent: Screening, neutralization, activated sludge

DESIGN OR OPERATING PARAMETERS

Unit configuration: 6.25 m^3 (1,650 gal) reactor/clarifier
Wastewater flow:
Chemical dosage(s): 40 mg/L alum (Al^{+3})
Mix detention time:
Mixing intensity (G):
Flocculation (GCt):
pH in clarifier: 6.9
Clarifier detention time:
Hydraulic loading: 16 $m^3/d/m^2$ (400 gpd/ft^2)

Weir loading:
Sludge underflow:
Percent solids in sludge:
Scum overflow:

REMOVAL DATA

Sampling period: 24 hr for toxic pollutants.

Pollutant/parameter	Concentration Influent	Concentration Effluent	Percent removal
Conventional pollutants, mg/L:			
BOD_5	9.3	3.6	61
COD	393	352	10
TOC	76	72	5
TSS	47	51	0[a]
Total phenol	0.023	0.016	30
Total phosphorus	2.7	2.3	15
Toxic pollutants, µg/L:			
Antimony	90	120	0[a]
Arsenic	1.6	<1	>37
Beryllium	1.5	2.2	0[a]
Cadmium	<2	2.9	0[a]
Chromium	5.5	17	0[a]
Copper	57	11	81
Lead	27	66	0[a]
Silver	80	72	10
Zinc	160	190	0[a]
Bis(2-ethylhexyl) phthalate	7.6	33	0[a]
Di-n-butyl phthalate	0.6	0.6	0
Phenol	0.4	<0.07	>82
1,2-Dichlorobenzene	<0.05	13	0[a]
Ethylbenzene	<0.2	1.3	0[a]
Toluene	15	1.0	93
Anthracene/phenanthrene	0.05	0.1	0[a]
Chlorodibromomethane	0.6	<0.3	>50
Methylene chloride[b]	160	70	56

[a]Actual data indicate negative removal.

[b]Presence may be due to sample contamination.

Note: Blanks indicate information was not specified.

TREATMENT TECHNOLOGY: Sedimentation with Chemical Addition (Alum)

Data source: Effluent Guidelines
Point source category: Textile mills
Subcategory: Wool finishing
Plant: B
References: A6, pp. VII-39 to 41

Data source status:
Engineering estimate ___
Bench scale ___
Pilot scale x
Full scale ___

Use in system: Tertiary
Pretreatment of influent: Screening, equalization, activated sludge

DESIGN OR OPERATING PARAMETERS

Unit configuration: Reactor/clarifier
Wastewater flow:
Chemical dosage(s): 27-35 mg/L alum (as Al^{+3})
Mix detention time:
Mixing intensity (G):
Flocculation (GCt):
pH in clarifier:
Clarifier detention time:
Hydraulic loading: 400-520 gpd/ft^2
Weir loading:
Sludge underflow:
Percent solids in sludge:
Scum overflow:

REMOVAL DATA

Sampling period: Average of 3 experimental runs; 21 samples for conventional pollutants and single 24-hr composite sample for toxics

Pollutant/parameter	Concentration Influent	Concentration Effluent	Percent removal
Conventional pollutants, mg/L:			
BOD_5	175	32	82
COD	962	212	78
TOC	321	72	78
TSS	244	28	89
Toxic pollutants, µg/L:			
Antimony	22	23	0[a]
Arsenic	60	62	0[a]
Chromium	120	41	65
Copper	23	16	30
Lead	30	30	0
Nickel	76	57	25
Silver	140	170	0[a]
Zinc	6,400	5,700	10
Bis(2-ethylhexyl) phthalate	32	44	0[a]
1,2-Dichlorobenzene	20	ND[b]	>50
Toluene	31	14	55
1,2,4-Trichlorobenzene	1,600	150	90

[a]Actual data indicate negative removal.
[b]Not detected; assumed to be <10 µg/L.

Note: Blanks indicate information was not specified.

TREATMENT TECHNOLOGY: Sedimentation with Chemical Addition (Alum)

Data source: Government report
Point source category:[a]
Subcategory:
Plant: Reichold Chemical, Inc.
References: B4, p. 46

Data source status:
Engineering estimate ___
Bench scale ___
Pilot scale x
Full scale ___

Use in system: Primary
Pretreatment of influent: Equalization

[a]Organic and inorganic wastes.

DESIGN OR OPERATING PARAMETERS

Unit configuration:
Wastewater flow:
Chemical dosage(s): 650 mg/L
Mix detention time:
Mixing intensity (G):
Flocculation (GCt):
pH in clarifier:
Clarifier detention time:
Hydraulic loading:

Weir loading:
Sludge underflow:
Percent solids in sludge:
Scum overflow:

REMOVAL DATA

Sampling period:

Pollutant/parameter	Concentration, mg/L Influent	Concentration, mg/L Effluent	Percent removal
Conventional pollutants:			
BOD_5	2,400	2,220	17
COD	3,610	3,470	4
TSS	136	28	79
Total phenol	325	225	31
Total phosphorous	49	43	12

Note: Blanks indicate information was not specified.

TREATMENT TECHNOLOGY: Sedimentation with Chemical Addition (Alum, Lime)

Data source: Effluent Guidelines
Point source category: Paint manufacturing
Subcategory:
Plant: 4
References: A4, Appendix G

Data source status:
Engineering estimate ___
Bench scale ___
Pilot scale ___
Full scale x

Use in system: Primary
Pretreatment of influent:

DESIGN OR OPERATING PARAMETERS

Unit configuration:
Wastewater flow:
Chemical dosage(s):
Mix detention time:
Mixing intensity (G):
Flocculation (GCt):
pH in clarifier:
Clarifier detention time:

Hydraulic loading:
Weir loading:
Sludge underflow:
Percent solids in sludge:
Scum overflow:

REMOVAL DATA

Sampling period:

Pollutant/parameter	Concentration[a] Influent	Effluent	Percent removal
Conventional pollutants, mg/L:			
BOD_5	3,300	3,900	0[b]
COD	147,000	7,970	95
TOC	13,000	2,300	82
TSS	14,000	480	97
Oil and grease	830	<16	>98
Total phenol	1.1	1.3	0[b]
Toxic pollutants, μg/L:			
Copper	500	60	88
Cyanide	150	30	80
Lead	370	<200	50
Mercury	7	2	71
Zinc	170,000	1,100	>99
Di-n-butyl phthalate	6,500	ND[c]	>99
Phenol	1,300	47	96
Benzene	92	46	50
Ethylbenzene	1,230	22	98
Toluene	1,900	72	96
Naphthalene	54	16	70
Carbon tetrachloride	12	ND	>17
Chloroform	16	74	0[b]
1,2-Dichloropropane	968	400	59
Methylene chloride	2,300	2,000	13
1,1,2,2-Tetrachloroethane	50	35	30
Tetrachloroethylene	270	13	95

[a]Average of several samples.
[b]Actual data indicate negative removal.
[c]Not detected; assumed to be <10 μg/L.

Note: Blanks indicate information not specified.

TREATMENT TECHNOLOGY: Sedimentation with Chemical Addition (Alum, Lime)

Data source: Effluent Guidelines
Point source category: Textile mills
Subcategory: Wool finishing
Plant: A
References: B3, pp. 39-44

Data source status:
Engineering estimate ___
Bench scale ___
Pilot scale _x_
Full scale ___

Use in system: Tertiary
Pretreatment of influent: Screening, equalization, activated sludge

DESIGN OR OPERATING PARAMETERS

Unit configuration: 6.25 m^3 (1,650 gal) reactor/clarifier
Wastewater flow:
Chemical dosage(s): 27-35 mg/L alum (as Al^{+3})
100 mg/L lime (as $Ca(OH)_2$)
Mix detention time:
Mixing intensity ($\bar{G}$):
Flocculation ($\bar{G}Ct$):
pH in clarifier: 6.1
Clarifier detention time:
Hydraulic loading: 16-21 $m^3/d/m^2$ (400-520 gpd/ft^2)

Weir loading:
Sludge underflow:
Percent solids in sludge:
Scum overflow:

REMOVAL DATA

Sampling period: 24-hr composite for toxic pollutants, volatile organics were grab-sampled

Pollutant/parameter	Concentration Influent	Effluent	Percent removal
Toxic pollutants, μg/L:			
Arsenic	60	62	0[a]
Chromium	110	31	72
Copper	20	13	35
Cyanide	10	<4	>60
Nickel	5.8	<1.0	>83
Zinc	6,400	5,700	11
Bis(2-ethylhexyl) phthalate	32	44	0[a]
Phenol	<0.07	3	0[a]
1,2-Dichlorobenzene	20	<0.05	>99
Ethylbenzene	5	<0.2	>96
Toluene	31	14	55
1,2,4-Trichlorobenzene	1,600	150	91
4,4'-DDT	2.1	<1.0	>52
Heptachlor	1.4	<1.0	>29

[a]Actual data indicate negative removal.

Note: Blanks indicate information was not specified.

TREATMENT TECHNOLOGY: Sedimentation with Chemical Addition (Alum, Lime, Ferric Chloride)

Data source: Effluent Guidelines
Point source category: Paint manufacturing
Subcategory:
Plant: 20
References: A4, Appendix G

Data source status:
- Engineering estimate ___
- Bench scale ___
- Pilot scale ___
- Full scale x

Use in system: Primary
Pretreatment of influent: None

DESIGN OR OPERATING PARAMETERS

Unit configuration:
Wastewater flow:
Chemical dosage(s):
Mix detention time:
Mixing intensity (G):
Flocculation (GCt):
pH in clarifier:
Clarifier detention time:
Hydraulic loading:
Weir loading:
Sludge underflow:
Percent solids in sludge:
Scum overflow:

REMOVAL DATA

Sampling period: Grab sample

Pollutant/parameter	Concentration		Percent removal
	Influent	Effluent	
Conventional pollutants, mg/L:			
BOD_5	4,670	1,110	76
COD	19,700	6,930	65
TOC	4,730	1,590	66
TSS	13,800	1,370	90
Oil and grease	393	91	77
Total phenol	0.115	0.046	60
Toxic pollutants, μg/L:			
Cadmium	30	<20	>33
Chromium	∿150	∿170	0[a]
Copper	300	170	44
Lead	∿300	∿250	17
Mercury	4,900	990	80
Nickel	100	<50	>50
Thallium	16	<10	>37
Zinc	870	∿1,400	0[a]
Di-n-butyl phthalate	360	<10	>97
Benzene	ND[b]	3,800	0[a]
Ethylbenzene	110	ND	>91
Toluene	3,800	4,200	0[a]
Carbon tetrachloride	19	ND	>47
Chloroform	55	4,700	0[a]
Methylene chloride	1	9,800	0[a]
Tetrachloroethylene	540	ND	>98
1,1,1-Trichloroethane	ND	120	0[a]
1,1,2-Trichloroethane	2,800	ND	>99
Trichloroethylene	250	300	0[a]

[a]Actual data indicate negative removal.

[b]Not detected; assumed to be <10 μg/L.

Note: Blanks indicate information was not specified.

TREATMENT TECHNOLOGY: Sedimentation with Chemical Addition (Alum, Lime, Polymer)

Data source: Effluent Guidelines
Point source category: Paint manufacturing
Subcategory:
Plant: 6
References: A4, Appendix G

Data source status:
Engineering estimate ___
Bench scale ___
Pilot scale ___
Full scale x

Use in system: Primary
Pretreatment of influent:

DESIGN OR OPERATING PARAMETERS

Unit configuration:
Wastewater flow:
Chemical dosage(s):
Mix detention time:
Mixing intensity (G):
Flocculation (GCt):
pH in clarifier:
Clarifier detention time:

Hydraulic loading:
Weir loading:
Sludge underflow:
Percent solids in sludge:
Scum overflow:

REMOVAL DATA

Sampling period:

Pollutant/parameter	Concentration[a] Influent	Effluent	Percent removal
Conventional pollutants, mg/L:			
BOD_5	7,100	9,000	0[b]
COD	32,000	12,000	62
TOC	9,800	2,500	74
TSS	23,900	100	>99
Oil and grease	980	22	98
Total phenol	0.27	0.14	48
Toxic pollutants, μg/L:			
Copper	400	97	76
Lead	800	≤200	≥75
Mercury	20	0.6	97
Zinc	300,000	17,000	94
Phenol	30	<10	>67
Benzene	2,020	195	90
Ethylbenzene	80	<10	>87
Toluene	8,700	1,400	84
Naphthalene	30	<10	>67
Carbon tetrachloride	93	ND[c]	>89
Chloroform	125	7	94
1,1-Dichloroethylene	28	ND	>64
Methylene chloride	275	90	67

[a]Average of several samples.
[b]Actual data indicate negative removal.
[c]Not detected; assumed to be <10 μg/L.

Note: Blanks indicate information was not specified.

TREATMENT TECHNOLOGY: Sedimentation with Chemical Addition (Alum, Polymer)

Data source: Effluent Guidelines
Point source category: Auto and other laundries
Subcategory: Power laundries
Plant: N
References: A28, Appendix C

Data source status:
- Engineering estimate ___
- Bench scale ___
- Pilot scale ___
- Full scale x

Use in system: Primary
Pretreatment of influent: Screening, equalization

DESIGN OR OPERATING PARAMETERS

Unit configuration: Circular clarifier 4.92 m^3 (1,300 gal) with mix tank
Wastewater flow: 15.2 m^3/d (4,000 gpd)
Chemical dosage(s): Alum - 2,800 mg/L
Polymer - 200 mg/L
Mix detention time:
Mixing intensity (G):
Flocculation (GCt):
pH in clarifier:
Clarifier detention time: 0.33 day

Hydraulic loading:
Weir loading:
Sludge underflow:
Percent solids in sludge:
Scum overflow:

REMOVAL DATA

Sampling period: 3 days total

Pollutant/parameter	Concentration Influent	Effluent	Percent removal
Conventional pollutants, mg/L:			
BOD_5	163	57	65
COD	240	125	48
TOC	63	40	37
TSS	40	46	0[a]
Oil and grease	15	4	73
Total phenol	0.038	0.028	26
Total phosphorus	7.0	1.6	77
Toxic pollutants, µg/L:			
Cadmium	51	12	76
Chromium	39	34	13
Copper	138	31	78
Lead	71	66	7
Nickel	55	50	9
Silver	14	11	21
Zinc	609	244	60
Phenol	18	2	89
Toluene	5	3	40
Tetrachloroethylene	2	100	0[a]
Trichloroethylene	0.5	12	0[a]

[a]Actual data indicate negative removal.

Note: Blanks indicate information was not specified.

TREATMENT TECHNOLOGY: Sedimentation with Chemical Addition (Alum, Polymer)

Data source: Effluent Guidelines
Point source category: Paint manufacturing
Subcategory:
Plant: 15
References: A4, Appendix G

Data source status:
Engineering estimate ___
Bench scale ___
Pilot scale ___
Full scale x

Use in system: Primary
Pretreatment of influent: None

DESIGN OR OPERATING PARAMETERS

Unit configuration:
Wastewater flow:
Chemical dosage(s):
Mix detention time:
Mixing intensity (G):
Flocculation (GCt):
pH in clarifier:
Clarifier detention time:

Hydraulic loading:
Weir loading:
Sludge underflow:
Percent solids in sludge:
Scum overflow:

REMOVAL DATA

Sampling period: Composite sample, period not specified

Pollutant/parameter	Concentration[a] Influent	Effluent	Percent removal
Conventional pollutants, mg/L:			
BOD_5	8,400	3,800	55
COD	48,000	30,000	38
TOC	9,000	4,800	47
TSS	14,200	6,000	58
Oil and grease	1,700	880	48
Total phenol	0.23	0.14	39
Toxic pollutants, μg/L:			
Cadmium	76	30	61
Chromium	1,600	83	95
Copper	800	500	38
Cyanide	37	74	0[b]
Lead	6,000	800	87
Mercury	55,000	14,000	74
Zinc	6,000	1,000	83
Di-n-butyl phthalate	40,000	ND[c]	>99
Carbon tetrachloride	30,000	1,800	94
Chloroform	ND	550	0[b]
1,1-Dichloroethylene	620	ND	>98
1,2-*Trans*-dichloroethylene	260	190	28
Methylene chloride	160,000	12,000	92

[a]Average of several samples.

[b]Actual data indicate negative removal.

[c]Not detected; assumed to be <10 μg/L.

Note: Blanks indicate information was not specified.

TREATMENT TECHNOLOGY: Sedimentation with Chemical Addition (Alum, Polymer)

Data source: Effluent Guidelines
Point source category: Paint manufacturing
Subcategory:
Plant: 8
References: A4, Appendix G

Data source status:
Engineering estimate ___
Bench scale ___
Pilot scale ___
Full scale x

Use in system: Primary
Pretreatment of influent: None

DESIGN OR OPERATING PARAMETERS

Unit configuration:
Wastewater flow:
Chemical dosage(s):
Mix detention time:
Mixing intensity (G):
Flocculation (GCt):
pH in clarifier:
Clarifier detention time:

Hydraulic loading:
Weir loading:
Sludge underflow:
Percent solids in sludge:
Scum overflow:

REMOVAL DATA

Sampling period: Grab samples

Pollutant/parameter	Concentration[a] Influent	Effluent	Percent removal
Conventional pollutants, mg/L:			
BOD_5	3,900	3,000	23
COD	41,000	9,500	77
TOC	8,500	2,500	71
TSS	16,000	140	99
Oil and grease	642	8	99
Total phenol	0.25	0.10	60
Toxic pollutants, μg/L:			
Chromium	300	30	90
Copper	3,700	27,000	0[b]
Lead	400	200	50
Mercury	13,000	1,500	88
Nickel	14,000	51,000	0[b]
Zinc	3,200	800	75
Benzene	290	310	0[b]
Ethylbenzene	180	ND[c]	>94
Toluene	73	350	0[b]
Chloroform	ND	36	0[b]
1,2-Dichloroethane	ND	90	0[b]
Methylene chloride	ND	3,100	0[b]
Tetrachloroethylene	400	700	0[b]
1,1,1-Trichloroethane	ND	120	0[b]

[a]Average of several samples.
[b]Actual data indicate negative removal.
[c]Not detected; assumed to be <10 μg/L.

Note: Blanks indicate information was not specified.

TREATMENT TECHNOLOGY: Sedimentation with Chemical Addition (Alum, Polymer)

Data source: Effluent Guidelines
Point source category: Paint manufacturing
Subcategory:
Plant: 24
References: A4, Appendix G

Data source status:
Engineering estimate ___
Bench scale ___
Pilot scale ___
Full scale x

Use in system: Primary
Pretreatment of influent:

DESIGN OR OPERATING PARAMETERS

Unit configuration:
Wastewater flow:
Chemical dosage(s):
Mix detention time:
Mixing intensity ($\bar{G}$):
Flocculation (GCt):
pH in clarifier:
Clarifier detention time:

Hydraulic loading:
Weir loading:
Sludge underflow:
Percent solids in sludge:
Scum overflow:

REMOVAL DATA

Sampling period: Grab samples

Pollutant/parameter	Concentration[a] Influent	Concentration[a] Effluent	Percent removal
Conventional pollutants, mg/L:			
BOD_5	16,000	1,100	25
COD	36,000	11,000	69
Total phenol	0.20	0.15	25
Toxic pollutants, μg/L:			
Ethylbenzene	1,800	460	75
Toluene	2,900	2,900	0
Chloroform	43	26	40
Methylene chloride	130,000	13,000	90
1,1,2-Trichloroethane	ND[b]	11	0[c]

[a]Average of several samples.

[b]Not detected.

[c]Actual data indicate negative removal.

Note: Blanks indicate information was not specified.

TREATMENT TECHNOLOGY: Sedimentation with Chemical Addition (Alum, Polymer)

Data source: Effluent Guidelines
Point source category: Paint manufacturing
Subcategory:
Plant: 1
References: A4, Appendix G

Data source status:
Engineering estimate ___
Bench scale ___
Pilot scale ___
Full scale x

Use in system: Primary
Pretreatment of influent: None

DESIGN OR OPERATING PARAMETERS

Unit configuration:
Wastewater flow:
Chemical dosage(s):
Mix detention time:
Mixing intensity (G):
Flocculation (GCt):
pH in clarifier:
Clarifier detention time:

Hydraulic loading:
Weir loading:
Sludge underflow:
Percent solids in sludge:
Scum overflow:

REMOVAL DATA

Sampling period: Grab samples

Pollutant/parameter	Concentration[a] Influent	Effluent	Percent removal
Conventional pollutants, mg/L:			
BOD_5	3,000	2,800	7
COD	51,000	10,000	80
TOC	10,000	3,200	68
TSS	11,000	2,600	76
Oil and grease	1,200	153	87
Total phenol	0.055	0.08	0[b]
Toxic pollutants, μg/L:			
Chromium	1,200	130	89
Copper	400	80	80
Lead	5,000	<200	>96
Mercury	60	30	50
Nickel	2,000	<50	>97
Zinc	1,700	600	65
Benzene	300	ND[c]	>97
Ethylbenzene	1,300	390	70
Toluene	2,700	720	73
Chloroform	160	ND	>94
1,2-Dichloroethane	25	ND	>60
Methylene chloride	4,800	110	98
Tetrachloroethylene	18	ND	>44
1,1,1-Trichloroethane	250	17	93

[a]Metals and conventional pollution concentrations represent an average of several samples, only one sample for organics.

[b]Actual data indicate negative removal.

[c]Not detected; assumed to be <10 μg/L.

Note: Blanks indicate information was not specified.

TREATMENT TECHNOLOGY: Sedimentation with Chemical Addition (Alum, Anionic Polymer)

Data source: Effluent Guidelines
Point source category: Textile mills
Subcategory: Knit fabric finishing
Plant: Q
References: A6, pp. VII-41 to 43

Data source status:
Engineering estimate ___
Bench scale ___
Pilot scale x
Full scale ___

Use in system: Tertiary
Pretreatment of influent: Screening, equalization, activated sludge

DESIGN OR OPERATING PARAMETERS

Unit configuration: 6.25 m^3 (1,650 gal) reactor/clarifier
Wastewater flow:
Chemical dosage(s): 20-30 mg/L alum (as Al^{+3}); 0.75-1.0 mg/L anionic polymer
Mix detention time:
Mixing intensity ($\bar{G}$):
Flocculation ($\bar{G}Ct$):
pH in clarifier:
Clarifier detention time:
Hydraulic loading: 320-400 gpd/ft^2
Weir loading:
Sludge underflow:
Percent solids in sludge:
Scum overflow:

REMOVAL DATA

Sampling period:

Pollutant/parameter	Concentration[a], mg/L Influent	Concentration[a], mg/L Effluent	Percent removal
Conventional pollutants:			
BOD_5	8.1	4.4	46
COD	270	185	31
TOC	30.3	21.5	29
TSS	45	66	0[b]

[a] Average of three experimental runs, 19 samples.
[b] Actual data indicate negative removal.

Note: Blanks indicate information was not specified.

TREATMENT TECHNOLOGY: Sedimentation with Chemical Addition ($BaCl_2$)

Data source: Effluent Guidelines
Point source category: Ore mining and dressing
Subcategory: Uranium mine/mill
Plant: 9411
References: A2, pp. V-62-63

Data source status:
Engineering estimate ___
Bench scale ___
Pilot scale ___
Full scale x

Use in system: Primary
Pretreatment of influent:

DESIGN OR OPERATING PARAMETERS

Unit configuration:
Wastewater flow:
Chemical dosage(s):
Mix detention time:
Mixing intensity ($\bar{G}$):
Flocculation ($\bar{G}Ct$):
pH in clarifier:
Clarifier detention time:

Hydraulic loading:
Weir loading:
Sludge underflow:
Percent solids in sludge:
Scum overflow:

REMOVAL DATA

Sampling period: 24-hr composite

Pollutant/parameter	Concentration		Percent removal
	Influent	Effluent	
Conventional pollutants, mg/L:			
COD	37	17	54
TOC	230	7	98
TSS	8	<1	>88
Toxic pollutants:			
Antimony, μg/L	50	<50	>0
Arsenic, μg/L	3	<2	>33
Asbestos, fibers/L	2.3×10^9	5.7×10^8	75
Chromium, μg/L	50	25	50
Copper, μg/L	40	<20	>50
Lead, μg/L	40	50	0[a]
Mercury, μg/L	3.8	0.5	87
Selenium, μg/L	5	10	0[a]
Zinc, μg/L	60	30	50
Bis(2-ethylhexyl) phthalate,[b] μg/L	47	2.4	95
Other pollutants, pCi/L:			
$Radium_{226}$ (total)	56.9	<2	>96

[a]Actual data indicate negative removal.

[b]Possibly due to tubing in sampling apparatus.

Note: Blanks indicate information was not specified.

TREATMENT TECHNOLOGY: Sedimentation with Chemical Addition ($BaCl_2$)

Data source: Effluent Guidelines
Point source category: Ore mining and dressing
Subcategory: Uranium mine
Plant: 9408
References: A2, pp. V-60-61

Data source status:
Engineering estimate ___
Bench scale ___
Pilot scale ___
Full scale x

Use in system: Primary
Pretreatment of influent:

DESIGN OR OPERATING PARAMETERS

Unit configuration:
Wastewater flow:
Chemical dosage(s):
Mix detention time:
Mixing intensity (G):
Flocculation (GCt):
pH in clarifier:
Clarifier detention time:
Hydraulic loading:

Weir loading:
Sludge underflow:
Percent solids in sludge:
Scum overflow:

REMOVAL DATA

Sampling period: 24-hr composite

Pollutant/parameter	Concentration		Percent
	Influent	Effluent	removal
Conventional pollutants, mg/L:			
COD	12	4	67
TOC	9	16	0[a]
TSS	270	26	90
Total phenol	0.01	0.01	0
Toxic pollutants:			
Arsenic, μg/L	8	15	0[a]
Asbestos, fibers/L	1.6×10^9	2.3×10^9	0[a]
Chromium, μg/L:	450	30	93
Copper, μg/L	110	30	73
Lead, μg/L	180	30	83
Silver, μg/L	<10	20	0[a]
Zinc, μg/L	150	30	80
Bis(2-ethylhexyl) phthalate,[b] μg/L	11	15	0[a]
Other pollutants, pCi/L:			
$Radium_{226}$ (total)	142	1.1	99
$Radium_{226}$ (dissolved)	120	<0.9	>99

[a]Actual data indicate negative removal.
[b]Possibly due to tubing used in sampling apparatus.

Note: Blanks indicate information was not specified.

TREATMENT TECHNOLOGY: Sedimentation with Chemical Addition ($BaCl_2$)

Data source: Effluent Guidelines
Point source category: Ore mining and dressing
Subcategory: Uranium Mine
Plant: See below
References: A1, pp. VI-49

Data source status:
Engineering estimate ___
Bench scale ___
Pilot scale ___
Full scale x

Use in system: Primary unless otherwise specified
Pretreatment of influent: None

DESIGN OR OPERATING PARAMETERS

Unit configuration:
Wastewater flow:
Chemical dosage(s): See below
Mix detention time:
Mixing intensity ($\bar{G}$):
Flocculation ($\bar{G}Ct$):
pH in clarifier:
Clarifier detention time:

Hydraulic loading:
Weir loading:
Sludge underflow:
Percent solids in sludge:
Scum overflow:

REMOVAL DATA

Sampling period:

Plant	Chemical dosage, mg/L $BaCl_2$	Total radium Concentration, picoCi/L Influent	Total radium Concentration, picoCi/L Effluent	Total radium Percent removal	Dissolved radium Concentration, picoCi/L Influent	Dissolved radium Concentration, picoCi/L Effluent	Dissolved radium Percent removal
9412	10.4	49	11	77	4.7	1.6	66
9405[a]	9.5	27.5	<3.0	>91	33.3	<2	>94
9403	7.4	110	4.0	96			

[a]Use in system: tertiary.

Note: Blanks indicate information was not specified.

TREATMENT TECHNOLOGY: Sedimentation with Chemical Addition (Fe^{+2}, lime)

Data source: Effluent Guidelines
Point source category: Steam electric power generating
Subcategory:
Plant: 1226
References: A31, p. 22 (Appendix E)

Data source status:
Engineering estimate ___
Bench scale ___
Pilot scale x
Full scale ___

Use in system: Primary
Pretreatment of influent:

DESIGN OR OPERATING PARAMETERS

Unit configuration: Cooling tower blowdown
Wastewater flow:
Chemical dosage(s):
Mix detention time:
Mixing intensity ($\bar{G}$):
Flocculation ($\bar{G}$Ct):
pH in clarifier:
Clarifier detention time:

Hydraulic loading:
Weir loading:
Sludge underflow:
Percent solids in sludge:
Scum overflow:

REMOVAL DATA

Sampling period:

Pollutant/parameter	Concentration, μg/L Influent	Effluent	Percent removal
Toxic pollutants:			
Antimony	7	9	0[a]
Arsenic	4	3	25
Cadmium	1.8	1.6	11
Chromium	5	3	40
Copper	47	4	91
Lead	3	<3	>0
Mercury	0.2	0.2	0
Nickel	6	6	0
Silver	0.7	0.4	43
Zinc	26	2	92

[a]Actual data indicate negative removal.

Note: Blanks indicate information was not specified.

TREATMENT TECHNOLOGY: Sedimentation with Chemical Addition (Fe^{2+}, lime)

Data source: Effluent Guidelines
Point source category: Steam electric power generating
Subcategory:
Plant: 1226
References: A31, p. 22 (Appendix E)

Data source status:
Engineering estimate ___
Bench scale ___
Pilot scale _x_
Full scale ___

Use in system: Secondary
Pretreatment of influent: Ash pond

DESIGN OR OPERATING PARAMETERS

Unit configuration:
Wastewater flow:
Chemical dosage(s):
Mix detention time:
Mixing intensity ($\bar{G}$):
Flocculation ($\bar{G}Ct$):
pH in clarifier:
Clarifier detention time:

Hydraulic loading:
Weir loading:
Sludge underflow:
Percent solids in sludge:
Scum overflow:

REMOVAL DATA

Sampling period:

Pollutant/parameter	Concentration, µg/L Influent	Effluent	Percent removal
Toxic pollutants:			
Antimony	7	9	0[a]
Arsenic	9	3	67
Cadmium	2.0	3.2	0[a]
Chromium	6	4	33
Copper	14	7	50
Lead	4	<3	>25
Selenium	8	7	12
Silver	0.5	0.6	0[a]
Zinc	7	6	14

[a]Actual data indicate negative removal.

Note: Blanks indicate information was not specified.

TREATMENT TECHNOLOGY: Sedimentation with Chemical Addition (Fe^{+2}, lime)

Data source: Effluent Guidelines
Point source category: Steam electric power generating
Subcategory:
Plant: 5604
References: A31, p. 22 (Appendix E)

Data source status:
Engineering estimate ___
Bench scale ___
Pilot scale x
Full scale ___

Use in system: Primary
Pretreatment of influent:

DESIGN OR OPERATING PARAMETERS

Unit configuration: Cooling tower blowdown
Wastewater flow:
Chemical dosage(s):
Mix detention time:
Mixing intensity ($\bar{G}$):
Flocculation ($\bar{G}Ct$):
pH in clarifier:
Clarifier detention time:

Hydraulic loading:
Weir loading:
Sludge underflow:
Percent solids in sludge:
Scum overflow:

REMOVAL DATA

Sampling period:

Pollutant/parameter	Concentration, µg/L Influent	Concentration, µg/L Effluent	Percent removal
Toxic pollutants:			
Arsenic	7	<1	>86
Copper	180	26	86
Nickel	6	3	50
Silver	3	10	0[a]
Zinc	780	36	95

[a]Actual data indicate negative removal.

Note: Blanks indicate information was not specified.

TREATMENT TECHNOLOGY: Sedimentation with Chemical Addition (Fe^{2+}, lime)

Data source: Effluent Guidelines
Point source category: Steam electric power generating
Subcategory:
Plant: 5604
References: A31, p.22 (Appendix E)

Data source status:
Engineering estimate ___
Bench scale ___
Pilot scale x
Full scale ___

Use in system: Secondary
Pretreatment of influent: Ash pond

DESIGN OR OPERATING PARAMETERS

Unit configuration:
Wastewater flow:
Chemical dosage(s):
Mix detention time:
Mixing intensity (G):
Flocculation (GCt):
pH in clarifier:
Clarifier detention time:

Hydraulic loading:
Weir loading:
Sludge underflow:
Percent solids in sludge:
Scum overflow:

REMOVAL DATA

Sampling period:

Pollutant/parameter	Concentration, µg/L Influent	Effluent	Percent removal
Toxic pollutants:			
Antimony	6	30	0[a]
Beryllium	2.5	0.5	80
Cadmium	1	<0.5	>50
Chromium	4	2	50
Copper	80	23	80
Nickel	9.5	<0.5	>95
Silver	5.5	5	9
Zinc	300	25	92

[a] Actual data indicate negative removal.

Note: Blanks indicate information was not specified.

TREATMENT TECHNOLOGY: Sedimentation with Chemical Addition (Ferrous sulfate, lime)

Data source: Effluent Guidelines
Point source category: Steam electric power generating
Subcategory:
Plant: 5409
References: A31, p. 24 (Appendix E)

Data source status:
- Engineering estimate ___
- Bench scale ___
- Pilot scale x
- Full scale ___

Use in system: Primary
Pretreatment of influent:

DESIGN OR OPERATING PARAMETERS

Unit configuration: Cooling tower blowdown
Wastewater flow:
Chemical dosage(s):
Mix detention time:
Mixing intensity ($\bar{G}$):
Flocculation ($\bar{G}$Ct):
pH in clarifier: 11.5
Clarifier detention time:

Hydraulic loading:
Weir loading:
Sludge underflow:
Percent solids in sludge:
Scum overflow:

REMOVAL DATA

Sampling period:

Pollutant/parameter	Concentration, µg/L Influent	Concentration, µg/L Effluent	Percent removal
Toxic pollutants:			
Beryllium	3.4	<0.5	>85
Cadmium	0.8	<0.5	>37
Chromium	37	<2.0	>95
Copper	620	48	92
Lead	70	<3.0	>96
Mercury	0.5	<0.2	>60
Nickel	4.0	3.6	10
Silver	14	1.0	93
Thallium	8.0	<1.0	>88
Zinc	61	<2	>97

Note: Blanks indicate information was not specified.

TREATMENT TECHNOLOGY: Sedimentation with Chemical Addition (Ferrous sulfate, lime)

Data source: Effluent Guidelines
Point source category: Steam electric power generating
Subcategory:
Plant: 5409
References: A31, p. 24 (Appendix E)

Data source status:
Engineering estimate ___
Bench scale ___
Pilot scale x
Full scale ___

Use in system: Secondary
Pretreatment of influent: Ash pond

DESIGN OR OPERATING PARAMETERS

Unit configuration:
Wastewater flow:
Chemical dosage(s):
Mix detention time:
Mixing intensity ($\bar{G}$):
Flocculation ($\bar{G}Ct$):
pH in clarifier: 11.5
Clarifier detention time:

Hydraulic loading:
Weir loading:
Sludge underflow:
Percent solids in sludge:
Scum overflow:

REMOVAL DATA

Sampling period:

Pollutant/parameter	Concentration, µg/L Influent	Effluent	Percent removal
Toxic pollutants:			
Antimony	5.0	3.5	30
Arsenic	74	<1	>99
Copper	26	18	31
Nickel	2.5	2.0	20
Selenium	42	32	24
Silver	1.0	1.1	0[a]
Thallium	9.0	7.0	22
Zinc	11	<2.0	>82

[a]Actual data indicate negative removal.

Note: Blanks indicate information was not specified.

TREATMENT TECHNOLOGY: Sedimentation with Chemical Addition (Lime)

Data source: Effluent Guidelines
Point source category: Inorganic chemicals
Subcategory: Hydrofluoric acid
Plant: 167
References: A29, p. 227

Data source status:
- Engineering estimate ___
- Bench scale ___
- Pilot scale ___
- Full scale x

Use in system: Primary
Pretreatment of influent:

DESIGN OR OPERATING PARAMETERS

Unit configuration: 47% of effluent is recycled
Wastewater flow:
Chemical dosage(s):
Mix detention time:
Mixing intensity ($\bar{G}$):
Flocculation (GCt):
pH in clarifier:
Clarifier detention time:

Hydraulic loading:
Weir loading:
Sludge underflow:
Percent solids in sludge:
Scum overflow:

REMOVAL DATA

Sampling period: Three 24-hr composite samples

Pollutant/parameter	Concentration,[a] µg/L Influent	Effluent	Percent removal
Toxic pollutants:			
Arsenic	150	<24	>84
Chromium	470	250	47
Copper	120	79	34
Lead	87	37	57
Mercury	27	<1.2	>96
Nickel	1,100	610	45
Selenium	63	87	0[b]
Zinc	240	180	25

[a]Values are combined for wastes from HF and AlF_3. Concentration data is calculated from pollutant flow in m^3/kkg and pollutant loading in kg/kkg.

[b]Actual data indicate negative removal.

Note: Blanks indicate information was not specified.

TREATMENT TECHNOLOGY: Sedimentation with Chemical Addition (Lime)

Data source: Effluent Guidelines
Point source category: Inorganic chemicals
Subcategory: Hydrofluoric acid
Plant: 705
References: A29, p. 227

Data source status:
- Engineering estimate ___
- Bench scale ___
- Pilot scale ___
- Full scale x

Use in system: Primary
Pretreatment of influent:

DESIGN OR OPERATING PARAMETERS

Unit configuration: 30-35% of effluent recycled, remaining effluent pH adjusted and discharged
Wastewater flow:
Chemical dosage(s):
Mix detention time:
Mixing intensity (G):
Flocculation (GCt):
pH in clarifier:
Clarifier detention time:

Hydraulic loading:
Weir loading:
Sludge underflow:
Percent solids in sludge:
Scum overflow:

REMOVAL DATA

Sampling period: Composite samples

Pollutant/parameter	Concentration,[a] μg/L Influent	Effluent	Percent removal
Toxic pollutants:			
Antimony	10	1.9	81
Arsenic	40	<9.7	>76
Cadmium	9.7	1.6	84
Chromium	390	47	88
Copper	290	19	93
Lead	50	23	54
Mercury	5.8	0.48	92
Nickel	560	<9.7	>98
Thallium	2.6	1.1	58
Zinc	240	53	78

[a]Values are for combined wastes from HF and AlF_3, concentrations are calculated from pollutant flow in m^3/kkg and pollutant loading in kg/kkg.

Note: Blanks indicate information was not specified.

TREATMENT TECHNOLOGY: Sedimentation with Chemical Addition (Lime)

Data source: Effluent Guidelines
Point source category: Iron and steel
Subcategory: Combination acid pickling-continuous
Plant: I
References: A42, p. 607

Data source status:
- Engineering estimate ___
- Bench scale ___
- Pilot scale ___
- Full scale x

Use in system: Primary
Pretreatment of influent:

DESIGN OR OPERATING PARAMETERS

Unit configuration: Settling lagoon
Wastewater flow:
Hydraulic detention time:
Hydraulic loading:
Weir loading:
Sludge underflow:
Percent solids in sludge:
Scum overflow:

REMOVAL DATA

Sampling period:

Pollutant/parameter	Concentration Influent	Concentration Effluent	Percent removal
Conventional pollutants, mg/L:			
TSS	560	130	77
Oil and grease	0.7	1.5	0[a]
Toxic pollutants, μg/L:			
Chromium	17,000	1,800	89
Copper	150	ND[b]	93
Nickel	6,000	5,200	13
Zinc	750	240	68
Other pollutants, μg/L:			
Fluoride	33,000	9,100	72

[a]Actual data indicate negative removal.

[b]Not detected, assumed to be <10 μg/L.

Note: Blanks indicate information was not specified.

TREATMENT TECHNOLOGY: Sedimentation with Chemical Addition (Lime)

Data source: Effluent Guidelines
Point source category: Iron and steel
Subcategory: Combination acid pickling-batch
Plant: U
References: A42, p. 604

Data source status:
Engineering estimate ___
Bench scale ___
Pilot scale ___
Full scale x

Use in system: Primary
Pretreatment of influent: Neutralization

DESIGN OR OPERATING PARAMETERS

Unit configuration: Three tanks in series
Wastewater flow:
Chemical dosage(s):
Mix detention time:
Mixing intensity ($\bar{G}$):
Flocculation ($\bar{G}Ct$):
pH in clarifier:
Clarifier detention time:
Hydraulic loading:
Weir loading:
Sludge underflow:
Percent solids in sludge:
Scum overflow:

REMOVAL DATA

Sampling period:

Pollutant/parameter	Concentration Influent	Concentration Effluent	Percent removal
Conventional pollutants, mg/L:			
TSS	4	12	0[a]
Oil and grease	3	1	66
Toxic pollutants, μg/L:			
Chromium (dissolved)	150,000	40	>99
Copper	1,400	30	98
Nickel (dissolved)	70,000	20	>99
Other pollutants, μg/L:			
Fluoride	500,000	12,000	98

[a]Actual data indicate negative removal.

Note: Blanks indicate information was not specified.

TREATMENT TECHNOLOGY: Sedimentation with Chemical Addition (Lime)

Data source: Effluent Guidelines
Point source category: Nonferrous metals
Subcategory: Columbium/Tantalum raw waste stream
Plant:
References: A36, p. 337

Data source status:
Engineering estimate ___
Bench scale ___
Pilot scale ___
Full scale x

Use in system: Primary
Pretreatment of influent:

DESIGN OR OPERATING PARAMETERS

Unit configuration:
Wastewater flow:
Chemical dosage(s):
Mix detention time:
Mixing intensity ($\bar{G}$):
Flocculation ($\bar{G}$Ct):
pH in clarifier:
Clarifier detention time:

Hydraulic loading:
Weir loading:
Sludge underflow:
Percent solids in sludge:
Scum overflow:

REMOVAL DATA

Sampling period:

Pollutant/parameter	Concentration		Percent
	Influent	Effluent	removal
Conventional pollutants, mg/L:			
COD	16	8	50
TSS	900	10	99
Toxic pollutants, µg/L:			
Cadmium	25	2	92
Copper	110,000	700	99
Nickel	60,000	500	99
Zinc	27,000	200	99
Other pollutants, µg/L:			
Fluoride	4,500	2,500	44
Aluminum	9,000	200	98
Calcium	550,000	230,000	58
Iron	120,000	300	>99
Manganese	17,000	200	99

Note: Blanks indicate information was not specified.

TREATMENT TECHNOLOGY: Sedimentation with Chemical Addition (Lime)

Data source: Effluent Guidelines
Point source category: Nonferrous metals
Subcategory: Tungsten raw waste stream
Plant:
References: A36, p. 337

Data source status:
Engineering estimate ___
Bench scale ___
Pilot scale ___
Full scale x

Use in system: Primary
Pretreatment of influent:

DESIGN OR OPERATING PARAMETERS

Unit configuration:
Wastewater flow:
Chemical dosage(s):
Mix detention time:
Mixing intensity (G):
Flocculation (GCt):
pH in clarifier:
Clarifier detention time:

Hydraulic loading:
Weir loading:
Sludge underflow:
Percent solids in sludge:
Scum overflow:

REMOVAL DATA

Sampling period:

Pollutant/parameter	Concentration Influent	Effluent	Percent removal
Conventional pollutants, mg/L:			
COD	300	53	84
TSS	300	150	28
Toxic pollutants, μg/L:			
Arsenic	7,000	80	99
Cadmium	200	80	60
Chromium	2,000	50	97
Copper	5,000	70	99
Lead	20,000	200	99
Nickel	1,000	100	90
Zinc	2,000	600	70
Other pollutants, μg/L:			
Chloride	25×10^6	19×10^6	26
Aluminum	3,000	500	83
Iron	50,000	2,000	96

Note: Blanks indicate information was not specified.

TREATMENT TECHNOLOGY: Sedimentation with Chemical Addition (Lime)

Data source: Effluent Guidelines
Point source category: Ore mining and dressing
Subcategory: Base-metal mine
Plant: Plant 3 of Canadian pilot plant study
References: A2, pp. VI-63-66

Data source status:
Engineering estimate ___
Bench scale ___
Pilot scale x
Full scale ___

Use in system: Primary
Pretreatment of influent:

DESIGN OR OPERATING PARAMETERS

Unit configuration: Two-stage lime addition
Wastewater flow:
Chemical dosage(s):
Mix detention time:
Mixing intensity ($\bar{G}$):
Flocculation ($\bar{G}$Ct):
pH in clarifier:
Clarifier detention time:
Hydraulic loading:
Weir loading:
Sludge underflow:
Percent solids in sludge:
Scum overflow:

REMOVAL DATA

Sampling period:

Pollutant/parameter	Concentration, μg/L		Percent removal
	Influent[a]	Effluent[b]	
Toxic pollutants:			
Copper	19,000	60	>99
Lead	1,300	150	88
Zinc	110,000	350	>99

[a]Average value for raw minewater influent to pilot plant.

[b]Effluent qualities during periods of optimized steady operation.

Note: Blanks indicate information was not specified.

TREATMENT TECHNOLOGY: Sedimentation with Chemical Addition (Lime)

Data source: Effluent Guidelines
Point source category: Ore mining and dressing
Subcategory: Copper mine/mill
Plant: 2120
References: A2, pp. V-78, 79

Data source status:
Engineering estimate ___
Bench scale ___
Pilot scale ___
Full scale x

Use in system: Primary
Pretreatment of influent: pH adjusted

DESIGN OR OPERATING PARAMETERS

Unit configuration:
Wastewater flow:
Chemical dosage(s):
Mix detention time:
Mixing intensity ($\bar{G}$):
Flocculation ($\bar{G}$Ct):
pH in clarifier:
Clarifier detention time:
Hydraulic loading:
Weir loading:
Sludge underflow:
Percent solids in sludge:
Scum overflow:

REMOVAL DATA

Sampling period: 24-hr composite

Pollutant/parameter	Concentration		Percent
	Influent	Effluent	removal
Conventional pollutants, mg/L:			
COD	10	18	0[a]
TOC	19	12	37
TSS	14	4	71
Total phenol	0.018[b]	0.012	33
Toxic pollutants, μg/L:			
Arsenic	4	3	25
Copper	500	80	84
Lead	40	40	0
Mercury	<1	1	0[a]
Nickel	<20	30	0[a]

[a]Actual data indicate negative removal.

[b]An ethoxylated phenol (Nalco 8800) is used as a wetting agent for dust suppression during secondary ore crushing.

Note: Blanks indicate information was not specified.

TREATMENT TECHNOLOGY: Sedimentation with Chemical Addition (Lime)

Data source: Effluent Guidelines
Point source category: Ore mining and dressing
Subcategory: Copper mine/mill/smelter
Plant: 2117
References: A2, pp. 29-22

Data source status:
Engineering estimate ___
Bench scale ___
Pilot scale ___
Full scale x

Use in system: Primary
Pretreatment of influent:

DESIGN OR OPERATING PARAMETERS

Unit configuration: Aerator also used for chemical oxidation
Wastewater flow:
Chemical dosage(s):
Mix detention time:
Mixing intensity ($\bar{G}$):
Flocculation ($\bar{G}$Ct):
pH in clarifier:
Clarifier detention time:

Hydraulic loading:
Weir loading:
Sludge underflow:
Percent solids in sludge:
Scum overflow:

REMOVAL DATA

Sampling period: Average of two 24 hour composites

Pollutant/parameter	Concentration		Percent removal
	Influent	Effluent	
Conventional pollutants, mg/L:			
COD	34.5	29.5	14
TOC	11	9	18
TSS	24	4.5	81
Total phenol	0.37	0.33	11
Toxic pollutants:			
Asbestos, fibers/L	1.3×10^8	6.1×10^6	95
Copper, μg/L	190	120	34
Cyanide, μg/L	<20	45	0[a]
Zinc, μg/L	760	120	85

[a]Actual data indicate negative removal.

Note: Blanks indicate information was not specified.

TREATMENT TECHNOLOGY: Sedimentation with Chemical Addition (Lime)

Data source: Effluent Guidelines
Point source category: Ore mining and dressing
Subcategory: Lead/zinc mine
Plant: 3113
References: A2, pp. VI-89-92

Data source status:
Engineering estimate ___
Bench scale ___
Pilot scale x
Full scale ___

Use in system: Primary
Pretreatment of influent:

DESIGN OR OPERATING PARAMETERS

Unit configuration:
Wastewater flow:
Chemical dosage(s):
Mix detention time:
Mixing intensity ($\bar{G}$):
Flocculation ($\bar{G}$Ct):
pH in clarifier: 9.1-9.7
Clarifier detention time:
Hydraulic loading:

Weir loading:
Sludge underflow:
Percent solids in sludge:
Scum overflow:

REMOVAL DATA

Sampling period:

Pollutant/parameter	Concentration Influent[a]	Concentration Effluent	Percent removal
Conventional pollutants, mg/L:			
TSS	112	33	71
Toxic pollutants, μg/L:			
Cadmium	230	25	89
Copper	1,500	100	93
Lead	88	100	0[b]
Zinc	71,000	<20	>99

[a]Average of seven observations.

[b]Actual data indicate negative removal.

Note: Blanks indicate information was not specified.

TREATMENT TECHNOLOGY: Sedimentation with Chemical Addition (Lime)

Data source: Effluent Guidelines
Point source category: Paint manufacturing
Subcategory:
Plant: 26
References: A4, Appendix G

Data source status:
Engineering estimate ___
Bench scale ___
Pilot scale ___
Full scale x

Use in system: Primary
Pretreatment of influent: None

DESIGN OR OPERATING PARAMETERS

Unit configuration:
Wastewater flow:
Chemical dosage(s):
Mix detention time:
Mixing intensity ($\bar{G}$):
Flocculation ($\bar{G}$Ct):
pH in clarifier:
Clarifier detention time:

Hydraulic loading:
Weir loading:
Sludge underflow:
Percent solids in sludge:
Scum overflow:

REMOVAL DATA

Sampling period: Grab samples

Pollutant/parameter	Concentration,[a] µg/L Influent	Effluent	Percent removal
Toxic pollutants:			
Antimony	1,000	180	83
Cadmium	40	30	25
Chromium	240	30	88
Copper	250	80	68
Lead	700	190	73
Mercury	5.8	8	0[b]
Nickel	210	310	0[b]
Zinc	270,000	8,200	97

[a]One sample.

[b]Actual data indicate negative removal.

Note: Blanks indicate information was not specified.

TREATMENT TECHNOLOGY: Sedimentation with Chemical Addition (Lime)

Data source: Effluent Guidelines
Point source category: Steam electric power generating
Subcategory:
Plant: 5604
References: A31, p. 20 (Appendix E)

Data source status:
Engineering estimate ___
Bench scale ___
Pilot scale _x_
Full scale ___

Use in system: Primary
Pretreatment of influent:

DESIGN OR OPERATING PARAMETERS

Unit configuration:
Wastewater flow:
Chemical dosage(s):
Mix detention time:
Mixing intensity ($\bar{G}$):
Flocculation ($\bar{G}$Ct):
pH in clarifier: 11.5
Clarifier detention time:

Hydraulic loading:
Weir loading:
Sludge underflow:
Percent solids in sludge:
Scum overflow:

REMOVAL DATA

Sampling period:

Pollutant/parameter	Concentration, µg/L Influent	Effluent	Percent removal
Toxic pollutants:			
Antimony	5	3	40
Arsenic	7	<1	>86
Chromium	2	<2	>0
Copper	180	48	73
Nickel	6	12	0[a]
Silver	3	4	0[a]
Zinc	780	140	82

[a]Actual data indicate negative removal.

Note: Blanks indicate information was not specified.

TREATMENT TECHNOLOGY: Sedimentation with Chemical Addition (Lime)

Data source: Effluent Guidelines
Point source category: Steam electric power generating
Subcategory:
Plant: 1226
References: A31, p. 20 (Appendix E)

Data source status:
Engineering estimate ___
Bench scale ___
Pilot scale x
Full scale ___

Use in system: Primary
Pretreatment of influent:

DESIGN OR OPERATING PARAMETERS

Unit configuration:
Wastewater flow:
Chemical dosage(s):
Mix detention time:
Mixing intensity ($\overline{G}$):
Flocculation ($\overline{G}$Ct):
pH in clarifier: 11.5
Clarifier detention time:

Hydraulic loading:
Weir loading:
Sludge underflow:
Percent solids in sludge:
Scum overflow:

REMOVAL DATA

Sampling period:

Pollutant/parameter	Concentration, µg/L Influent	Concentration, µg/L Effluent	Percent removal
Toxic pollutants:			
Antimony	7	4	43
Arsenic	4	3	25
Beryllium	<0.5	0.9	0[a]
Cadmium	1.8	3.0	0[a]
Chromium	4	9	0[a]
Copper	47	18	62
Lead	3	5	0[a]
Mercury	0.2	0.7	0[a]
Nickel	6.0	2.9	52
Silver	0.7	0.9	0[a]
Zinc	26	2	92

[a] Actual data indicate negative removal.

Note: Blanks indicate information was not specified.

TREATMENT TECHNOLOGY: Sedimentation with Chemical Addition (Lime)

Data source: Effluent Guidelines
Point source category: Steam electric power generating
Subcategory:
Plant: 1226
References: A31, p. 20 (Appendix E)

Data source status:
Engineering estimate ___
Bench scale ___
Pilot scale x
Full scale ___

Use in system: Secondary
Pretreatment of influent: Ash pond

DESIGN OR OPERATING PARAMETERS

Unit configuration:
Wastewater flow:
Chemical dosage(s):
Mix detention time:
Mixing intensity (G):
Flocculation (GCt):
pH in clarifier: 11.5
Clarifier detention time:

Hydraulic loading:
Weir loading:
Sludge underflow:
Percent solids in sludge:
Scum overflow:

REMOVAL DATA

Sampling period:

Pollutant/parameter	Concentration, µg/L Influent	Effluent	Percent removal
Toxic pollutants:			
Antimony	7	10	0[a]
Arsenic	9	1	89
Cadmium	2.0	2.0	0
Chromium	6	11	0[a]
Copper	14	10	29
Lead	4	<3	>25
Mercury	<0.2	0.3	0[a]
Nickel	5.5	6.0	0[a]
Selenium	8	8	0
Silver	0.5	0.4	20
Zinc	7	2	57

[a]Actual data indicate negative removal.

Note: Blanks indicate information was not specified.

TREATMENT TECHNOLOGY: Sedimentation with Chemical Addition (Lime)

Data source: Effluent Guidelines
Point source category: Steam electric power generating
Subcategory:
Plant: 5409
References: A31, p. 22 (Appendix E)

Data source status:
Engineering estimate ___
Bench scale ___
Pilot scale x
Full scale ___

Use in system: Primary
Pretreatment of influent:

DESIGN OR OPERATING PARAMETERS

Unit configuration: Ash pond
Wastewater flow:
Chemical dosage(s):
Mix detention time:
Mixing intensity ($\bar{G}$):
Flocculation ($\bar{G}Ct$):
pH in clarifier: 11.5
Clarifier detention time:

Hydraulic loading:
Weir loading:
Sludge underflow:
Percent solids in sludge:
Scum overflow:

REMOVAL DATA

Sampling period:

Pollutant/parameter	Concentration, μg/L Influent	Effluent	Percent removal
Toxic pollutants:			
Antimony	5	4	20
Arsenic	74	<1	>99
Copper	26	12	54
Nickel	2.5	2.2	12
Selenium	42	52	0[a]
Silver	1	1.1	0[a]
Thallium	9	8	11
Zinc	11	<2	>82

[a]Actual data indicate negative removal.

Note: Blanks indicate information was not specified.

TREATMENT TECHNOLOGY: Sedimentation with Chemical Addition (Lime)

Data source: Effluent Guidelines
Point source category: Steam electric power generating
Subcategory:
Plant: 5409
References: A31, p. 22 (Appendix E)

Data source status:
Engineering estimate ___
Bench scale ___
Pilot scale x
Full scale ___

Use in system: Primary
Pretreatment of influent:

DESIGN OR OPERATING PARAMETERS

Unit configuration:
Wastewater flow:
Chemical dosage(s):
Mix detention time:
Mixing intensity ($\bar{G}$):
Flocculation ($\bar{G}Ct$):
pH in clarifier: 11.5
Clarifier detention time:

Hydraulic loading:
Weir loading:
Sludge underflow:
Percent solids in sludge:
Scum overflow:

REMOVAL DATA

Sampling period:

Pollutant/parameter	Concentration Influent	Effluent	Percent removal
Conventional pollutants, mg/L:			
TOC	21	<20	>5
Toxic pollutants, μg/L:			
Antimony	<1	4	0[a]
Arsenic	<1	2.5	0[a]
Beryllium	3.4	0.8	76
Cadmium	0.8	<0.5	>38
Chromium	37	8.8	76
Copper	620	70	89
Lead	70	<3	>96
Mercury	0.5	<0.2	>60
Nickel	4	2.3	43
Selenium	<2	2.3	0[a]
Silver	14	7.8	44
Thallium	8	<1	>88
Zinc	61	<2	>97

[a]Actual data indicate negative removal.

Note: Blanks indicate information was not specified.

TREATMENT TECHNOLOGY: Sedimentation with Chemical Addition (Lime)

Data source: Effluent Guidelines
Point source category: Steam electric power generating
Subcategory:
Plant: See below
References: A31, pp. 219, 220, 222

Data source status:
- Engineering estimate ___
- Bench scale ___
- Pilot scale ___
- Full scale x

Use in system: Primary
Pretreatment of influent: None

DESIGN OR OPERATING PARAMETERS

Unit configuration: Ash pond
Wastewater flow:
Chemical dosage(s):
Mix detention time:
Mixing intensity (G):
Flocculation (GCt):
pH in clarifier:
Clarifier detention time:

Hydraulic loading:
Weir loading:
Sludge underflow:
Percent solids in sludge:
Scum overflow:

REMOVAL DATA

Sampling period:

	COD			TSS			Arsenic		
	Concentration, mg/L		Percent	Concentration, mg/L		Percent	Concentration, μg/L		Percent
Plant	Influent	Effluent	removal	Influent	Effluent	removal	Influent	Effluent	removal
Shawnee power plant, pond A	20	37	0[a]	16	12.5	22	20	7.5	63
Shawnee power plant, pond B				160	6	96			
Shawnee power plant, pond D							240	110	54

	Lead			Mercury			Selenium		
	Concentration, μg/L		Percent	Concentration, μg/L		Percent	Concentration, μg/L		Percent
Plant	Influent	Effluent	removal	Influent	Effluent	removal	Influent	Effluent	removal
Shawnee power plant, pond A	170	46	73	0.76	0.23	70	3	3.2	0[a]
Shawnee power plant, pond B									
Shawnee power plant, pond D	260	39	85	0.1	0.3	0[a]			

[a]Actual data indicate negative removal.

Note: Blanks indicate information was not specified.

TREATMENT TECHNOLOGY: Sedimentation with Chemical Addition (Lime)

Data source: Effluent Guidelines
Point source category: Textile mills
Subcategory: Knit fabric finishing
Plant:
References: A6, p. VII-48

Data source status:
- Engineering estimate ___
- Bench scale x
- Pilot scale ___
- Full scale ___

Use in system: Sample taken from aeration basin at plant (primary)
Pretreatment of influent:

DESIGN OR OPERATING PARAMETERS

Unit configuration:
Wastewater flow:
Chemical dosage(s):
Mix detention time:
Mixing intensity (G):
Flocculation (GCt):
pH in clarifier:
Clarifier detention time:
Hydraulic loading:
Weir loading:
Sludge underflow:
Percent solids in sludge:
Scum overflow:

REMOVAL DATA

Sampling period:

Pollutant/parameter	Concentration, μg/L Influent	Effluent	Percent removal
Toxic pollutants:			
Cadmium	10	ND[a]	>0
Chromium	930	80	91
Copper	500	30	94
Lead	100	ND	>90
Nickel	50	ND	>80
Silver	50	ND	>80
Zinc	3,200	110	97

[a]Not detected; assumed to be <10 μg/L.

Note: Blanks indicate information was not specified.

TREATMENT TECHNOLOGY: Sedimentation with Chemical Addition (Lime, Polymer)

Data source: Effluent Guidelines
Point source category: Foundry industry
Subcategory: Aluminum foundries - die casting
Plant: 574C
References: A27, pp. V-13, VI-49-56

Data source status:
Engineering estimate ___
Bench scale ___
Pilot scale ___
Full scale x

Use in system: Secondary
Pretreatment of influent: Emulsion break

DESIGN OR OPERATING PARAMETERS

Unit configuration: Flocculation with polymer
Wastewater flow:
Chemical dosage(s):
Mix detention time:
Mixing intensity (G):
Flocculation (GCt):
pH in clarifier:
Clarifier detention time:

Hydraulic loading:
Weir loading:
Sludge underflow:
Percent solids in sludge:
Scum overflow:

REMOVAL DATA

Sampling period:

Pollutant/parameter	Concentration, μg/L Influent	Effluent	Percent removal
Toxic pollutants:			
Cyanide	5	2.3	54
Lead	200	150	24
Zinc	1,300	40	97
Bis(2-ethylhexyl) phthalate	5,500	32	99
Butyl benzyl phthalate	690	BDL[b]	>99
Di-n-butyl phthalate	74	1	99
Diethyl phthalate	730	BDL	>99
2,4-Dimethylphenol	41	BDL	>76
Phenol	16	BDL	>37
p-Chloro-*m*-cresol	110	62	44
Anthracene/phenanthrene	10	BDL	>0
Benzo(a)pyrene	53	BDL	>81
Chrysene	780	10	99
Fluoranthene	370	BDL	>97
Fluorene	800	BDL	>99
Naphthalene	160	3	98
Pyrene	80	BDL	>87
Chloroform	4	7	0[a]
Methylene chloride	2	39	0[a]
1,1,1-Trichloroethane	0	51	0[a]

[a]Actual data indicate negative removal.

[b]Below detection limits; was detected but not in sufficient amounts to be quantified, assumed to be <10 μg/L.

Note: Blanks indicate information was not specified.

TREATMENT TECHNOLOGY: Sedimentation with Chemical Addition (Lime, Coagulant Aids)

Data source: Effluent Guidelines
Point source category: Iron and steel
Subcategory: Combination acid pickling-batch
Plant: C
References: A42, p. 604

Data source status:
- Engineering estimate ___
- Bench scale ___
- Pilot scale ___
- Full scale x

Use in system: Primary
Pretreatment of influent: Equalization

DESIGN OR OPERATING PARAMETERS

Unit configuration:
Wastewater flow: 0.378 L/s (6 gpm)
Chemical dosage(s):
Mix detention time:
Mixing intensity ($\bar{G}$):
Flocculation ($\bar{G}Ct$):
pH in clarifier:
Clarifier detention time:
Hydraulic loading:

Weir loading:
Sludge underflow:
Percent solids in sludge:
Scum overflow:

REMOVAL DATA

Sampling period:

Pollutant/parameter	Concentration Influent	Concentration Effluent	Percent removal
Conventional pollutants, mg/L:			
TSS	106	31	71
Oil and grease	5	0.3	94
Toxic pollutants, μg/L:			
Chromium (dissolved)	140,000	1,300	99
Nickel (dissolved)	240,000	2,500	99
Other pollutants, μg/L:			
Fluoride	1,700,000	130,000	92

Note: Blanks indicate information was not specified.

TREATMENT TECHNOLOGY: Sedimentation with Chemical Addition (Lime, Polymer)

Data source: Effluent Guidelines
Point source category: Iron and steel
Subcategory: Combination acid pickling-bath
Plant: 123
References: A42, p. 605

Data source status:
Engineering estimate ___
Bench scale ___
Pilot scale ___
Full scale x

Use in system: Primary
Pretreatment of influent: Equalization neutralization

DESIGN OR OPERATING PARAMETERS

Unit configuration: Clarifier
Wastewater flow:
Chemical dosage(s):
Mix detention time:
Mixing intensity (G):
Flocculation (GCt):
pH in clarifier:
Clarifier detention time:
Hydraulic loading:
Weir loading:
Sludge underflow:
Percent solids in sludge:
Scum overflow:

REMOVAL DATA

Sampling period:

Pollutant/parameter	Concentration Influent	Concentration Effluent	Percent removal
Conventional pollutants, mg/L:			
Oil and grease	5	8.5	0[a]
Toxic pollutants, µg/L:			
Arsenic	10	<10	>0
Chromium	3,300	360	89
Copper	260	30	88
Cyanide	110	39	65
Nickel	7,700	330	96
Zinc	90	120	0[b]
Chloroform	46	42	9

[a]Actual data indicate negative removal.

Note: Blanks indicate information was not specified.

TREATMENT TECHNOLOGY: Sedimentation with Chemical Addition (Lime, Polymer)

Data source: Effluent Guidelines
Point source category: Iron and steel
Subcategory: Hydrochloric acid pickling
Plant: 093
References: A42, pp. 369-370

Data source status:
Engineering estimate ___
Bench scale ___
Pilot scale ___
Full scale x

Use in system: Primary
Pretreatment of influent: Neutralization

DESIGN OR OPERATING PARAMETERS

Unit configuration: Clarifier
Wastewater flow: 17.4 L/s (276 gpm)
Chemical dosage(s):
Mix detention time:
Mixing intensity ($\bar{G}$):
Flocculation ($\bar{G}Ct$):
pH in clarifier:
Clarifier detention time:
Hydraulic loading:
Weir loading:
Sludge underflow:
Percent solids in sludge:
Scum overflow:

REMOVAL DATA

Sampling period:

Pollutant/parameter	Concentration Influent	Concentration Effluent	Percent removal
Conventional pollutants, mg/L:			
TSS	490	43	91
Oil and grease	650	5.5	99
Toxic pollutants, µg/L:			
Arsenic	<10	<10	0
Cadmium	<20	<20	0
Chromium	790	75	90
Copper	680	170	75
Cyanide	18	2	89
Lead	420	575	0[a]
Nickel	440	270	39
Selenium	<20	<10	∿50
Silver	<24	90	0[a]
Zinc	1,480	245	83
Bis(2-ethylhexyl) phthalate	4,630	150	97
Methylene chloride	<13	13	0[a]

[a]Actual data indicate negative removal.

Note: Blanks indicate information was not specified.

TREATMENT TECHNOLOGY: Sedimentation with Chemical Addition (Lime, Polymer)

Data source: Effluent Guidelines
Point source category: Iron and steel
Subcategory: Pipe and tube-welded
Plant: 087
References: A41, p. 75

Data source status:
Engineering estimate ___
Bench scale ___
Pilot scale ___
Full scale x

Use in system: Primary
Pretreatment of influent:

DESIGN OR OPERATING PARAMETERS

Unit configuration:
Wastewater flow: 1,750 L/s (27,700 gpm)
Chemical dosage(s):
Mix detention time:
Mixing intensity ($\bar{G}$):
Flocculation ($\bar{G}$Ct):
pH in clarifier:
Clarifier detention time:
Hydraulic loading:

Weir loading:
Sludge underflow:
Percent solids in sludge:
Scum overflow:

REMOVAL DATA

Sampling period:

Pollutant/parameter	Concentration Influent	Concentration Effluent	Percent removal
Conventional pollutants, mg/L:			
TSS	66	38	42
Oil and grease	5	4	20
Toxic pollutants, µg/L:			
Copper	90	12	87
Mercury			
Selenium			

[a]Actual data indicate negative removal.

Note: Blanks indicate information was not specified.

TREATMENT TECHNOLOGY: Sedimentation with Chemical Addition (Lime, Polymer)

Data source: Effluent Guidelines
Point source category: Ore mining and dressing
Subcategory: Copper mill
Plant: 2122
References: A2, pp. 84-87

Data source status:
Engineering estimate ___
Bench scale ___
Pilot scale x
Full scale ___

Use in system: Secondary
Pretreatment of influent: Tailing pond

DESIGN OR OPERATING PARAMETERS

Unit configuration:
Wastewater flow:
Chemical dosage(s):
Mix detention time:
Mixing intensity ($\bar{G}$):
Flocculation ($\bar{G}$Ct):
pH in clarifier: 9.3-9.9
Clarifier detention time: 2.6 hr
Hydraulic loading:

Weir loading:
Sludge underflow:
Percent solids in sludge:
Scum overflow:

REMOVAL DATA

Sampling period:

Pollutant/parameter	Concentration[a] Influent	Effluent	Percent removal
Conventional pollutants, mg/L:			
TSS	2,550	36.5	99
Toxic pollutants, µg/L:			
Chromium	190	32	83
Copper	2,000	38	98
Lead	160	75	53
Nickel	190	45	76
Zinc	100	25	75

[a]Average values: TSS (54 observations)
Metals (46 observations).

Note: Blanks indicate information was not specified.

TREATMENT TECHNOLOGY: Sedimentation with Chemical Addition (Lime, Polyelectrolyte)

Data source: Effluent Guidelines
Point source category: Ore mining and dressing
Subcategory: Copper mine/mill/smelter/refinery
Plant: 2121
References: A2, pp. V-18-19

Data source status:
Engineering estimate ___
Bench scale ___
Pilot scale ___
Full scale x

Use in system: Primary
Pretreatment of influent:

DESIGN OR OPERATING PARAMETERS

Unit configuration:
Wastewater flow:
Chemical dosage(s):
Mix detention time:
Mixing intensity (G):
Flocculation (GCt):
pH in clarifier:
Clarifier detention time:
Hydraulic loading:
Weir loading:
Sludge underflow:
Percent solids in sludge:
Scum overflow:

REMOVAL DATA

Sampling period: 24-hr composite

Pollutant/parameter	Concentration		Percent removal
	Influent	Effluent	
Conventional pollutants, mg/L:			
COD	960	2	>99
TOC	9	7	22
TSS	211,000	5	>99
Toxic pollutants:			
Asbestos, fibers/L	3.0×10^{11}	8.2×10^{6}	>99
Copper, µg/L	190,000	90	>99
Zinc, µg/L	28,000	40	>99
Bis(2 ethylhexyl) phthalate,[a] µg/L	0.1	12	0[b]

[a]Possibly from the tubing in sampling apparatus.

[b]Actual data indicate negative removal.

Note: Blanks indicate information was not specified.

TREATMENT TECHNOLOGY: Sedimentation with Chemical Addition (Lime, Polymer)

Data source: Effluent Guidelines
Point source category: Ore mining and dressing
Subcategory: See below
Plant: See below
References: A2, pp. VI-89-92, 76-79, 80-83, 63-66

Data source status:
Engineering estimate ___
Bench scale ___
Pilot scale x
Full scale ___

Use in system: Primary unless otherwise specified
Pretreatment of influent: None unless otherwise specified

DESIGN OR OPERATING PARAMETERS

Unit configuration:
Wastewater flow:
Chemical dosage(s):
Mix detention time:
Mixing intensity (G):
Flocculation (GCt):
pH in clarifier: See below
Clarifier detention time:

Hydraulic loading:
Weir loading:
Sludge underflow:
Percent solids in sludge:
Scum overflow:

REMOVAL DATA

Subcategory	Plant	TSS Concentration (mg/ℓ) Influent	TSS Effluent	TSS Percent Removal	Cadmium Concentration (mg/ℓ) Influent	Cadmium Effluent	Cadmium Percent Removal	Copper Concentration (mg/ℓ) Influent	Copper Effluent	Copper Percent Removal
Base metal mine	Mine No. 1*	–	–	–	–	–	–	10,000	40	>99
Base metal mine	Mine No. 2*	–	–	–	–	–	–	47,000	50	>99
Lead/zinc mine	3113**	112	10	91	230	15	93	1,500	50	97
Lead/zinc mine/mill	3121***	4.5	17	0†	–	–	–	100	40	60
Lead/zinc mine/mill/smelter/refinery	3107††	16	6	62	120	60	50	31	15	52

Subcategory	Plant	Lead Concentration (mg/ℓ) Influent	Lead Effluent	Lead Percent Removal	Zinc Concentration (mg/ℓ) Influent	Zinc Effluent	Zinc Percent Removal
Base metal mine	Mine No. 1*	3,900	180	95	1,200,000	330	>99
Base metal mine	Mine No. 2*	1,200	440	63	540,000	450	>99
Lead/zinc mine	3113**	88	<20	>77	71,000	1,400	98
Lead/zinc mine/mill	3121***	210	65	69	740	260	66
Lead/zinc mine/mill/smelter refinery	3107††	130	70	46	2,900	1,000	66

*Of Canadian pilot plan study.
**pH in clarifier is 8.8 to 9.8.
***Use in system: secondary; tailing pond was used in pretreatment of influent; ph in clarifier is 9.2 to 11.3.
†Actual data indicate negative removal.
††Use in system: tertiary, pretreatment of influent included tailing pond, lime precipitation, aeration, flocculation, and clarification; pH in clarifier is 8.1 to 8.7.

Note: Blanks indicate information was not specified.

TREATMENT TECHNOLOGY: Sedimentation with Chemical Addition (Polymer)

Data source: Effluent Guidelines
Point source category: Paint manufacturing
Subcategory:
Plant: 14
References: A4, Appendix G

Data source status:
Engineering estimate ___
Bench scale ___
Pilot scale ___
Full scale x

Use in system: Primary
Pretreatment of influent: None

DESIGN OR OPERATING PARAMETERS

Unit configuration: Batch operation
Wastewater flow:
Chemical dosage(s):
Mix detention time:
Mixing intensity (G):
Flocculation (GCt):
pH in clarifier:
Clarifier detention time:

Hydraulic loading:
Weir loading:
Sludge underflow:
Percent solids in sludge:
Scum overflow:

REMOVAL DATA

Sampling period: Grab samples

Pollutant/parameter	Concentration[a] Influent	Effluent	Percent removal
Conventional pollutants, mg/L:			
BOD_5	4,800	4,700	2
COD	28,000	8,000	71
TOC	9,000	1,600	82
TSS	12,400	39	>99
Oil and grease	1,100	22	98
Total phenol	0.705	0.3	58
Toxic pollutants, μg/L:			
Cadmium	45	100	0[b]
Chromium	950	25	97
Copper	550	400	27
Lead	5,000	140	97
Mercury	9,400	140	99
Zinc	55,000	6,000	89
Bis(2-ethylhexyl)phthalate	390	<10	>97
Di-n-butyl phthalate	4,000	<10	>99
Phenol	ND[c]	74	0[b]
Ethylbenzene	690	130	81
Toluene	3,100	1,900	39
Chloroform	ND	11	0[b]
1,2-*Trans*-dichloroethylene	ND	21	0[b]
Methylene chloride	ND	130	0[b]
Trichloroethylene	ND	14	0[b]

[a]Average of several samples.

[b]Actual data indicate negative removal.

[c]Not detected; assumed to be less than the corresponding effluent concentration.

Note: Blanks indicate information was not specified.

TREATMENT TECHNOLOGY: Sedimentation with Chemical Addition (Polymer)

Data source: Effluent Guidelines
Point source category: Textile mills
Subcategory: Knit fabric finishing
Plant: E, P (different references)
References: A6, p. VII-45; B3, pp. 60-64

Data source status:
Engineering estimate ___
Bench scale ___
Pilot scale x
Full scale ___

Use in system: Tertiary
Pretreatment of influent: Screening, activated sludge

DESIGN OR OPERATING PARAMETERS

Unit configuration: 6.25 m^3 (1,650 gal) reactor/clarifier
Wastewater flow:
Chemical dosage(s): 20 mg/L 572 C polymer (American Cyanimid-Cationic)
Mix detention time:
Mixing intensity ($\bar{G}$):
Flocculation ($\bar{G}Ct$):
pH in clarifier: 6.9
Clarifier detention time:
Hydraulic loading:
Weir loading:
Sludge underflow:
Percent solids in sludge:
Scum overflow:

REMOVAL DATA

Sampling period: 24-hr composite samples, volatile organics were composites of 3 grab samples

Pollutant/parameter	Concentration Influent	Concentration Effluent	Percent removal
Conventional pollutants, mg/L:			
Total phenol	0.072	0.082	0[a]
Toxic pollutants, μg/L:			
Antimony	77	43	44
Chromium	98	<4	>96
Copper	36	<4	>89
Lead	25	<22	>12
Mercury	0.4	<0.3	>25
Nickel	66	43	35
Zinc	5,200	160	97
Bis(2-ethylhexyl) phthalate	10	10	0
Di-n-butyl phthalate	2.1	2.8	0[a]
Diethyl phthalate	1.3	<0.03	>98
Phenol	0.7	0.5	29
Benzene	<0.2	0.4	0[a]
Toluene	0.4	0.4	0
Anthracene/Phenanthrene	0.8	0.9	0[a]
Methylene chloride[b]	0.4	2.5	0[a]
Trichloroethylene	<0.5	0.8	0[a]

[a]Actual data indicate negative removal.
[b]Presence may be due to sample contamination.

Note: Blanks indicate information was not specified.

TREATMENT TECHNOLOGY: Sedimentation with Chemical Addition (Sulfide)

Data source: Effluent Guidelines
Point source category: Textile mills
Subcategory: Knit fabric finishing
Plant:
References: A6, p. III-48

Data source status:
Engineering estimate ___
Bench scale x
Pilot scale ___
Full scale ___

Use in system: Sample taken from aeration basin at plant
Pretreatment of influent:

DESIGN OR OPERATING PARAMETERS

Unit configuration:
Wastewater flow:
Chemical dosage(s):
Mix detention time:
Mixing intensity (G):
Flocculation (GCt):
pH in clarifier:
Clarifier detention time:
Hydraulic loading:
Weir loading:
Sludge underflow:
Percent solids in sludge:
Scum overflow:

REMOVAL DATA

Sampling period:

Pollutant/parameter	Concentration, μg/L Influent	Effluent	Percent removal
Toxic pollutants:			
Cadmium	10	ND[a]	>0
Chromium	930	50	95
Copper	500	10	98
Lead	100	ND	>90
Nickel	50	ND	>80
Silver	50	ND	>80
Zinc	3,200	90	97

[a]Not detected; assumed to be <10 μg/L.

Note: Blanks indicate information was not specified.

4.4 GAS FLOTATION (Dissolved Air Flotation) [1]

4.4.1 Function

Gas flotation is used to remove suspended solids by flotation.

4.4.2 Description

Gas flotation or dissolved air flotation (DAF) is used to remove suspended solids by using flotation (rising) to decrease their apparent density. DAF consists of saturating a portion or all of the wastewater feed, or a portion of recycled effluent, with air at a pressure of 25 to 70 lb/in.^{2} (gage). The pressurized wastewater is held at this pressure for 0.5 to 3.0 minutes in a retention tank and then released to atmospheric pressure in the flotation chamber. The sudden reduction in pressure results in the release of microscopic air bubbles, which attach themselves to oil and suspended particles in the wastewater in the flotation chamber. This results in agglomeration which, due to the entrained air, results in greatly increased vertical rise rates of about 0.5 to 2.0 ft/min. The floated materials rise to the surface to form a froth layer. Specially designed flight scrapers or other skimming devices continuously remove the froth. The retention time in the flotation chambers is usually about 20 to 60 minutes. The effectiveness of dissolved air flotation depends on the attachment of bubbles to the suspended oil and other particles that are to be removed from the waste stream. The attraction between the air bubble and particle is primarily a result of the particle surface charges and bubble-size distribution.

The more uniform the distribution of water and microbubbles, the shallower the flotation unit can be. Generally, the depth of effective flotation units is between 4 and 9 feet.

In certain cases, the surface sludge layer can attain a thickness of many inches and can be relatively stable for a short period. The layer thickens with time, but undue delays in removal will cause a release of particulates back to the liquid.

4.4.3 Common Modifications

Significant modifications are systems utilizing pressurized raw waste and pressurized recycle. DAF units can be round, square, or rectangular. In addition, gases other than air can be used. The petroleum industry has used nitrogen, with closed vessels, to reduce the possibilities of fire.

4.4.4 Technology Status

DAF has been used for many years to treat industrial wastewaters. It has been commonly used to thicken sludges including those generated by municipal wastewater treatment; however, it is not widely used to treat municipal wastewaters.

4.4.5 Applications

Used to remove lighter suspended materials whose specific gravity is only slightly in excess of 1.0; usually used to remove oil and grease materials; sometimes used when existing clarifiers are overloaded hydraulically because converting to DAF requires less surface area.

4.4.6 Limitations

Will only be effective on particles with densities near or less than water.

4.4.7 Chemicals Required

The use of chemical addition is covered in the section entitled "Gas Flotation with Chemical Addition", Section 4.5 of this manual.

4.4.8 Residuals Generated

A froth layer is generated, which is skimmed off the top of the unit. It is generally denser than sludge from clarifiers.

4.4.9 Reliability

DAF systems have been found to be reliable. DAF units are subject to variable influent conditions, resulting in widely varying performance.

4.4.10 Environmental Impact

Requires very little use of land; air released in unit is unlikely to strip volatile organic material into air; air compressors will need silencers to control the noise generated; sludge generated will need methods for disposal.

4.4.11 Design Criteria

Criteria	Units	Range/value
Pressure	lb/in.^{2} (gauge)	25 - 70
Air-to-solids ratio	lb/lb	0.01 - 0.1
Detention time	min	20 - 60
Surface hydraulic loading	gpd/ft^2	500 - 8,000
Recycle (where employed)	percent	5 - 120
Solids loading	lb/ft^2/hr	0.5 - 5

4.4.12 Flow Diagram

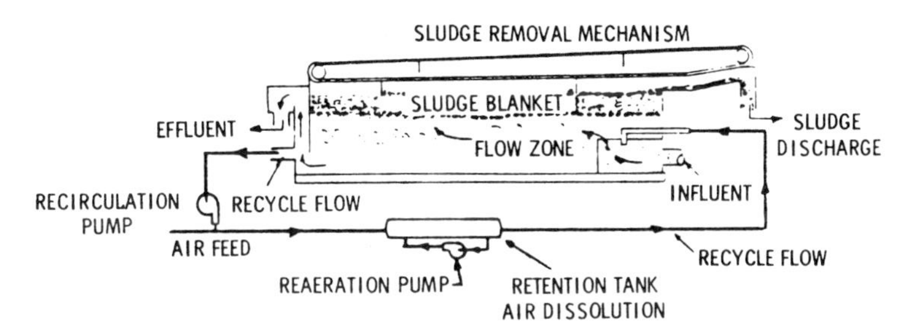

4.4.13 Performance

Subsequent data sheets provide performance data from studies on the following industries and/or wastestreams:

Petroleum refining

Pulp, paper, and paperboard production
Nonintegrated tissue

4.4.14 References

1. Innovative and Alternative Technology Assessment Manual. EPA-430/9-78-009 (draft), U.S. Environmental Protection Agency, Cincinnati, Ohio, 1978. 252 pp.

CONTROL TECHNOLOGY SUMMARY FOR GAS FLOTATION

Pollutant	Number of data points	Effluent concentration				Removal efficiency, %			
		Minimum	Maximum	Median	Mean	Minimum	Maximum	Median	Mean
Conventional pollutants, mg/L:									
BOD_5	1				250				4
COD	2	18	1,000		509	0[a]	95		48
TOC	1				280				0[a]
TSS	1				131				6
Oil and grease	1				220				0[a]
Total phenol	1				23				4
Toxic pollutants, µg/L:									
Chromium	2	2	570		290	21	87		54
Copper	2	5	19		12	58	69		64
Cyanide	1				2,300				0[a]
Lead	2	2	210		160	16	82		49
Mercury	1				0.6				0[a]
Nickel	2	2	52		27	0[a]	0[a]		0[a]
Selenium	1				8.5				0[a]
Zinc	2	83	53,000		27,000	0[a]	22		11
Bis(2-ethylhexyl) phthalate	2	30	1,100		570	0[a]	0[a]		0[a]
Butyl benzyl phthalate	1				<10[b]				>99
Diethyl phthalate	1				<10[b]				>17
Phenol	2	5	2,400		1,200	0[a]	51		26
Ethylbenzene	1				<10[b]				>99
Toluene	1				<10[b]				>92
Anthracene/phenanthrene	1				~600				~45
Naphthalene	2	60	~700		380	0[a]	~36		18
Aroclor 1016	1				7.9				0[a]
Aroclor 1242	1				0.5				0
Chloroform	1				<3[c]				>0
Other pollutants, µg/L:									
Xylenes	1				<10[b]				>99

Note: Blanks indicate data not applicable.

[a]Actual data indicate negative removal.

[b]Not detected, assumed to be <10 µg/L.

[c]Not detected; assumed to be less than the corresponding influent concentration.

TREATMENT TECHNOLOGY: Gas Flotation

Data source: Effluent Guidelines
Point source category: Petroleum refining
Subcategory:
Plant: G
References: A3, pp. IV-36-63

Data source status:
Engineering estimate ___
Bench scale ___
Pilot scale ___
Full scale x

Use in system: Secondary
Pretreatment of influent: API design gravity oil separator

DESIGN OR OPERATING PARAMETERS

Process type: Dissolved air flotation
Unit configuration:
Wastewater flow: 3.2 MGD
Float detention time:
Hydraulic loading:
Percent recycle:
Solids loading:
Gas requirement:
Gas-to-solids ratio:
Pressure:
Sludge overflow:
Percent solids in sludge:

REMOVAL DATA

Sampling period:

Pollutants/parameter	Concentration Influent	Concentration Effluent	Percent removal
Conventional pollutants, mg/L:			
BOD_5	260	250	4
COD	840	1,000	0[a]
TOC	230	280	0[a]
TSS	140	131	6
Oil and grease	93	220	0[a]
Total phenol	24	23	4
Toxic pollutants, μg/L:			
Chromium	720	570	21
Copper	16	5	69
Cyanide	1,300	2,300	0[a]
Lead	250	210	16
Mercury	0.2	0.6	0[a]
Nickel	47	52	0[a]
Selenium	7.8	8.5	0[a]
Zinc	110	83	22
Bis(2-ethylhexyl) phthalate	770	1,100	0[a]
Phenol	4,900	2,400	51
Anthracene/phenanthrene[b]	∿1,100	∿600	∿45
Naphthalene	∿1,100	∿700	∿36
Aroclor 1016	1.8	7.9	0[a]
Aroclor 1242	0.5	0.5	0[a]

[a]Actual data indicate negative removal.

[b]Concentrations represent sums for these two compounds which elute simultaneously and have the same major ions for GC/MS.

Note: Blanks indicate information was not specified.

TREATMENT TECHNOLOGY: Gas Flotation

Data source: Effluent Guidelines
Point source category: Pulp, paper and paperboard
Subcategory: Nonintegrated tissue
Plant:
References: A26, pp. A-104-107
Use in system: Primary
Pretreatment of influent:

Data source status:
- Engineering estimate ___
- Bench scale ___
- Pilot scale ___
- Full scale x

DESIGN OR OPERATING PARAMETERS

Process type: Dissolved air flotation
Unit configuration:
Wastewater flow:
Detention time:
Hydraulic loading:
Percent recycle:
Solids loading:

Gas requirements:
Gas-to-solids ratio:
Pressure:
Sludge overflow:
Percent solids in sludge:

REMOVAL DATA

Sampling period:

Pollutant/parameter	Concentration[a] Influent	Effluent	Percent removal
Concentional pollutants, mg/L:			
COD	395	18	95
Toxic pollutants, μg/L:			
Chromium	15	2	87
Copper	45	19	58
Lead	11	2	82
Nickel	1	2	0[b]
Zinc	92	53,000	0[b]
Bis(2-ethylhexyl) phthalate	8	30	0[b]
Butyl benzyl phthalate	800	ND[c]	>99
Diethyl phthalate	12	ND[c]	>17
Phenol	1	5	0[b]
Ethylbenzene	13,000	ND[c]	>99
Toluene	130	ND[c]	>92
Napthalene	46	60	0[b]
Chloroform	3	ND[d]	>0
Other pollutants, μg/L:			
Xylenes	14,000	ND[c]	>99

[a]Average concentration.

[b]Actual data indicate negative removal.

[c]Not detected, assumed to be <10 μg/L.

[d]Not detected, assumed to be less than the corresponding influent concentration.

Note: Blanks indicate information was not specified.

4.5 GAS FLOTATION WITH CHEMICAL ADDITION [1]

4.5.1 Function

Gas flotation with chemical addition is utilized to remove colloidal and suspended solids.

4.5.2 Description

The use of chemical addition in conjuction with gas flotation is similar to the treatment technology described for sedimentation with chemical addition, except that gas flotation is utilized instead of sedimentation. The reader is referred to Section 4.3 for a thorough discussion of chemical addition and Section 4.4 for a discussion of gas flotation.

4.5.3 Technology Status

Gas flotation with chemical addition is a well-developed technology; installed equipment is currently operating in many industrial applications.

4.5.4 Applications

Any industrial wastestream where land/space availability is limited and/or sedimentation is not practical.

4.5.5 Limitations

Effluent from gas flotation with chemical addition may require additional solids removal (e.g., multimedia filtration).

4.5.6 Residuals Generated/Environmental Impact

Solids must be disposed of properly; odor may be a problem with certain wastestreams. Energy requirements are greater than those for sedimentation; noise pollution due to air compressors.

4.5.7 Design Criteria

Design criteria for gas flotation with chemical addition are the same as those described in Section III.4.3.10 for chemical addition and Section III.4.4 for gas flotation.

4.5.8 Performance

Subsequent data sheets provide performance data from studies on the following industries and/or wastestreams:

- Auto and other laundries industry
  - Industrial laundries
  - Linen supplies
  - Power laundries

- Textile milling
  - Woven fabric finishing

4.5.9 References

1. Innovative and Alternative Technology Assessment Manual. EPA-430/9-78-009 (draft), U.S. Environmental Protection Agency, Cincinnati, Ohio, 1978. 252 pp.

CONTROL TECHNOLOGY SUMMARY FOR GAS FLOTATION WITH CHEMICAL ADDITION ($CaCl_2$, POLYMER)

Pollutant	Number of data points	Effluent concentration				Removal efficiency, %			
		Minimum	Maximum	Median	Mean	Minimum	Maximum	Median	Mean
Toxic pollutants, µg/L:									
Antimony	5	<10	310	<20	<78	0[a]	>89	>20	47
Arsenic	4	2	12	<10	<8.5	8	80	>13	>29
Cadmium	6	<2	72	<3	<17	0[a]	>98	>96	79
Chromium	6	100	620	280	≤330	42	67	≥50	≥52
Copper	5	150	500	330	300	67	91	79	78
Cyanide	4	54	530	260	280	0[a]	98	3	26
Lead	6	67	300	120	150	94	98	98	97
Mercury	3	<0.2	2	<0.2	0.8	33	90	>80	68
Nickel	5	<5	250	<50	<73	>0	>94	>67	>55
Selenium	1				2				0[a]
Silver	2	<15	19		<17	0[a]	>48		64
Thallium	1				50				0[a]
Zinc	6	<10	310	≤130	<140	94	>99	97	>95
Bis(2-ethylhexyl) phthalate	2	220	1,000		610	62	82		72
Butyl benzyl phthalate	1				<0.03				>99
Di-n-butyl phthalate	2	19	290		160	0[a]	79		40
Di-n-octyl phthalate	1				33				78
N-nitrosodiphenyl amine	1				620				66
2,4-Dimethylphenol	1				<0.1				>99
Pentachlorophenol	1				27				0[a]
Phenol	3	42	120	100	87	0	80	57	69
2,4,6-Trichlorophenol	1				3				0[a]
Benzene	2	5	200		100	0[a]	0[a]		0[a]
1,2-Dichlorobenzene; 1,3-Dichlorobenzene; 1,4-Dichlorobenzene	1				260				76
Ethylbenzene	4	<10[b]	970	77	280	0[a]	>99	31	40
Toluene	4	380	2,100	850	1,000	0[a]	65	6	19
Anthracene/phenanthrene	1				66				83
Naphthalene	3	480	840	790	700	0[a]	82	80	54
Carbon tetrachloride	1				1				50
Chloroform	3	0.8	9	8	6	0[a]	74	20	31
Methylene chloride	3	2	6,000	500	2,200	0[a]	7	0	2
Tetrachloroethylene	4	5	1,000	660	580	0[a]	94	47	24
1,1,1-Trichloroethane	1				14				22
Trichloroethylene	2	6	30		18	0[a]	86		43
N-nitrosodimethylamine	1				620				66
Isophorone	1				<10[b]				>95

Note: Blanks indicate data not applicable.

[a]Actual data indicate negative removal.

[b]Not detected, assumed to be <10 µg/L.

CONTROL TECHNOLOGY SUMMARY FOR GAS FLOTATION WITH CHEMICAL ADDITION (POLYMER)

Pollutant	Number of data points	Effluent concentration				Removal efficiency, %			
		Minimum	Maximum	Median	Mean	Minimum	Maximum	Median	Mean
Toxic pollutants, µg/L:									
Antimony	1				64				0[a]
Cadmium	1				5				0[a]
Chromium	1				28				0[a]
Copper	2	50	81		66	9	75		42
Cyanide	1				25				14
Lead	2	<10[a]	29		>20	0[a]	>29		15
Nickel	2	32	63		48	0[a]	0[b]		0
Silver	1				29				0[a]
Zinc	2	<10	240		130	17	>60		39
Bis(2-ethylhexyl) phthalate	2	45	74		60	10	92		51
Butyl benzyl phthalate	1				<0.03				>99
Di-n-butyl phthalate	2	<0.02	<10[b]		<5	>23	>99		>61
Di-n-octyl phthalate	1				11				61
2-Chlorophenol	1				2				0[a]
2,4-Dichlorophenol	1				6				0[a]
2,4-Dimethylphenol	1				28				0[a]
Pentachlorophenol	2	8	30		19	0[a]	19		10
Phenol	2	9	26		18	0[a]	72		36
Anthracene/phenanthrene	1				2				0[a]
Fluoranthene	1				0.5				0[a]
Naphthalene	2	0.6	<10[b]		<5	33	>96		>65
Pyrene	1				0.3				0
Chloroform	1				24				41
Methylene chloride	1				22				61
Tetrachloroethylene	1				2				0
1,1,1-Trichloroethane	2	<2	<10		<6	>0	>9		>5
Thallium	1				14				0[b]
Benzene	1				12				33
Ethylbenzene	1				160				65
Toluene	1				130				59
Methyl chloride	1				30				0[b]

Note: Blanks indicate data not applicable.

[a]Actual data indicate negative removal.

[b]Not detected, assumed to be <10 µg/L.

TREATMENT TECHNOLOGY: Gas Flotation with Chemical Addition (Alum, Polymer)

Data source: Effluent Guidelines
Point source category: Auto and other laundries
Subcategory: Industrial laundries
Plant: K
References: A28, Appendix C

Data source status:
- Engineering estimate ___
- Bench scale ___
- Pilot scale ___
- Full scale x

Use in system: Primary
Pretreatment of influent: Screening, equalization

DESIGN OR OPERATING PARAMETERS

Process type: Dissolved air flotation (DAF)
Unit configuration: Circular DAF unit; no recycle
Wastewater flow: 45 m^3/d (12,000 gpd), 159 m^3/d (42,000 gpd), design
Chemical dosage(s): Alum - 1,200 mg/L, Polymer - 80 mg/L
pH in flotation chamber: 5 - 6
Detention time:
Hydraulic loading:
Percent recycle: 0
Solids loading:
Gas requirement:
Gas-to-solids ratio:
Pressure: 552 kPa (80 psi)
Sludge overflow:
Percent solids in sludge:

REMOVAL DATA

Sampling period: Average of two, one-day composites

Pollutant/parameter	Concentration		Percent
	Influent	Effluent	removal
Conventional pollutants, mg/L:			
BOD_5	346	178	49
COD	2,550	2,110	17
TOC	728	544	25
TSS	498	742	0[a]
Oil and grease	205	76	63
Total phenol	0.108	0.094	13
Total phosphorus	24.0	12.2	49
Toxic pollutants, µg/L:			
Antimony	2,400	2,200	6
Arsenic	8.0	3.5	56
Cadmium	40	40	0
Chromium	450	360	19
Copper	810	660	19
Cyanide	26	≤10	≥61
Lead	1,000	1,000	0
Mercury	1.5	1.0	33
Nickel	460	270	41
Selenium	≤1	≤1	-0
Silver	120	66	44
Zinc	2,600	2,300	10
Bis(2-ethylhexyl) phthalate	120	90	25
Butyl benzyl phthalate	<0.03	81	0[a]
Di-n-butyl phthalate	300	300	0
Di-n-octyl phthalate	<0.9	21	0[a]
Phenol	20	28	0[a]
Ethylbenzene	1.5	3.0	0[a]
Toluene	5.0	4.5	10
Anthracene/phenanthrene	7.5	10	0[a]
Naphthalene	23	11	52
2-Chloronaphthalene	17	16	3
Carbon tetrachloride	1,700	410	76
Chloroform	6.0	19	0[a]
Dichlorobromomethane	6.0	<0.9	>85
Methylene chloride	48	8.0	84
Tetrachloroethylene	1.0	<0.9	>10
1,1,1-Trichloroethane	3,300	860	74
Trichlorofluoromethane	4.0	<2.0	>50
Acrolein	<100	720	0[a]

[a]Actual data indicate negative removal.

Note: Blanks indicate information was not specified.

TREATMENT TECHNOLOGY: Gas Flotation with Chemical Addition (Calcium Chloride, Polymer)

Data source: Effluent Guidelines
Point source category: Auto and other laundries
Subcategory: Industrial laundries
Plant: A
References: A28, Appendix C

Data source status:
- Engineering estimate ___
- Bench scale ___
- Pilot scale ___
- Full scale x

Use in system: Secondary
Pretreatment of influent: Screening, equalization, gravity oil separation

DESIGN OR OPERATING PARAMETERS

Process type: Dissolved air flotation (DAF)
Unit configuration: Rectangular DAF unit; recycle pressurization
Wastewater flow: 0.27 m^3/min (70 gpm), 0.57 m^3/min (150 gpm), design
Chemical dosage(s): $CaCl_2$ - 1,800 mg/L, Polymers - 2 mg/L
pH in flotation chamber: 11.6
Detention time:
Hydraulic loading: 0.038 $m^3/min/m^2$ (0.93 gpm/ft^2)
Percent recycle: 50
Solids loading:
Gas requirement:
Gas-to-solids ratio: 0.0097
Pressure: 476 kPa (4.7 atm)
Sludge overflow: 0.082 m^3/min (2 gpm)
Percent solids in sludge: 5

REMOVAL DATA

Sampling period: 2 days

Pollutant/parameter	Concentration Influent	Concentration Effluent	Percent removal
Conventional pollutants, mg/L:			
COD	6,400	3,200	50
TOC	1,700	690	59
TSS	390	98	75
Oil and grease	703	143	80
Total phenol	0.78	0.76	3
Total phosphorus	41.6	1.7	96
Toxic pollutants, μg/L:			
Antimony	94	<10	>89
Arsenic	10	2	80
Cadmium	110	<2	>98
Chromium	480	270	44
Copper	1,500	500	67
Cyanide	57	54	5
Lead	4,800	130	97
Nickel	350	250	29
Selenium	1	2	0[a]
Thallium	<40	50	0[a]
Zinc	3,700	230	94
Bis(2-ethylhexyl) phthalate	1,200	220	82
Butyl benzyl phthalate	310	<0.03	>99
Di-n-butyl phthalate	92	19	79
Di-n-octyl phthalate	150	33	78
2,4-Dimethylphenol	460	<0.1	>99
Pentachlorophenol	<0.4	27	0[a]
Phenol	98	42	57
2,4,6-Trichlorophenol	<0.2	3	0[a]
Benzene	3	5	0[a]
Dichlorobenzene	1,100	260	76
Ethylbenzene	25	44	0[a]
Toluene	360	380	0[a]
Anthracene/phenanthrene	380	66	83
Naphthalene	4,800	840	82
Carbon tetrachloride	2	1	50
Chloroform	0.7	0.8	0[a]
Methylene chloride	2	2	0
Tetrachloroethylene	320	330	0[a]
1,1,1-Trichloroethane	18	14	22
Trichloroethylene	4	6	0[a]

[a]Actual data indicate negative removal.

Note: Blanks indicate information was not specified.

TREATMENT TECHNOLOGY: Gas Flotation with Chemical Addition (Calcium Chloride, Polymer)

Data source: Effluent Guidelines
Point source category: Auto and other laundries
Subcategory: Industrial laundries
Plant: B
References: A28, Appendix C

Data source status:
Engineering estimate ___
Bench scale ___
Pilot scale ___
Full scale x

Use in system: Primary
Pretreatment of influent: Screening, equalization

DESIGN OR OPERATING PARAMETERS

Process type: Dissolved air flotation (DAF)
Unit configuration: Rectangular DAF unit; recycle pressurization
Wastewater flow:
Chemical dosage(s):
pH in flotation chamber: 11.6
Detention time:
Hydraulic loading:
Percent recycle:
Solids loading:

Gas requirement:
Gas-to-solids ratio:
Pressure:
Sludge overflow:
Percent solids in sludge:

REMOVAL DATA

Sampling period: 1 day

Pollutant/parameter	Concentration Influent	Concentration Effluent	Percent removal
Conventional pollutants, mg/L:			
COD	3,800	1,300	66
TSS	700	48	93
Oil and grease	440	190	57
Total phenol	0.016	<0.001	>94
Toxic pollutants, μg/L:			
Antimony	41	<20	>51
Arsenic	12	<10	>17
Cadmium	170	23	86
Chromium	270	≤130	≥52
Copper	1,600	330	79
Lead	9,400	230	98
Mercury	2	<0.2	>90
Nickel	150	<50	>67
Zinc	4,500	200	96
Di-n-butyl phthalate	ND[a]	290	0[b]
N-nitrosodiphenylamine	1,800	620	66
Phenol	600	120	80
Ethylbenzene	260	110	58
Toluene	750	790	0[b]
Naphthalene	4,000	790	80
Chloroform	10	8	20
Methylene chloride	540	500	7
Tetrachloroethylene	880	1,000	0[b]
Trichloroethylene	210	30	86
Isophorone	190	ND	>95

[a]Not detected; assumed to be <10 μg/L.
[b]Actual data indicate negative removal.

Note: Blanks indicate information was not specified.

TREATMENT TECHNOLOGY: Gas Flotation with Chemical Addition (Calcium Chloride, Polymer)

Data source: Effluent Guidelines
Point source category: Auto and other laundries
Subcategory: Industrial laundries
Plant: C
References: A28, Appendix C

Data source status:
Engineering estimate ___
Bench scale ___
Pilot scale ___
Full scale x

Use in system: Primary
Pretreatment of influent: Screening, equalization

DESIGN OR OPERATING PARAMETERS

Process type: Dissolved air flotation (DAF)
Unit configuration: Rectangular DAF unit; recycle pressurization
Wastewater flow:
Chemical dosage(s):
pH in flotation chamber: 11.3
Detention time:
Hydraulic loading:
Percent recycle:
Solids loading:

Gas requirement:
Gas-to-solids ratio:
Pressure:
Sludge overflow:
Percent solids in sludge:

REMOVAL DATA

Sampling period: 1 day

Pollutant/parameter	Concentration Influent	Effluent	Percent removal
Conventional pollutants, mg/L:			
COD	3,200	1,200	62
TSS	520	64	88
Oil and grease	760	170	78
Total phenol	0.028	0.56	0[a]
Toxic pollutants, µg/L:			
Antimony	<25	<20	>20
Arsenic	13	12	8
Cadmium	54	<2	>96
Chromium	1,200	620	48
Copper	1,200	340	72
Lead	4,400	67	98
Mercury	1	<0.2	>80
Nickel	50	<50	>0
Silver	<29	<15	>48
Zinc	2,600	≤68	≥97
Phenol	100	100	0
Ethylbenzene	1,000	970	3
Toluene	2,400	2,100	12
Naphthalene	ND[b]	480	0[a]
Chloroform	35	9	74
Methylene chloride	110	6,000	0[a]
Tetrachloroethylene	84	5	94

[a]Actual data indicate negative removal.

[b]Not detected.

Note: Blanks indicate information was not specified.

TREATMENT TECHNOLOGY: Gas Flotation with Chemical Addition (Calcium Chloride, Polymer)

Data source: Effluent Guidelines
Point source category: Auto and other laundries
Subcategory: Industrial laundries
Plant: D
References: A28, Appendix C

Data source status:
Engineering estimate ___
Bench scale ___
Pilot scale ___
Full scale x

Use in system: Primary
Pretreatment of influent: Screening, equalization

DESIGN OR OPERATING PARAMETERS

Process type: Dissolved air flotation (DAF)
Unit configuration: Rectangular DAF unit; recycle pressurization
Wastewater flow:
Chemical dosage(s):
pH in flotation chamber: 11.7
Detention time:
Hydraulic loading:
Percent recycle:
Solids loading:
Gas requirement:
Gas-to-solids ratio:
Pressure:
Sludge overflow:
Percent solids in sludge:

REMOVAL DATA

Pollutant/Parameter	....Concentration.... Influent	Effluent	Percent Removal
Conventional pollutants, mg/ℓ			
BOD_5	2,400	1,000	58
COD	7,100	2,000	72
TOC	1,800	500	72
TSS	940	100	89
Oil and grease	1,600	230	86
Toxic pollutants, μg/ℓ			
Antimony	160	310	0*
Cadmium	70	3	96
Chromium	980	570	42
Copper	1,700	150	91
Cyanide	280	290	0*
Lead	5,400	110	98
Nickel	80	<10	>87
Zinc	2,700	ND**	>99
Bis(2-ethylhexyl) phthalate	2,600	1,000	62
Benzene	130	200	0
Ethylbenzene	18,000	ND	>99
Toluene	2,600	900	65
Tetrachloroethylene	30	980	0*

*Actual data indicate negative removal.
**Not detected, assumed to be <10 μg/ℓ.

Note: Blanks indicate information was not specified.

TREATMENT TECHNOLOGY: Gas Flotation with Chemical Addition (Calcium Chloride, Polymer)

Data source: Effluent Guidelines
Point source category: Auto and other laundries
Subcategory: Industrial laundries
Plant: E
References: A28, Appendix C

Data source status:
- Engineering estimate ___
- Bench scale ___
- Pilot scale ___
- Full scale x

Use in system: Primary
Pretreatment of influent: Screening, equalization

DESIGN OR OPERATING PARAMETERS

Process type: Dissolved air flotation (DAF)
Unit configuration: Rectangular DAF unit; recycle pressurization
Wastewater flow:
Chemical dosage(s):
pH in flotation chamber:
Detention time:
Hydraulic loading:
Percent recycle:
Solids loading:
Gas requirement:
Gas-to-solids ratio:
Pressure:
Sludge overflow:
Percent solids in sludge:

REMOVAL DATA

Sampling period: 1 day

Pollutant/parameter	Concentration Influent	Concentration Effluent	Percent removal
Conventional pollutants, mg/L:			
BOD_5	1,700	540	68
COD	4,900	1,100	78
TOC	460	270	41
TSS	900	18	98
Oil and grease	230	84	63
Total phenol	0.10	0.32	0[a]
Total phosphorus	13	23	0[a]
Toxic pollutants, μg/L:			
Antimony	120	29	76
Arsenic	11	ND[b]	>9
Cadmium	60	<2	>97
Chromium	300	100	67
Copper	1,000	200	80
Cyanide	240	530	0[a]
Lead	3,000	70	98
Mercury	-3	2	33
Nickel	80	<5	>94
Silver	8	19	0[a]
Zinc	2,000	60	97

[a]Actual data indicate negative removal.
[b]Not detected; assumed to be <10 μg/L.

Note: Blanks indicate information was not specified.

TREATMENT TECHNOLOGY: Gas Flotation with Chemical Addition (Calcium Chloride, Polymer)

Data source: Effluent Guidelines
Point source category: Auto and other laundries
Subcategory: Industrial luandries
Plant: F
References: A28, Appendix C

Data source status:
Engineering estimate ___
Bench scale ___
Pilot scale ___
Full scale x

Use in system: Primary
Pretreatment of influent: Screening, equalization

DESIGN OR OPERATING PARAMETERS

Process type: Dissolved air flotation (DAF)
Unit configuration: Rectangular DAF unit; recycle pressurization
Wastewater flow: 0.38 m^3/min (101 gpm), 0.78 m^3/min (200 gpm), design
Chemical dosage(s): Calcium chloride - 1,600 mg/L, polymer - 2 mg/L
pH in flotation chamber:
Detention time:
Hydraulic loading: 0.0027 m^3/min/m^2 (0.66 gpm/ft^2)
Percent recycle:
Solids loading:
Gas requirement:
Gas-to-solids ratio:
Pressure:
Sludge overflow:
Percent solids in sludge: 3-5

REMOVAL DATA

Sampling period: 5 days

Pollutant/parameter	Concentration Influent	Concentration Effluent	Percent removal
Conventional pollutants, mg/L:			
BOD_5	877	318	64
TOC	139	155	0[a]
TSS	792	142	82
Oil and grease	513	53	90
Toxic pollutants, µg/L:			
Cadmium	48	72	0[a]
Chromium	650	290	56
Lead	5,400	300	94
Zinc	2,900	310	89

[a] Actual data indicate negative removal.

Note: Blanks indicate information was not specified.

TREATMENT TECHNOLOGY: Gas Flotation with Chemical Addition (Ferrous Sulfate, Lime, Polymer)

Data source: Effluent Guidelines
Point source category: Auto and other laundries
Subcategory: Industrial laundries
Plant: L
References: A28, Appendix C

Data source status:
Engineering estimate ___
Bench scale ___
Pilot scale ___
Full scale x

Use in system: Primary
Pretreatment of influent: Screening

DESIGN OR OPERATING PARAMETERS

Process type: Dissolved air flotation (DAF)
Unit configuration: Rectangular DAF unit; recycle pressurization
Wastewater flow: 83 m^3/d (22,000 gpd), design
Chemical dosage(s): $FeSO_4$ - 300 mg/L, Cationic polymer - 2 mg/L
pH in flotation chamber:
Detention time:
Hydraulic loading:
Percent recycle:
Solids loading:
Gas requirement:
Gas-to-solids ratio:
Pressure:
Sludge overflow:
Percent solids in sludge:

REMOVAL DATA

Sampling Period: Toxic Organics—3 Days; Other Pollutants—8 Days

Pollutant/Parameter	....Concentration.... Influent	Effluent	Percent Removal
Conventional pollutants, mg/ℓ			
BOD_5	1,310	209	84
COD	4,770	600	87
TOC	771	177	77
TSS	711	86	88
Oil and grease	915	28	97
Total phenol	0.367	1.09	0*
Total phosphorus	21.7	0.14	99
Toxic pollutants, μg/ℓ			
Antimony	95	18	81
Arsenic	32	11	65
Cadmium	97	≤15	≥84
Chromium	410	≤27	≥93
Copper	3,600	73	98
Cyanide	46	≤32	≥30
Lead	7,200	≤140	≥98
Mercury	2.7	≤0.97	≥64
Nickel	130	<5	>96
Silver	-4	<1	>75
Zinc	2,500	130	95
Bis(2-ethylhexyl) phthalate	5,100	110	98
Butyl benzyl phthalate	1,500	42	97
Di-n-butyl phthalate	600	21	97
Di-n-octyl phthalate	410	ND**	>98
N-nitrosodiphenylamine	ND	84	0*
Pentachlorophenol	ND	13	0*
Phenol	ND	190	0*
Benzene	ND	120	0*
Chlorobenzene	ND	57	0*
Dichlorobenzene	ND	18	0*
Anthracene/phenanthrene	470	≤10	≥98
Fluoranthene	ND	≤10	0*
Fluorene	ND	14	0*
Naphthalene	410	96	77
Pyrene	ND	18	0*
Carbon tetrachloride	ND	36	0*
Dichlorobromomethane	ND	290	0*
1,1-Dichloroethylene	ND	1,000	0*
1,2-Dichloropropane	ND	930	0*

*Actual data indicate negative removal.
**Not detected.

Note: Blanks indicate information was not specified.

TREATMENT TECHNOLOGY: Gas Flotation with Chemical Addition (Ferric Sulfate, Polymer)

Data source: Effluent Guidelines
Point source category: Auto and other laundries
Subcategory: Linen supply
Plant: M
References: A28, Appendix C

Data source status:
Engineering estimate ___
Bench scale ___
Pilot scale ___
Full scale x

Use in system: Primary
Pretreatment of influent: Screening, equalization

DESIGN OR OPERATING PARAMETERS

Process type: Dissolved air flotation (DAF)
Unit configuration: Rectangular DAF unit; full flow pressurization
Wastewater flow: 170 m^3/d (45,000 gpd), design
Chemical dosage(s): $Fe_2(SO_4)_3$ - 1,200 mg/L, Anionic polymer - 25 mg/L
pH in flotation chamber: 6
Detention time: 29 min
Hydraulic loading:
Percent recycle: 0
Solids loading:
Gas requirement:
Gas-to-solids ratio:
Pressure:
Sludge overflow:
Percent solids in sludge:

REMOVAL DATA

Sampling period: Eight days unless otherwise specified

Pollutant/parameter	Concentration		Percent removal
	Influent	Effluent	
Conventional pollutants, mg/L:			
BOD_5	1,420	486	66
COD	3,600	410	89
TOC	599	160	73
TSS	536	61	89
Oil and grease	341	101	70
Total phenol	0.065	0.034	48
Total phosphorus	19	0.3	98
Toxic pollutants, µg/L:			
Antimony[a]	8	3	62
Arsenic[a]	3	9	0[b]
Chromium	140	58	59
Copper[a]	230	400	0[b]
Lead	330	≤87	≥74
Mercury	2	1.2	40
Zinc	670	910	0[b]

[a]Average of three samples.

[b]Actual data indicate negative removal.

Note: Blanks indicate information was not specified.

TREATMENT TECHNOLOGY: Gas Flotation with Chemical Addition (Polymer)

Data source: Effluent Guidelines
Point source category: Auto and other luandries
Subcategory: Power Laundries
Plant: J
References: A28, Appendix C

Data source status:
- Engineering estimate ___
- Bench scale ___
- Pilot scale ___
- Full scale x

Use in system: Primary
Pretreatment of influent: Screening, equalization

DESIGN OR OPERATING PARAMETERS

Process type: Dissolved air flotation (DAF)
Unit configuration: Rectangular DAF unit; recycle pressurization
Wastewater flow: 341 m^3/d (90,000 gpd), 379 m^3/d (100,000 gpd), design
Chemical dosage(s): 60 mg/L polyelectrolyte
pH in flotation chamber: 10.3 - 10.6
Detention time:
Hydraulic loading: 0.11 $m^3/min/m^2$ (2.6 gpm/ft^2)
Percent recycle: 50
Solids loading:

Gas requirement:
Gas-to-solids ratio: 0.5
Pressure: 517 kPa (5.1 atm)
Sludge overflow: 0.11 m^3/d (30 gpd)
Percent solids in sludge: 7.5

REMOVAL DATA

Sampling period: 2 days

Pollutant/parameter	Concentration Influent	Concentration Effluent	Percent removal
Conventional pollutants, mg/L:			
BOD_5	113	142	0[a]
COD	497	459	8
TOC	135	87	36
TSS	50	32	36
Oil and grease	39	16	59
Total phenol	0.432	0.385	11
Total phosphorus	0.8	1.0	0[a]
Toxic pollutants, µg/L:			
Antimony	<10	64	0[a]
Cadmium	<2	5	0[a]
Chromium	26	28	0[a]
Copper	55	50	9
Cyanide	29	25	14
Lead	<22	70	0[a]
Nickel	<36	63	0[a]
Silver	<5	29	0[a]
Zinc	290	240	17
Bis(2-ethylhexyl) phthalate	82	74	10
Butyl benzyl phthalate	17	<0.03	>99
Di-n-butyl phthalate	2	<0.02	>99
Di-n-octyl phthalate	28	11	61
2-Chlorophenol	0.3	2	0[a]
2,4-Dichlorophenol	1	6	0[a]
2,4-Dimethylphenol	2	28	0[a]
Pentachlorophenol	3	8	0[a]
Phenol	2	9	0[a]
Anthracene/phenanthrene	0.9	2	0[a]
Fluoranthene	0.3	0.5	0[a]
Naphthalene	0.9	0.6	33
Pyrene	0.3	0.3	0
Chloroform	41	24	41
Methylene chloride	57	22	61
Tetrachloroethylene	2	2	0
1,1,1-Trichloroethane	2	<2	>0

[a]Actual data indicate negative removal.

Note: Blanks indicate information was not specified.

TREATMENT TECHNOLOGY: Gas Flotation with Chemical Addition (Cationic polymer)

Data source: Effluent Guidelines
Point source category: Textile mills
Subcategory: Woven fabric finishing
Plant:
References: A6, p. VII-80

Data source status:
Engineering estimate ___
Bench scale ___
Pilot scale ___
Full scale _x_

Use in system: Primary
Pretreatment of influent: Equalization, grit removal, coarse screening, chemical addition (alum and caustic), and fine screening

DESIGN OR OPERATING PARAMETERS

Process type: Dissolved air flotation
Unit configuration:
Wastewater flow: 1.2 m^3/min (300 gpm)
Chemical dosage(s):
pH in flotation chamber:
Detention time:
Hydraulic loading:
Percent recycle:
Solids loading:

Gas requirement:
Gas-to-solids ratio:
Pressure:
Sludge overflow:
Percent solids in sludge:

REMOVAL DATA

Sampling period: Average of two 24-hr samples

Pollutant/parameter	Concentration Influent	Effluent	Percent removal
Conventional pollutants, mg/L:			
BOD_5	400	<200	>50
COD	1,050	725	31
TSS	195	32	84
Total phenol	0.092	0.026	72
Toxic pollutants, μg/L:			
Copper	320	81	75
Lead	14	ND[a]	>29
Nickel	28	32	0[b]
Thallium	<10	14	0[b]
Zinc	25	<10	>60
Bis(2-ethylhexyl) phthalate	570	45	92
Di-n-butyl phthalate	13	ND	>23
Pentachlorophenol	37	30	19
Phenol	94	26	72
Benzene	18	12	33
Ethylbenzene	460	160	65
Toluene	320	130	59
Naphthalene	250	ND	>96
Methyl chloride	26	30	0[b]
1,1,1-Trichloroethane	11	<10	>9

[a]Not detected; assumed to be <10 μg/L.
[b]Actual data indicate negative removal.

Note: Blanks indicate information was not specified.

4.6 GRANULAR MEDIA FILTRATION [1, 2]

4.6.1 Function

Granular media filtration is used to remove suspended solids from a liquid wastestream.

4.6.2 Description

Granular media filtration, one of the oldest and most widely applied types of filtration for the removal of suspended solids from aqueous liquid streams, utilizes a bed of granular particles (typically sand or sand with coal) as the filter medium. The bed is typically contained within a basin or tank and is supported by an underdrain system which allows the filtered liquid to be drawn off while retaining the filter medium in place. The underdrain system typically consists of metal or plastic strainers located at intervals on the bottom of the filter. As suspended particle-laden water passes through the bed of the filter medium, particles are trapped on top of and within the bed, thus reducing its porous nature and either reducing the filtration rate at constant pressure or increasing the amount of pressure needed to force the water through the filter. If left to continue in this manner, the filter would eventually plug up with solids; the solids, therefore, must be removed. This is done by forcing a wash water stream through the bed of granular particles in the reverse direction of the original fluid flow. The wash water is sent through the bed at a velocity sufficiently high so that the filter bed becomes fluidized and turbulent. In this turbulent condition, the solids are dislodged from the granular particles and are discharged in the spent wash water. This whole process is referred to as "back-washing." When the backwashing cycle is completed, the filter is returned to service.

The spent backwash water contains the suspended solids removed from the liquid, and, therefore, presents a liquid disposal problem in itself. The volume of the backwash water stream, however, is normally only a small fraction (2% to 10%) of the volume of the liquid being filtered. Consequently, the suspended solids concentration of the backwash water is far greater than that of the liquid filtered. Granular media filtration essentially removes suspended solids from one liquid stream and concentrates them in another, but much smaller, liquid stream. Depending on the specific process configuration, backwash water itself can be treated to remove suspended solids by flocculation and/or sedimentation or by returning it to the portion of the process from whence the liquid stream subjected to filtration originated; e.g., a settling pond.

4.6.3 Common Modifications

Dual-media filtration involves the use of both sand and anthracite as filter media, with anthracite being placed on top of the sand. In multimedia filtration, three or more layers of different granular materials are employed. A common design incorporates sequential layers of anthracite, sand and garnet (top to bottom). Precoat pressure filters utilizing diatomaceous earth is another type of filtration system used, usually for relatively low volume streams requiring exceptionally high clarity.

Gravity filters operate by either using the available head from the previous treatment unit, or by pumping to a flow-split box after which the wastewater flows by gravity to the filter cells. Pressure filters utilize pumping to increase the available head. Flow control methods include constant head declining rate, constant rate control etc.

Filtration systems can be constructed of concrete or steel, with single or multiple compartment units. Steel units can be either horizontal or vertical and are generally used for pressure filters. Systems can be manually or automatically operated. High rate (10-20 gpm/ft^2) deep bed (4-10 ft) filters are utilized for certain industrial applications for removal of dense particulates in high concentrations.

Backwash sequences can include air scour or surface wash steps. Backwash water can be stored separately or in chambers that are integral parts of the filter unit. Backwash water can be pumped through the unit or can be supplied through gravity head tanks.

4.6.4 Technology Status

Granular media filtration has been used for many years in the potable water industry and for 10 to 15 years in the wastewater treatment field.

4.6.5 Applications

Removal of residual biological floc in settled effluents from secondary treatment, and removal of residual chemical-biological floc after alum, iron, or lime precipitation in tertiary or independent physical-chemical waste treatment; in these applications, filtration may serve both as an intermediate process to prepare wastewater for further treatment (such as carbon adsorption, clinoptilolite ammonia exchange columns, or reverse osmosis) or as a final polishing step following other processes.

4.6.6 Limitations

Economics are highly dependent on consistent pretreatment quality and flow modulations; increasing suspended solids loading will reduce run lengths, and large flow variations will deleteriously affect effluent quality in chemical treatment sequences; depending on suspended solids concentration of wastewater streams, it may be necessary to install other liquid/solid separation processes such as flocculation and/or sedimentation ahead of granular media filtration to take the bulk of the suspended solids load off the filters.

4.6.7 Chemicals Required

Alum salts, iron salts, and polymers can be added as coagulants or coagulant aids directly ahead of filtration units; however, this will generally reduce run lengths.

4.6.8 Residuals Generated

Backwash water, which generally approximates two to ten percent of the throughput; backwash water can be returned to the head of the plant, for removal of solids in a clarification step.

4.6.9 Reliability

Granular filtration systems are very reliable from both a process and unit standpoint.

4.6.10 Environmental Impact

Requires relatively little use of land; backwash water will need further treatment, with an ultimate production of solids that will need disposal; air scour blowers usually need silencers to control noise; no air pollution generated.

4.6.11 Design Criteria (for Dual-Media Filtration)

Criteria	Units	Range/value
Filtration rate	gpm/ft^2	2 to 8
Bed depth	in.	24 to 48
Depth ratio (sand to anthracite)		1:1 to 1:4
Backwash rate	gpm/ft^2	15 to 25
Air scour rate	standard $ft^3/min/ft^2$	3 to 5
Filter run length	hr	8 to 48
Terminal head loss	ft (gravity type)	6 to 15
Terminal head loss	ft (pressure type)	∿30

Note: Precoat, dual, and multi-media filtration utilize similar criteria; however, the depth ratios will differ.

4.6.12 Flow Diagram

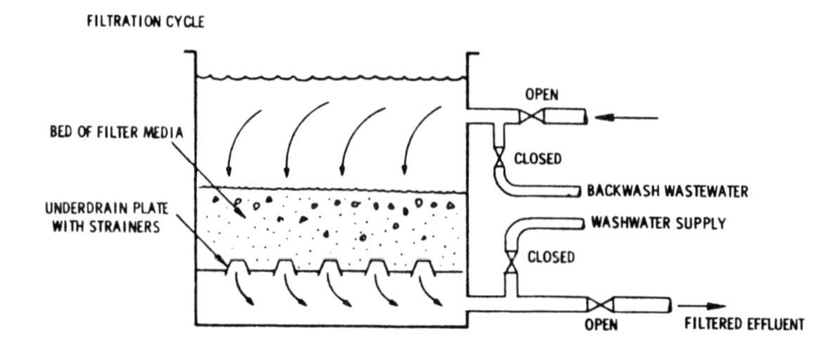

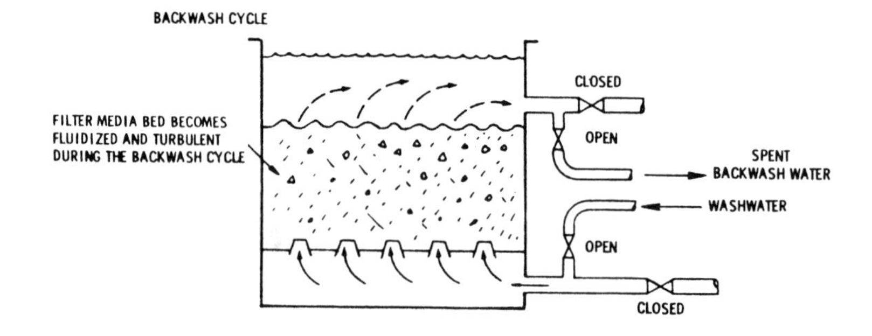

4.6.13 Performance

Subsequent data sheets provide performance data from studies on the following industries and/or waste streams:

Auto and other laundries industry
- Industrial laundries
- Power laundries

Electroplating

Foundry industry
- Aluminum foundry - die lube operation

Inorganic chemicals production
- Chlorine - diaphragm cell plant operations
- Chrome pigment production
- Copper sulfate production

Iron and steel industry
- Continuous casting
- Hot forming - primary
- Vacuum degassing

Nonferrous metals industry

Ore mining and dressing
- Asbestos - cement processing
- Asbestos mining
- Base metal mining
- Copper milling
- Lead/zinc mining/milling/smelting/refining
- Molybdenum mining/milling

Paint manufacturing

Petroleum refining

Pulp, paper, and paperboard production
- Man-made fiber processing
- Pulp milling

Textile milling
- Knit fabric finishing
- Stock and yarn finishing
- Wool finishing
- Wool scouring
- Woven fabric finishing

4.6.14 References

1. Innovative and Alternative Technology Assessment Manual. EPA-430/9-78-009 (draft), U.S. Environmental Protection Agency, Cincinnati, Ohio, 1978. 252 pp.

2. Physical, Chemical, and Biological Treatment Techniques for Industrial Wastes, PB 275 287, U.S. Environmental Protection Agency, Washington, D.C., November 1976. pp. 22-1 - 22-25.

CONTROL TECHNOLOGY SUMMARY FOR FILTRATION

Pollutant	Number of data points	Effluent concentration				Removal efficiency, %			
		Minimum	Maximum	Median	Mean	Minimum	Maximum	Median	Mean
Conventional pollutants, mg/L:									
BOD_5	13	2.4	23,400	19	2,280	0[a]	51	22	24
COD	22	29	260,000	195	13,400	0[a]	75	20.5	26
TOC	17	10	25,000	40	2,000	0[a]	49	12	15
TSS	40	<0.01	7,330	12	247	0[a]	>99	75	68
Oil and grease	13	<0.5	9,940	11	899	0[a]	>98	20	30
Total phenol	19	0.0011	64.4	0.032	3.46	0[a]	65	8	16
Total phosphorous	7	0.23	13	2.0	3.0	7	83	30	40
Toxic pollutants, μg/L:									
Antimony	15	<10	1,800	42	270	0[a]	>84	24	>33
Arsenic	8	<1	100	7	28	0[a]	>99	0	31
Asbestos (total), fibers/L	8	8×10^4	3.2×10^9	2.5×10^6	4.7×10^8	36	>99	>99	90
Chrysotile, fibers/L	3	1×10^5	1×10^9	3×10^6	3.3×10^8	>99	>99	>99	>99
Beryllium	4	1.2	2	1.6	1.6	0[a]	71	23	58
Cadmium	19	<1	97	5	16	0[a]	>99	55	44
Chromium	21	<4	320	34	65	0[a]	>99	31	41
Chromium (+3)	1				610				95
Chromium (+6)	3	10	20	20	17	0[a]	98	0	33
Copper	35	<4	4,500	32	210	0[a]	>99	35	42
Cyanide	11	10	260	20	47	0[a]	>99	0[a]	14
Lead	31	5	2,100	61	150	0[a]	>99	31	39
Mercury	8	0.3	2,900	0.6	380	0[a]	86	48	41
Nickel	16	<5	240	47	62	0[a]	>99	14	40
Selenium	6	≤1.0	100	41	48	0[a]	10	0	1.7
Silver	12	<5	77	9	21	0[a]	>83	0	17
Zinc	38	<10	5,900	150	550	0[a]	>99	43	43
Bis(2-ethylhexyl) phthalate	14	3.3	16,000	15	1,200	0[a]	98	38	46
Butyl benzyl phthalate	4	<0.03	19	6	7.8	0[a]	>99	49	49
Di-n-butyl phthalate	12	<0.02	9,300	2.8	1,900	0[a]	>99	5.5	16
Diethyl phthalate	5	<0.03	10,000	0.8	2,000	0[a]	>99	38	37
Dimethyl phthalate	1				<0.03				>98
Di-n-octyl phthalate	3	<0.9	4	2	2.3	50	>96	64	70
N-nitrosodiphenylamine	1				0.4				0[a]
2-Chlorophenol	1				2				0
2,4-Dichlorophenol	2	0.2	2		1.1	0[a]	67		34
2,4-Dimethylphenol	3	0.4	29	0.9	10	0[a]	0[a]	0[a]	0[a]
Pentachlorophenol	4	<0.4	12	10	8.1	0[a]	>87	0[a]	22
Phenol	10	<0.07	34,000	2.8	3,400	0[a]	>93	11	25
2,4,6-Trichlorophenol	1				69				80

(continued)

CONTROL TECHNOLOGY SUMMARY FOR FILTRATION (continued)

Pollutant	Number of data points	Effluent concentration				Removal efficiency, %			
		Minimum	Maximum	Median	Mean	Minimum	Maximum	Median	Mean
p-Chloro-m-cresol	2	0.3	0.6		0.45	0[a]	0[a]		0[a]
Benzene	6	0.5	200	<8.45	45	0[a]	>99	<15	28
Chlorobenzene	2	4.8	460		230	0[a]	0[a]		0[a]
1,2-Dichlorobenzene	3	<0.05	5.8	5.4	<3.8	0[a]	>94	55	>50
Ethylbenzene	6	<0.2	<10	<0.2	<2.1	33	>99	82	>75
Toluene	14	<0.1	200	2.7	30	0[a]	>99	17	31
1,2,4-Trichlorobenzene	1				94				37
Acenaphthene	1				0.6				73
Anthracene/phenanthrene	9	0.03	≤3,200	0.5	<400	0[a]	70	40	31
Benzo(a)pyrene	2	0.2	0.8		0.5	0[a]	0[a]		0[a]
Benzo(k)fluoranthene	1				0.1				0[a]
Fluoranthene	4	0.05	0.4	0.14	0.18	0[a]	50	29	27
Fluorene	1				10,000				0[a]
Naphthalene	3	0.9	65	<10	<25	0[a]	86	>70	52
Pyrene	3	0.1	0.3	0.3	0.23	0[a]	0	0[a]	0
Aroclor 1232, Aroclor 1242, Aroclor 1248, Aroclor 1260	1				480				16
Aroclor 1254	1				650				20
2-Chloronaphthalene	1				17				0[a]
Carbon tetrachloride	3	<10	55	30	<32	>37	93	89	>73
Chloroform	6	12	500	59	160	0[a]	50	0[a]	8
1,1-Dichloroethane	1				180				0[a]
1,2-Dichloroethane	1				170				0[a]
1,1-Dichloroethylene	1				<2				>52
1,2-Trans-dichloroethylene	1				47				0[a]
1,2-Dichloropropane	1				1.0				0[a]
Methylene chloride	15	<0.4	31,000	19	2,300	0[a]	>87	0[a]	15
1,1,2,2-Tetrachloroethane	2	0.7	0.9		0.8	0[a]	0[a]		0[a]
Tetrachloroethylene	7	1.0	210	21	49	0[a]	>99	1.5	30
1,1,1-Trichloroethane	4	$<10^6$	2,200	300	710	0[a]	94	>88	67
1,1,2-Trichloroethane	1				2,100				0[a]
Trichloroethylene	6	<0.5	140	3	<29	0[a]	>90	>37	34
Trichlorofluoromethane	2	5	12		8.5	0[a]	0[a]		0[a]
Acrolein	1				<100				>86
α-BHC	2	1.4	6		3.7	0[a]	77		39
β-BHC	1				55				21
Chlordane	1				24				37

Note: Blanks indicate data not applicable.

[a] Actual data indicate negative removal.

TREATMENT TECHNOLOGY: Filtration

Data source: Effluent Guidelines	Data source status:	
Point source category: Auto and other laundries	Engineering estimate	
Subcategory: Industrial laundries	Bench scale	
Plant: K	Pilot scale	
References: A28, Appendix C	Full scale	x

Use in system: Secondary
Pretreatment of influent: Screening equalization, dissolved air flotation (alum, polymer)

DESIGN OR OPERATING PARAMETERS

Unit configuration: Downflow multimedia filter
Media (top to bottom): Plastic chips, anthracite, sand, garnet, gravel
Bed depth - total:
Effective size of media:
Uniformity coefficient of media:
Wastewater flow: 45 m^3/d (12,000 gpd); 159 m^3/d design (42,000 gpd)
Filtration rate (Hydraulic loading):
Backwash rate:
Air scour rate:
Filter run length:
Terminal head loss:

REMOVAL DATA

Sampling Period—2 Days	..... Concentration		Percent
Pollutant/Parameter	Influent	Effluent	Removal
Conventional pollutants, mg/ℓ			
BOD_5	178	92	48
COD	2,110	1,080	49
TOC	544	286	47
TSS	742	71	90
Oil and grease	76	46	39
Total phenol	0.094	0.076	19
Total phosphorus	12.2	2.0	83
Toxic pollutants, μg/ℓ			
Antimony	2,300	1,800	22
Arsenic	3.5	≤1.0	≥71
Cadmium	40	9.5	76
Chromium	360	200	44
Copper	660	350	47
Cyanide	~10	12	0*
Lead	1,000	180	83
Mercury	1.0	<1.0	>0
Nickel	270	≤38	≥86
Selenium	≤1.0	≤1.0	≥0
Silver	66	52	21
Zinc	2,300	1,200	50
Bis(2-ethylhexyl) phthalate	90	98	0*
Butyl benzyl phthalate	81	<0.03	>99
Di-n-butyl phthalate	300	210	12
Di-n-octyl phthalate	21	<0.9	>96
Phenol	28	18	33
Ethylbenzene	3.0	2.0	33
Toluene	4.5	5.0	0*
Anthracene/phenanthrene	10	3.5	65
Naphthalene	11	65	86
2-Chloronaphthalene	16	17	0*
Carbon tetrachloride	410	30	93
Chloroform	~12	20	0*
Methylene chloride	8.0	113	0*
Tetrachloroethylene	<0.9	1.0	0*
1,1,1-Trichloroethane	860	54	94
Trichlorofluoromethane	<2.0	12	0*
Acrolein	720	<100	>86

*Actual data indicates negative removal.

Note: Blanks indicate information was not specified.

TREATMENT TECHNOLOGY: Filtration

Data source: Effluent Guidelines
Point source category: Auto and other laundries
Subcategory: Power laundries
Plant: N
References: A28, Appendix C

Data source status:
Engineering estimate ___
Bench scale ___
Pilot scale ___
Full scale x

Use in system: Tertiary
Pretreatment of influent: Screening, equalization, sedimentation with alum and polymer addition, carbon adsorption

DESIGN OR OPERATING PARAMETERS

Unit configuration:
Media (top to bottom):
Bed depth - total:
Effective size of media:
Uniformity coefficient of media:
Wastewater flow: 15.2 m^3/d (4,000 gpd)
Filtration rate (Hydraulic loading):
Backwash rate:
Air scour rate:
Filter run length:
Terminal head loss:

REMOVAL DATA

Sampling period: 2 days

Pollutant/parameter	Concentration Influent	Concentration Effluent	Percent removal
Conventional pollutants, mg/L:			
BOD_5	35.5	23	36
COD	136	59	57
TOC	38	21	45
TSS	78	37	53
Oil and grease	8	1	87
Total phenol	0.029	0.013	55
Total phosphorus	2.0	0.9	55
Toxic pollutants, µg/L:			
Antimony	44	<10	>77
Cadmium	15	14	7
Chromium	36	25	31
Copper	42	32	24
Lead	65	31	52
Nickel	<36	37	0[a]
Silver	7	7	0
Zinc	210	240	0[a]
Bis(2-ethylhexyl) phthalate	23	16	30
Butyl benzyl phthalate	17	4	76
Di-n-butyl phthalate	5	3	40
Diethyl phthalate	3	<0.03	>99
Di-n-octyl phthalate	4	2	50
Pentachlorophenol	3	<0.4	>87
Phenol	1	<0.07	>93
Toluene	4	6	0[a]
Chloroform	18	95	0[a]
Methylene chloride	3	<0.4	>87
1,1,2,2-Tetrachloroethane	<0.6	0.7	0[a]
Tetrachloroethylene	32	31	3
Trichloroethylene	5	3	40

[a]Actual data indicates negative removal.

Note: Blanks indicate information was not specified.

TREATMENT TECHNOLOGY: Filtration

Data source: Effluent Guidelines
Point source category: Auto and other laundries
Subcategory: Power laundries
Plant: J
References: A28, Appendix C

Data source status:
Engineering estimate ___
Bench scale ___
Pilot scale ___
Full scale x

Use in system: Secondary
Pretreatment of influent: Screening, equalization, dissolved air flotation with polymer addition

DESIGN OR OPERATING PARAMETERS

Unit configuration: Downflow, multimedia filter
Media (top to bottom):
Bed depth - total:
Effective size of media:
Uniformity coefficient of media:
Wastewater flow: 341 m^3/d (90,000 gpd) 379 m^3/d design (100,000 gpd)
Filtration rate (Hydraulic loading):
Backwash rate:
Air scour rate:
Filter run length:
Terminal head loss:

REMOVAL DATA

Sampling period: Two days

Pollutant/parameter	Concentration Influent	Concentration Effluent	Percent removal
Conventional pollutants, mg/L:			
BOD_5	142	118	17
COD	459	378	18
TOC	87	94	0[a]
TSS	32	40	0[a]
Oil and grease	16	33	0[a]
Total phenol	0.385	0.264	31
Total phosphorus	1.0	0.7	30
Toxic pollutants, µg/L:			
Antimony	64	<10	>84
Cadmium	5	<2	>60
Chromium	28	16	43
Copper	50	52	0[a]
Cyanide	25	11	56
Lead	70	<22	69
Nickel	63	<36	>43
Silver	29	<5	>83
Zinc	240	100	56
Bis(2-ethylhexyl) phthalate	74	54	27
Butyl benzyl phthalate	<0.03	8	0[a]
Di-n-butyl phthalate	<0.02	0.9	0[a]
Di-n-octyl phthalate	11	4	64
2-Chlorophenol	2	2	0
2,4-Dichlorophenol	6	2	67
2,4-Dimethylphenol	28	29	0[a]
Pentachlorophenol	8	10	0[a]
Phenol	9	7	22
Anthracene/Phenanthrene	2	2	0
Fluoranthene	0.5	0.4	20
Naphthalene	0.6	0.9	0[a]
Pyrene	0.3	0.3	0
Chloroform	24	12	50
Methylene chloride	22	520	0[a]
1,1,2,2-Tetrachloroethane	<0.6	0.9	0[a]
Tetrachloroethylene	2	2	0
Trichlorofluoromethane	<2	5	0[a]

[a]Actual data indicate negative removal.

Note: Blanks indicate information was not specified.

TREATMENT TECHNOLOGY: Filtration

Data source: Effluent Guidelines
Point source category: Electroplating
Subcategory:
Plant: 36041
References: A14, p. 187

Data source status:
Engineering estimate ___
Bench scale ___
Pilot scale ___
Full scale x

Use in system: Primary
Pretreatment of influent:

DESIGN OR OPERATING PARAMETERS

Unit configuration:
Media (top to bottom):
Bed depth - total:
Effective size of media:
Uniformity coefficient of media:
Filtration rate (Hydraulic loading):
Backwash rate:
Air scour rate:
Filter run length:
Terminal head loss:

REMOVAL DATA

Sampling period:

Pollutant/parameter	Concentration		Percent
	Influent	Effluent	removal
Conventional pollutants, mg/L:			
TSS	524	10	98
Toxic pollutants, µg/L:			
Chromium (+3)	12,000	610	95
Copper	7,500	440	94
Nickel	2,600	44	98
Zinc	13,000	140	99

Note: Blanks indicate information was not specified.

TREATMENT TECHNOLOGY: Filtration

Data source: Effluent Guidelines
Point source category: Electroplating
Subcategory:
Plant: 19066
References: A14, p. 203

Data source status:
- Engineering estimate ___
- Bench scale ___
- Pilot scale ___
- Full scale x

Use in system: Tertiary
Pretreatment of influent: Chemical reduction, pH adjustment

DESIGN OR OPERATING PARAMETERS

Unit configuration: Membrane filtration
Media (top to bottom):
Bed depth - total:
Effective size of media:
Uniformity coefficient of media:
Filtration rate (Hydraulic loading):
Backwash rate:
Air scour rate:
Filter run length:
Terminal head loss:

REMOVAL DATA

Sampling period:

Pollutant/parameter	Concentration Influent	Concentration Effluent	Percent removal
Conventional pollutants, mg/L:			
TSS	630	<0.01	>99
Toxic pollutants, µg/L:			
Cadmium	7	<5	>29
Chromium (+6)	460	10	98
Chromium (total)	4,100	18	>99
Copper	19,000	43	>99
Lead	650	<10	>98
Nickel	9,600	17	>99
Zinc	2,100	46	98

Note: Blanks indicate information was not specified.

TREATMENT TECHNOLOGY: Filtration[a]

Data source: Effluent Guidelines
Point source category: Foundry Industry
Subcategory: Aluminum Foundry - Die Lube Operation
Plant: 715C
References: A27, p. VII-1-13, VI-57-62, p. VII-27

Data source status:
Engineering estimate ___
Bench scale ___
Pilot scale ___
Full scale x

Use in system: Tertiary
Pretreatment of influent: Skimmer on holding tank, cyclone separator

[a]100% recycle, none of waste is discharged.

DESIGN OR OPERATING PARAMETERS

Unit configuration: Paper filter
Media (top to bottom):
Bed depth - total:
Effective size of media:
Uniformity coefficient of media:
Filtration rate (Hydraulic loading):
Backwash rate:
Air scour rate:
Filter run length:
Terminal head loss:

REMOVAL DATA

Sampling period:

Pollutant/parameters	Concentration		Percent removal
	Influent[a]	Effluent	
Conventional pollutants, mg/L:			
TSS	1,740	1,560	8
Oil and grease	8,500	9,940	0[b]
Total phenol	66.3	64.4	3
Toxic pollutants, µg/L:			
Cyanide	8	10	0[b]
Lead	2,000	2,100	0[b]
Zinc	1,600	1,500	6
Bis(2-ethylhexyl) phthalate	820,000	16,000	98
Di-n-butyl phthalate	5,400	9,300	0[b]
Diethyl phthalate	600	10,000	0[b]
Phenol	26,000	34,000	0[b]
2,4,6-Trichlorophenol	350	69	80
Benzene	84	50	40
Chlorobenzene	250	460	0[b]
Toluene	540	180	64
Anthracene	≤470	≤3,200	0[b]
Fluorene	32	10,000	0[b]
Phenathrene	≤470	≤3,200	0[b]
Aroclor 1232-Aroclor 1242-Aroclor 1248-Aroclor 1260	570	480	16
Aroclor 1254	810	650	20
Carbon tetrachloride	480	55	89
Chloroform	450	500	0[b]
Methylene chloride	2,400	2,500	0[b]
Tetrachloroethylene	160	210	0[b]
1,1,1-Trichloroethane	16,000	2,200	86
Trichloroethylene	280	140	50
α-BHC	26	6	77
β-BHC	70	55	21
Chlordane	38	24	37

[a]Influent concentration is the concentration in the raw waste.
[b]Actual data indicate negative removal.

Note: Blanks indicate information was not specified.

TREATMENT TECHNOLOGY: Filtration

Data source: Effluent Guidelines
Point source category: Inorganic chemicals
Subcategory: Chlorine-Diaphragm Cell plant
Plant: 261
References: A29, pp. 158-162

Data source status:
Engineering estimate ___
Bench scale ___
Pilot scale ___
Full scale x

Use in system: Primary
Pretreatment of influent:

DESIGN OR OPERATING PARAMETERS

Unit configuration:
Media (top to bottom):
Bed depth - total:
Effective size of media:
Uniformity coefficient of media:
Filtration rate (Hydraulic loading):
Backwash rate:
Air scour rate:
Filter run length:
Terminal head loss:

REMOVAL DATA

Sampling period: Three 24 hr composite samples

Pollutant/parameter	Concentration Influent[a]	Concentration Effluent	Percent removal
Conventional pollutants, mg/L:			
TSS	476	9	98
Toxic pollutants, µg/L:			
Lead	260,000	75	>99

[a]Influent concentration is calculated from flow in $m^3/kkgCl_2$ and pollutant load in $kg/kkgCl_2$.

Note: Blanks indicate information was not specified.

TREATMENT TECHNOLOGY: Filtration

Data source: Effluent Guidelines
Point source category: Inorganic chemicals
Subcategory: Chrome pigment
Plant: 894
References: A29, pp. 395-396

Data source status:
- Engineering estimate ___
- Bench scale ___
- Pilot scale ___
- Full scale x

Use in system: Secondary
Pretreatment of influent: Equalization, neutralization, sedimentation with chemical addition

DESIGN OR OPERATING PARAMETERS

Unit configuration: 2 sand filters
Media (top to bottom):
Bed depth - total:
Effective size of media:
Uniformity coefficient of media:
Filtration rate (Hydraulic loading):
Backwash rate:
Air scour rate:
Filter run length:
Terminal head loss:

REMOVAL DATA

Sampling period: Three 24 hr composite samples

Pollutant/parameter	Concentration		Percent
	Influent	Effluent	removal
Conventional pollutants, mg/L:			
TSS	780	3.9	>99
Toxic pollutants, μg/L:			
Antimony	740	300	59
Cadmium	900	8.4	99
Chromium	78,000	320	>99
Copper	3,600	40	99
Cyanide	5,100	<66	>99
Lead	15,000	110	99
Zinc	4,200	58	99

Note: Blanks indicate information was not specified.

TREATMENT TECHNOLOGY: Filtration

Data source: Effluent Guidelines
Point source category: Inorganic chemicals
Subcategory: Copper sulfate
Plant: 034
References: A29, pp. 501-502, 508

Data source status:
- Engineering estimate ___
- Bench scale ___
- Pilot scale ___
- Full scale x

Use in system: Primary
Pretreatment of influent:

DESIGN OR OPERATING PARAMETERS

Unit configuration: Pressure filter preceded by lime neutralization and coagulation
Media (top to bottom):
Bed depth - total:
Effective size of media:
Uniformity coefficient of media:
Filtration rate (Hydraulic loading):
Backwash rate:
Air scour rate:
Filter run length:
Terminal head loss:

REMOVAL DATA

Sampling period: 72 hr composite sample and three 24 hr composite samples

Pollutant/parameter	Concentration Influent[a]	Effluent	Percent removal
Conventional mg/L:			
TSS	38.6[b]	34.5[b]	11
Toxic pollutants, μg/L:			
Antimony	330	36	89
Arsenic	3,500	<20	>99
Cadmium	870	1	>99
Chromium	140	5	96
Copper	1,800,000[b]	4,500[b]	>99
Lead	180	5	97
Nickel	110,000[b]	240[b]	>99
Selenium	<11	100	0[c]
Zinc	11,000	16	>99
Phenol	18	12	33

[a]Infiltration of gound water into the collection sump was suspected at the time of sampling.

[b]Concentration is calculated from pollutant flow in m^3/kkg and pollutant loading in kg/kkg.

[c]Actual data indicate negative removal.

Note: Blanks indicate information was not specified.

TREATMENT TECHNOLOGY: Filtration

Data source: Effluent Guidelines
Point source category: Iron and steel
Subcategory: Continuous casting
Plant: AF
References: A33, p. 345

Data source status:
Engineering estimate ___
Bench scale ___
Pilot scale ___
Full scale x

Use in system: Secondary
Pretreatment of influent: Scale pit

DESIGN OR OPERATING PARAMETERS

Unit configuration: High flow rate pressure filters
Media (top to bottom):
Bed depth - total:
Effective size of media:
Uniformity coefficient of media: 113.6 L/s (1,800 gpm)
Filtration rate (Hydraulic loading):
Backwash rate: 176.7 L/s (2,800 gpm)
Air scour rate:
Filter run length:
Terminal head loss:

REMOVAL DATA

Sampling period:

Pollutant/parameter	Concentration		Percent
	Influent	Effluent	removal
Conventional pollutants, mg/L:			
TSS	110	22	80
Oil and grease	22	<0.5	>98
Toxic pollutants, μg/L:			
Copper	370	250	32
Zinc	2,600	1,600	38

Note: Blanks indicate information was not specified.

TREATMENT TECHNOLOGY: Filtration

Data source: Effluent Guidelines
Point source category: Iron and steel
Subcategory: Hot forming - primary
Plant: C-2
References: A35, p. 161

Data source status:
Engineering estimate ___
Bench scale ___
Pilot scale ___
Full scale _x_

Use in system: Primary
Pretreatment of influent:

DESIGN OR OPERATING PARAMETERS

Unit configuration: Deep bed filter
Media (top to bottom):
Bed depth - total:
Effective size of media:
Uniformity coefficient of media:
Filtration rate (Hydraulic loading):
Backwash rate:
Air scour rate:
Filter run length:
Terminal head loss:
Filter effluent flow rate: 145 L/s (2,300 gpm)

REMOVAL DATA

Sampling period:

Pollutant/parameter	Concentration, mg/L Influent	Concentration, mg/L Effluent	Percent removal
Conventional pollutants:			
TSS	26	5	81
Oil and grease	8.8	6.7	24

Note: Blanks indicate information was not specified.

TREATMENT TECHNOLOGY: Filtration

Data source: Effluent Guidelines
Point source category: Iron and steel
Subcategory: Vacuum degassing
Plant: AF
References: A33, p. 217

Data source status:
Engineering estimate ___
Bench scale ___
Pilot scale ___
Full scale x

Use in system: Secondary
Pretreatment of influent: Scale pit

DESIGN OR OPERATING PARAMETERS

Unit configuration: High flow rate pressure filters
Media (top to bottom):
Bed depth - total:
Effective size of media:
Uniformity coefficient of media:
Filtration rate (Hydraulic loading):
Backwash rate:
Air scour rate:
Filter run length:
Terminal head loss:
Flow rate: 114 L/sec (1,800 gpm)

REMOVAL DATA

Sampling period:

Pollutant/parameter	Concentration		Percent removal
	Influent	Effluent	
Conventional pollutants, mg/L:			
TSS	110	22	80
Toxic pollutants, µg/L:			
Lead	1,140	320	72
Zinc	8,700	1,600	82

Note: Blanks indicate information was not specified.

TREATMENT TECHNOLOGY: Filtration

Data source: Effluent Guidelines
Point source category: Nonferrous metals
Subcategory:
Plant:
References: A52, p. 340

Data source status:
- Engineering estimate ___
- Bench scale ___
- Pilot scale ___
- Full scale x

Use in system: Primary
Pretreatment of influent:

DESIGN OR OPERATING PARAMETERS

Unit configuration:
Media (top to bottom):
Bed depth - total:
Effective size of media:
Uniformity coefficient of media:
Filtration rate (Hydraulic loading):
Backwash rate:
Air scour rate:
Filter run length:
Terminal head loss:

REMOVAL DATA

Sampling period:

Pollutant/parameter	Concentration, µg/L Influent	Concentration, µg/L Effluent[a]	Percent removal
Toxic pollutants:			
Fluoranthene	0.08	0.05	38
Methylene chloride	46	37	20

[a]Calculate from influent and percent removal.

Note: Blanks indicate information was not specified.

TREATMENT TECHNOLOGY: Filtration

Data source: Effluent Guidelines
Point source category: Ore mining and dressing
Subcategory: Base metal mine
Plant: See below
References: A2

Data source status:
- Engineering estimate ___
- Bench scale ___
- Pilot scale x
- Full scale ___

Use in system: Tertiary unless otherwise specified
Pretreatment of influent: Sedimentation with lime and polymer addition, secondary settling (unless otherwise specified)

DESIGN OR OPERATING PARAMETERS

Unit configuration:
Media (top to bottom): Sand
Bed depth - total:
Effective size of media:
Uniformity coefficient of media:
Filtration rate (Hydraulic loading):
Backwash rate:
Air scour rate:
Filter run length:
Terminal head loss:

REMOVAL DATA

Sampling period:

	Copper			Lead			Zinc		
	Concentration, μg/L		Percent	Concentration, μg/L		Percent	Concentration, μg/L		Percent
Plant	Influent	Effluent	removal	Influent	Effluent	removal	Influent	Effluent	removal
Mine 1 of Canadian pilot plant study	40	30	25	210	150	29	290	390	0[a]
Mine 2 of Canadian pilot plant study	30	30	0	290	290	0	220	150	32
Mine 3 of Canadian pilot plant study	70	30	57	110	80	27	220	120	45
In Canada[b]	50	40	20	250	120	52	370	190	49

[a]Actual data indicate negative removal.

[b]Pretreatment of influent included lime addition, flocculation, clarification; use in system: secondary.

Note: Blanks indicate information was not specified.

TREATMENT TECHNOLOGY: Filtration

Data source: Effluent Guidelines
Point source category: Ore mining and dressing
Subcategory: Copper mill
Plant: 2122
References: A2, pp. VI-83-87

Data source status:
Engineering estimate ___
Bench scale ___
Pilot scale x
Full scale ___

Use in system: Secondary
Pretreatment of influent: Tailing pond

DESIGN OR OPERATING PARAMETERS

Unit configuration: Three dual media, downflow pressure filters
Media (top to bottom):
Bed depth - total:
Effective size of media:
Uniformity coefficient of media:
Filtration rate (Hydraulic loading):
Backwash rate:
Air scour rate:
Filter run length:
Terminal head loss:
pH: 7.9-8.2

REMOVAL DATA

Sampling period:

Pollutant/parameter	Concentration		Percent removal
	Influent[a]	Effluent[b]	
Conventional pollutants, mg/L:			
TSS	2,550	7.1	>99
Toxic pollutants, µg/L:			
Chromium	190	30	84
Copper	2,000	32	98
Lead	160	75	53
Nickel	190	50	74
Zinc	100	60	40

[a]Average concentration TSS (27 values), metals (23 values).
[b]Average concentration.

Note: Blanks indicate information was not specified.

TREATMENT TECHNOLOGY: Filtration

Data source: Effluent Guidelines
Point source category: Ore mining and dressing
Subcategory: Lead/zinc mine
Plant: 3113
References: A2

Data source status:
Engineering estimate ___
Bench scale ___
Pilot scale x
Full scale ___

Use in system: Secondary
Pretreatment of influent: See below

DESIGN OR OPERATING PARAMETERS

Unit configuration: Dual media filter
Media (top to bottom):
Bed depth - total:
Effective size of media:
Uniformity coefficient of media:
Filtration rate (Hydraulic loading):
Backwash rate:
Air scour rate:
Filter run length:
Terminal head loss:
pH: See below

REMOVAL DATA

Sampling period:

Pretreatment of influent	TSS Concentration, mg/L Influent	TSS Concentration, mg/L Effluent	TSS Percent removal	Cadmium Concentration, µg/L Influent	Cadmium Concentration, µg/L Effluent	Cadmium Percent removal	Copper Concentration, µg/L Influent	Copper Concentration, µg/L Effluent	Copper Percent removal
Sedimentation with lime and polymer, aeration, and flocculation[a]	15	<1	>93	5	<5	>0	20	13	35
Sedimentation with lime and polymer, aeration, and flocculation[b]	6	<1	>83	20	12	40	20	<10	>50
Sedimentation with lime addition, aeration[c]	35	1	97	20	20	75	110	20	82

Pretreatment of influent	Lead Concentration, µg/L Influent	Lead Concentration, µg/L Effluent	Lead Percent removal	Zinc Concentration, µg/L Influent	Zinc Concentration, µg/L Effluent	Zinc Percent removal
Sedimentation with lime and polymer, aeration, and flocculation[a]				670	27	96
Sedimentation with lime and polymer, aeration, and flocculation[b]	80	<20	>75	1,900	150	92
Sedimentation with lime addition, aeration[c]	20	<20	>0	4,100	150	96

[a]pH: 9.5.
[b]pH: 8.5.
[c]pH: 9.5.

Note: Blanks indicate information was not specified.

TREATMENT TECHNOLOGY: Filtration

Data source: Effluent Guidelines
Point source category: Ore mining and dressing
Subcategory: Lead/zinc mine/mill and molybdenum mine/mill
Plant: See below
References: A2

Data source status:
- Engineering estimate ___
- Bench scale ___
- Pilot scale x
- Full scale ___

Use in system: Tertiary
Pretreatment of influent: See below

DESIGN OR OPERATING PARAMETERS

Unit configuration:
Media (top to bottom):
Bed depth - total:
Effective size of media:
Uniformity coefficient of media:
Filtration rate (Hydraulic loading):
Backwash rate:
Air scour rate:
Filter run length:
Terminal head loss:
pH: See below

REMOVAL DATA

Sampling period:

		TSS			Cadmium			Copper		
		Concentration, mg/L		Percent	Concentration, µg/L		Percent	Concentration, µg/L		Percent
Plant	pH	Influent	Effluent	removal	Influent	Effluent	removal	Influent	Effluent	removal
3121[a]	9.2	17	1	94				50	20	60
3121[a]	11.3							30	20	33
3107[b]	3.1 - 3.7	16	<1	>93	120	35	71	31	16	48
6102[c]		62	≤5	≥92						

		Lead			Zinc		
		Concentration, µg/L		Percent	Concentration, µg/L		Percent
Plant	pH	Influent	Effluent	removal	Influent	Effluent	removal
3121[a]	9.2	80	40	50	380	160	58
3121[a]	11.3	50	60	0[d]	130	80	42
3107[b]	3.1 - 3.7	130	61	53	2,900	42	99
6102[c]					80	60	25

[a]Pretreatment of influent: tailing pond, lime and polymer addition, flocculation, secondary settling

[b]Pretreatment of influent: tailing pond, lime addition, aeration, flocculation, and clarification; unit configuration: dual media granular pressure filter.

[c]Pretreatment of influent: Sedimentation, ion exchange, lime precipitation, electrocoagulation, alkaline chlorination; unit configuration: four individual filters; anthracite, garnet, and pea gravel were used as media.

[d]Actual data indicate negative removal.

Note: Blanks indicate information was not specified.

TREATMENT TECHNOLOGY: Filtration

Data source: Effluent Guidelines
Point source category: Ore mining and dressing
Subcategory: See below
Plant: See below
References: A2, pp. VI-39, 41, 43

Data source status:
- Engineering estimate ___
- Bench scale ___
- Pilot scale[a] x
- Full scale ___

Use in system: Secondary unless otherwise specified
Pretreatment of influent: See below

[a]Unless otherwise specified

DESIGN OR OPERATING PARAMETERS

Unit configuration: See below
Media (top to bottom):
Bed depth - total:
Effective size of media:
Uniformity coefficient of media:
Filtration rate (Hydraulic loading):
Backwash rate:
Air scour rate:
Filter run length:
Terminal head loss:

REMOVAL DATA

Sampling period:

			Asbestos					
			Total			Chrysotile		
		Unit	Conc., fibers/L		Percent	Conc., fibers/L		Percent
Subcategory	Plant	configuration	Influent	Effluent	removal	Influent	Effluent	removal
Asbestos-cement processing plant[a]			5×10^9	3.2×10^9	36			
Asbestos mine[b]	In Baie Verte, Newfoundland	Alum-coated-diatomaceous earth filter	1×10^9	$<1 \times 10^5$	>99			
Asbestos mine[b]	In Baie Verte, Newfoundland	Dual media filter	1×10^{10}	5×10^8	95			
Asbestos mine[b]	In Baie Verte Newfoundland	Uncoated diatomaceous earth filter	1×10^9	2×10^6	>99			
Asbestos mine	In Asbestos, Quebec[c]	Mixed media filter	1×10^9	3×10^7	97			
Asbestos mine	In Asbestos, Quebec[c]	Coated diatomaceous earth	1×10^9	8×10^4	>99			
Asbestos mine	In Asbestos, Quebec[c]	Uncoated diatomaceous earth	1×10^9	3×10^6	>99			
Chlorine/caustic facility	In Michigan[d]	Pressure leaf filter used with flocculants	$>5 \times 10^9$	$\sim 3 \times 10^5$	>99			
-[b]		Alum-coated diatomaceous earth filter				4×10^{12}	1×10^5	>99
	-[b,e]	Dual media filtration				4×10^{12}	1×10^9	>99
	-[b]	Uncoated diatomaceous earth filter				4×10^{12}	3×10^6	>99

[a]Pretreatment of influent: Sedimentation (24 hr).
[b]Pretreatment of influent: Sedimentation.
[c]Use in system: tertiary.
[d]Use in system: tertiary; full scale; flow rate: 0.095 m^3/min.
[e]Bed depth: 34.3 cm (13.5 in.).

Note: Blanks indicate information was not specified.

TREATMENT TECHNOLOGY: Filtration

Data source: Effluent Guidelines
Point source category: Paint manufacturing
Subcategory:
Plant: 17
References: A4, Appendix G

Data source status:
Engineering estimate ___
Bench scale ___
Pilot scale ___
Full scale x

Use in system: Primary
Pretreatment of influent:

DESIGN OR OPERATING PARAMETERS

Unit configuration: Lime precoagulation
Media (top to bottom):
Bed depth - total:
Effective size of media:
Uniformity coefficient of media:
Filtration rate (Hydraulic loading):
Backwash rate:
Air scour rate:
Filter run length:
Terminal head loss:

REMOVAL DATA

Sampling period: Composite sample

Pollutant/parameter	Concentration[a] Influent	Effluent	Percent removal
Conventional pollutants, mg/L:			
BOD_5	6,370	5,870	8
COD	28,700	29,300	0[b]
TOC	7,100	8,130	0[b]
TSS	14,500	7,330	49
Oil and grease	1,000	1,140	0[b]
Total phenol	0.347	0.267	23
Toxic pollutants, µg/L:			
Antimony	40	<30	>25
Cadmium	~25	~30	0[b]
Chromium	130	130	0
Copper	530	370	31
Lead	100	300	0[b]
Mercury	20,000	2,900	86
Nickel	~67	80	0[b]
Silver	20	<10	>50
Thallium	22	<10	>55
Zinc	~9,200	18,000	0[b]
Di-n-butyl phthalate	ND[c]	1,300	0[b]
Benzene	1,300	ND	>99
Toluene	1,700	ND	>99
Naphthalene	33	ND	>70
Carbon tetrachloride	16	ND	>37
Chloroform	200	300	0[b]
1,1-Dichloroethane	ND	180	0[b]
1,2-Dichloroethane	ND	170	0[b]
1,2-*Trans*-dichloroethylene	ND	47	0[b]
Methylene chloride	15	ND	>33
Tetrachloroethylene	730	ND	>99
1,1,1-Trichloroethane	90	ND	>89
1,1,2-Trichloroethane	ND	2,100	0[b]
Trichloroethylene	100	ND	>90

[a]Average of several samples.
[b]Actual data indicate negative removal.
[c]Not detected

Note: Blanks indicate information was not specified.

TREATMENT TECHNOLOGY: Filtration

Data source: Effluent Guidelines
Point source category: Paint manufacturing
Subcategory:
Plant: 27
References: A4, Appendix G

Data source status:
Engineering estimate ___
Bench scale ___
Pilot scale ___
Full scale x

Use in system: Primary
Pretreatment of influent: None

DESIGN OR OPERATING PARAMETERS

Unit configuration: Polymer precoagulation
Media (top to bottom):
Bed depth - total:
Effective size of media:
Uniformity coefficient of media:
Filtration rate (Hydraulic loading):
Backwash rate:
Air scour rate:
Filter run length:
Terminal head loss:

REMOVAL DATA

Sampling period: Grab sample

Pollutant/parameter	Concentration		Percent removal
	Influent	Effluent	
Conventional pollutants, mg/L:			
BOD_5	25,000	23,400	6
COD	70,000	260,000	0[a]
TOC	7,500	25,000	0[a]
TSS	46,000	400	99
Total phenol	0.0012	0.0011	8
Toxic pollutants, μg/L:			
Beryllium	7	2	71
Cadmium	130	58	55
Chromium	1,400	100	93
Copper	260	120	56
Lead	12,000	98	99
Mercury	1,000	140	86
Nickel	450	<5	>99
Zinc	60,000	4,200	93
Benzene	280	200	29
Ethylbenzene	730	ND[b]	>99
Toluene	290	200	31
Chloroform	ND	23	0[a]
Methylene chloride	6,300	31,000	0[a]
Tetrachloroethylene	110	25	77
1,1,1-Trichloroethane	120	560	0[a]

[a]Actual data indicate negative removal.

[b]Not detected.

Note: Blanks indicate information was not specified.

TREATMENT TECHNOLOGY: Filtration

Data source: Effluent Guidelines
Point source category: Petroleum refining
Subcategory:
Plant: B
References: A3, pp. VI-36 to 42

Data source status:
Engineering estimate ___
Bench scale ___
Pilot scale x
Full scale ___

Use in system:
Pretreatment of influent: Dissolved air flotation plus unspecified secondary treatment

DESIGN OR OPERATING PARAMETERS

Unit configuration: Multimedia filter
Media (top to bottom):
Bed depth - total:
Effective size of media:
Uniformity coefficient of media:
Filtration rate (Hydraulic loading):
Backwash rate:
Air scour rate:
Filter run length:
Terminal head loss:

REMOVAL DATA

Sampling period: Average of three daily samples and a composite sample

Pollutant/parameter	Concentration		Percent removal
	Influent	Effluent	
Conventional pollutants, mg/L:			
COD	110	101	8
TOC	43	40	7
TSS	29	21	28
Oil and grease	8	8	0
Total phenol	0.024	0.022	8
Toxic pollutants, μg/L:			
Beryllium	2	2	0
Cadmium	3	<1	>67
Chromium	37	30	19
Chromium (+6)	20	20	0
Cyanide	50	50	0
Selenium	62	56	10
Zinc	25	65	0[a]

[a]Actual data indicate negative removal.

Note: Blanks indicate information was not specified.

TREATMENT TECHNOLOGY: Filtration

Data source: Effluent Guidelines
Point source category: Petroleum refining
Subcategory:
Plant: K
References: A3, pp. VI-36 to 42

Data source status:
Engineering estimate ___
Bench scale ___
Pilot scale x
Full scale ___

Use in system: Tertiary
Pretreatment of influent: Dissolved air flotation plus unspecified secondary treatment

DESIGN OR OPERATING PARAMETERS

Unit configuration: Multimedia filter
Media (top to bottom):
Bed depth - total:
Effective size of media:
Uniformity coefficient of media:
Filtration rate (Hydraulic loading):
Backwash rate:
Air scour rate:
Filter run length:
Terminal head loss:

REMOVAL DATA

Sampling period: Average of three daily samples and a composite sample

Pollutant/parameter	Concentration		Percent
	Influent	Effluent	removal
Conventional pollutants, mg/L:			
COD	135	56	59
TOC	43	22	49
TSS	50	4	92
Oil and grease	35	6	83
Total phenol	0.024	0.023	4
Toxic pollutants, μg/L:			
Chromium	200	34	83
Copper	28	7	75
Mercury	0.8	<0.5	>37
Zinc	200	92	55

Note: Blanks indicate information was not specified.

TREATMENT TECHNOLOGY: Filtration

Data source: Effluent Guidelines
Point source category: Petroleum refining
Subcategory:
Plant: H
References: A3, pp. VI-36 to 42

Data source status:
- Engineering estimate ___
- Bench scale ___
- Pilot scale x
- Full scale ___

Use in system: Tertiary
Pretreatment of influent: API-design oil separator plus unspecified secondary treatment

DESIGN OR OPERATING PARAMETERS

Unit configuration: Multimedia filter
Media (top to bottom):
Bed depth - total:
Effective size of media:
Uniformity coefficient of media:
Filtration rate (Hydraulic loading):
Backwash rate:
Air scour rate:
Filter run length:
Terminal head loss:

REMOVAL DATA

Sampling period: Average of three daily samples and a composite sample

Pollutant/parameter	Concentration		Percent removal
	Influent	Effluent	
Conventional pollutants, mg/L:			
COD	34	29	15
TOC	22	19	14
TSS	7	4	43
Oil and grease	10	8	20
Toxic pollutants, μg/L:			
Cadmium	5	<1	>80
Chromium	7	7	0
Chromium (+6)	<20	20	0[a]
Copper	21	12	43
Lead	17	23	0[a]
Zinc	15	20	0[a]

[a]Actual data indicate negative removal.

Note: Blanks indicate information was not specified.

TREATMENT TECHNOLOGY: Filtration

Data source: Effluent Guidelines
Point source category: Petroleum refining
Subcategory:
Plant: M
References: A3, pp. VI-36 to 42

Data source status:
- Engineering estimate ___
- Bench scale ___
- Pilot scale x
- Full scale ___

Use in system: Tertiary
Pretreatment of influent: Dissolved air flotation plus unspecified secondary treatment

DESIGN OR OPERATING PARAMETERS

Unit configuration: Multimedia filter
Media (top to bottom):
Bed depth - total:
Effective size of media:
Uniformity coefficient of media:
Filtration rate (Hydraulic loading):
Backwash rate:
Air scour rate:
Filter run length:
Terminal head loss:

REMOVAL DATA

Sampling period: Average of three 1-day composites and a 3-day composite sample

Pollutant/parameter	Concentration		Percent
	Influent	Effluent	removal
Conventional pollutants, mg/L:			
COD	107	55	49
TOC	18	17	6
TSS	9	3	67
Oil and grease	12	12	0
Toxic pollutants, μg/L:			
Cadmium	4	<1	>75
Chromium	62	48	23
Copper	12	7	42
Cyanide	40	42	0[a]
Lead	37	22	41
Mercury	0.8	<0.5	>37
Nickel	8	9	0[a]
Selenium	25	26	0[a]
Silver	5	5	0
Zinc	92	200	0[a]

[a]Actual data indicate negative removal.

Note: Blanks indicate information was not specified.

TREATMENT TECHNOLOGY: Filtration

Data source: Effluent Guidelines
Point source category: Petroleum refining
Subcategory:
Plant: O
References: A3, pp. VI-36 to 42

Data source status:
- Engineering estimate ___
- Bench scale ___
- Pilot scale _x_
- Full scale ___

Use in system: Tertiary
Pretreatment of influent: Dissolved air flotation plus unspecified secondary treatment

DESIGN OR OPERATING PARAMETERS

Unit configuration: Multimedia filter
Media (top to bottom):
Bed depth - total:
Effective size of media:
Uniformity coefficient of media:
Filtration rate (Hydraulic loading):
Backwash rate:
Air scour rate:
Filter run length:
Terminal head loss:

REMOVAL DATA

Sampling period: Average of three 1-day composites and a 3-day composite sample

Pollutant/parameter	Concentration		Percent removal
	Influent	Effluent	
Conventional pollutants, mg/L:			
BOD_5	11	19	0[a]
COD	125	120	4
TOC	38	44	0[a]
TSS	32	18	44
Oil and grease	18	11	39
Total phenol	0.028	0.032	0[a]
Toxic pollutants, µg/L:			
Chromium	70	60	14
Copper	9	7	22

[a]Actual data indicate negative removal.

Note: Blanks indicate information was not specified.

TREATMENT TECHNOLOGY: Filtration

Data source: Effluent Guidelines
Point source category: Petroleum refining
Subcategory:
Plant: P
References: A3, pp. VI-36 to 42

Data source status:
Engineering estimate ___
Bench scale ___
Pilot scale x
Full scale ___

Use in system: Tertiary
Pretreatment of influent: API-design gravity oil separator plus unspecified secondary treatment

DESIGN OR OPERATING PARAMETERS

Unit configuration: Multimedia filter
Media (top to bottom):
Bed depth - total:
Effective size of media:
Uniformity coefficient of media:
Filtration rate (Hydraulic loading):
Backwash rate:
Air scour rate:
Filter run length:
Terminal head loss:

REMOVAL DATA

Sampling period: Average of three 1-day composites and a 3-day composite sample

Pollutant/parameter	Concentration Influent	Concentration Effluent	Percent removal
Conventional pollutants, mg/L:			
BOD_5	12	13	0[a]
COD	100	130	0[a]
TOC	38	45	0[a]
TSS	17	14	18
Oil and grease	27	17	37
Total phenol	0.047	0.051	0[a]
Toxic pollutants, μg/L:			
Antimony	470	430	9
Cadmium	1	1	0
Chromium	32	27	16
Copper	9	8	11
Cyanide	45	42	7
Nickel	10	10	0
Zinc	17	30	0[a]

[a] Actual data indicate negative removal.

Note: Blanks indicate information was not specified.

TREATMENT TECHNOLOGY: Filtration

Data source: Effluent Guidelines
Point source category: Pulp, paper and paperboard
Subcategory: See below
Plant: See below
References: A26, p. VII-8

Data source status: See below
Engineering estimate ___
Bench scale ___
Pilot scale ___
Full scale ___

Use in system: Tertiary
Pretreatment of influent: Activated sludge unless otherwise specified

DESIGN OR OPERATING PARAMETERS

Unit configuration: See below
Media (top to bottom): See below
Bed depth - total: See below
Effective size of media:
Uniformity coefficient of media:
Filtration rate (Hydraulic loading): See below
Backwash rate:
Air scour rate:
Filter run length:
Terminal head loss:

REMOVAL DATA

Subcategory	Plant	Scale	Unit configuration	Media (top to bottom)	Bed depth-total	Filtration rate
Man-made fiber processing	A-4	Full	3 filters	Coal, sand, garnet	914 mm	0.0877 $m^3/min/m^2$ (2.15 gpm/ft^2)
Pulp mill	-[a]	Pilot		Coarse coal, medium sand, coarse sand	381 mm	0.10 - 0.47 $m^3/min/m^2$ (2.4 - 3.6 gpm/ft^2)
	A-1	Full	3 filters	Coal and sand	686 mm	0.13 $m^3/min/m^2$ (3.2 gpm/ft^2)

	TSS Concentration, mg/L Influent	Effluent	Percent removal
Man-made fiber processing	49.5	16.2	67
Pulp mill	40	21	48
	10.8	5.9	45

[a] Aerated lagoon was used in the pretreatment of influent.

Note: Blanks indicate information was not specified.

TREATMENT TECHNOLOGY: Filtration

Data source: Effluent Guidelines
Government report
Point source category: Textile mills
Subcategory: Knit fabric finishing
Plant: E, P (different references)
References: A6, pp. VII-74 to 75; B3, pp. 60-64

Data source status:

Engineering estimate	
Bench scale	
Pilot scale	x
Full scale	

Use in system: Tertiary
Pretreatment of influent: Screening, activated sludge, sedimentation with chemical addition (polymer)

DESIGN OR OPERATING PARAMETERS

Unit configuration: Downflow multimedia filter
Media (top to bottom): Anthracite, sand, gravel
Bed depth - total: 1,000 mm (40 in.)
anthracite: 300 mm (12 in.)
sand: 300 mm (12 in.)
gravel: 400 mm (16 in.)
Effective size of media: anthracite: 0.9-1.5 mm
sand: 0.4-0.8 mm
gravel: 6-16 mm
Uniformity coefficient of media:
Wastewater flow:
Filtration rate (Hydraulic loading):
Backwash rate:
Air scour rate:
Filter run length:
Terminal head loss:

REMOVAL DATA

Sampling period: 24-hr composite samples, volatile organics were grab sampled

Pollutant/parameter	Concentration Influent	Effluent	Percent removal
Conventional pollutants, mg/L:			
Total phenol	0.082	0.13	0[a]
Toxic pollutants, µg/L:			
Antimony	43	34	21
Mercury	<0.3	0.4	0[a]
Nickel	43	36	16
Zinc	160	160	0
Bis(2-ethylhexyl) phthalate	10	3.3	67
Di-n-butyl phthalate	2.8	2.5	11
Diethyl phthalate	0.03	1.0	0[a]
Phenol	0.5	2.6	0[a]
Benzene	0.4	0.5	0[a]
Toluene	0.4	2.6	0[a]
Anthracene/phenanthrene	0.9	0.5	44
Methylene chloride[b]	2.5	4.7	0[a]
Trichloroethylene	0.8	<0.5	>37

[a]Actual data indicate negative removal.
[b]Presence may be due to sample contamination.

Note: Blanks indicate information was not specified.

TREATMENT TECHNOLOGY: Filtration

Data source: Effluent Guidelines
Government report
Point source category: Textile mills
Subcategory: Knit fabric finishing
Plant: E, P (different references)
References: A6, pp. VII-74 to 75; B3, pp. 60-64

Data source status:

Engineering estimate	
Bench scale	
Pilot scale	x
Full scale	

Use in system: Tertiary
Pretreatment of influent: Screening, activated sludge

DESIGN OR OPERATING PARAMETERS

Unit configuration: Downflow multimedia filter
Media (top to bottom): Anthracite, sand, gravel
Bed depth - total: 1,000 mm (40 in.)
anthracite: 300 mm (12 in.)
sand: 300 mm (12 in.)
gravel: 400 mm (16 in.)
Effective size of media:
anthracite: 0.9-1.5 mm
sand: 0.4-0.8 mm
gravel: 6-16 mm
Uniformity coefficient of media:
Wastewater flow:
Filtration rate (Hydraulic loading):
Backwash rate:
Air scour rate:
Filter run length:
Terminal head loss:

REMOVAL DATA

Sampling Period: 24 hr Composite, Volatile Organics Were Grab Sampled

Pollutant/Parameter	Concentration Influent	Concentration Effluent	Percent Removal
Conventional pollutants, mg/ℓ			
Total phenol	0.072	0.068	6
Toxic pollutants, μg/ℓ			
Antimony	77	48	38
Chromium	98	<4	>96
Copper	36	<4	>89
Lead	25	<22	>12
Mercury	0.4	0.3	25
Nickel	66	58	12
Silver	<5	5	0*
Zinc	5,200	150	97
Bis(2-ethylhexyl) phthalate	10	3.9	61
Di-n-butyl phthalate	2.1	1.6	24
Diethyl phthalate	1.3	0.8	38
Phenol	0.7	1.8	0*
Benzene	<0.2	1.0	0*
Toluene	0.4	2.7	0*
Anthracene/phenanthrene	0.8	0.5	37
Methylene chloride**	0.4	4.1	0*

*Actual data indicate negative removal.
**Presence may be due to sample contamination.

Note: Blanks indicate information was not specified.

TREATMENT TECHNOLOGY: Filtration

Data source: Effluent Guidelines
Government report
Point source category: Textile mills
Subcategory: Wool finishing
Plant: O, N (different references)
References: A6, p. VII-76; B3, pp. 65-69

Data source status:

Engineering estimate	___
Bench scale	___
Pilot scale	x
Full scale	___

Use in system: Tertiary
Pretreatment of influent: Screening, activated sludge

DESIGN OR OPERATING PARAMETERS

Unit configuration: Downflow multimedia filter
Media (top to bottom): Anthracite, sand, gravel
Bed depth - total: 1,000 mm (40 in.)
anthracite: 300 mm (12 in.)
sand: 300 mm (12 in.)
gravel: 400 mm (16 in.)
Effective size of media:
anthracite: 0.9-1.5 mm
sand: 0.4-0.8 mm
gravel: 6-16 mm
Uniformity coefficient of media:
Wastewater flow:
Filtration rate (Hydraulic loading):
Backwash rate:
Air scour rate:
Filter run length:
Terminal head loss:

REMOVAL DATA

Sampling Period: 73 hr for Conventional Pollutants; 24 hr Composite Samples for the Toxic Pollutants Grab Samples for Volatile Organics

Pollutant/Parameter	.... Concentration..... Influent	Effluent	Percent Removal
Conventional pollutants, mg/ℓ			
COD	128	210	0*
TSS	75	<1	>99
Total phenol	0.031	0.017	45
Total phosphorus	2.5	2.3	8
Toxic pollutants, μg/ℓ			
Antimony	18	<10	>44
Arsenic	3	3	0
Chromium	170	95	44
Copper	14	130	0*
Silver	5.5	<5	>9
Zinc	1,300	590	55
Bis(2-ethylhexyl) phthalate	230	29	87
Di-n-butyl phthalate	0.6	1.1	0*
Diethyl phthalate	0.8	0.4	50
Dimethyl phthalate	1.4	<0.03	>98
1,2-Dichlorobenzene	0.9	<0.05	>94
Ethylbenzene	0.9	<0.2	>78
Toluene	0.4	0.6	0*
Anthracene/phenanthrene	0.4	0.5	0*
Fluoranthene	0.07	0.08	0*
Pyrene	0.1	0.1	0
1,2-Dichloropropane	<0.7	1.0	0*
Methylene chloride**	46	47	0*

*Actual data indicate negative removal.
**Presence may be due to sample contamination.

Note: Blanks indicate information was not specified.

TREATMENT TECHNOLOGY: Filtration

Data source: Effluent Guidelines
Point source category: Textile mills
Subcategory: Knit fabric finishing
Plant: Q
References: A6, p. VII-58

Data source status:
Engineering estimate ___
Bench scale ___
Pilot scale ___
Full scale x

Use in system: Tertiary
Pretreatment of influent: Screening, equalization, activated sludge

DESIGN OR OPERATING PARAMETERS

Unit configuration: Downflow multimedia pressure filter
Media (top to bottom):
Bed depth - total:
Effective size of media:
Uniformity coefficient of media:
Wastewater flow: 9,500 m^3/d (2.5 MGD)
Filtration rate (Hydraulic loading): 0.0012 $m^3/min/m^2$ (3.5 gpm/ft^2), design
Backwash rate:
Air scour rate:
Filter run length:
Terminal head loss:

REMOVAL DATA

Sampling period: Conventional pollutant influent is a 48-hr composite sample, toxic pollutant influent is an average of two 24-hr grab samples, effluents are the average of two, 24-hr composite samples

Pollutant/parameter	Concentration		Percent
	Influent	Effluent	removal
Conventional pollutants, mg/L:			
COD	312	233	25
TSS	28	6	79
Oil and grease	303	476	0[a]
Total phenol	0.059	0.048	19
Toxic pollutants, μg/L:			
Antimony	670	700	0[a]
Chromium	32	32	0
Copper	100	79	24
Cyanide	ND[b]	10	0[a]
Lead	48	33	31
Selenium	41	100	0[a]
Silver	13	8	38
Zinc	48	84	0[a]
Bis(2-ethylhexyl) phthalate	15	12	20
Tetrachloroethylene	17	17	0

[a]Actual data indicate negative removal.

[b]Not detected; assumed to be <10 μg/L.

Note: Blanks indicate information was not specified.

TREATMENT TECHNOLOGY: Filtration

Data source:	Effluent Guidelines Government report	Data source status:	
Point source category:	Textile mills	Engineering estimate	___
Subcategory:	Knit fabric finishing	Bench scale	___
Plant:	W, S (different references)	Pilot scale	x
References:	A6, p. VII-71; B3, pp. 55-59	Full scale	___

Use in system: Tertiary
Pretreatment of influent: Primary sedimentation equalization, activated sludge

DESIGN OR OPERATING PARAMETERS

Unit configuration: Downflow multimedia filter
Media (top to bottom): Anthracite, sand, gravel
Bed depth - total: 1,000 mm (40 in.)
anthracite: 300 (12 in.)
sand: 300 mm (12 in.)
gravel: 400 mm (16 in.)
Effective size of media:
anthracite: 0.9-1.5 mm
sand: 0.4-0.8 mm
gravel: 6-16 mm
Uniformity coefficient of media:
Wastewater flow: 0.03 m^3/min (7 gpm)
Filtration rate (hydraulic loading): 0.002 $m^3/min/m^2$ (7 gpm/ft^2)
Backwash rate:
Air scour rate:
Filter run length:
Terminal head loss:

REMOVAL DATA

Sampling Period: 24 hr Composite, Volatile Organics Were Grab Sampled

Pollutant/Parameter	Concentration Influent	Concentration Effluent	Percent Removal
Conventional pollutants, mg/ℓ			
BOD_5	4.6	2.4	48
COD	73	48	34
TOC	14	10	29
TSS	26	13	50
Total phenol	0.011	0.016	0*
Toxic pollutants, μg/ℓ			
Antimony	610	600	2
Arsenic	<10	11	0*
Cadmium	5	6	0*
Copper	26	24	8
Lead	75	85	0*
Mercury	1.7	0.7	59
Nickel	83	98	0*
Zinc	41	55	0*
Bis(2-ethylhexyl) phthalate	25	16	36
Di-n-butyl phthalate	2.8	<0.02	>99
2,4-Dichlorophenol	<0.1	0.2	0*
2,4-Dimethylphenol	<0.1	0.4	0*
Phenol	0.6	0.2	67
p-Chloro-m-cresol	<0.1	0.3	0*
Toluene	1.8	1.4	22
Acenaphthene	2.2	0.6	73
Methylene chloride**	12	7.9	34

*Actual data indicate negative removal.
**Presence may be due to sample contamination.

Note: Blanks indicate information was not specified.

TREATMENT TECHNOLOGY: Filtration

Data source: Effluent Guidelines
Government report
Point source category: Textile mills
Subcategory: Wool finishing
Plant: B, A
References: A6, pp. VII-64, 65

Data source status:

Engineering estimate	___
Bench scale	___
Pilot scale	x
Full scale	___

Use in system: Tertiary
Pretreatment of influent: Screening, equalization, activated sludge, sedimentation with chemical addition (alum, lime)

DESIGN OR OPERATING PARAMETERS

Unit configuration: Downflow multimedia filter
Media (top to bottom): Anthracite, sand, gravel
Bed depth - total: 1,000 mm (40 in.)
anthracite: 300 mm (12 in.)
sand: 300 mm (12 in.)
gravel: 400 mm (16 in.)
Effective size of media: anthracite: 0.9-1.5 mm
sand: 0.4-0.8 mm
gravel: 6-16 mm
Uniformity coefficient of media:
Wastewater flow: 0.020-0.027 m^3/min (5.4-7.0 gpm)
Filtration rate (Hydraulic loading): 0.002-0.0025 $m^3/min/m^2$ (5.4-7.0 gpm/ft^2)
Backwash rate:
Air scour rate:
Filter run length:
Terminal head loss:
Chemical dosage: 35 mg/L alum

REMOVAL DATA

Sampling Period: 24 hr for Priority Pollutants

Pollutant/Parameter	Concentration Influent	Concentration Effluent	Percent Removal
Conventional pollutants, mg/ℓ			
BOD_5	32	25	22
COD	212	184	13
TOC	72	60	17
TSS	28	12	57
Total phenol	0.047	0.055	0*
Toxic pollutants, μg/ℓ			
Arsenic	62	100	0*
Beryllium	<0.04	1.2	0*
Cadmium	<2	97	0*
Chromium	31	34	0*
Copper	13	110	0*
Cyanide	<4	10	0*
Lead	<22	79	0*
Zinc	5,700	5,900	0*
Bis(2-ethylhexyl) phthalate	44	14	68
N-nitrosodiphenylamine	<0.07	0.4	0*
2,4-Dimethylphenol	<0.1	0.9	0*
Pentachlorophenol	<0.4	10	0*
Phenol	3	3	0
1,2-Dichlorobenzene	<0.05	5.4	0*
Toluene	14	12	14
1,2,4-Trichlorobenzene	150	94	37
Benzo(a)pyrene	<0.02	0.8	0*
α-BHC	<1.0	1.9	0*

*Actual data indicate negative removal.

Note: Blanks indicate information was not specified.

TREATMENT TECHNOLOGY: Filtration

Data source: Effluent Guidelines
Government report
Point source category: Textile mills
Subcategory: Wool scouring
Plant: A, W (different references)
References: A6, p. VII-75; B3, pp. 50-54

Data source status:

Engineering estimate	___
Bench scale	___
Pilot scale	x
Full scale	___

Use in system: Tertiary
Pretreatment of influent: Grit removal, activated sludge, tertiary sedimentation

DESIGN OR OPERATING PARAMETERS

Unit configuration: Downflow multimedia filter
Media (top to bottom): Anthracite, sand, gravel
Bed depth - total: 1,000 mm (40 in.)
anthracite: 300 mm (12 in.)
sand: 300 mm (12 in.)
gravel: 400 mm (16 in.)
Effective size of media: anthracite: 0.9-1.5 mm
sand: 0.4-0.8 mm
gravel: 6-16 mm
Uniformity coefficient of media:
Wastewater flow:
Filtration rate (Hydraulic loading):
Backwash rate:
Air scour rate:
Filter run length:
Terminal head loss:

REMOVAL DATA

Sampling Period: 24 hr Composite, Volatile Organics Were Grab Sampled

Pollutant/Parameter	Concentration Influent	Concentration Effluent	Percent Removal
Conventional pollutants, mg/ℓ			
Total phenol	0.049	0.017	65
Toxic pollutants, μg/ℓ			
Arsenic	39	83	0*
Copper	110	120	0*
Cyanide	240	260	0*
Zinc	190	400	0*
Bis(2-ethylhexyl) phthalate	23	14	39
Ethylbenzene	3.0	<0.2	>93
Toluene	9.5	<0.1	>99
Anthracene/phenanthrene	0.4	0.2	50
Benzo(a)pyrene	<0.02	0.2	0*
Benzo(k)fluoranthene	<0.02	0.1	0*
Fluoranthene	0.4	0.2	50
Pyrene	0.2	0.3	0*
Methylene chloride**	2.2	4.8	0*

*Actual data indicate negative removal.
**Presence may be due to sample contamination.

Note: Blanks indicate information was not specified.

TREATMENT TECHNOLOGY: Filtration

Data source: Government report
Point source category: Textile mills
Subcategory: Woven fabric finishing
Plant: T
References: B3, pp. 76-82

Data source status:
- Engineering estimate ___
- Bench scale ___
- Pilot scale x
- Full scale ___

Use in system: Tertiary
Pretreatment of influent: Equalization, activated sludge

DESIGN OR OPERATING PARAMETERS

Unit configuration: Downflow multimedia filter
Media (top to bottom): Anthracite, sand, gravel
Bed depth - total: 1,000 mm (40 in.)
anthracite: 300 mm (12 in.)
sand: 300 mm (12 in.)
gravel: 400 mm (16 in.)
Effective size of media:
anthracite: 0.9-1.5 mm
sand: 0.4-0.8 mm
gravel: 6-16 mm
Uniformity coefficient of media:
Wastewater flow:
Filtration rate (Hydraulic loading):
Backwash rate:
Air scour rate:
Filter run length:
Terminal head loss:

REMOVAL DATA

Sampling period: 24-hr composite samples, volatile organics were grab sampled

Pollutant/parameter	Concentration Influent	Influent	Percent removal
Conventional pollutants, mg/L:			
COD	630	160	75
TSS	20	14	30
Total phenol	0.026	0.16	0[a]
Total phosphorus	14	13	7
Toxic pollutants, μg/L:			
Antimony	54	58	0[a]
Arsenic	3	3	0
Cadmium	2	<2	>0
Chromium	97	95	2
Copper	110	100	9
Cyanide	11	20	0[a]
Lead	22	26	0[a]
Nickel	93	100	0[a]
Selenium	2	2	0
Silver	23	32	0[a]
Zinc	150	97	35
Bis(2-ethylhexyl) phthalate	24	19	21
Butyl benzyl phthalate	5.2	2.5	52
Di-n-butyl phthalate	4.4	7.0	0[a]
Phenol	0.4	1.1	0[a]
p-Chloro-m-cresol	<0.1	0.6	0[a]
Benzene	5.7	6.9	0[a]
Chlorobenzene	4.1	4.8	0[a]
Ethylbenzene	0.5	0.2	60
Toluene	1.0	0.8	20
1,1-Dichloroethylene	4.2	<2.0	>52
Methylene chloride[b]	20	19	5

[a]Actual data indicate negative removal.
[b]Presence may be due to sample contamination.

Note: Blanks indicate information was not specified.

TREATMENT TECHNOLOGY: Filtration

Data source: Government report	Data source status:	
Point source category: Textile mills	Engineering estimate	___
Subcategory: Woven fabric finishing	Bench scale	___
Plant: V	Pilot scale	x
References: B3, pp. 70-75	Full scale	___

Use in system: Tertiary
Pretreatment of influent: Screening, activated sludge

DESIGN OR OPERATING PARAMETERS

Unit configuration: Downflow multimedia filter with $FeCl_3$ precoagulation (16 mg/L)
Media (top to bottom): Anthracite, sand, gravel
Bed depth - total: 1,000 mm (40 in.)
anthracite: 300 mm (12 in.)
sand: 300 mm (12 in.)
gravel: 400 mm (16 in.)
Effective size of media:
anthracite: 0.9-1.5 mm
sand: 0.4-0.8 mm
gravel: 6-16 mm
Uniformity coefficient of media:
Wastewater flow:
Filtration rate (Hydraulic loading):
Backwash rate:
Air scour rate:
Filter run length:
Terminal head loss:

REMOVAL DATA

Sampling Period: 24 hr Composite, Volatile Organics Were Grab Sampled

Pollutant/Parameter	Concentration Influent	Concentration Effluent	Percent Removal
Conventional pollutants, mg/ℓ			
COD	9.3	36	61
TSS	12	20	0*
Total phenol	0.029	0.022	24
Total phosphorus	1.2	0.23	81
Toxic pollutants, μg/ℓ			
Antimony	<10	24	0*
Arsenic	4	<1	>75
Chromium	4.3	6.7	0*
Copper	85	100	0*
Cyanide	23	27	0*
Lead	<22	37	0*
Nickel	<36	73	0*
Silver	<5	12	0*
Zinc	240	330	0*
Bis(2-ethylhexyl) phthalate	9.5	46	0*
Di-n-butyl phthalate	5.7	5.4	5
Toluene	1.1	1.1	0
Anthracene/phenanthrene	0.2	0.1	50
Methylene chloride**	24	14	42
Trichloroethylene	0.7	2.1	0*

*Actual data indicate negative removal.
**Presence may be due to sample contamination.

Note: Blanks indicate information was not specified.

TREATMENT TECHNOLOGY: Filtration

Data source: Effluent Guidelines
Government report
Point source category: Textile mills
Subcategory: Woven fabric finishing
Plant: V, C (different references)
References: A6, pp. VII-70-71; B3, pp. 45-49

Data source status:

Engineering estimate	___
Bench scale	___
Pilot scale	x
Full scale	___

Use in system: Tertiary
Pretreatment of influent: Screening, neutralization, activated sludge, sedimentation with chemical addition (alum)

DESIGN OR OPERATING PARAMETERS

Unit configuration: Downflow multimedia filter
Media (top to bottom): Anthracite, sand, gravel
Bed depth - total: 1,000 mm (40 in.)
anthracite: 300 mm (12 in.)
sand: 300 mm (12 in.)
gravel: 400 (16 in.)
Effective size of media: anthracite: 0.9-1.5 mm
sand: 0.4-0.8 mm
gravel: 6-16 mm
Uniformity coefficient of media: 0.12 $m^3/min/m^2$
Wastewater flow: 0.011 m^3/min (3.0 gpm)
Filtration rate (Hydraulic loading): 0.0011 $m^3/min/m^2$ (3.0 gpm/ft^2)
Backwash rate:
Air scour rate:
Filter run length:
Terminal head loss:
Chemical dosage: 40 mg/L alum

REMOVAL DATA

Sampling period: 24 hr; toxic pollutants were composite sampled, volatile organics were grab sampled

Pollutant/parameter	Concentration Influent	Concentration Effluent	Percent removal
Conventional pollutants, mg/L:			
BOD_5	3.6	2.5	31
COD	352	331	6
TOC	72	62	14
TSS	51	20	61
Total phenol	0.016	0.019	0[a]
Total phosphorus	2.3	2.0	13
Toxic pollutants, μg/L:			
Antimony	120	140	0[a]
Beryllium	2.2	1.2	45
Cadmium	2.9	2.7	7
Chromium	17	14	18
Copper	11	25	0[a]
Lead	66	64	3
Silver	72	77	0[a]
Zinc	190	230	0[a]
Bis(2-ethylhexyl) phthalate	33	5.3	84
Di-n-butyl phalate	0.6	0.6	0
Pentachlorophenol	<0.4	12	0[a]
1,2-Dichlorobenzene	13	5.8	55
Ethylbenzene	1.3	<0.2	>85
Toluene	1.0	<0.1	>90
Anthracene/phenanthrene	0.1	0.03	70
Methylene chloride[b]	70	210	0[a]

[a]Actual data indicate negative removal.
[b]Presence may be due to sample contamination.

Note: Blanks indicate information was not specified.

TREATMENT TECHNOLOGY: Filtration

Data source: Effluent Guidelines
Point source category: Textile mills
Subcategory: Woven fabric and knit fabric finishing
Plant: See below
References: A6, pp. IV-62, 66-68, 68-69

Data source status:
Engineering estimate ___
Bench scale ___
Pilot scale x
Full scale ___

Use in system: Tertiary
Pretreatment of influent: Screening, equalization, activated sludge (unless otherwise specified)

DESIGN OR OPERATING PARAMETERS

Unit configuration: Downflow multimedia filter
Media (top to bottom): Anthracite, sand, gravel
Bed depth - total: 1,000 mm (40 in.)
anthracite: 300 mm (12 in.)
sand: 300 mm (12 in.)
gravel: 400 mm (16 in.)
Effective size of media: anthracite: 0.9-1.5 mm
sand: 0.4-0.8 mm
gravel: 6-16 mm
Uniformity coefficient of media:
Wastewater flow: See below
Filtration rate (Hydraulic loading): See below
Backwash rate:
Air scour rate:
Filter run length:
Terminal head loss:

REMOVAL DATA

	BOD_5			COD		
	Concentration, mg/L		Percent	Concentration, mg/L		Percent
Plant	Influent	Effluent	removal	Influent	Effluent	removal
D[a]	24	19	21	814	630	23
P[b]	12	12	0	107	106	1
Q[c]	10	7	30	338	258	24
Q[d]	8.2	4	51	272	206	27

	TOC			TSS		
	Concentration, mg/L		Percent	Concentration, mg/L		Percent
Plant	Influent	Effluent	removal	Influent	Effluent	removal
D[a]	179	157	12	294	85	71
P[b]	27	25	1	63	18	71
Q[c]	18	18	0	77	28	64
Q[d]	27	22	19	46	4	91

[a]Filtration rate: 0.18 $m^3/min/m^2$; wastewater flow: 0.017 m^3/min; neutralization was used in pretreatment of influent in lieu of equalization.

[b]Filtration rate: 0.1-0.3 $m^3/min/m^2$; wastewater flow: 0.01-0.03 m^3/min; neutralization was also included in pretreatment of effluent.

[c]Filtration rate: 0.1 $m^3/min/m^2$; wastewater flow: 0.0095 m^3/min.

[d]Filtration rate: 0.08 $m^3/min/m^2$; wastewater flow: 0.0076 m^3/min.

Note: Blanks indicate information was not specified.

TREATMENT TECHNOLOGY: Filtration

Data source: Effluent Guidelines
Point source category: Textile mills
Subcategory: Woven fabric/stock yarn finishing
Plant: DD
References: A6, p. VII-63

Data source status:
Engineering estimate ___
Bench scale ___
Pilot scale x
Full scale ___

Use in system: Tertiary
Pretreatment of influent: Screening, neutralization, activated sludge

DESIGN OR OPERATING PARAMETERS

Unit configuration: Downflow multimedia filter with alum precoagulation (20 mg/L or Al^{+3})
Media (top to bottom): Anthracite, sand, gravel
Bed depth - total: 1,000 mm (40 in.)
anthracite: 300 mm (12 in.)
sand: 300 mm (12 in.)
gravel: 400 mm (16 in.)
Effective size of media: anthracite: 0.9-1.5 mm
sand: 0.4-0.8 mm
gravel: 6-16 mm
Uniformity coefficient of media:
Wastewater flow: 0.004-0.02 m^3/min (1-4 gpm)
Filtration rate (hydraulic loading): 0.0004-0.002 $m^3/min/m^2$ (1-4 gpm/ft^2)
Backwash rate:
Air scour rate:
Filter run length:
Terminal head loss:

REMOVAL DATA

Sampling period: 8 hr

Pollutant/parameter	Concentration, µg/L Influent	Concentration, µg/L Effluent	Percent removal
Toxic pollutants:			
Chromium	58	110	0[a]
Copper	59	28	53
Lead	37	31	16
Nickel	72	67	7
Silver	25	28	0[a]
Zinc	190	280	0[a]

[a]Actual data indicate negative removal.

Note: Blanks indicate information was not specified.

TREATMENT TECHNOLOGY: Filtration

Data source: Government report
Point source category:[a]
Subcategory:
Plant: Reichhold Chemical Inc.
References: B4, p. 57

Data source status:
Engineering estimate ___
Bench scale ___
Pilot scale x
Full scale ___

Use in system: Tertiary
Pretreatment of influent:

[a]Organic and inorganic waste

DESIGN OR OPERATING PARAMETERS

Unit configuration: Diameter - 50.8 mm (2 in.)
Media (top to bottom): Sand
Bed depth - total: Sand: 0.61 m (2 ft)
Effective size of media:
Uniformity coefficient of media:
Filtration rate (hydraulic loading): 7×10^{-5} $m^3/min/m^2$ (0.2 gpm/ft^2)
Backwash rate:
Air scour rate:
Filter run length:
Terminal head loss:

REMOVAL DATA

Sampling period: 24 hr composite

Pollutant/parameter	Concentration[a]		Percent
	Influent	Effluent	removal
Conventional pollutant, mg/L:			
COD	853	703	18

[a]Average of seven samples.

Note: Blanks indicate information was not specified.

4.7 ULTRAFILTRATION [1]

4.7.1 Function

Ultrafiltration is used to segregate dissolved or suspended solids from a liquid stream on the basis of molecular size.

4.7.2 Description

Ultrafiltration is a membrane filtration process that separates high-molecular-weight solutes or colloids from a solution or suspension. The process has been successfully applied to both homogeneous solutions and colloidal suspensions, which are difficult to separate practically by other techniques. To date, commercial applications have been entirely focused on aqueous media.

The basic principle of operation of ultrafiltration can be explained as follows. Flowing by a porous membrane is a solution containing two solutes: one of a molecular size too small to be retained by the membrane, and the other of a larger size allowing 100% retention. A hydrostatic pressure, typically 10 to 100 psig, is applied to the upstream side of the supported membrane, and the large-molecule solute or colloid is retained (rejected) by the membrane. A fluid concentrated in the retained solute is collected as a product from the upstream side, and a solution of small-molecule solute and solvent is collected from the downstream side of the membrane. Of course, where only a single solute is present and is rejected by the membrane, the liquid collected downstream is (ideally) pure solvent.

Retained solute (or particle) size is one characteristic distinquishing ultrafiltration from other filtration processes. Viewed on a spectrum of membrane separation processes, ultrafiltration is only one of a series of membrane methods that can be used. For example, reverse osmosis, a membrane process capable of separating dissolved ionic species from water, falls further down the same scale of separated partical size.

Ultrafiltration membranes are asymmetric structures, possessing an extremely thin selective layer (0.1 to 1.0 μm thick) supported on a thicker spongy substructure. Controlled variation of fabrication methods can produce membranes with desirable rentitive characteristics for a number of separation applications. It has become possible to tailor membranes with a wide range of selective properties. For example, tight membranes can retain organic solutes of 500 to 1,000 molecular weight while allowing passage of most inorganic salts; conversely, loose membranes can discriminate between solutes of 1,000,000 vs. 250,000 molecular weight.

Ultrafiltration membranes are different from so-called "solution-diffusion" membranes, which have been studied for a wide variety of gas and liquid-phase separations. The latter group possesses a permselective structure that is nonporous, and separation is effected on the basis of differences in solubility and molecular diffusivity within the actual polymer matrix. Reverse osmosis membranes generally fall into this category.

Membranes can be made from various synthetic or natural polymeric materials. These range from hydrophilic polymers such as cellulose, to very hydrophobic materials such as fluorinated polymers. Polyaryl sulfones and inorganic materials have been introduced to deal with high temperatures and pH values.

Membranes of this type are in many respects similar to reverse osmosis membranes except for the openness of their pores. Other forms and materials are available as well, including porous zirconia, deposited on a porous carbon substrate and on a porous ceramic tube. The latter two systems, while more expensive than the former, are capable of use at very high pH values and temperatures.

4.7.3 Technology Status

Ultrafiltration has demonstrated unique capabilities in oil/water separation, electropaint recovery, textile industry, and the dairy processing industry. It is certain that new applications will continue to be developed.

4.7.4 Applications

Can be used for 1) concentration, where the desired component is rejected by the membrane and taken off as a fluid concentrate; 2) fractionation, for systems where more than one solute are to be recovered, and products are taken from both the rejected concentrate and permeate; and 3) purification, where the desired product is purified solvent. Major existing ultrafiltration applications (commercial and developmental) are summarized below; the function of ultrafiltration processing for each specific application is also provided; developmental applications listed are likely to be commercial within the next 5 years.

COMMERCIAL APPLICATIONS OF ULTRAFILTRATIONS

Application	Function
Electrocoat	Fractionation
Paint rejuvenation and rinse water recovery	
Protein recovery from cheese whey	Concentration and fractionation
Metal machining, rolling, and drawing - oil emulsion treatment	Purification
Textile sizing (PVA) waste treatment	Fractionation
Electronics component	
Manufacturing wash water treatment	Purification
Pharmaceuticals manufacturing sterile water production	Purification

DEVELOPMENTAL APPLICATIONS OF ULTRAFILTRATION

Application	Function
Dye waste treatment	Concentration and purification
Pulp-mill waste treatment	Concentration and purification
Industrial laundry waste treatment	Purification and fractionation
Protein recovery from soy whey	Concentration
Hot alkaline cleaner treatment	Fractionation and purification
Power-plant boiler feedwater treatment	Purification
Sugar recovery from orange-juice pulp	Fractionation
Product recovery in pharmaceutical and fermentation industries	Concentration
Colloid-free water pollution for beverages	Purification

4.7.5 Limitations

Uniquely capable of making certain separations especially from concentrated streams; however, each installation must be carefully piloted as the system design and determination of operating parameters is critical.

4.7.6 Reliability

Process continually being refined; individual process reliability will depend on the specific application and past performance of process in that application.

4.7.7 Residuals Generated/Environmental Impact

Because ultrafiltration involves no chemical conversion, residues from process are typically a concentrate of the undesirable or hazardous components. The process generally serves to provide a greatly reduced volume of hazardous waste, but does not inherently provide any elimination of waste. Noteworthy exceptions are those cases where a pollutant can be recovered as a valuable by-product, such as soluble whey proteins or PVA. Otherwise, organic concentrates require further processing for ultimate disposal, such as additional concentration and incineration. In some fractionation applications, the concentrate and ultrafiltrate require further processing before end disposal occurs; for example, in cheese whey treatment, the lactose content of the ultrafiltrate is far too high to permit sewering, and additional processing steps must be taken before the stream is ready for disposal.

4.7.8 Design Criteria

Criteria	Units	Value/range
Pore size	µm	0.001-1.0
Flux	gal/ft²/d/Δpsi	0.5-1,000
Operating pressure	psi	60-200
Pressure drop	psi	5-30

Selection of membrane type configuration is dependent upon specific applications and results of pilot tests.

4.7.9 Flow Diagram

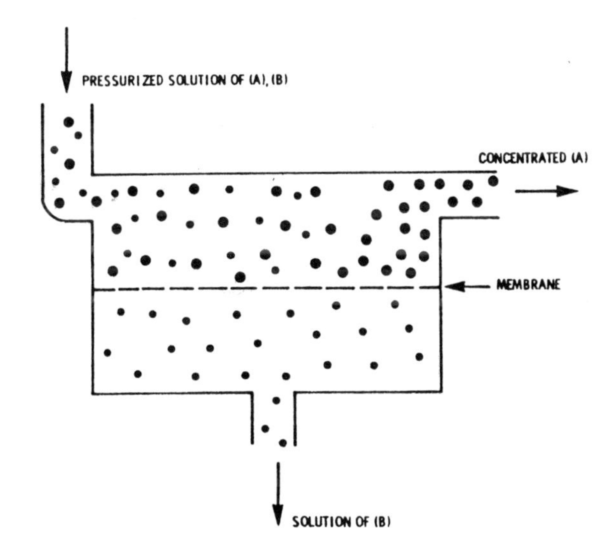

4.7.10 Performance

Subsequent data sheets provide performance data from studies on the following industries and/or wastestreams:

Adhesives and sealants production

Auto and other laundries industry
  Industrial laundries

Synthetic rubber manufacturing
  Emulsion crumb process
  Solution crumb process
  Styrene-butadiene latex production

Timber products processing
  Pentachlorophenol wastewater

4.7.11 References

1. Physical, Chemical, and Biological Treatment Techniques for Industrial Wastes, PB 275 287, U.S. Environmental Protection Agency, Washington, D.C. November 1976. pp. 43-1 - 43-12.

CONTROL TECHNOLOGY SUMMARY FOR ULTRAFILTRATION

Pollutant	Number of data points	Effluent concentration				Removal efficiency, %			
		Minimum	Maximum	Median	Mean	Minimum	Maximum	Median	Mean
Conventional pollutants, mg/L:									
BOD_5	12	12	8,890	457	2,850	0[a]	88	66	53
COD	11	444	36,600	830	9,130	12	99	72	57
TOC	15	66	939	246	366	15	97	79	63
TSS	12	2.4	539	54	105	60	>99	99	>92
Oil and grease	11	5	195	38	72	23	>99	96	>78
Total phenol	4	44.6	131	79.1	83.4	0[a]	82	32	37
Toxic pollutants, μg/L:									
Cadmium	3	<5	<10	<10	8.3	>67	>93	>90	>83
Chromium	1				2,900				67
Copper	3	<500	1,100	<500	<700	>58	90	>71	>73
Cyanide	1				5,000				0[a]
Lead	3	<1,000	<1,000	<1,000	<1,000	>52	>95	>74	>74
Mercury	2	0.4	0.8		0.6	11	20		16
Nickel	1				<500				>32
Zinc	6	180	40,000	<1,000	8,600	22	98	94	>78

[a]Actual data indicate negative removal.

TREATMENT TECHNOLOGY: Ultrafiltration

Data source: Government report
Point source category: Adhesives and sealants
Subcategory:
Plant: San Leandro
References: B10, pp. 62, 64

Data source status:
Engineering estimate ___
Bench scale ___
Pilot scale _x_
Full scale ___

Use in system: Secondary
Pretreatment of influent: Settling, equalization

DESIGN OR OPERATING PARAMETERS[a]

Product flow rate:
Flux rate:
Membrane configuration: 21 tubular assemblies, 3 parallel banks of seven tubes in series
Membrane type: See below
Retentate (concentrate) flow rate:
Recycle flow rate:
Operating temperature: 32.2°C
Rated production capacity:
Membrane inlet pressure: 280-340 kPa (40-50 psig)
Feed circulation rate: 164 m^3/d (30 gpm)
Tube diameter: 0.025 m (1 in.)
Tube length: 1.52 m (5 ft)

[a]Standard operating parameters for the study.

REMOVAL DATA

Sampling period:

Membrane type	BOD_5 Concentration, mg/L Influent	BOD_5 Concentration, mg/L Effluent	BOD_5 Percent removal	COD Concentration, mg/L Influent	COD Concentration, mg/L Effluent	COD Percent removal	TSS Concentration, mg/L Influent	TSS Concentration, mg/L Effluent	TSS Percent removal
Abcor, Inc., type HFD							2,470	10	>99
Abcor, Inc., type HFM							2,060	18	99
Abcor, Inc., type HFM or HFD[a]	8,820	7,180	19	21,200	18,200	14	1,590	66	96

Membrane type	Oil and grease Concentration, mg/L Influent	Oil and grease Concentration, mg/L Effluent	Oil and grease Percent removal	Total phenol Concentration, mg/L Influent	Total phenol Concentration, mg/L Effluent	Total phenol Percent removal
Abcor, Inc., type HFD						
Abcor, Inc., type HFM						
Abcor, Inc., type HFM or HFD[a]	252	195	23	113	131	0[b]

[a]Ultrafiltration with surfactant addition.
[b]Actual data indicate negative removal.

Note: Blanks indicate information was not specified.

TREATMENT TECHNOLOGY: Ultrafiltration

Data source: Government report
Point source category: Adhesives and sealants
Subcategory:
Plant: San Leandro
References: B10, pp. 62, 64

Data source status:
- Engineering estimate ___
- Bench scale ___
- Pilot scale x
- Full scale ___

Use in system: Secondary
Pretreatment of influent: Settling, equalization

DESIGN OR OPERATING PARAMETERS[a]

Product flow rate:
Flux rate:
Membrane configuration: 21 tubular assemblies, 3 parallel banks of seven tubes in series
Membrane type: Abcor, Inc., type HFM
Retentate (concentrate) flow rate:
Recycle flow rate:
Operating temperature: 32.2°C
Rated production capacity:
Membrane inlet pressure: 280-340 kPa(40-50 psig)
Feed circulation rate: 164 m^3/d (30 gpm)
Tube diameter: 0.025 m (1 in.)
Tube length: 1.52 m (5 ft)

[a]Standard operating parameters for the study.

REMOVAL DATA[a]

Sampling period:

Pollutant/parameter	Concentration Influent	Concentration Effluent	Percent removal
Conventional pollutants, mg/L:			
BOD_5	6,670	7,070	0[b]
COD	25,300	22,200	12
TSS	2,260	539	76
Oil and grease[c]	522	162	69
Total phenol[b]	84	56.1	33
Toxic pollutants, μg/L:			
Cyanide[c]	4,500	5,000	0[b]
Zinc	49,000	40,000	22

[a]Summary of 11 tests including tests with surfactants added and certain steams not included.
[b]Actual data indicate negative removal.
[c]Interference in assays suspected.

Note: Blanks indicate information was not specified.

TREATMENT TECHNOLOGY: Ultrafiltration[a]

Data source: Government report
Point source category: Adhesives and sealants
Subcategory:
Plant: San Leandro
References: B10, p. 69

Data source status:
Engineering estimate ___
Bench scale ___
Pilot scale x
Full scale ___

Use in system: Secondary
Pretreatment of influent: Settling, equalization

[a]With surfactant addition, summary of two tests.

DESIGN OR OPERATING PARAMETERS

Product flow rate:
Flux rate:
Membrane configuration: 21 tubular assemblies, 3 parallel banks of 7 tubes in series.
Membrane type: Abcor, Inc., type HFM or HFD
Retentate (concentrate) flow rate:
Recycle flow rate:
Operating temperature: 32.2°C
Rated production capacity:
Membrane inlet pressure: 280-340 kPa (40-50 psig)
Feed circulation rate: 164 m^3/d (30 gpm)
Tube diameter: 0.025 m (1 in)
Tube length: 1.52 m (5 ft)

REMOVAL DATA

Sampling period: Equal volume grab samples collected throughout an 8-hr day.

Pollutant/parameter	Concentration Influent	Concentration Effluent	Percent removal
Conventional pollutants, mg/L:			
BOD_5	8,700	8,570	1
COD	23,000	16,900	27
TSS	4,230	61.3	99
Oil and grease[a]	478	184	62
Total phenol[a]	148	102	31
Toxic pollutants, µg/L:			
Zinc	120,000	9,300[b]	92

[a]Interference in assay suspected.

[b]Excludes one reading of 1,100 mg/L.

Note: Blanks indicate information was not specified.

TREATMENT TECHNOLOGY: Ultrafiltration

Data source: Government report
Point source category: Adhesives and sealants
Subcategory:
Plant: San Leandro
References: B10, p. 67

Data source status:
Engineering estimate ___
Bench scale ___
Pilot scale x
Full scale ___

Use in system: Secondary
Pretreatment of influent: Settling, equalization

DESIGN OR OPERATING PARAMETERS

Product flow rate:
Flux rate: 1.38 $m^3/m^2/d$
Membrane configuration: 21 tubular assemblies, 3 parallel banks of 7 tubes in series.
Membrane type: Abcor, Inc., type HFM
Retentate (concentrate) flow rate:
Recycle flow rate:
Operating temperature: 32.2°C
Rated production capacity:
Membrane inlet pressure: 280-340 kPa (40-50 psig)
Feed circulation rate: 164 m^3/d (30 gpm)
Tube diameter: 0.025 m (1 in)
Tube length: 1.52 m (5 ft)

REMOVAL DATA

Sampling period: Equal volumes grab samples collected throughout an 8-hr day.

Pollutant/parameter	Concentration[a] Influent	Effluent	Percent removal
Conventional pollutants, mg/L:			
BOD_5	11,300	8,890	21
COD	56,100	36,600	35
TSS	13,400	<27.0[b]	>99
Oil and grease[c]	3,250	100	97
Total phenol[c]	244	44.6	82
Toxic pollutants, µg/L:			
Zinc	100,000	1,500[d]	98

[a]Average concentration for six tests with no additives.
[b]Most readings were <5 mg/L.
[c]Interferences in analysis suspected.

Note: Blanks indicate information was not specified.

TREATMENT TECHNOLOGY: Ultrafiltration

Data source: Government report
Point source category: Auto and other laundries
Subcategory: See below
Plant:
References: B9, p. 41

Data source status:
Engineering estimate ___
Bench scale ___
Pilot scale x
Full scale ___

Use in system: See below
Pretreatment of influent: Gravity oil separation or none listed

DESIGN OR OPERATING PARAMETERS

Product flow rate:
Flux rate: See below
Membrane configuration: See below
Membrane type: Abcor, Inc., type HFD or HFM
Retentate (concentrate) flow rate:
Recycle flow rate: Total recycle of permeate and concentrate
Operating temperature: 52°C (125°F)
Rated production capacity:
Average feed flow rate: See below
Average pressure drop: See below

REMOVAL DATA

Sampling period: See below

Subcategory	Flux rate, $m^3/d/m^2$ (gpd/ft^2)	Average feed flow rate, m^3/min (gpm)	Average pressure drop, kPa (psi)	TOC Concentration, mg/L		Percent removal
				Influent	Effluent	
Industrial laundries[a]	∿0.69(17)	0.17(45)	102(15)	2,510	409	84
Industrial laundries[b]	1.5(38)	0.23(60)	41(6)	2,510	371	85
Industrial laundries[c]	1.6(40)	0.23(90)	83(12)	34,500	939	97
Industrial laundries[d]	1.84(45)	0.26(95)	89(13)	34,500	918	97

[a]Lightly polluted industrial laundry wastewater; use in system: primary; membrane configuration: spiral wound corrugated; sampling period: sampled after 53 and 239 hr of operation.

[b]Lightly polluted industrial laundry wastewater; use in system: primary; membrane configuration: spiral wound, open mesh; sampling period: sampled after 53 and 239 hr of operation.

[c]Heavily polluted industrial laundry wastewater; use in system: secondary; membrane configuration: spiral wound, open spacer; sampling period: sampled after 19.4 and 242 hr.

[d]Heavily polluted industrial laundry wastewater; use in system: secondary; membrane configuration: spiral wound, corrugated spacer; sampling period: sampled after 19.4 and 242 hr.

Note: Blanks indicate information was not specified.

TREATMENT TECHNOLOGY: Ultrafiltration

Data source: Government report
Point source category: Auto and other laundries
Subcategory: Industrial laundries
Plant: Standard Uniform Rental Service
(Dorchester, Mass.)
References: B9, p. 70

Data source status:
Engineering estimate ___
Bench scale ___
Pilot scale x
Full scale ___

Use in system: Secondary
Pretreatment of influent: Foam filtration

DESIGN OR OPERATING PARAMETERS

Product flow rate:
Flux rate: 0.011-$m^3/d/m^2$ (3-40 gpd/ft^2)
Membrane configuration: Spiral wound
Membrane type: Abcor, Inc., type HFM/open mesh spacer, Vexar spacer, corrugated spacer
Retentate (concentrate) flow rate:
Recycle flow rate:
Operating temperature: 57°C, 135°F
Rated production capacity:
Feed flow rate: 3.4-17 m^3/d (900-4,500 gpd)
Inlet pressure: 324-345 kPa (47-50 psig)

REMOVAL DATA

Sampling period:

Pollutant/parameter	Concentration,[a] mg/L Influent	Effluent	Percent removal
Conventional pollutants:			
BOD_5	1,030	190	82
COD	2,960	350	88
TOC	760	120	84
TSS	480	<17	>96
Oil and grease	600	<9	>98

[a]Average of concentrations for six different conversion periods; 67-99%.

Note: Blanks indicate information was not specified.

TREATMENT TECHNOLOGY: Ultrafiltration

Data source: Government report
Point source category: Auto and other laundries
Subcategory: Industrial laundries[a]
Plant:
References: B9, p. 89

Data source status:
Engineering estimate ___
Bench scale ___
Pilot scale x
Full scale ___

Use in system: Primary
Pretreatment of influent:

[a] "Lightly polluted" industrial laundry laundry wastewater.

DESIGN OR OPERATING PARAMETERS

Product flow rate: 0.16-021 m^3/min (42-56 gpm)
Flux rate: 1.2-1.6 $m^3/d/m^2$ (30-40 gpd/ft^2)
Membrane configuration: Spiral wound
Membrane type: Abcor, Inc., types HFD/corrugated spacer and HFM/open spacer
Retentate (concentrate) flow rate: 0.01-0.02 m^3/min (3-4 gpm)
Recycle flow rate: All concentrate recycled
Operating temperature: 52°C (125°F)
Rated production capacity:
Feed flow rate: 0.17-0.23 m^3/m (45-60 gpm)
Pressure drop: 41-75 kPa (6-11 psi)
Water recovery: 92.8%
Inlet pressure: 410-440 kPa (45-50 psig)

REMOVAL DATA

Sampling period:

Pollutant/parameter	Concentration Influent	Effluent	Percent removal
Conventional pollutants, mg/L:			
BOD_5	2,800	360	87
COD	3,780	672	82
TOC	1,100	202	82
TSS	700	<4	>99
Oil and grease	749	27.7	96
Toxic pollutants, µg/L:			
Cadmium	50	<5	>90
Copper	1,700	<500	>71
Lead	3,900	<1,000	>74
Zinc	3,900	200	95

Note: Blanks indicate information was not specified.

TREATMENT TECHNOLOGY: Ultrafiltration

Data source: Government report
Point source category: Auto and other laundries
Subcategory: Industrial laundries[a]
Plant:
References: B9, p. 90

Data source status:
Engineering estimate ___
Bench scale ___
Pilot scale x
Full scale ___

Use in system: Primary
Pretreatment of influent:

[a]"Medium polluted" industrial laundry wastewater

DESIGN OR OPERATING PARAMETERS

Product flow rate: 0.23-0.26 m^3/m (60-70 gpm)
Flux rate: 1.2-1.4 $m^3/d/m^2$ (30-35 gpd/ft^2)
Membrane configuration: Spiral wound
Membrane type: Abcor, Inc., types HFD/corrugated spacer and HFM/open spacer
Retentate (concentrate) flow rate: 0.02 m^3/min (5 gpm)
Recycle flow rate: All concentrate recycled
Operating temperature: 52°C (125°F)
Rated production capacity:
Feed flow rate: 0.25-0.28 m^3/m (65-75 gpm)
Pressure drop: 61-68 kPa (9-10 psi)
Water recovery: 92.8%
Inlet pressure: 410-440 kPa (45-50 psig)

REMOVAL DATA

Sampling period:

Pollutant/parameter	Concentration Influent	Concentration Effluent	Percent removal
Conventional pollutants, mg/L:			
BOD_5	1,650	553	66
COD	5,480	796	86
TOC	1,240	196	84
TSS	675	2.4	>99
Oil and grease	795	10	99
Toxic pollutants, μg/L:			
Cadmium	30	<10	>67
Copper	1,200	<500	>58
Lead	2,100	<1,000	>52
Mercury	0.5	0.4	20
Zinc	1,400	<500	>64

Note: Blanks indicate information was not specified.

TREATMENT TECHNOLOGY: Ultrafiltration

Data source: Government report
Point source category: Auto and other laundries
Subcategory: Industrial laundries[a]
Plant:
References: B9, p. 91

Data source status:
Engineering estimate ___
Bench scale ___
Pilot scale x
Full scale ___

Use in system: Secondary
Pretreatment of influent: Gravity oil separation

[a]"Heavily polluted" industrial laundry wastewater

DESIGN OR OPERATING PARAMETERS

Product flow rate: 0.32-0.33 m^3/min (84-88 gpm)
Flux rate: 1.8-2.0 $m^3/d/m^2$ (45-50 gpd/ft^2)
Membrane configuration: Spiral wound
Membrane type: Abcor, Inc., types HFD/open-spacer and HFM/corrugated spacer
Retentate (concentrate) flow rate: 0.02-0.03 m^3/m (6-7 gpm)
Recycle flow rate: All concentrate recycled
Operating temperature: 52°C (125°F)
Rated production capacity:
Feed flow rate: 0.34-036 m^3/m (90-95 gpm)
Pressure drop:
Water recovery; 92.8%
Inlet pressure: 410-440 kPa (45-50 psig)

REMOVAL DATA

Pollutant/Parameter	Concentration Influent	Concentration Effluent	Percent Removal
Conventional pollutants, mg/ℓ			
BOD_5	7,850	930	88
COD	27,400	2,370	91
TOC	6,750	642	90
TSS	4,500	<5	>99
Oil and grease	7,890	38	>99
Toxic pollutants, μg/ℓ			
Cadmium	150	<10	>93
Chromium	8,800	2,900	67
Copper	11,000	1,100	90
Lead	22,000	<1,000	>95
Mercury	0.9	0.8	11
Nickel	740	<500	>32
Zinc	9,000	180	98

Note: Blanks indicate information was not specified.

TREATMENT TECHNOLOGY: Ultrafiltration

Data source: Government report
Point source category: Synthetic rubber manufacturing
Subcategory: See below
Plant: See below
References: B1, pp. 63, 68, 79, 122, 159

Data source status:
- Engineering estimate ___
- Bench scale ___
- Pilot scale[a] X
- Full scale ___

Use in system: Primary
Pretreatment of influent: Screening

[a]Unless otherwise specified.

DESIGN OR OPERATING PARAMETERS

Product flow rate:
Flux rate:
Membrane configuration: Tubular unless otherwise specified
Membrane type: Abcor, Inc., type HFM unless otherwise specified
Retentate (concentrate) flow rate:
Recycle flow rate:
Operating temperature: See below
Rated production capacity:
Membrane inlet pressure: 345 kPa
Feed Circulation rate: See below

REMOVAL DATA

Sampling period:

Subcategory	Operating temperature, °C	BOD_5 Concentration, mg/L Influent	BOD_5 Concentration, mg/L Effluent	BOD_5 Percent removal	COD Concentration, mg/L Influent	COD Concentration, mg/L Effluent	COD Percent removal	TOC Concentration, mg/L Influent	TOC Concentration, mg/L Effluent	TOC Percent removal
Emulsion crumb[a]	38	98	12	88	917	830	88	334	246	26
Latex[b]	50	100	47	53				320	66	79
Latex[c]	50	1,400	230	84	99,200	775	99			
Solution crumb[d]	38	86	30	65	625	444	29	144	122	15

Subcategory	TSS Concentration, mg/L Influent	TSS Concentration, mg/L Effluent	TSS Percent removal	Oil and grease Concentration, mg/L Influent	Oil and grease Concentration, mg/L Effluent	Oil and grease Percent removal
Emulsion crumb[a]	191	48	75	12	5	58
Latex[b]						
Latex[c]	23,800	222	99			
Solution crumb[e]				28[f]	11	61

(Operating temperature, °C for the second table: Emulsion crumb 38; Latex[b] 50; Latex[c] 50; Solution crumb 38.)

[a]Wastewater was adjusted with sulfuric acid to a pH of 4.0 before shipment in order to maintain sample integrity; membrane configuration: tubular and spiral; feed circulation rate: tubular model - 6.8 m^3/hr; spiral model - 22.7 m^3/hr.

[b]Plant: styrene-butadiene latex manufacturing; feed circulation rate: 7.9 to 8.4 m^3/hr.

[c]Plant: styrene-butadiene latex manufacturing; feed circulation rate: 7.9 to 8.4 m^3/hr; wastewater is 3.6% latex washwater, in full scale operation this would represent 70% to 90% of plant effluent; bench scale.

[d]Wastewater is from production of solution crumb rubbers, adhesives, and antioxidants. Approximately 70% of wastewater is attributed to solution crumb rubber manufacture, of this volume, two-thirds comes from the production of polyisoprene rubber. Feed circulation rate: 6.9 m^3/hr.

[e]Since the majority of production at the time of sampling was geared to "nonextended" rubbers, the relatively low oil and grease content in the sampled wastewater would be expected.

Note: Blanks indicate information was not specified.

TREATMENT TECHNOLOGY: Ultrafiltration

Data source: Government report
Point source category: Synthetic rubber manufacturing
Subcategory:
Plant:
References: B1, p. 159

Data source status:
Engineering estimate ___
Bench scale ___
Pilot scale x
Full scale ___

Use in system: Primary
Pretreatment of influent: See below

DESIGN OR OPERATING PARAMETERS

Product flow rate:
Flux rate:
Membrane configuration:
Membrane type: Abcor, Inc., type HFM unless otherwise specified
Retentate (concentrate) flow rate:
Recycle flow rate:
Operating temperature:
Rated production capacity:

REMOVAL DATA

	TOC		
	Concentration, mg/L		Percent
Pretreatment of influent	Influent	Effluent	removal
Screening[a]	649	385	41
-[a]	649	408	37
Screening[b]	266	198	26
-	266	186	31

[a]1% triton x-100 (a nonionic surfactant) was added.
[b]Membrane type used is Abcor, Inc., type HFA.

Note: Blanks indicate information was not specified.

TREATMENT TECHNOLOGY: Ultrafiltration

Data source: Effluent Guidelines
Point source category: Timber products (pentachlorophenol wastewater)
Subcategory:
Plant:
References: A1, p. E-3

Data source status:
Engineering estimate ___
Bench scale ___
Pilot scale x
Full scale ___

Use in system: Primary
Pretreatment of influent: None reported

DESIGN OR OPERATING PARAMETERS

Product flow rate:
Flux rate:
Membrane configuration:
Membrane type:
Retentate (concentrate) flow rate:
Recycle flow rate:
Operating temperature:
Rated production capacity:
Wastewater flow: 0.095 m^3/min (25 gpm)
Pressure: 331 kPa (48 psi)
Flux: 4,030·m^3/hr/m^2 (35 gpd/ft^2)
Water recovery: 96.2%

REMOVAL DATA

Sample period:

Pollutant/parameter	Concentration, mg/L Influent	Concentration, mg/L Effluent	Percent removal
Conventional pollutants:			
Oil and grease	2,160	55	97

Note: Blanks indicate information was not specified.

5. Secondary Wastewater Treatment

5.1 ACTIVATED SLUDGE [1]

5.1.1 Function

Activated sludge treatment is used to remove dissolved and collodial biodegradable organics.

5.1.2 Description

Conventional Activated Sludge. Activated sludge is a continous flow, biological treatment process characterized by a suspension of aerobic microorganisms, maintained in a relatively homogeneous state by the mixing and turbulence induced by aeration. The microorganisms are used to oxidize soluble and colloidal organics to CO_2 and H_2O in the presence of molecular oxygen. The process is generally, but not always, preceded by primary sedimentation. The mixture of microorganisms and wastewater (called mixed liquor) formed in the aeration basins is transferred to gravity clarifiers following treatment for liquid-solids separation. The major portion of the microorganisms settling out in the clarifiers is recycled to the aeration basins to be mixed with incoming wastewater, while the excess, which constitutes the waste sludge, is sent to the sludge handling facilities or lost as effluent solids. The rate and concentration of activated sludge returned to the aeration basins determines the mixed liquor suspended solids (MLSS) level developed and maintained in the basins. During the oxidation process, a certain amount of the organic material is synthesized into new cells, some of which then undergoes auto-oxidation (self-oxidation, or endogenous respiration) in the aeration basins, the remainder forming net growth or excess sludge. Oxygen is required in the process to support the oxidation and synthesis reactions. Volatile compounds are driven off to a certain extent in the aeration process. Metals will also be partially removed, with accumulation in the sludge.

Various aeration methods employed in transferring oxygen to wastewater are described below:

Diffused Aeration. In the conventional activated sludge plant, the municipal wastewater is commonly aerated for a period of four to eight hours (based on average daily flow) in a plug-flow hydraulic mode. Industrial wastes may require greater aeration times depending on waste strength. Long narrow aeration tanks are commonly employed to attain pseudo plug-flow conditions with a significant degree of backmixing occurring in each aeration basin. Diffusers are employed to transfer oxygen from air to wastewater. Compressors are used to supply air to the submerged systems, normally through a network of diffusers, although newer submerged devices which do not come under the general category of diffusers (e.g., static aerators and jet aerators) are being developed and applied. Diffused air systems may be classified fine bubble or coarse bubble. Diffusers commonly used in activated sludge service include porous cermaic plates laid in the basin bottom (fine bubble), porous ceramic domes or ceramic or plastic tubes connected to a pipe header and lateral system (fine bubble), tubes covered with synthetic fabric or wound filaments (fine or coarse bubble), and specially designed spargers with multiple openings (coarse bubble). The fine bubble units typically attain higher oxygen adsorption efficiencies and require a cleaner air supply to prevent clogging. With their higher efficiencies they use significantly less power than coarse bubble units.

Spiral roll aeration has normally been utilized in long narrow aeration tanks. This type of system which uses diffusers along one wall of the tank produces lower oxygen adsorption efficiencies than whole plane areation. In the latter type diffusers are installed over the whole tank bottom or headers are installed perpendicular to the tank wall instead of parallel.

Mechanical Aeration. Mechanical aeration methods include the submerged turbine with compressed air spargers (agitator/sparger system) and the surface-type mechanical entrainment aerators. The surface-type aerators entrain atmospheric air by producing a region of intense turbulence at the surface around their periphery. They are designed to pump large quantities of liquid, thus dispersing the entrained air and agitating and mixing the basin contents. The agitator/sparger system consists of a radial-flow turbine located below the mid-depth of the basin with compressed air supplied to the turbine through a sparger.

The submerged turbine aeration system affords a convenient and relatively economical method for upgrading overloaded activated sludge plants. To attain optimum flexibility of oxygen input, the surface aerator can be combined with the submerged turbine aerator. Several manufacturers supply such equipment, with both aerators mounted on the same vertical shaft. Such an arrangement might be advantageous if space limitations require the use of deep aeration basins. In addition, mechanical aerators may be either the floating or fixed installation type, using either high-speed small diameter units or low speed large diameter units with gear reducers. When utilizing surface aerators, both oxygen transfer and mixing requirements must be considered. The greater the separation distance betweeen aerators, the greater horsepower required to maintain solids suspension. In addition to the conventional activated sludge process, various modified ones are also commonly utilized in wastewater treatment.

Pure Oxygen (covered and uncovered). The use of pure oxygen for activated sludge treatment has become competitive with the use of air due to the development of efficient oxygen dissolution systems. The covered oxygen system can be a high-rate activated sludge system. The main benefits cited for the process include reduced power requirements for dissolving oxygen in the wastewater developed by Union Carbide, reduced areation tank requirements, and improved biokinetics of the activated sludge system. Lower excess sludge may be generated while the activated sludge thickening capability is generally greater than air activated sludge. In the covered "UNOX" system, oxygenation is performed in a staged, covered reactor. High-purity oxygen gas (90 to 100 percent volume/volume) either from direct on-site generation, off-site generation combined with pipeline delivery, or trucked-in and on-site stored liquid oxygen followed by vaporization enters the first stage of the system and flows concurrently with the wastewater being treated through the oxygenation basin. Gas is vented only from the last stage after approximately 90% of the oxygen has been utilized. Pressure under the tank covers is essentially atmospheric, being held at 2 to 4 inches water column, sufficient to maintain oxygen gas feed control and prevent backmixing from stage to stage. Effluent mixed liquor is separated in conventional gravity clarifiers, and the thickened sludge is recycled to the first stage for contact with influent wastewater.

Mass transfer and mixing within each stage are accomplished either with surface aerators or with a submerged-turbine rotating-sparge system. In the first case, mass transfer occurs at the gas/liquid interface; in the latter, oxygen is sparged into the mixed liquor where mass transfer occurs from the oxygen bubbles to the bulk liquid. In both cases, the mass-transfer process is enhanced by the high oxygen partial pressure maintained under the tank covers in each stage.

The UNOX and OASES processes are examples of patented and licensed systems, respectively, for pure oxygen activated sludge based on the description presented here.

Although flexibility is claimed to permit operation in any of the normally used flow regimes, i.e., plug-flow, complete mix, step aeration, and contact stabilization, the method of oxygen contact employed favors the plug-flow mode.

In the uncovered system, oxygenation is performed in an open reactor in which extremely fine porous diffusers are utilized to develop small oxygen gas bubbles that are completely dissolved before breaking surface in normal-depth tanks. The principles that apply in the transfer of oxygen in conventional diffused air systems also apply to the open-tank, pure-oxygen system.

The pure-oxygen, open-tank system currently available is the FMC system (formerly referred to as the "Marox" system) in which ultrafine bubbles are produced, with a correspondingly high gas-surface area. These ultrafine bubbles are of micron size, whereas "fine bubbles" normally produced in diffused air systems are in millimeter sizes. The complete oxygenation system is composed of an oxygen dissolution system comprised of rotating diffusers; a source of high-purity oxygen gas (normally, an on-site oxygen generator); and an oxygen control system, which balances oxygen supply with oxygen demand through use of basin-located dissolved-oxygen probes and control valves.

The influent to the system enters the oxygenation tank and is mixed with return activated sludge. The mixed liquor is continuously and thoroughly mixed using low-energy mechanical agitation near the tank bottom. Mixing is produced by radial turbine impellers located on both surfaces (top and bottom) of the rotating diffusion discs. Pure oxygen gas in the form of micron-size bubbles is simultaneously introduced into the tank to accomplish oxygen transfer. The rotating diffuser is a gear-driven disc-shaped device equipped with a porous medium to assist in the diffusion process. As the diffuser rotates at constant speed in the mixed liquor, hydraulic shear wipes bubbles from the medium before they have an opportunity to coalesce and enlarge.

High Rate Aeration. The term modified aeration has been adopted to apply to those high-rate air-activated sludge systems with design F/M loadings in the range of 0.75 to 1.5 lb BOD_5/d/lb MLVSS (mixed liquor volatile suspended solids). Modified aeration systems are characterized by low MLSS concentrations, short aeration detention times, high volumetric loadings, low air usage rates, and intermediate levels of BOD_5 and suspended solids removal efficiencies. Prior to enactment of nationwide secondary treatment regulations, modified aeration was utilized as an independent treatment system for plants where BOD_5 removals of 50 to 70 percent would suffice. With present-day treatment requirements, modified aeration no longer qualifies as a "stand-alone" activated sludge option.

Modified aeration basins are normally designed to operate in either complete-mix or plug-flow hydraulic configurations. Either surface or submerged aeration systems can be employed to transfer oxygen from air to wastewater, although submerged equipment is specified more frequently for this process. Compressors are used to supply air to submerged aeration systems.

Due to land limitations, interest has been recently expressed in the development of high-rate, diffused aeration systems that would produce a high quality secondary effluent. As with modified aeration, aeration detention times would remain low and volumetric loadings high. In contrast to modified aeration systems, high MLSS concentrations would have to be utilized to permit F/M loadings to be maintained at reasonable levels. The key to development of efficient high-rate air systems is the availability of submerged aeration equipment that could satisfy the high oxygen demand rates that accompany high MLSS levels and short aeration times. New innovations in fine bubble diffuser and jet aeration technology offer potential for uniting high-efficiency oxygen transfer with high-rate, air-activated sludge-flow regimes to achieve acceptable secondary treatment as independent "stand-alone" processes. Research evaluations and field studies currently underway should provide performance and cost data on this subject in the next several years.

Step Aeration. To balance the oxygen demand throughout the aeration tankage, step aeration is often practiced. This involves distributing the raw waste load to the 1/4 or 1/3 points in an aeration tank while return activated sludge is recycled to the head of the tank. Sequential dilution of the MLSS occurs after each waste addition. The oxygen demand is highest near the head of the plant and lowest near the effluent, however the gradation is much less severe than a conventional system. Thus the aeration equiment is normally spread uniformly throughout the aeration basin while in a conventional system it is often tapered.

Contact Stabilization. In this modification, the adsorptive capacity of the floc is utilized in the contact tank to adsorb suspended, colloidal, and some dissolved organics. The hydraulic detention time in the contact tank is only 30 to 60 minutes (based on average daily flow). After the biological sludge is separated from the wastewater in the secondary clarifier, the concentrated sludge is separately aerated in the stabilization tank with a detention time of 2 to 6 hours (based on sludge recycle flow). The adsorbed organics undergo oxidation in the stabilization tank and are synthesized into microbial cells. If the detention time is long enough in the stabilization tank, endogenous respiration will occur, along with a concomitant decrease in excess biological sludge production. Following stabilization, the reaerated sludge is mixed with incoming wastewater in the contact tank, and the cycle starts anew.

This process requires smaller total aeration volume than the conventional activated sludge process. It also can handle greater organic shock and toxic loadings because of the biological buffering capacity of the stabilization tank and the fact that at any given time the majority of the activated sludge is isolated from the main stream of the plant flow. Generally, the total aeration basin volume (contact plus stabilization basins) is only 50% to 75% of that required in the conventional activated sludge system.

Extended Aeration. Extended aeration is the "low-rate" modification of the activated sludge process. The F/M loading is in the range of 0.05 to 0.15 lb BOD_5/d/lb MLVSS, and the detention time is about 24 hours for municipal wastewater. Primary clarification is rarely used. The extended aeration system operates in the endogenous respiration phase of the bacterial growth cycle, because of the low BOD_5 loading. The organisms are starved and forced to undergo partial auto-oxidation.

In the complete mix version of the extended aeration process, all portions of the aeration basin are essentially homogeneous, resulting in a uniform oxygen demand throughout the aeration tank. This condition can be accomplished fairly simply in a symmetrical (square or circular) basin with a single mechanical aerator or by diffused aeration. The raw wastewater and return sludge enter at a point (e.g., under a mechanical aerator) where they are quickly dispersed through the basin. In rectangular basins with mechanical aerators or diffused air, the incoming waste and return sludge are distributed along one side of the basin, and the mixed liquor is withdrawn from the opposite side. The extended aeration activated-sludge process is commonly used to treat small wastewater flows from schools, housing developments, trailer parks, institutions, and small communities. This process can accept periodic (intermittent) loadings without becoming upset.

Oxidation Ditch. An oxidation ditch is an activated sludge biological treatment process, which is commonly operated in the extended aeration mode, although conventional activated sludge treatment is also possible. Typical oxidation ditch treatment systems consist of a single or closed loop channel, 4 to 6 feet deep, with 45° sloping sidewalls.

Some form of preliminary treatment such as screening, comminution or grit removal normally precedes the process. After pretreatment (primary clarification is usually not practiced) the wastewater is aerated in the ditch using mechanical aerators that are mounted across the channel. Horizontal brush, cage or disc-type aerators, as well as jet aerators, specially designed for oxidation ditch applications, are normally used. The aerators provide mixing and circulation in the ditch, as well as sufficient oxygen transfer. Mixing in the channels is uniform, but zones of low dissolved oxygen concentration can develop. Aerators operate in the 60 to 110 RPM range and provide sufficient velocity to maintain solids in suspension. A high degree of nitrification occurs in the process without special modification because of the long detention times and high solid retention times (10 to 50 days) utilized. Secondary settling of the aeration ditch effluent is provided in a separate clarifier.

Ditches may be constructed of various materials, including concrete, gunite, asphalt, or impervious membranes; concrete is the most common. Ditch loops may be oval or circular in shape. "Ell" and "horseshoe" configurations have been constructed to maximize land usage. Conventional activated sludge treatment, in contrast to extended aeration, may be practiced. Oxidation ditch systems with depths of 10 feet or more with vertical sidewalls and vertical shaft aerators may also be used.

5.1.3 Technology Status

Conventional Activated Sludge. Conventional activated sludge is the most versatile and widely used biological wastewater treatment process. Since 1950, the mechanical aeration method has been utilized more often, particularly in the industrial wastewater treatment field.

High Rate Aeration. High rate aeration was more widely used in the 1950's and 1960's than it is today, because of the less stringent effluent standards in effect during these periods.

Contact Stabilization. Contact stabilization has evolved as an outgrowth of activated sludge technology since 1950. The technology has seen common usage in package plants and some usage for on-site constructed plants.

Extended Aeration. Extended aeration plants have evolved since the latter part of the 1940's. Pre-engineered, package plants have been widely utilized for this process.

Oxidation Ditch. There are nearly 650 shallow oxidation ditch installations in the United States and Canada. Numerous shallow and deep oxidation ditch systems are in operation in Europe. The overall process is fully demonstrated for carbon removal, as a secondary treatment process.

5.1.4 Applications

Conventional Activated Sludge. Conventional activated sludge is employed in domestic and biodegradable industrial wastewater treatment. The main advantage of this process is the lower cost compared to that of other processes, especially when a high quality effluent is required. Industrial wastewater which is amenable to biological treatment and degradation may be jointly treated with domestic wastewater in a conventional activated sludge system.

High Rate Aeration. Since the early 1970's, employed generally as a pretreatment or roughing process in a two-stage activated sludge system, where the second stage is used for biological nitrification; alum or one of the iron salts is sometimes added to modified aeration basins preceding second-stage nitrification units for phosphorus removal.

Contact Stabilization. The contact stabilization activated sludge process is usually employed to treat wastewaters that have an appreciable amount of BOD_5. The main advantages of this process are the small aeration volume requirement and the ability to handle greater organic shock and toxic loadings compared to that of other processes.

Oxidation Ditch. Applicable in any situation where activated sludge treatment is appropriate; process cost of treatment is competitive with other biological processes in the range of wastewater flows between 0.1 and 10 Mgal/d.

5.1.5 Limitations

Conventional Activated Sludge. The conventional activated sludge process has a limited BOD_5 loading capacity and poor organic load distribution. It also requires an aeration time of four to eight hours. Plant can be upset with extreme variations in hydraulic, organic, and toxic loadings. Other disadvantages are high operating costs, operational complexity, and energy consumption. Maintenance is necessary for diffusers or mechanical aerators. When pure oxygen is utilized in the activated sludge process, the oxygen generation also makes a high contribution to the cost.

High Rate Aeration. High-rate activated sludge alone does not produce an effluent with BOD_5 and suspended solids concentrations suitable for discharge into most surface water in the United States.

Contact Stabilization. Unlikely that effluent standards can be met in plants smaller than 50,000 gal/d without some prior flow equalization; operational complexity; high operating costs; high energy consumption, high diffuser maintenance; fraction of soluble BOD_5 in the influent wastewater increases, the required total aeration volume of contact stabilization process approaches that of the conventional process.

Extended Aeration. High power costs, operation costs, and capital costs (for large permanent installations where pre-engineered plants would not be appropriate), and high land requirements. Periodic ditching of suspended solids requires sand filtration to consistently maintain high effluent quality.

Oxidation Ditch. Limitations of oxidation ditch are the same as extended aeration.

5.1.6 Residuals Generated

Conventional Activated Sludge. Anticipated increase in excess sludge, volatile suspended solids (VSS) production from the conventional activated sludge process as settled wasewater food-to-microorganism (F/M) loadings increase is shown below:

F/M	Excess VSS
0.3	0.3 - 0.6 lb/lb BOD_5 removed
0.5	0.4 - 0.7 lb/lb BOD_5 removed

The actual sludge production will be a function of the type of wastewater, the greater the colloidal fraction in the waste, the greater the sludge production.

High Rate Aeration. Same as reported for conventional activated sludge.

Contact Stabilization. Same as reported for conventional activated sludge.

Extended Aeration. Because of low F/M loadings and long hydraulic detention times employed, excess sludge production for the extended aeration process (and the closely related oxidation ditch process) is the lowest of any of the activated sludge process alternatives, generally in the range of 0.15 to 0.3 lb excess sludge suspended solids/lb BOD_5 removed at F/M of 0.1.

Oxidation Ditch. Sludge produced is less volatile due to higher oxidation efficiency and increased solids retention times.

5.1.7 Reliability

Conventional Activated Sludge. All activated sludge plants require close operator attention. Reliability is based on the capability of the secondary clarifiers to perform adequately. In-plant recycle streams from sludge handling processes often cause poor settling and "fines" discharged to the effluent.

High Rate Aeration. Requires close operator attention.

Contact Stabilization. Requires close operator attention.

Extended Aeration. Good.

Oxidation Ditch. Average reliability of 12 shallow oxidation ditch plants is summarized below:

	Percent of time effluent concentration less than					
	10 mg/L		20 mg/L		30 mg/L	
	TSS	BOD	TSS	BOD	TSS	BOD
Average of all plants	65	65	85	90	94	96

5.1.8 Environmental Impact

Conventional Activated Sludge. Sludge disposal; odor potential; and energy consumption.

High Rate Aeration. Same as conventional activated sludge.

Contact Stabilization. Same as conventional activated sludge.

Oxidation Ditch. Solid waste, odor and air pollution impacts are similar to those encountered with standard activated sludge processes.

5.1.9 Design Criteria

Conventional Activated Sludge (using diffused aeration or mechanical aeration method). A table of partial listing of design criteria for the conventional activated sludge process is shown below:

Criteria	Unit	Range/value
Volumetric loading	lb BOD_5/d/1,000 ft^3	25 - 50
Aeration detention time (based on average daily flow)	hr	4 - 8
MLSS	mg/L	1,500 - 3,000
F/M	lb BOD_5/d/lb MLSS	0.25 - 0.5
Air requirement	std ft^3/lb BOD_5 removed	800 - 1,500
Sludge retention time	d	5 - 10

High Rate Aeration. A partial listing of design criteria for two high rate activated sludge processes is as follows:

Criteria	Unit	Modified aerators	High solids, high rate aeration
Volumetric loading	lb BOD_5/d/1,000 ft^3	50 - 100	50 - 125
MLSS	mg/L	800 - 2,000	3,000 - 5,000
F/M	lb BOD_5/d/lb MLSS	0.75 - 1.5	0.4 - 0.8
Aeration detention time (based on influent flow)	hr	2 - 3	2 - 4
Air requirement	std ft^3/lb BOD_5 removed	400 - 800	800 - 1,200
Sludge retention time	d	0.75 - 2	2 - 5
Recycle ratio	-	0.25 - 1.0	0.25 - 0.5
Volatile fraction of MLSS	-	0.75 - 0.85	0.7 - 0.8

Contact Stabilization. A partial listing of design criteria for the contact stabilization process is shown below:

Criteria	Unit	Range/value	
Volumetric loading (based on contact and stabilization volume)	lb BOD_5/d/1,000 ft^3	30 - 50	
MLSS	mg/L	1,000 - 2,500	for contact tank
		4,000 - 10,000	for stabilization tank
F/M	lb BOD_5/d/lb MLSS	0.2 - 0.6	
Aeration time	hr	0.5 - 1.0	for contact tank (based on average daily flow)
		2 - 6	for stabilization tank (based on sludge recycle flow)
Sludge retention time	d	5 - 10	
Air requirement	std ft^3/lb BOD_5 removed	800 - 1,200	
Volatile fraction of MLSS	-	0.6 - 0.8	
Recycle ratio	-	0.25 - 1.0	

Extended Aeration. A partial listing of design criteria for extended aeration activated sludge process is shown below:

Criteria	Unit	Range/value
Volumetric loading	lb BOD_5/d/1,000 ft^3	5 - 10
MLSS	mg/L	3,000 - 6,000
F/M	lb BOD_5/d/lb MLSS	0.05 - 0.15
Aeration time (based on average daily flow)	hr	18 - 36
Air requirement	std ft^3/lb BOD_5 applied	3,000 - 4,000
Sludge retention time	d	20 - 40
Recycle ratio	-	0.75 - 1.5
Volatile fraction of MLSS	-	0.6 - 0.7

Oxidation Ditch.

Criteria/factor	Unit	Range/value
BOD_5 loading	lb BOD_5/1,000 ft^3 of aeration volume/d	10 - 15
Sludge age	d	10 - 50
Channel dept	ft	4 - 6
Channel geometry	-	45 degree or vertical walls
Aeration channel detention time	d	1

5.1.10 Flow Diagram

Conventional Activated Sludge Using Diffused Aeration.

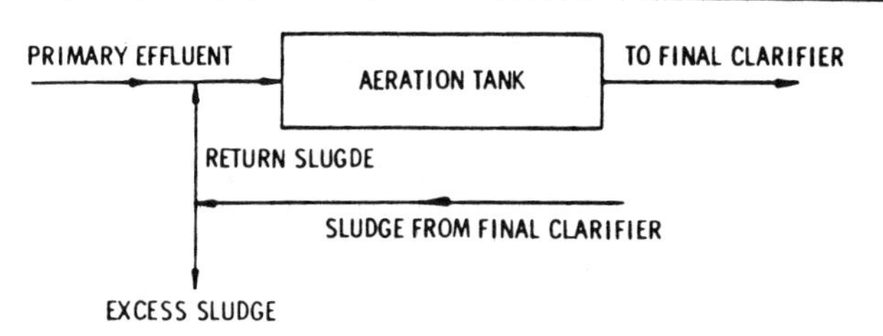

Conventional Activated Sludge Using Mechanical Aeration. See Diffused Aeration for typical flow diagram.

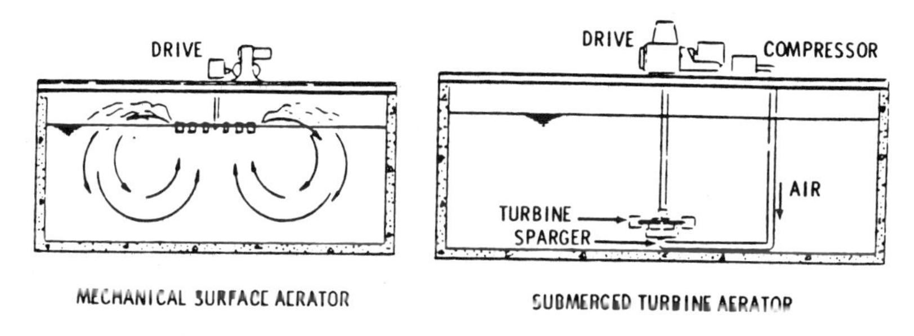

MECHANICAL SURFACE AERATOR SUBMERGED TURBINE AERATOR

Activated Sludge Using Pure Oxygen (covered).

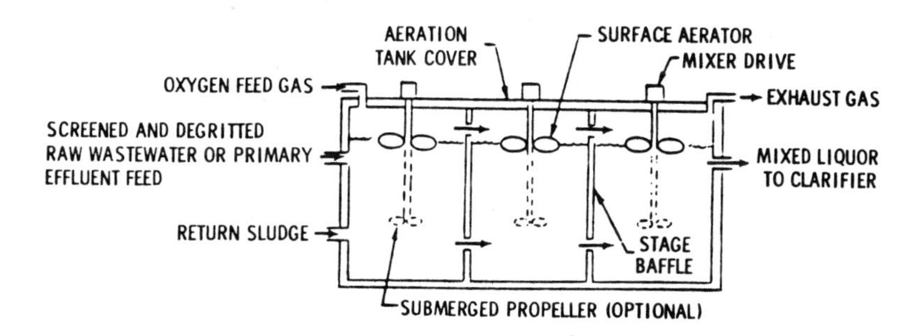

Activated Sludge Using Pure Oxygen (uncovered).

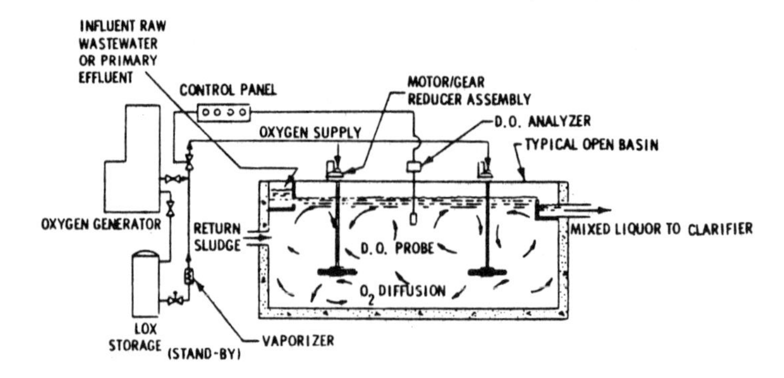

High Rate Aeration.

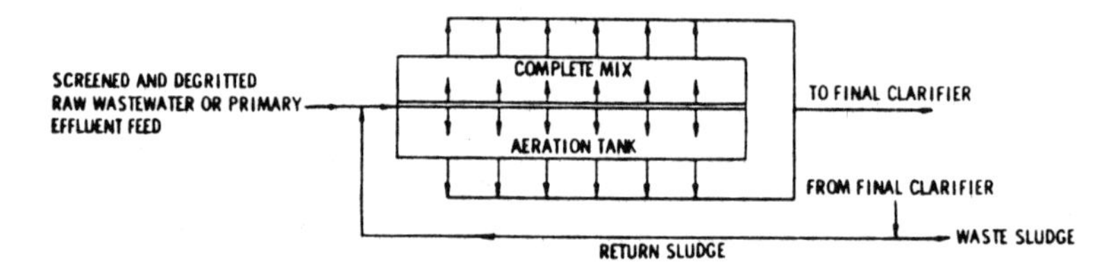

Contact Stabilization.

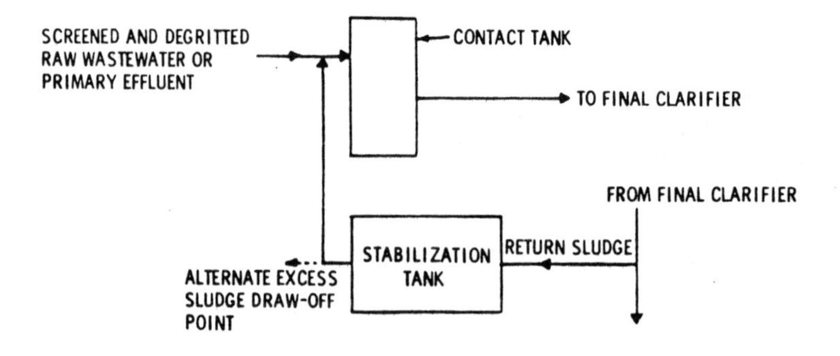

Extended Aeration.

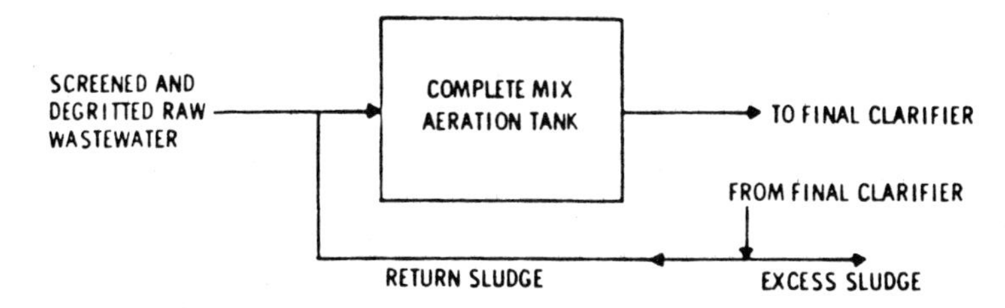

Oxidation Ditch.

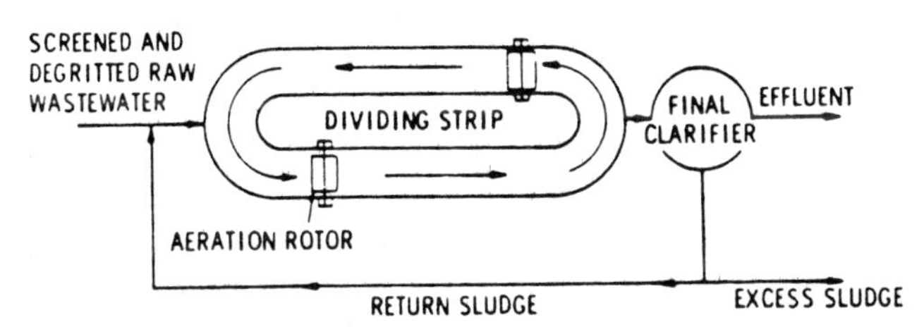

5.1.11 Performance

Performance data presented on the following data sheets include information from studies on the listed industries or wastestreams:

Coal gas washing

Coal-tar distillation

Coke gasification

Hospital wastewater

Iron and steel industry
  By-product coke manufacturing

Leather tanning and finishing
  Cattle, hair save, chrome tanning
  Cattle, hair pulp, chrome tanning
  Cattle, hair pulp, combination tanning
  Hair save, chrome tanning, retanning - wet finishing
  Hair save, nonchrome (primarily vegetable) tanning, retanning - wet finishing
  Shearling

Municipal wastewater
  Mixed industrial and domestic wastewaters

Organic chemicals production
  Aqueous liquid-phase reaction systems
  Batch and semicontinuous process
  Processes with process water contact as steam diluent or absorbent
  Organosilicones production

Pharmaceuticals production
  Biological and natural products
  Chemical synthesis products
  Fermentation products
  Formulation products
  Miscellaneous pharmaceuticals and fine organic chemicals

Pulp, paper, and paperboard production
  Sulfite - papergrade
  Wastepaper - board

Rubber processing

Synthetic resin production
  Cellophane
  Cellulosics

Textile mills
  Carpet finishing
  Knit fabric finishing
  Stock and yarn finishing
  Woven fabric finishing

Timber products processing
  Hardboard processing
  Plywood processing
  Wood preserving

5.1.12 References

1. Innovative and Alternative Technology Assessment Manual. EPA-430/9-78-009 (Draft), U.S. Environmental Protection Agency, Cincinnati, Ohio, 1978. 252 pp.

2. Echendelder, W. W. "Industrial Water Pollution Control," McGraw-Hill Book Company, 1966.

CONTROL TECHNOLOGY SUMMARY FOR ACTIVATED SLUDGE

Pollutant	Number of data points	Effluent concentration				Removal efficiency, %			
		Minimum	Maximum	Median	Mean	Minimum	Maximum	Median	Mean
Conventional pollutants, mg/L:									
BOD_5	92	5	4,640	49	184	17	>99	91	86
COD	84	45	7,420	425	890	0	96	67	63
TOC	14	35	1,700	280	427	8	95	69	63
TSS	74	6	4,050	92	283	0[a]	96	25	34
Oil and grease	7	<5	303	25	70.8	6	>98	92	74
Total phenol	31	0.007	500	0.028	18.7	0	>99	64	60
TKN	6	27	322	174	174	26	63	44	43
Total phosphorus	27	0.14	46.8	3.46	6.7	0	97	27	32
Toxic pollutants, μg/L:									
Antimony	18	0.3	670	3.5	46	0[a]	90	15	30
Arsenic	8	<5	160	13	48	0	96	39	>43
Cadmium	17	<0.5	13	4	4	0[a]	>99	0	>31
Chromium	34	<0.2	20,000	28	910	0[a]	99	48	45
Copper	37	<0.2	170	30	43	0[a]	>99	56	52
Cyanide	24	<1	38,000	28	2,000	0[a]	>90	0[a]	>18
Lead	26	0.6	160	61	38	0[a]	99	50	>49
Mercury	9	<0.5	1.6	0.7	<0.98	0[a]	>97	>29	>31
Nickel	32	4	400	40	78	0[a]	92	7	>29
Selenium	1				41				0[a]
Silver	17	<5	95	33	32	0[a]	>96	20	31
Thallium	1				29				38
Zinc	36	48	150,000	180	5,800	0[a]	92	27	35
Bis(chloromethyl) ether	1				<10				>83
Bis(2-chloroethyl) ether	1				<10				>47
4-Bromophenyl phenyl ether	1				18				95
Bis(2-ethylhexyl) phthalate	38	<0.04	1,300	13	64				
Butyl benzyl phthalate	1				11				0
Di-n-butyl phthalate	9	<0.02	58	3.6	9.2	0[a]	>99	>84	>60
Diethyl phthalate	17	<0.03	69	<0.03	6.6	0[a]	>99	>99	>56
Dimethyl phthalate	9	<0.03	200	<0.03	24	0[a]	>99	>99	>60
Di-n-octyl phthalate	1				5,000				0
Benzidine	1				4				0
1,2-Diphenylhydrazine	1				340				0[a]
N-nitrosodiphenylamine	2	<0.07	1.6		0.84	69	>99		>84
N-nitroso-di-n-propylamine	2	2	19		11				0[a]
2-Chlorophenol	2	0.9	10		5.5	0	92		46
2,4-Dichlorophenol	2	<10	<10		<10	>0	>50		>25
2,4-Dimethylphenol	3	8	<10	9	<27	0[a]	>95	0[a]	>32
2-Nitrophenol	1				<0.4				>99
4-Nitrophenol	1				<0.9				>99
Pentachlorophenol	15	<0.4	3,100	<0.4	1,600	0[a]	>99	>98	>70

(continued)

CONTROL TECHNOLOGY SUMMARY FOR ACTIVATED SLUDGE (continued)

Pollutant	Number of data points	Effluent concentration				Removal efficiency, %			
		Minimum	Maximum	Median	Mean	Minimum	Maximum	Median	Mean
Toxic pollutants, µg/L (continued):									
Phenol	30	<0.07	440	<0.07	<77	0[a]	>99	99	>77
2,4,6-Trichlorophenol	10	<0.2	4,300	41	450	0[a]	98	0	>36
p-Chloro-m-cresol	4	<0.1	<10	0.85	<3.9	0[a]	>98	>49	>65
Benzene	9	<0.2	37,000	<0.2	4,100	0[a]	>99	>96	>60
Chlorobenzene	6	<0.2	26	<0.2	5	0[a]	>99	>99	>67
1,2-Dichlorobenzene	12	<0.05	69	<0.05	<7.5	0[a]	>99	>99	>74
1,4-Dichlorobenzene	8	<0.04	21	0.13	<5.3	>47	>99	95	>82
2,6-Dinitrotoluene	1				390				0[a]
Ethylbenzene	24	<0.2	3,000	<0.2	150	0[a]	>99	>95	>83
Hexachlorobenzene	4	<0.05	0.8	0.28	<0.35	0[a]	>97	>49	>47
Toluene	31	<0.1	1,400	24	57	0[a]	>99	18	>52
1,2,4-Trichlorobenzene	11	<0.09	920	<6.3	94	0[a]	>99	>99	>67
Acenaphthene	10	<0.04	<10	<1.0	2.4	0[a]	>99	>99	>84
Anthracene/Phenanthrene	7	<0.01	<10	1.4	<3.4	0[a]	>98	83	>60
Fluoranthene	1				2				0
Fluorene	2	<0.02	<0.02		<0.02				>99
Indeno(1,2,3-cd)pyrene	1				<0.02				>99
Naphthalene	26	<0.007	260	<2.3	<18	0[a]	>99	>99	>64
Pyrene	5	0.1	9	0.2	1.9	0[a]	78	0[a]	16
2-Chloronaphthalene	1				1				50
Bromoform	1				3				0
Carbon tetrachloride	2	0.1	<10		<5.1	98	>99		>99
Chloroform	16	<1	58	32	<13	0[a]	>99	>2	>61
Dichlorobromomethane	2	1.5	<10		<5.8	0[a]	>0		0
1,1-Dichloroethane	2	<3	<10		<6.5	>0	>18		>9
1,2-Dichloropropane	2	<0.7	<10		<5.3	>53	>82		>68
1,3-Dichloropropane	1				0.89				0[a]
Methylene chloride	5	0.9	250	9	95	0[a]	99	0[a]	34
1,1,2,2-Tetrachloroethane	2				<10	>0	>44		>22
Tetrachloroethylene	11	<0.1	40	17	<6.4	0[a]	>99	0[a]	>75
1,1,1-Trichloroethane	6	<2	3.3	<2.0	<2.4	0[a]	>99	>85	>74
1,1,2-Trichloroethane	1				<10				>9
Trichloroethylene	12	<0.5	84	<0.5	10	0[a]	>99	>98	>68
Trichlorofluoromethane	5	1.7	2,100	2,100	450	0[a]	96	0[a]	19
Heptachlor	1				1.5				76
Isophorone	2	<10	<10		<10	>0	>0		>0

Note: Blanks indicate data not applicable.

[a]Actual data indicate negative removal.

TREATMENT TECHNOLOGY: Activated Sludge

Data source: Effluent Guidelines
Point source category: See below
Subcategory:
Plant:
References: A1, Appendix D-1

Data source status:
Engineering estimate ___
Bench scale ___
Pilot scale ___
Full scale ___

Use in system:
Pretreatment of influent:

DESIGN OR OPERATING PARAMETERS

Process modification:
Wastewater flow:
Hydraulic aeration detention time:
Volumetric loading:
MLSS:
Volatile fraction of MLSS:
F/M:
Mean cell residence time:
Sludge recycle ratio:
Mixed liquor dissolved oxygen:
Oxygen consumption:
Aerator power requirement:

Secondary clarifier configuration:
Depth:
Hydraulic loading (overflow rate):
Solids loading:
Weir loading:
Sludge underflow:
Percent solids in sludge:

REMOVAL DATA

Sampling period:

Point source category	Total phenol Concentration, μg/L Influent	Effluent[a]	Percent removal
Coal gas washing process[b]	1,200	<12	<99
Coke gasification plant[c,d,e,f,g]	5,000	<500	<90
Coal-tar distillation plant[c]	500	<5	<99

[a]Calculated from influent concentration and percent removal.

[b]F/M ratio: 0.116 kg phenol/kg MLSS/d.

[c]Use in system; secondary.

[d]Hydraulic aeration detention time: 2 d.

[e]Volumetric loading: 1,600 kg to 2,400 kg phenol/1,000 m^3/d.

[f]MLSS: 2,000 mg/L.

[g]Unit configuration: continuous flow through, bench-scale system.

Note: Blanks indicate information was not specified.

TREATMENT TECHNOLOGY: Activated Sludge

Data source: Government report
Point source category: Combined waste from petrochemical plants and paper mills
Subcategory:
Plant: Washburn tunnel facility
References: B16, pp. 288-289

Data source status:
- Engineering estimate ___
- Bench scale ___
- Pilot scale ___
- Full scale x

Use in system: Secondary
Pretreatment of influent: Bar screening, grit removal, primary clarification, nutrient addition, pH control

DESIGN OR OPERATING PARAMETERS

Process modification: High rate
Wastewater flow: ~1.7 x 10^5 m^3/d (45.0 mgd)
Hydraulic aeration detention time:
Volumetric loading:
MLSS:
Volatile fraction of MLSS:
F/M:
Mean cell residence time:
Sludge recycle ratio:
Mixed liquor dissolved oxygen:
Oxygen consumption:
Aerator power requirement:

Secondary clarifier configuration:
Depth:
Hydraulic loading (overflow rate):
Solids loading:
Weir loading:
Sludge underflow:
Percent solids in sludge:

REMOVAL DATA

Sampling period: Four days

Pollutant/parameter	Concentration		Percent removal
	Influent[a]	Effluent	
Conventional pollutants, mg/L:			
BOD_5	4,951	684	86
COD	2,670	761	72
TOC	323	216	33
TSS	1,460	191	93
Total phenol	19.5	0.116	92
Toxic pollutants, μg/L:			
Bis(2-chloroethyl) ether	19	BDL[b]	>47
4-Bromophenyl phenyl ether	358	18	95
Bis(2-ethylhexyl) phthalate	1	2	0[c]
Di-n-butyl phthalate	2	BDL[d]	>0
Diethyl phthalate	0.6	6	0[c]
Benzidine	4	4	0
1 2-Diphenylhydrazine	250	340	0[c]
2-Chlorophenol	0.1	0.9	0[c]
2,4-Dichlorophenol	4	BDL[d]	>0
Phenol	43	8	81
2,4,6-Trichlorophenol	4	BDL[d]	>0
p-Chloro-*m*-cresol	68	4	94
2,6-Dinitrotoluene	0.9	390	0[c]
Acenaphthene	1	1	0
Acenaphthylene	0.4	1	0[c]
Fluoranthene	2	2	0
Naphthalene	1.2	4.0	0[c]
Phenanthrene	0.9	1	0[c]
Pyrene	3	9	0[c]
2-Chloronaphthalene	2	1	50
Isophorone	0.2	BDL[d]	>0

[a] Influent to aeration basin.

[b] Below detectable limits; assumed to be <10 μg/L.

[c] Actual data indicate negative removal.

[d] Below detectable limits; assumed to be less than corresponding influent concentration.

Note: Blanks indicate information was not specified.

TREATMENT TECHNOLOGY: Activated Sludge

Data source: Effluent Guidelines
Point source category: Hospital
Subcategory:
Plant: See below
References: A22, p. 52

Data source status:
- Engineering estimate ___
- Bench scale ___
- Pilot scale ___
- Full scale x

Use in system: Secondary
Pretreatment of influent:

DESIGN OR OPERATING PARAMETERS

Process modification:
Wastewater flow:
Hydraulic aeration detention time:
Volumetric loading:
MLSS:
Volatile fraction of MLSS:
F/M:
Mean cell residence time:
Sludge recycle ratio:
Mixed liquor dissolved oxygen:
Oxygen consumption:
Aerator power requirement:

Secondary clarifier configuration:
Depth:
Hydraulic loading (overflow rate):
Solids loading:
Weir loading:
Sludge underflow:
Percent solids in sludge:

REMOVAL DATA

Sampling period:

	BOD_5 [a]		
	Concentration, mg/L		Percent
Plant	Influent	Effluent	removal
102	206[b]	16	92
93	335[b]	16	95

[a]Average of three samples.

[b]Calculated from effluent and percent removal.

Note: Blanks indicate information was not specified.

TREATMENT TECHNOLOGY: Activated Sludge

Data source: Effluent Guidelines
Point source category: Iron and steel
Subcategory: By-product coke manufacturing
Plant: B
References: A38, pp.60, 68, 70

Data source status:
Engineering estimate ___
Bench scale ___
Pilot scale ___
Full scale x

Use in system: Secondary
Pretreatment of influent:

DESIGN OR OPERATING PARAMETERS

Process modification:
Wastewater flow:
Hydraulic aeration detention time: 12-15 hr
Volumetric loading:
MLSS:
Volatile fraction of MLSS:
F/M:
Mean cell residence time:
Sludge recycle ratio:
Mixed liquor dissolved oxygen:
Oxygen consumption:
Aerator power requirement:

Secondary clarifier configuration:
Depth:
Hydraulic loading (overflow rate):
Solids loading:
Weir loading:
Sludge underflow:
Percent solids in sludge:

REMOVAL DATA

Sampling period:

Pollutant/parameter	Concentration Influent[a]	Concentration Effluent	Percent removal
Conventional pollutants, mg/L:			
TSS	5.74	16.3	0[b]
Oil and grease	240	5	>98
Total phenol	350	0.064	>99
Toxic pollutants, μg/L:			
Cyanide	135,000	37,700	72

[a]Calculated from effluent and percent removal.
[b]Actual data indicate negative removal.

Note: Blanks indicate information was not specified.

TREATMENT TECHNOLOGY: Activated Sludge

Data source: Effluent Guidelines
Point source category: Leather tanning and finishing
Subcategory: See below
Plant: See below
References: A15, p. 90

Data source status:
Engineering estimate ___
Bench scale ___
Pilot scale ___
Full scale x

Use in system: Secondary
Pretreatment of influent:

DESIGN OR OPERATING PARAMETERS

Process modification:
Wastewater flow: See below
Hydraulic aeration detention time: See below
Volumetric loading: See below
MLSS:
Volatile fraction of MLSS:
F/M:
Mean cell residence time:
Sludge recycle ratio:
Mixed liquor dissolved oxygen:
Oxygen consumption:
Aerator power requirement:

Secondary clarifier configuration:
Depth:
Hydraulic loading (overflow rate):
Solids loading:
Weir loading:
Sludge underflow:
Percent solids in sludge:

REMOVAL DATA

Subcategory	Plant	BOD_5 Concentration (mg/ℓ) Influent	Effluent	Percent Removal*	TSS Concentration (mg/ℓ) Influent	Effluent	Percent Removal*
Cattle, save chrome	Moench Tanning Co. (Gowanda, NY)**	1,700	343	80	2,400	190	92
Cattle, pulp, chrome	S.B. Foot Tanning Co. (Red Wing, MN)***	1,360	325	76	2,970	325	89
Cattle, pulp, combination tanning	Caldwell Lace Leather, (Auburn, KY)†	1,440	96	93	3,140	223	93

Subcategory	Plant	COD Concentration (mg/ℓ) Influent	Effluent	Percent Removal*	TKN Concentration (mg/ℓ) Influent	Effluent	Percent Removal*
Cattle, save chrome	Moench Tanning Co. (Gowanda, NY)**	–	–	–	–	–	–
Cattle, pulp chrome	S.B. Foot Tanning Co. (Red Wing, MN)***	–	–	–	–	–	–
Cattle, pulp, combination tanning	Caldwell Lace Leather (Auburn, KY)†	4,020	481	88	490	322	34

*Percent removal for entire plant.
**Wastewater flow: 1,510 m^3/day; hydraulic aeration detention time: 12 hr; volumetric loading: 3,600 kg BOD_5/day/1,000 m^3.
***Wastewater flow: 3,780 m^3/day; pretreatment influent: screening, primary sedimentation.
†Wastewater flow: 61 m^3/day; hydraulic aeration detention time: 1.6 days; volumetric loading: 908 kg BOD_5/day/1,000 m^3.

Note: Blanks indicate information was not specified.

TREATMENT TECHNOLOGY: Activated Sludge

Data source: Effluent Guidelines
Point source category: Leather tanning and finishing
Subcategory: Hair save, chrome tan, retan-wet finish
Plant: 248
References: A34, p. 208

Data source status:
Engineering estimate ___
Bench scale ___
Pilot scale ___
Full scale x

Use in system:
Pretreatment of influent:

DESIGN OR OPERATING PARAMETERS

Process modification:
Wastewater flow:
Hydraulic aeration detention time:
Volumetric loading:
MLSS:
Volatile fraction of MLSS:
F/M:
Mean cell residence time:
Sludge recycle ratio:
Mixed liquor dissolved oxygen:
Oxygen consumption:
Aerator power requirement:

Secondary clarifier configuration:
Depth:
Hydraulic loading (overflow rate):
Solids loading:
Weir loading:
Sludge underflow:
Percent solids in sludge:

REMOVAL DATA

Sampling period:

Pollutant/parameter	Concentration		Percent
	Influent	Effluent	removal
Conventional pollutants, mg/L:			
BOD_5	1,240	917	26
COD	2,560	1,780	31
TSS	1,100	557	49
Oil and grease	171	91	47
TKN	252	186	26
Toxic pollutants, μg/L:			
Chromium	31,000	20,000	35
Copper	57	37	35
Cyanide	20	40	0[a]
Lead	100	30	70
Nickel	5	34	0[a]
Zinc	230	140	39
Pentachlorophenol	9,500	3,100	67
Phenol	480	440	9
2,4,6-Trichlorophenol	10,500	4,300	59
1,2-Dichlorobenzene	220	69	68
1,4-Dichlorobenzene	99	21	79
Anthracene/phenanthrene	56	TR[b]	>98
Naphthalene	49	15	69
Chloroform	41	TR	>98

[a]Actual data indicate negative removal.

[b]Trace; <1 μg/L based on reported influent concentration and percent removal.

Note: Blanks indicate information was not specified.

TREATMENT TECHNOLOGY: Activated Sludge

Data source: Effluent Guidelines
Point source category: Leather tanning and finishing
Subcategory: Hair save, nonchrome (primarily vegetable) tan, retan-wet finish
Plant: 47
References: A34, p. 208

Data source status:
Engineering estimate ___
Bench scale ___
Pilot scale ___
Full scale x

Use in system:
Pretreatment of influent:

DESIGN OR OPERATING PARAMETERS

Process modification:
Wastewater flow:
Hydraulic aeration detention time:
Volumetric loading:
MLSS:
Volatile fraction of MLSS:
F/M:
Mean cell residence time:
Sludge recycle ratio:
Mixed liquor dissolved oxygen:
Oxygen consumption:
Aerator power requirement:

Secondary clarifier configuration:
Depth:
Hydraulic loading (overflow rate):
Solids loading:
Weir loading:
Sludge underflow:
Percent solids in sludge:

REMOVAL DATA

Pollutant/Parameter	.... Concentration..... Influent	Effluent	Percent Removal
Conventional pollutants, mg/ℓ			
BOD_5	1,530	49	97
COD	5,950	553	91
TSS	6,380	227	96
Oil and grease	247	35	86
TKN	750	277	63
Toxic pollutants, μg/ℓ			
Chromium	6,400	170	97
Copper	200	25	88
Cyanide	100	400	0*
Lead	100	50	50
Nickel	60	30	50
Zinc	460	59	87
Bis(2-ethylhexyl) phthalate	ND**	26	0*
Pentachlorophenol	2,900	200	93
Phenol	840	ND**	>99
2,4,6-Trichlorophenol	1,700	38	98
1,2-Dichlorobenzene	49	ND**	>80
1,4-Dichlorobenzene	19	ND**	>47
Ethylbenzene	43	TR***	>98
Anthracene/phenanthrene	7.6	ND†	>0
Naphthalene	19	ND**	>47

*Actual data indicate negative removal.
**Not detected; assumed to be <10 μg/ℓ.
***Trace; <1 μg/ℓ based on reported influent concentration and percent reported.
†Not detected; assumed to be less than the corresponding influent concentration.

Note: Blanks indicate information was not specified.

TREATMENT TECHNOLOGY: Activated Sludge

Data source: Effluent Guidelines
Point source category: Leather tanning and finishing
Subcategory: Hair save, chrome tan, retan-wet finish
Plant: 320
References: A34, p. 208

Data source status:
Engineering estimate ___
Bench scale ___
Pilot scale ___
Full scale x

Use in system:
Pretreatment of influent:

DESIGN OR OPERATING PARAMETERS

Process modification:
Wastewater flow:
Hydraulic aeration detention time:
Volumetric loading:
MLSS:
Volatile fraction of MLSS:
F/M:
Mean cell residence time:
Sludge recycle ratio:
Mixed liquor dissolved oxygen:
Oxygen consumption:
Aerator power requirement:

Secondary clarifier configuration:
Depth:
Hydraulic loading (overflow rate):
Solids loading:
Weir loading:
Sludge underflow:
Percent solids in sludge:

REMOVAL DATA

Sampling period:

Pollutant/parameter	Concentration Influent	Concentration Effluent	Percent removal
Conventional pollutants, mg/L:			
BOD_5	2,000	297	86
COD	4,030	893	87
TSS	2,250	129	94
Oil and grease	553	17	97
TKN	287	163	43
Toxic pollutants, μg/L:			
Chromium	170,000	1,700	99
Copper	220	8	96
Cyanide	50	40	20
Lead	3,100	60	98
Nickel	75	30	60
Zinc	2,100	170	92
Bis(2-ethylhexyl) phthalate	32	6	83
Pentachlorophenol	ND[a]	12	0[b]
Phenol	5,500	1,400	75
2,4,6-Trichlorophenol	ND	12	0[b]
Ethylbenzene	>100	TR[c]	>99
Toluene	>100	TR	>99
Anthracene/phenanthrene	2.9	1.4	52
Napthalene	ND	2.3	0[b]

[a]Not detected.

[b]Actual data indicate negative removal.

[c]Trace; <1 μg/L based on reported influent concentrations and percent removal.

Note: Blanks indicate information was not specified.

TREATMENT TECHNOLOGY: Activated Sludge

Data source: Effluent Guidelines
Point source category: Leather tanning and finishing
Subcategory: Shearing
Plant: 253
References: A34, p. 208

Data source status:
Engineering estimate ___
Bench scale ___
Pilot scale ___
Full scale x

Use in system:
Pretreatment of influent:

DESIGN OR OPERATING PARAMETERS

Process modification:
Wastewater flow:
Hydraulic aeration detention time:
Volumetric loading:
MLSS:
Volatile fraction of MLSS:
F/M:
Mean cell residence time:
Sludge recycle ratio:
Mixed liquor dissolved oxygen:
Oxygen consumption:
Aerator power requirement:

Secondary clarifier configuration:
Depth:
Hydraulic loading (overflow rate):
Solids loading:
Weir loading:
Sludge underflow:
Percent solids in sludge:

REMOVAL DATA

Sampling period:

Pollutant/parameter	Concentration Influent	Effluent	Percent removal
Conventional pollutants, mg/L:			
BOD_5	1,020	27	97
COD	2,370	488	79
TSS	768	108	86
Oil and grease	413	25	94
TKN	49	27	45
Toxic pollutants, μg/L:			
Chromium	53,000	2,200	96
Copper	120	7	94
Lead	80	30	63
Nickel	27	19	30
Zinc	500	68	86
Bis(2ethylhexyl) phthalate	93	34	63
Pentachlorophenol	400	130	68
Phenol	91	ND[a]	>89
Benzene	5	ND[b]	>0
1,4-Dichlorobenzene	20	ND[a]	>50
Toluene	9	ND[b]	>0
Anthracene/phenanthrene	36	6	83
Naphthalene	35	ND[a]	>71
Chloroform	12	10	16
1,1,2,2-Tetrachloroethane	18	ND[a]	>44

[a]Not detected; assumed to be <10 μg/L.

[b]Not detected; assumed to be less than the corresponding influent concentration.

Note: Blanks indicate information was not specified.

TREATMENT TECHNOLOGY: Activated Sludge

Data source: Government report
Point source category: Mixed industrial (mainly textile)/ domestic wastes
Subcategory:
Plant: Deep shaft treatment plant (Paris, Ontario)
References: B16, pp. 297-301

Data source status:
Engineering estimate ___
Bench scale ___
Pilot scale x
Full scale ___

Use in system: Secondary
Pretreatment of influent: Bar screening, comminutor, acid neutralization

DESIGN OR OPERATING PARAMETERS

Process modification: Deep shaft biooxidator, air flotation
Wastewater flow: 4.5 x 10^2 m^3/day
Volumetric loading:
MLSS:
Volatile fraction of MLSS:
F/M:
Mean cell residence time:
Sludge recycle ratio:
Mixed liquor dissolved oxygen:
Oxygen consumption:
Aerator power requirement:

Secondary clarifier configuration: Air flotation tank
Depth:
Hydraulic loading (overflow rate):
Solids loading:
Weir loading:
Sludge underflow:
Percent solids in sludge:
Hydraulic aeration detention time: 30 min

REMOVAL DATA

Sampling period: Five days

Pollutant/parameter	Concentration Influent	Concentration Effluent	Percent removal
Conventional pollutants,[a] mg/L:			
BOD_5	182	33	82
COD	895	200	78
TSS	311	60	81
Toxic pollutants, μg/L:[b]			
Dimethyl phthalate	70	200	0[c]
Di-n-octyl phthalate	1,000	5,000	0[c]
Phenol	18	BDL[d]	>44
Benzene	340	BDL[d]	>97
Toluene	30	BDL[d]	>67
1,2,4-Trichlorobenzene	5	BDL[e]	>C
Acenaphthene	180	BDL[d]	>94
Carbon tetrachloride	2,200	BDL[d]	>99
Chloroform	22,000	BDL[d]	>99
1,1,2,2-Tetrachloroethane	8	BDL[e]	>0
Tetrachloroethene	5	BDL[e]	>0
1,1,2-Trichloroethane	11	BDL[d]	>9
Isophorone	7	BDL[e]	>0

[a]Average of daily composites.

[b]Grab samples.

[c]Actual data indicate negative removal.

[d]Below detectable limits; assumed to be <10 μg/L.

[e]Below detectable limits; assumed to be less than corresponding influent concentration.

Note: Blanks indicate information was not specified.

TREATMENT TECHNOLOGY: Activated Sludge

Data source: Effluent Guidelines unless otherwise noted
Point source category: Organic chemicals
Subcategory: See below
Plant: See below
References: A24, A25, p. 300 unless otherwise noted

Data source status:
- Engineering estimate ___
- Bench scale ___
- Pilot scale ___
- Full scale x

Use in system: Secondary
Pretreatment of influent:

DESIGN OR OPERATING PARAMETERS

Process modification:
Wastewater flow:
Hydraulic aeration detention time:
Volumetric loading:
MLSS:
Volatile fraction of MLSS:
F/M:
Mean cell residence time:
Sludge recycle ratio:
Mixed liquor dissolved oxygen:
Oxygen consumption:
Aerator power requirement:

Secondary clarifier configuration:
Depth:
Hydraulic loading (overflow rate):
Solids loading:
Weir loading:
Sludge underflow:
Percent solids in sludge:

REMOVAL DATA

Subcategory/ Plant	BOD_5 Concentration (mg/ℓ) Influent	BOD_5 Concentration (mg/ℓ) Effluent	BOD_5 Percent Removal	COD Concentration (mg/ℓ) Influent	COD Concentration (mg/ℓ) Effluent	COD Percent Removal	TOC Concentration (mg/ℓ) Influent	TOC Concentration (mg/ℓ) Effluent	TOC Percent Removal	TSS Concentration (mg/ℓ) Influent	TSS Concentration (mg/ℓ) Effluent	TSS Percent Removal
Aqueous liquid-phase reaction system, plant 9	938*	75	92	2,380*	595	75	781*	242	69	<50	50	0**
Aqueous liquid-phase reaction system, plant 20	1,900*	19	<99	7,920*	317	96	3,800	114	97	<100	100	0**
Union Carbide (in Sistersirele, WV)†	450	36***	92	–	–	–	–	–	–	–	–	–

*Calculated from effluent and percent removal.
**Actual data indicate negative removal.
***Calculated from influent and percent removal.
†Data source, Government Report; reference, B16, p 70; aeration method, pure oxygen activated sludge system; F/M, 0.5 to 1.5.

Note: Blanks indicate information was not specified.

TREATMENT TECHNOLOGY: Activated Sludge

Data source: Effluent Guidelines
Point source category: Organic chemicals
Subcategory: Process with process water contact as steam diluent or absorbent unless otherwise noted
Plant: See below
References: A25, p. 300

Data source status:
Engineering estimate ___
Bench scale ___
Pilot scale ___
Full scale x

Use in system: Secondary
Pretreatment of influent:

DESIGN OR OPERATING PARAMETERS

Process modification:
Wastewater flow:
Hydraulic aeration detention time:
Volumetric loading:
MLSS:
Volatile fraction of MLSS:
F/M:
Mean cell residence time:
Sludge recycle ratio:
Mixed liquor dissolved oxygen:
Oxygen consumption:
Aerator power requirement:

Secondary clarifier configuration:
Depth:
Hydraulic loading (overflow rate):
Solids loading:
Weir loading:
Sludge underflow:
Percent solids in sludge:

REMOVAL DATA

	BOD_5			COD			TOC			TSS		
	Concentration (mg/ℓ)			Concentration (mg/ℓ)			Concentration (mg/ℓ)			Concentration (mg/ℓ)		
Plant	Influent*	Effluent	Percent Removal	Influent*	Effluent	Percent Removal	Influent*	Effluent	Percent Removal	Influent*	Effluent	Percent Removal
13**	1,770	177	90	2,690	940	65	1,310	470	64	154	338	0***
4	72	13	82	498	214	57	123	80	35	23	14	40
22	404	210	48	1,630	1,370	16	598	550	8	174	82	53
23	586	41	93	3,200	147	95	760	35	95	<37	37	0***

*Calculated from effluent and percent removal.
**Subcategory: process with process water contact as steam diluent or absorbent and aqueous liquid phase reaction systems.
***Actual data indicate negative removal.

Note: Blanks indicate information was not specified.

TREATMENT TECHNOLOGY: Activated Sludge

Data source: Effluent Guidelines
Point source category: Organic chemicals
Subcategory: Batch and semicontinuous process
Plant: See below
References: A25, p. 300

Data source status:
- Engineering estimate ___
- Bench scale ___
- Pilot scale ___
- Full scale x

Use in system: Secondary
Pretreatment of influent:

DESIGN OR OPERATING PARAMETERS

Process modification:
Wastewater flow:
Hydraulic aeration detention time:
Volumetric loading:
MLSS:
Volatile fraction of MLSS:
F/M:
Mean cell residence time:
Sludge recycle ratio:
Mixed liquor dissolved oxygen:
Oxygen consumption:
Aerator power requirement:

Secondary clarifier configuration:
Depth:
Hydraulic loading (overflow rate):
Solids loading:
Weir loading:
Sludge underflow:
Percent solids in sludge:

REMOVAL DATA

	BOD_5			COD			TOC			TSS		
	Concentration (mg/ℓ)			Concentration (mg/ℓ)			Concentration (mg/ℓ)			Concentration (mg/ℓ)		
Plant	Influent*	Effluent	Percent Removal	Influent*	Effluent	Percent Removal	Influent*	Effluent	Percent Removal	Influent*	Effluent	Percent Removal
3	274	74	73	979	284	71	455	132	71	<62	62	0**
16	1,670	300	82	3,670	1,650	55	1,470	280	81	986	552	44
17	1,920	275	86	3,870	1,200	69	3,700	385	70	<1,300	0**	–
18	783	650	17	3,440	2,680	22	–	–	–	2,050	1,170	43
19	6,000	1,800	70	12,800	5,100	60	3,860	1,700	56	<2,500	2,500	0**

*Calculated from effluent and percent removal.
**Actual data indicate negative removal.

Note: Blanks indicate information was not specified.

TREATMENT TECHNOLOGY: Activated Sludge

Data source: Effluent Guidelines
Point source category: Pharmaceuticals manufacturing
Subcategory:
Plant: B
References: A32, Supplement 2

Data source status:
- Engineering estimate ___
- Bench scale ___
- Pilot scale ___
- Full scale x

Use in system: Secondary
Pretreatment of influent: Equalization

DESIGN OR OPERATING PARAMETERS

Process modification:
Wastewater flow: 1,890 m^3/d (0.50 mgd)
Hydraulic aeration detention time:
Volumetric loading:
MLSS:
Volatile fraction of MLSS:
F/M:
Mean cell residence time:
Sludge recycle ratio: 200 to 500%
Mixed liquor dissolved oxygen:
Oxygen consumption:
Aerator power requirement: 7.5-37.3 kW (10-50 hp)

Secondary clarifier configuration: Multiple settling tanks, 5,200 m^2 (56,000 ft^2) surface area
Depth:
Hydraulic loading (overflow rate):
Solids loading:
Weir loading:
Sludge underflow:
Percent solids in sludge:

REMOVAL DATA

Sampling period:

Pollutant/parameter	Concentration Influent	Effluent	Percent removal
Conventional pollutants, mg/L:			
BOD_5	3,000	120	96
TSS	950	500	47
Toxic pollutants, µg/L:			
Arsenic	70	20	71
Chromium	680	190	72
Copper	180	31	83
Cyanide	580	7,700	0[a]
Lead	15	24	0[a]
Nickel	630	190	70
Thallium	47	29	38
Zinc	540	160	70
Bis(2-ethylhexyl) phthalate	24	33	0[a]

[a] Actual data indicate negative removal.

Note: Blanks indicate information was not specified.

TREATMENT TECHNOLOGY: Activated Sludge

Data source: Effluent Guidelines
Point source category: Pharmaceuticals
Subcategory: Biological and natural extraction products, formulation products
Plant: H
References: A32, Supplement 2

Data source status:
- Engineering estimate ___
- Bench scale ___
- Pilot scale ___
- Full scale x

Use in system: Secondary
Pretreatment of influent:

DESIGN OR OPERATING PARAMETERS

Process modification:
Wastewater flow: 644 m^3/d (0.17 mgd)
Hydraulic aeration detention time: 2.56 days
Volumetric loading:
MLSS: 3,500 mg/L
Volatile fraction of MLSS:
F/M: 0.30
Mean cell residence time: 6.85 days
Sludge recycle ratio:
Mixed liquor dissolved oxygen:
Oxygen consumption:
Aerator power requirement: 344.7 kW (60 hp)
Sludge recycle flow rate: 992 m^3/d (262,000 gpd)

Secondary clarifier configuration:
Depth:
Hydraulic loading (overflow rate): 21.4 $m^3/d/m^2$ (525 gal/d/ft^2)
Solids loading: 107 kg TSS/d/m^2 (22 lb TSS/d/ft^2)
Weir loading:
Sludge underflow:
Percent solids in sludge:

REMOVAL DATA

Sampling period:

Pollutant/parameter	Concentration Influent	Concentration Effluent	Percent removal
Conventional pollutants, mg/L:			
BOD_5	7,520	4,640	38
COD	12,000	7,420	38
TSS	4,920	4,050	18
Toxic pollutants, µg/L:			
Benzene	40	10	75
Methylene chloride	130	210	0[a]

[a]Actual data indicate negative removal.

Note: Blanks indicate information was not specified.

TREATMENT TECHNOLOGY: Activated Sludge

Data source: Effluent Guidelines
Point source category: Pharmaceuticals
Subcategory: Formulation products
Plant: S
References: A32, Supplement 2

Data source status:
Engineering estimate ___
Bench scale ___
Pilot scale ___
Full scale x

Use in system: Secondary
Pretreatment of influent:

DESIGN OR OPERATING PARAMETERS

Process modification: Four 1,290 m^3 (340,000 gal) aeration tanks
Wastewater flow: 606 m^3/d (0.16 mgd)
Hydraulic aeration detention time:
Volumetric loading:
MLSS:
Volatile fraction of MLSS:
F/M:
Mean cell residence time:
Sludge recycle ratio:
Mixed liquor dissolved oxygen:
Oxygen consumption:
Aerator power requirement:

Secondary clarifier configuration:
Depth:
Hydraulic loading (overflow rate):
Solids loading:
Weir loading:
Sludge underflow:
Percent solids in sludge:

REMOVAL DATA

Sampling period:

Pollutant/parameter	Concentration, µg/L Influent	Concentration, µg/L Effluent	Percent removal
Toxic pollutants:			
Chromium	30	10	66
Copper	80	20	75
Bis(2-ethylhexyl) phthalate	50	10	80
Methylene chloride	800	250	69

Note: Blanks indicate information was not specified.

TREATMENT TECHNOLOGY: Activated Sludge

Data source: Effluent Guidelines unless otherwise noted
Point source category: Pharmaceutical
Subcategory: See below
Plant: See below
References: See below

Data source status:
Engineering estimate ___
Bench scale ___
Pilot scale ___
Full scale x

Use in system: Secondary
Pretreatment of influent:

DESIGN OR OPERATING PARAMETERS

Process modification:
Wastewater flow: See below
Hydraulic aeration detention time: See below
Volumetric loading:
MLSS:
Volatile fraction of MLSS:
F/M:
Mean cell residence time:
Sludge recycle ratio:
Mixed liquor dissolved oxygen:
Oxygen consumption:
Aerator power requirement:

Secondary clarifier configuration:
Depth:
Hydraulic loading (overflow rate):
Solids loading:
Weir loading:
Sludge underflow:
Percent solids in sludge:

REMOVAL DATA

		BOD$_5$			COD			TOC		
		Concentration (mg/ℓ)		Percent	Concentration (mg/ℓ)		Percent	Concentration (mg/ℓ)		Percent
Subcategory	Plant	Influent	Effluent	Removal	Influent	Effluent	Removal	Influent	Effluent	Removal
Fermentation products	20*	1,380	110	92	4,380	1,300	70	1,520	218	86
Fermentation products and synthesis products	25**	3,830	280	93	7,740	4,070	47	1,900	1,260	34
Fermentation products, chemical synthesis products, and mixing/compounding and formulation	19***	3,110	134	96	6,800	680	90	2,220	292	87
	Texas†	7,470	75††	99	14,800	592††	96	–	–	–

		TSS			TKN			Total Phosphorus		
		Concentration (mg/ℓ)		Percent	Concentration (mg/ℓ)		Percent	Concentration (mg/ℓ)		Percent
Subcategory	Plant	Influent	Effluent	Removal	Influent	Effluent	Removal	Influent	Effluent	Removal
Fermentation products	20	–	–	–	–	–	–	–	–	–
Fermentation products and synthesis products	25	858	1,340	0†††	–	–	–	–	–	–
Fermentation products, chemical synthesis products, and mixing/compounding and formulation	19	1,700	210	88	196	60	69	32	3.5	89
	Texas†	–	–	–	690	593††	14	–	–	–

*Wastewater flow: 950 m^3/day; hydraulic aeration detention time: 4.8 days; secondary clarifier configuration: circular 10 m diameter; aerator power requirement: three 34 kW floating 50 hp aerators utilized (Reference A12, pp 113-114).

**Wastewater flow: 1,000 m^3/day; hydraulic aeration detention time: 3.5 days; process modification: four aeration tanks (Reference A12, p 123).

***Wastewater flow: 2,850 m^3/day; hydraulic aeration detention time: 24 hr; pretreatment of influent: equalization (Reference A12, p 113).

†Process modification: Two-stage activated sludge system.

††Calculated from influent concentration and percent removal.

†††Actual data indicate negative removal.

Note: Blanks indicate information was not specified.

TREATMENT TECHNOLOGY: Activated Sludge

Data source: Effluent Guidelines
Point source category: Pulp, paper and paperboard
Subcategory: Sulfite-papergrade
Plant:
References: A26, pp. A-34 to 41

Data source status:
Engineering estimate ___
Bench scale ___
Pilot scale ___
Full scale x

Use in system: Secondary
Pretreatment of influent:

DESIGN OR OPERATING PARAMETERS

Process modification:
Wastewater flow:
Hydraulic aeration detention time:
Volumetric loading:
MLSS:
Volatile fraction of MLSS:
F/M:
Mean cell residence time:
Sludge recycle ratio:
Mixed liquor dissolved oxygen:
Oxygen consumption:
Aerator power requirement:

Secondary clarifier configuration:
Depth:
Hydraulic loading (overflow rate):
Solids loading:
Weir loading:
Sludge underflow:
Percent solids in sludge:

REMOVAL DATA

Pollutant/Parameter	..... Concentration*..... Influent	Effluent	Percent Removal
Conventional pollutants, mg/ℓ			
COD	4,790	2,890	40
Toxic pollutants, μg/ℓ			
Chromium	13	10	23
Copper	81	20	75
Lead	13	10	23
Nickel	16	17	0**
Zinc	91	58	36
Bis(2-ethylhexyl) phthalate	38	3	92
Pentachlorophenol	4	ND***	>0
Phenol	53	2	96
2,4,6-Trichlorophenol	4	ND***	>0
Benzene	53	ND†	>81
Toluene	15	ND†	>33
Naphthalene	72	53	26
Chloroform	3,200	56	98
Dichlorobromomethane	9	ND***	>0
1,1-Dichloroethane	4	ND***	>0
Methylene chloride	460	5	99
1,1,1-Trichloroethane	410	3	99
Trichloroethylene	5	ND***	>0

*Average values.
**Actual data indicate negative removal.
***Not detected.
†Not detected; assumed to be <10 μg/ℓ.

Note: Blanks indicate information was not specified.

TREATMENT TECHNOLOGY: Activated Sludge

Data source: Effluent Guidelines
Point source category: Pulp, paper and paperboard
Subcategory: Wastepaper-board
Plant:
References: A26, pp. A-78 to 85

Data source status:
Engineering estimate ___
Bench scale ___
Pilot scale ___
Full scale x

Use in system: Tertiary
Pretreatment of influent: Lagooning, trickling filter

DESIGN OR OPERATING PARAMETERS

Process modification:
Wastewater flow:
Hydraulic aeration detention time:
Volumetric loading:
MLSS:
Volatile fraction of MLSS:
F/M:
Mean cell residence time:
Sludge recycle ratio:
Mixed liquor dissolved oxygen:
Oxygen consumption:
Aerator power requirement:

Secondary clarifier configuration:
Depth:
Hydraulic loading (overflow rate):
Solids loading:
Weir loading:
Sludge underflow:
Percent solids in sludge:

REMOVAL DATA

Pollutant/Parameter	Concentration* Influent	Concentration* Effluent	Percent Removal
Conventional pollutants, mg/ℓ			
COD	622	967	0**
Toxic pollutants, μg/ℓ			
Chromium	17	33	0**
Copper	42	37	12
Cyanide	16	14	13
Lead	49	31	37
Bis(2-ethylhexyl) phthalate	6	73	0**
Butyl benzyl phthalate	<1	11	0**
Di-n-butyl phthalate	6	7	0**
Diethyl phthalate	139	69	50
Pentachlorophenol	3	200	0**
Phenol	37	72	0**
2,4,6-Trichlorophenol	2	72	0**
Toluene	13	2	85
Naphthalene	55	54	2
Bromoform	ND***	3	0**
Chloroform	19	2	89
Methylene chloride	1	9	0**
Trichloroethylene	1	<1	>0
Other pollutants, μg/ℓ			
Xylenes	2	ND	>0

*Average values.
**Actual data indicate negative removal.
***Not detected; assumed to be less than corresponding influent concentration.

Note: Blanks indicate information was not specified.

TREATMENT TECHNOLOGY: Activated Sludge

Data source: Effluent Guidelines
Point source category: Rubber processing
Subcategory:
Plant: 000012
References: A30, p. 121

Data source status:
Engineering estimate ___
Bench scale ___
Pilot scale ___
Full scale x

Use in system: Secondary
Pretreatment of influent:

DESIGN OR OPERATING PARAMETERS

Process modification:
Wastewater flow:
Hydraulic aeration detention time:
Volumetric loading:
MLSS:
Volatile fraction of MLSS:
F/M:
Mean cell residence time:
Sludge recycle ratio:
Mixed liquor dissolved oxygen:
Oxygen consumption:
Aerator power requirement:

Secondary clarifier configuration:
Depth:
Hydraulic loading (overflow rate):
Solids loading:
Weir loading:
Sludge underflow:
Percent solids in sludge:

REMOVAL DATA

Sampling period: 24 hr

Pollutant/parameter	Concentration,[a] μg/L Influent	Effluent	Percent removal
Toxic pollutants:			
Cadmium	1	<1	>0
Mercury	2.5	1.6	36
Nickel	610	400	34
Bis(2-ethylhexyl) phthalate	260	220	15
N-nitrosodiphenylamine	5.2	1.6	69
Phenol	41	19	54
Toluene	250	<0.1	>99
Carbon tetrachloride	4.7	0.1	98
Chloroform	27	4.1	85
Methylene chloride	<0.1	0.9	0[b]
Tetrachloroethylene	1.4	<0.1	>93
1,1,1-Trichloroethane	1.0	3.3	0[b]

[a]Values presented are averages three of composite samples.

[b]Actual data indicate negative removal.

Note: Blanks indicate information was not specified.

TREATMENT TECHNOLOGY: Activated Sludge

Data source: Effluent Guidelines
Point source category: Synthetic resins
Subcategory: See below
Plant:
References: A23, p. 105

Data source status:
Engineering estimate ___
Bench scale ___
Pilot scale ___
Full scale x

Use in system: Secondary
Pretreatment of influent:

DESIGN OR OPERATING PARAMETERS

Process modification:
Wastewater flow: See below
Hydraulic aeration detention time: See below
Volumetric loading: See below
MLSS:
Volatile fraction of MLSS:
F/M:
Mean cell residence time:
Sludge recycle ratio:
Mixed liquor dissolved oxygen:
Oxygen consumption:
Aerator power requirement: See below

Secondary clarifier configuration:
Depth:
Hydraulic loading (overflow rate):
Solids loading:
Weir loading:
Sludge underflow:
Percent solids in sludge:

REMOVAL DATA

Sampling period:

	BOD_5			COD		
	Concentration, mg/L		Percent	Concentration, mg/L		Percent
Subcategory	Influent	Effluent	removal	Influent	Effluent	removal
Cellophane[a]	90	20	78	228	197	14
Cellulosic[b]	1,320	37	97			

[a]Wastewater flow: 26,000 m^3/d; hydraulic aeration detention time: 1.5 hr; volumetric loading: 1.0 kg $BOD_5/d/m^3$; aerator power requirement: 130 W/m^3 (660 hp/Mgal).

[b]Wastewater flow: 12,900 m^3/d; hydraulic aeration detention time: 64 hr; volumetric loading: 0.48 kg $BOD_5/d/m^3$; aerator power requirement: 18.4 W/m^3 (93.5 hp/Mgal).

Note: Blanks indicate information was not specified.

TREATMENT TECHNOLOGY: Activated Sludge

Data source: Effluent Guidelines
Point source category: Textile mills
Subcategory: Knit fabric finishing unless otherwise noted
Plant:
References: A6, p. VII-25

Data source status:
- Engineering estimate ___
- Bench scale ___
- Pilot scale ___
- Full scale x

Use in system:
Pretreatment of influent:

DESIGN OR OPERATING PARAMETERS

Process modification: Extended aeration, surface aeration
Wastewater flow:
Hydraulic aeration detention time: See below
Volumetric loading:
MLSS:
Volatile fraction of MLSS:
F/M:
Mean cell residence time:
Sludge recycle ratio:
Mixed liquor dissolved oxygen:
Oxygen consumption:
Aerator power requirement: See below

Secondary clarifier configuration:
Depth:
Hydraulic loading (overflow rate):
Solids loading:
Weir loading:
Sludge underflow:
Percent solids in sludge:

REMOVAL DATA

Sampling period:

Hydraulic aeration detention time[a] hr	Aerator power requirement, W/m^3 (hp/Mgal)	BOD_5 Concentration, mg/L Influent	BOD_5 Concentration, mg/L Effluent	BOD_5 Percent removal	COD Concentration, mg/L Influent	COD Concentration, mg/L Effluent	COD Percent removal	TSS Concentration, mg/L Influent	TSS Concentration, mg/L Effluent	TSS Percent removal
130[b]	9(44)	207	29	86	614	272	63	93	50	46
417	8(40)	198	13	93	745	226	70	49	62	0[c]
48	12(60)	272	45	83	694	354	49	28	55	0[c]
76	32(160)	1,100	11	99				281	45	84
82	15(74)	190	19	90	342	164	52	97	63	35
110	15(75)	181	5	97				18	18	0[c]

[a] Based on average flow and full basin volume.
[b] Subcategory: Carpet finishing.
[c] Actual data indicate negative removal.

Note: Blanks indicate information was not specified.

TREATMENT TECHNOLOGY: Activated Sludge

Data source: Effluent Guidelines
Point source category: Textile mills
Subcategory: Stock and yarn finishing
Plant:
References: A6, p. VII-25

Data source status:
- Engineering estimate ___
- Bench scale ___
- Pilot scale ___
- Full scale ___

Use in system:
Pretreatment of influent:

DESIGN OR OPERATING PARAMETERS

Process modification: Extended aeration, surface aeration
Wastewater flow:
Hydraulic aeration detention time: See below
Volumetric loading:
MLSS:
Volatile fraction of MLSS:
F/M:
Mean cell residence time:
Sludge recycle ratio:
Mixed liquor dissolved oxygen:
Oxygen consumption:
Aerator power requirement: See below

Secondary clarifier configuration:
Depth:
Hydraulic loading (overflow rate):
Solids loading:
Weir loading:
Sludge underflow:
Percent solids in sludge:

REMOVAL DATA

Sampling period:

Hydraulic aeration detention time,[a] hr	Aerator power requirement, W/m^3 (hp/Mgal)	BOD_5			COD			TSS		
		Concentration, mg/L		Percent removal	Concentration, mg/L		Percent removal	Concentration, mg/L		Percent removal
		Influent	Effluent		Influent	Effluent		Influent	Effluent	
50	16(80)	125	5	96				46	21	54
33	16(80)	150	6	96	496	124	75	36	27	25
44	98(500)	1,630	233	86	4,760	1,840	61	136	195	0[b]

[a]Based on average flow and full basin volume.
[b]Actual data indicate negative removal.

Note: Blanks indicate information was not specified.

TREATMENT TECHNOLOGY: Activated Sludge

Data source: Effluent Guidelines
Point source category: Textile mills
Subcategory: Wool scouring
Plant:
References: A6, p. VII-25

Data source status:
- Engineering estimate ___
- Bench scale ___
- Pilot scale ___
- Full scale x

Use in system:
Pretreatment of influent:

DESIGN OR OPERATING PARAMETERS

Process modification: Extended aeration, surface aeration
Wastewater flow:
Hydraulic aeration detention time:[a] 99 hr
Volumetric loading:
MLSS:
Volatile fraction of MLSS:
F/M:
Mean cell residence time:
Sludge recycle ratio:
Mixed liquor dissolved oxygen:
Oxygen consumption:
Aerator power requirement: 32 W/m^3 (160 hp/Mgal)

Secondary clarifier configuration:
Depth:
Hydraulic loading (overflow rate):
Solids loading:
Weir loading:
Sludge underflow:
Percent solids in sludge:

[a]Based on average flow and full basin volume.

REMOVAL DATA

Sampling period: Data are average values for 1976

Pollutant/parameter	Concentration, mg/L Influent	Effluent	Percent removal
Conventional pollutants:			
BOD_5	1,560	125	92
COD	16,200	2,600	84
TSS	3,970	1,230	69

Note: Blanks indicate information was not specified.

TREATMENT TECHNOLOGY: Activated Sludge

Data source: Effluent Guidelines
Point source category: Textile mills
Subcategory: Woven fabric finishing
Plant:
References: A6, p. VII-25

Data source status:
- Engineering estimate ___
- Bench scale ___
- Pilot scale ___
- Full scale x

Use in system:
Pretreatment of influent:

DESIGN OR OPERATING PARAMETERS

Process modification: Extended aeration, surface aeration
Wastewater flow:
Hydraulic aeration detention time: See below
Volumetric loading:
MLSS:
Volatile fraction of MLSS:
F/M:
Mean cell residence time:
Sludge recycle ratio:
Mixed liquor dissolved oxygen:
Oxygen consumption:
Aerator power requirement: See below

Secondary clarifier configuration:
Depth:
Hydraulic loading (overflow rate):
Solids loading:
Weir loading:
Sludge underflow:
Percent solids in sludge:

REMOVAL DATA

Sampling period: Data are average values for each mill for the year 1976

Hydraulic aeration detention time,[a] hr	Aerator power requirement, W/m^3 (hp/Mgal)	BOD_5			COD			TSS		
		Concentration, mg/L		Percent removal	Concentration, mg/L		Percent removal	Concentration, mg/L		Percent removal
		Influent	Effluent		Influent	Effluent		Influent	Effluent	
78	16(80)	640	105	84	1,240	664	46	173	176	0[b]
131	11(58)	400	8	98				80	8	90
75	8.1(41)	267	24	91	840	336	60			
120	12(60)	180	9	95	468	159	66	26	18	31
80	18(90)	250	5	98				218	48	78
97	49(250)	329	23	93	2,970	594	80			
106	24(120)	475	19	96						
24	12(60)	133	22	83	472	307	35	34	38	0[b]

[a] Based on average flow and full basin volume.
[b] Actual data indicate negative removal.

Note: Blanks indicate information was not specified.

TREATMENT TECHNOLOGY: Activated Sludge

Data source: Government report
Point source category: Textile mills
Subcategory:
Plant: A
References: B5, pp. 32-53

Data source status:
Engineering estimate ___
Bench scale ___
Pilot scale ___
Full scale x

Use in system: Secondary
Pretreatment of influent:

DESIGN OR OPERATING PARAMETERS

Aeration method: Surface aeration
Wastewater flow:
Hydraulic aeration detention time:
Volumetric loading:
MLSS:
Volatile fraction of MLSS:
F/M:
Mean cell residence time:
Sludge recycle ratio:
Mixed liquor dissolved oxygen:
Oxygen consumption:
Aerator power requirement:

Secondary clarifier configuration:
Depth:
Hydraulic loading (overflow rate):
Solids loading:
Weir loading:
Sludge underflow:
Percent solids in sludge:

REMOVAL DATA

Sampling period: 1 day

Pollutant/parameter	Concentration Influent	Effluent	Percent removal
Conventional pollutants, mg/L:			
BOD_5	459	168	63
COD	1,740	1,650	5
TSS	165	228	0[a]
Total phenol	0.092	0.065	29
Total phosphorus	1.2	0.50	58
Toxic pollutants, µg/L:			
Antimony	<0.5	30	0[a]
Chromium	190	180	5
Copper	21	27	0[a]
Cyanide	<4	15	0[a]
Mercury	4	<0.5	>87
Nickel	1,300	6,400	0[a]
Zinc	1,300	6,400	0[a]
Bis(2-ethylhexyl) phthalate	0.5	6	0[a]
Diethyl phthalate	1	<0.03	>97
Dimethyl phthalate	3	<0.03	>99
Pentachlorophenol	71	<0.4	>99
Phenol	1.2	<0.07	>94
1,2,-Dichlorobenzene	<0.05	1	0[a]
1,4-Dichlorobenzene	11	0.05	>99
Toluene	<0.1	8.4	0[a]
1,2,4-Trichlorobenzene	90	46	49
Naphthalene	0.1	<0.007	>93
Heptachlor	6.4	1.5	76

[a]Actual data indicate negative removal.

Note: Blanks indicate information was not specified.

TREATMENT TECHNOLOGY: Activated Sludge

Data source: Government report
Point source category: Textile mills
Subcategory:
Plant: B
References: B25, pp. 32-53

Data source status:
Engineering estimate ___
Bench scale ___
Pilot scale ___
Full scale X

Use in system: Secondary
Pretreatment of influent:

DESIGN OR OPERATING PARAMETERS

Aeration method: Surface aeration
Wastewater flow:
Hydraulic aeration detention time:
Volumetric loading:
MLSS:
Volatile fraction of MLSS:
F/M:
Mean cell residence time:
Sludge recycle ratio:
Mixed liquor dissolved oxygen:
Oxygen consumption:
Aerator power requirement:

Secondary clarifier configuration:
Depth:
Hydraulic loading (overflow rate):
Solids loading:
Weir loading:
Sludge underflow:
Percent solids in sludge:

REMOVAL DATA

Sampling period: 1 day

Pollutant/parameter	Concentration Influent	Concentration Effluent	Percent removal
Conventional pollutants, mg/L:			
BOD_5	1,050	<5	>99
COD	1,260	99	92
TSS	32	8	75
Total phenol	0.042	0.015	64
Total phosphorus	12	6.5	46
Toxic pollutants, µg/L:			
Cadmium	0.7	<0.5	>29
Chromium	12	4	67
Copper	74	30	59
Cyanide	17	<4	>76
Mercury	0.9	0.6	33
Zinc	300	170	43
Bis(2-ethylhexyl) phthalate	5.7	3	47
Diethyl phthalate	3.3	<0.03	>99
N-nitroso-di-n-propylamine	<0.2	2	0[a]
Toluene	3.7	<0.1	>97
Anthracene/phenanthrene	0.1	<0.01	>90
Naphthalene	41	<0.007	>99
Pyrene	<0.01	0.3	0[a]
Trichlorofluoromethane	<2.0	2.6	0[a]

[a]Actual data indicate negative removal.

Note: Blanks indicate information was not specified.

TREATMENT TECHNOLOGY: Activated Sludge

Data source: Government report
Point source category: Textile mills
Subcategory:
Plant: C
References: B5, pp. 32-53

Use in system: Secondary
Pretreatment of influent:

Data source status:
Engineering estimate ___
Bench scale ___
Pilot scale ___
Full scale x

DESIGN OR OPERATING PARAMETERS

Process modification: Surface aeration
Wastewater flow:
Hydraulic aeration detention time:
Volumetric loading:
MLSS:
Volatile fraction of MLSS:
F/M:
Mean cell residence time:
Sludge recycle ratio:
Mixed liquor dissolved oxygen:
Oxygen consumption:
Aerator power requirement:

Secondary clarifier configuration:
Depth:
Hydraulic loading (overflow rate):
Solids loading:
Weir loading:
Sludge underflow:
Percent solids in sludge:

REMOVAL DATA

Sampling period: 1 day

Pollutant/parameter	Concentration Influent	Effluent	Percent removal
Conventional pollutants, mg/L:			
BOD_5	445	25	94
COD	802	396	51
TSS	49	300	0[a]
Total phenol	0.074	0.088	0[a]
Total phosphorus	4.0	4.1	0[a]
Toxic pollutants, µg/L:			
Antimony	7	4	43
Cadmium	5	6	0[a]
Chromium	35	31	11
Copper	8	20	0[a]
Cyanide	7	13	0[a]
Lead	120	120	0
Mercury	<0.5	0.7	0[a]
Nickel	150	140	7
Zinc	74	120	0[a]
Bis(2-ethylhexyl) phthalate	140	3.0	98
Diethyl phthalate	4.1	<0.03	>99
Phenol	0.5	<0.07	>86
1,2-Dichlorobenzene	1.1	0.3	73
Ethylbenzene	110	2.0	98
Toluene	240	2.6	99
1,2,4-Trichlorobenzene	<0.09	10	0[a]
Acenaphthene	<0.04	0.5	0[a]
Anthracene/phenanthrene	<0.01	4.4	0[a]
Tetrachloroethylene	26	<0.9	>97
Trichloroethylene	18	<0.5	>97

[a]Actual data indicate negative removal.

Note: Blanks indicate information was not specified.

TREATMENT TECHNOLOGY: Activated Sludge

Data source: Government report
Point source category: Textile mills
Subcategory:
Plant: D
References: B5, pp. 32-53

Data source status:
- Engineering estimate ___
- Bench scale ___
- Pilot scale ___
- Full scale x

Use in system: Secondary
Pretreatment of influent:

DESIGN OR OPERATING PARAMETERS

Aeration method: Surface aeration
Wastewater flow:
Hydraulic aeration detention time:
Volumetric loading:
MLSS:
Volatile fraction of MLSS:
F/M:
Mean cell residence time:
Sludge recycle ratio:
Mixed liquor dissolved oxygen:
Oxygen consumption:
Aerator power requirement:

Secondary clarifier configuration:
Depth:
Hydraulic loading (overflow rate):
Solids loading:
Weir loading:
Sludge underflow:
Percent solids in sludge:

REMOVAL DATA

Sampling period: 1 day

Pollutant/parameter	Concentration		Percent removal
	Influent	Effluent	
Conventional pollutants, mg/L:			
BOD_5	71	6.6	91
COD	224	64	71
TSS	16	154	0[a]
Total phenol	0.024	0.0.8	25
Total phosphorus	1.6	1.0	37
Toxic pollutants, μg/L:			
Antimony	3	2	33
Arsenic	17	6	65
Copper	31	<0.2	>99
Cyanide	210	210	0
Nickel	30	<10	>67
Silver	11	<5	>55
Zinc	210	210	0
Bis(2-ethylhexyl) phthalate	8.9	5	44
Di-n-butyl phthalate	16	<0.02	>99
Diethyl phthalate	<0.03	1	0[a]
Pentachlorophenol	22	<0.4	>98
Ethylbenzene	57	<0.2	>99
Toluene	2.3	1.3	27
Naphthalene	0.3	<0.007	>98

[a]Actual data indicates negative removal.

Note: Blanks indicate information was not specified.

TREATMENT TECHNOLOGY: Activated Sludge

Data source: Government report
Point source category: Textile mills
Subcategory:
Plant: E
References: B5, pp. 32-53

Data source status:
Engineering estimate ___
Bench scale ___
Pilot scale ___
Full scale x

Use in system: Secondary
Pretreatment of influent:

DESIGN OR OPERATING PARAMETERS

Aeration method: Surface aeration
Wastewater flow:
Hydraulic aeration detention time:
Volumetric loading:
MLSS:
Volatile fraction of MLSS:
F/M:
Mean cell residence time:
Sludge recycle ratio:
Mixed liquor dissolved oxygen:
Oxygen consumption:
Aerator power requirement:

Secondary clarifier configuration:
Depth:
Hydraulic loading (overflow rate):
Solids loading:
Weir loading:
Sludge underflow:
Percent solids in sludge:

REMOVAL DATA

Sampling Period: 1 Day

Pollutant/Parameter	**..... Concentration .....**		**Percent**
	Influent	**Effluent**	**Removal**
Conventional pollutants, mg/ℓ			
BOD_5	18	<5	>72
COD	2,660	79	97
TSS	52	19	63
Total phenol	0.069	0.014	80
Total phosphorus	1.9	1.4	26
Toxic pollutants, μg/ℓ			
Antimony	8	0.8	90
Cadmium	6	1	83
Chromium	11	4	64
Copper	840	30	96
Lead	8	<1	>87
Nickel	40	40	0
Silver	7	<5	>29
Zinc	7,900	5,100	35
Bis(2-ethylhexyl) phthalate	5	18	0*
Diethyl phthalate	<0.03	0.5	0*
Dimethyl phthalate	<0.03	1	0*
Pentachlorophenol	30	<0.4	>99
Phenol	5.7	<0.07	>99
Benzene	5.4	<0.2	>96
Chlorobenzene	1.0	<0.2	>80
1,2-Dichlorobenzene	<0.05	0.2	0*
1,4-Dichlorobenzene	2	0.2	90
Ethylbenzene	21	<0.2	>99
Toluene	61	5.5	91
Naphthalene	1	<0.007	>99
Pyrene	<0.01	0.1	0*
Chloroform	22	<5	>77
1,1,1-Trichloroethane	17	<2.0	>88
Trichloroethylene	2.0	<0.5	>75

*Actual data indicate negative removal.

Note: Blanks indicate information was not specified.

TREATMENT TECHNOLOGY: Activated Sludge

Data source: Government report
Point source category: Textile mills
Subcategory:
Plant: F
References: B5, pp. 32-53

Data source status:
Engineering estimate ___
Bench scale ___
Pilot scale ___
Full scale x

Use in system: Secondary
Pretreatment of influent:

DESIGN OR OPERATING PARAMETERS

Aeration method: Surface aeration
Wastewater flow:
Hydraulic aeration detention time:
Volumetric loading:
MLSS:
Volatile fraction of MLSS:
F/M:
Mean cell residence time:
Sludge recycle ratio:
Mixed liquor dissolved oxygen:
Oxygen consumption:
Aerator power requirement:

Secondary clarifier configuration:
Depth:
Hydraulic loading (overflow rate):
Solids loading:
Weir loading:
Sludge underflow:
Percent solids in sludge:

REMOVAL DATA

Sampling Period: 1 Day

Pollutant/Parameter	Concentration Influent	Concentration Effluent	Percent Removal
Conventional pollutants, mg/ℓ			
BOD_5	194	69	64
COD	583	276	53
TSS	23	44	0*
Total phenol	0.74	0.028	96
Total phosphorus	24	9.5	60
Toxic pollutants, μg/ℓ			
Antimony	1	0.3	70
Cadmium	10	10	0
Chromium	6	4	33
Copper	590	130	78
Lead	80	0.6	99
Mercury	<0.5	0.9	0*
Nickel	100	60	40
Silver	100	80	20
Zinc	260	570	0*
Bis(2-ethylhexyl) phthalate	<0.04	23	0*
Diethyl phthalate	34	<0.03	>99
2,4-Dimethylphenol	<0.1	9	0*
Pentachlorophenol	2.4	<0.4	>83
Phenol	8.2	<0.07	>99
1,2-Dichlorobenzene	35	<0.05	>99
1,4-Dichlorobenzene	6.5	<0.04	>99
Ethylbenzene	<0.2	2.7	0*
Toluene	12	0.85	93
1,2,4-Trichlorobenzene	120	6.3	95
Acenaphthene	12	<0.04	>99
Fluorene	15	<0.02	>99
1,2-Dichloropropane	1.5	<0.7	>53
1,1,1-Trichloroethane	11	<2.0	>82
Trichlorofluoromethane	45	1.7	96

*Actual data indicate negative removal.

Note: Blanks indicate information was not specified.

TREATMENT TECHNOLOGY: Activated Sludge

Data source: Government report
Point source category: Textile mills
Subcategory:
Plant: J
References: B5, pp. 32-53

Data source status:
Engineering estimate ___
Bench scale ___
Pilot scale ___
Full scale x

Use in system: Secondary
Pretreatment of influent:

DESIGN OR OPERATING PARAMETERS

Aeration method: Surface aeration
Wastewater flow:
Hydraulic aeration detention time:
Volumetric loading:
MLSS:
Volatile fraction of MLSS:
F/M:
Mean cell residence time:
Sludge recycle ratio:
Mixed liquor dissolved oxygen:
Oxygen consumption:
Aerator power requirement:

Secondary clarifier configuration:
Depth:
Hydraulic loading (overflow rate):
Solids loading:
Weir loading:
Sludge underflow:
Percent solids in sludge:

REMOVAL DATA

Sampling period: 1 day

Pollutant/parameter	Concentration Influent	Concentration Effluent	Percent removal
Conventional pollutants, mg/L:			
BOD_5	210	25	88
COD	810	376	54
Total phenol	0.063	0.024	62
Total phosphorus	3.3	0.6	82
Toxic pollutants, μg/L:			
Antimony	0.7	<0.5	>29
Chromium	48	25	48
Copper	2,400	100	96
Lead	29	<1	>97
Nickel	97	90	7
Silver	60	<5	>92
Zinc	2,100	800	62
Bis(2-ethylhexyl) phthalate	160	35	78
Di-n-butyl phthalate	23	3.6	84
Diethyl phthalate	6.5	<0.03	>99
Ethylbenzene	<0.2	51	0[a]
Toluene	36	8.0	78
Naphthalene	80	<0.007	>99
Pyrene	<0.01	0.1	0[a]

[a]Actual data indicate negative removal.

Note: Blanks indicate information was not specified.

TREATMENT TECHNOLOGY: Activated Sludge

Data source: Government report
Point source category: Textile mills
Subcategory:
Plant: JJ
References: B5, pp. 32-53

Data source status:
Engineering estimate ___
Bench scale ___
Pilot scale ___
Full scale x

Use in system: Secondary
Pretreatment of influent:

DESIGN OR OPERATING PARAMETERS

Aeration method: Surface aeration
Wastewater flow:
Hydraulic aeration detention time:
Volumetric loading:
MLSS:
Volatile fraction of MLSS:
F/M:
Mean cell residence time:
Sludge recycle ratio:
Mixed liquor dissolved oxygen:
Oxygen consumption:
Aerator power requirement:

Secondary clarifier configuration:
Depth:
Hydraulic loading (overflow rate):
Solids loading:
Weir loading:
Sludge underflow:
Percent solids in sludge:

REMOVAL DATA

Sampling period: 1 day

Pollutant/parameter	Concentration		Percent removal
	Influent	Effluent	
Conventional pollutants, mg/L:			
COD	1,540	510	67
Total phenol	0.144	0.055	62
Total phosphorus	3.5	2.3	34
Toxic pollutants, μg/L:			
Arsenic	200	160	20
Cadmium	5	5	0
Chromium	160	80	50
Copper	32	31	3
Cyanide	5	28	0[a]
Lead	84	65	23
Nickel	100	120	0[a]
Silver	47	49	0[a]
Zinc	130	320	0[a]
Phenol	41	<0.07	>99
1,2-Dichlorobenzene	11	<0.05	>99
Ethylbenzene	14	<0.2	>99
1,2,4-Trichlorobenzene	440	32	93
Tetrachloroethylene	1,100	<0.9	>99
Trichloroethylene	190	84	55

[a]Actual data indicate negative removal.

Note: Blanks indicate information was not specified.

TREATMENT TECHNOLOGY: Activated Sludge

Data source: Government report
Point source category: Textile mills
Subcategory:
Plant: G
References: B5, pp. 32-53

Data source status:
Engineering estimate ___
Bench scale ___
Pilot scale ___
Full scale x

Use in system: Secondary
Pretreatment of influent:

DESIGN OR OPERATING PARAMETERS

Aeration method: Surface aeration
Wastewater flow:
Hydraulic aeration detention time:
Volumetric loading:
MLSS:
Volatile fraction of MLSS:
F/M:
Mean cell residence time:
Sludge recycle ratio:
Mixed liquor dissolved oxygen:
Oxygen consumption:
Aerator power requirement:

Secondary clarifier configuration:
Depth:
Hydraulic loading (overflow rate):
Solids loading:
Weir loading:
Sludge underflow:
Percent solids in sludge:

REMOVAL DATA

Sampling period:

Pollutant/parameter	Concentration		Percent removal
	Influent	Effluent	
Conventional pollutants, mg/L:			
BOD_5	203	42	79
COD	1,340	502	63
TSS	37	6	84
Total phenol	0.028	0.054	0[a]
Total phosphorus	6.4	6.1	5
Toxic pollutants, µg/L:			
Antimony	52	11	79
Chromium	4	2	25
Copper	63	28	56
Cyanide	<4	6	0[a]
Lead	6	<1	>83
Nickel	28	13	54
Silver	8.5	<5	>41
Zinc	450	260	42
Bis(2-ethylhexyl) phthalate	19	10	46
Diethyl phthalate	<0.03	11	0[a]
Phenol	0.8	2	0[a]
Hexachlorobenzene	<0.05	0.8	0[a]
Toluene	<0.1	0.8	0[a]
Acenaphthene	270	2.0	99
Fluorene	5	<0.02	>99
Naphthalene	95	<0.007	>99
Chloroform	5.2	<5	>4

[a]Actual data indicate negative removal.

Note: Blanks indicate information was not specified.

TREATMENT TECHNOLOGY: Activated Sludge

Data source: Government report
Point source category: Textile mills
Subcategory:
Plant: H
References: B5, pp. 32-53

Data source status:
Engineering estimate ___
Bench scale ___
Pilot scale ___
Full scale x

Use in system: Secondary
Pretreatment of influent:

DESIGN OR OPERATING PARAMETERS

Aeration method: Surface aeration
Wastewater flow:
Hydraulic aeration detention time:
Volumetric loading:
MLSS:
Volatile fraction of MLSS:
F/M:
Mean cell residence time:
Sludge recycle ratio:
Mixed liquor dissolved oxygen:
Oxygen consumption:
Aerator power requirement:

Secondary clarifier configuration:
Depth:
Hydraulic loading (overflow rate):
Solids loading:
Weir loading:
Sludge underflow:
Percent solids in sludge:

REMOVAL DATA

Sampling period: 1 day

Pollutant/parameter	Concentration		Percent
	Influent	Effluent	removal
Conventional pollutants, mg/L:			
BOD_5	288	14	95
COD	320	300	6
TSS	39	43	0[a]
Total phenol	0.047	0.019	60
Total phosphorus	0.99	0.20	80
Toxic pollutants, µg/L:			
Antimony	4	6	0[a]
Chromium	4	<0.2	>95
Copper	22	<0.2	>99
Nickel	14	<10	>29
Silver	41	<5	>88
Bis(2-ethylhexyl) phthalate	14	230	0[a]
Di-n-butyl phthalate	2	<0.02	>99
2-Nitrophenol	60	<0.4	>99
4-Nitrophenol	65	<0.9	>99
Phenol	63	<0.07	>99
p-Chloro-*m*-cresol	4.5	<0.1	>98
1,2-Dichlorobenzene	0.5	<0.05	>90
Ethylbenzene	5.7	<0.2	>96
Toluene	26	12	54
Acenaphthene	27	<0.04	>99
Naphthalene	3	<0.007	>99
Trichlorofluoromethane	<2.0	2,100	0[a]

[a]Actual data indicate negative removal.

Note: Blanks indicate information was not specified.

TREATMENT TECHNOLOGY: Activated Sludge

Data source: Government report
Point source category: Textile mills
Subcategory:
Plant: K
References: B5, pp. 32-53

Data source status:
Engineering estimate ___
Bench scale ___
Pilot scale ___
Full scale x

Use in system: Secondary
Pretreatment of influent:

DESIGN OR OPERATING PARAMETERS

Process modification: Surface aeration
Wastewater flow:
Hydraulic aeration detention time:
Volumetric loading:
MLSS:
Volatile fraction of MLSS:
F/M:
Mean cell residence time:
Sludge recycle ratio:
Mixed liquor dissolved oxygen:
Oxygen consumption:
Aerator power requirement:

Secondary clarifier configuration:
Depth:
Hydraulic loading (overflow rate):
Solids loading:
Weir loading:
Sludge underflow:
Percent solids in sludge:

REMOVAL DATA

Sampling period: 1 day

Pollutant/parameter	Concentration		Percent
	Influent	Effluent	removal
Conventional pollutants, mg/L:			
BOD_5	564	<5	>99
COD	1,720	131	92
TSS	69	21	70
Total phenol	0.067	0.018	73
Total phosphorus	1.9	0.93	51
Toxic pollutants, µg/L:			
Antimony	3	0.018	73
Arsenic	6	<5	>17
Cadmium	4	<0.5	>87
Chromium	19	4	79
Copper	26	15	42
Lead	30	<1	>97
Nickel	100	<10	>90
Silver	130	<5	>96
Zinc	150	110	27
Bis(2-ethylhexyl) phthalate	<0.04	8	0[a]
Diethyl phthalate	0.2	<0.03	>85
Pentachlorophenol	3.9	<0.4	>90
2,4,6-Trichlorophenol	0.7	<0.2	>71
Ethylbenzene	64	0.7	99
Toluene	29	24	18
Naphthalene	0.03	0.5	0[a]
Chloroform	4.8	58	0[a]
Trichloroethylene	<0.5	4.6	0[a]

[a]Actual data indicate negative removal.

Note: Blanks indicate information was not specified.

TREATMENT TECHNOLOGY: Activated Sludge

Data source: Government report
Point source category: Textile mills
Subcategory:
Plant: KK
References: B5, pp. 32-53

Data source status:
Engineering estimate ___
Bench scale ___
Pilot scale ___
Full scale x

Use in system: Secondary
Pretreatment of influent:

DESIGN OR OPERATING PARAMETERS

Aeration method: Surface aeration
Wastewater flow:
Hydraulic aeration detention time:
Volumetric loading:
MLSS:
Volatile fraction of MLSS:
F/M:
Mean cell residence time:
Sludge recycle ratio:
Mixed liquor dissolved oxygen:
Oxygen consumption:
Aerator power requirement:

Secondary clarifier configuration:
Depth:
Hydraulic loading (overflow rate):
Solids loading:
Weir loading:
Sludge underflow:
Percent solids in sludge:

REMOVAL DATA

Sampling period: 1 day

Pollutant/parameter	Concentration Influent	Concentration Effluent	Percent removal
Conventional pollutants, mg/L:			
COD	1,950	447	77
Total phosphorus	6.3	6.4	0[a]
Total phenol	0.150	0.052	65
Toxic pollutants, μg/L:			
Arsenic	120	<5	>96
Cadmium	2	4	0[a]
Chromium	16	13	19
Copper	86	37	57
Lead	49	44	10
Nickel	77	110	0[a]
Silver	22	44	0[a]
Zinc	1,100	390	64
Bis(2-ethylhexyl) phthalate	9.3	4.1	56
Diethyl phthalate	2.5	<0.03	>99
Dimethyl phthalate	11.6	<0.03	>99
2-Chlorophenol	130	10	92
Pentachlorophenol	20	<0.4	>98
2,4,6-Trichlorophenol	20	21	0[a]
Benzene	0.2	64	0[a]
Chlorobenzene	42	26	38
Ethylbenzene	26	<0.2	>99
Toluene	28	<0.1	>99
Pyrene	0.9	0.2	78
Trichloroethylene	52	<0.5	>99

[a]Actual data indicate negative removal.

Note: Blanks indicate information was not specified.

TREATMENT TECHNOLOGY: Activated Sludge

Data source: Government report
Point source category: Textile mills
Subcategory:
Plant: L
References: B5, pp. 32-53

Data source status:
Engineering estimate ___
Bench scale ___
Pilot scale ___
Full scale x

Use in system: Secondary
Pretreatment of influent:

DESIGN OR OPERATING PARAMETERS

Aeration method: Surface aeration
Wastewater flow:
Hydraulic aeration detention time:
Volumetric loading:
MLSS:
Volatile fraction of MLSS:
F/M:
Mean cell residence time:
Sludge recycle ratio:
Mixed liquor dissolved oxygen:
Oxygen consumption:
Aerator power requirement:

Secondary clarifier configuration:
Depth:
Hydraulic loading (overflow rate):
Solids loading:
Weir loading:
Sludge underflow:
Percent solids in sludge:

REMOVAL DATA

Sampling period: 1 day

Pollutant/parameter	Concentration		Percent
	Influent	Effluent	removal
Conventional pollutants, mg/L:			
BOD_5	379	13	97
COD	1,120	234	79
TSS	19	78	0[a]
Total phenol	0.038	0.026	32
Total phosphorus	2.2	1.6	27
Toxic pollutants, μg/L:			
Antimony	5	3	40
Chromium	3	30	0[a]
Copper	300	96	68
Cyanide	<4	170	0[a]
Lead	36	<1	>97
Nickel	54	35	35
Zinc	1,000	720	28
Bis(2-ethylhexyl) phthalate	3	2	33
Dimethyl phthalate	110	<0.03	>99
Benzene	<0.2	0.5	0[a]
1,4-Dichlorobenzene	1	<0.04	>96
Ethylbenzene	2.0	<0.2	>90
Toluene	5.2	<0.2	>98
Acenaphthene	30	<0.04	>99

[a]Actual data indicate negative removal.

Note: Blanks indicate information was not specified.

TREATMENT TECHNOLOGY: Activated Sludge

Data source: Government report
Point source category: Textile mills
Subcategory:
Plant: LL
References: B5, pp. 32-53

Data source status:
Engineering estimate ___
Bench scale ___
Pilot scale ___
Full scale x

Use in system: Secondary
Pretreatment of influent:

DESIGN OR OPERATING PARAMETERS

Aeration method: Surface aeration
Wastewater flow:
Hydraulic aeration detention time:
Volumetric loading:
MLSS:
Volatile fraction of MLSS:
F/M:
Mean cell residence time:
Sludge recycle ratio:
Mixed liquor dissolved oxygen:
Oxygen consumption:
Aerator power requirement:

Secondary clarifier configuration:
Depth:
Hydraulic loading (overflow rate):
Solids loading:
Weir loading:
Sludge underflow:
Percent solids in sludge:

REMOVAL DATA

Sampling period: 1 day

Pollutant/parameter	Concentration Influent	Effluent	Percent removal
Conventional pollutants, mg/L:			
COD	727	155	79
Total phenol	0.001	0.094	0[a]
Total phosphorus	18.8	28.8	0[a]
Toxic pollutants, µg/L:			
Arsenic	100	70	30
Cadmium	4	2	50
Chromium	11	20	0[a]
Copper	38	92	0[a]
Cyanide	8	6	25
Lead	60	48	20
Nickel	130	150	0[a]
Silver	58	56	3
Zinc	67	68	0[a]
Bis(2-ethylhexyl) phthalate	0.04	5.2	0[a]
Dimethyl phthalate	<0.03	0.2	0[a]
Phenol	16	<0.07	>99
1,2-Dichlorobenzene	0.6	<0.05	>92
Ethylbenzene	480	<0.2	>99
1,2,4-Trichlorobenzene	320	<0.09	>99
Naphthalene	51	<0.007	>99
Chloroform	500	<5	>99
Tetrachloroethylene	1,100	<0.9	>99

[a]Actual data indicate negative removal.

Note: Blanks indicate information was not specified.

TREATMENT TECHNOLOGY: Activated Sludge

Data source: Government report
Point source category: Textile mills
Subcategory:
Plant: M
References: B5, pp. 32-53

Data source status:
Engineering estimate ___
Bench scale ___
Pilot scale ___
Full scale x

Use in system: Secondary
Pretreatment of influent:

DESIGN OR OPERATING PARAMETERS

Aeration method: Surface aeration
Wastewater flow:
Hydraulic aeration detention time:
Volumetric loading:
MLSS:
Volatile fraction of MLSS:
F/M:
Mean cell residence time:
Sludge recycle ratio:
Mixed liquor dissolved oxygen:
Oxygen consumption:
Aerator power requirement:

Secondary clarifier configuration:
Depth:
Hydraulic loading (overflow rate):
Solids loading:
Weir loading:
Sludge underflow:
Percent solids in sludge:

REMOVAL DATA

Sampling period:

Pollutant/parameter	Concentration Influent	Effluent	Percent removal
Conventional pollutants, mg/L:			
BOD_5	830	<5	>99
COD	2,260	255	89
TSS	210	21	90
Total phenol	0.037	0.025	32
Total phosphorus	3.99	3.46	13
Toxic pollutants, µg/L:			
Antimony	0.8	4	0[a]
Copper	9	5	44
Zinc	1,200	410	66
Bis(2-ethylhexyl) phthalate	300	<0.04	>99
Di-n-butyl phthalate	<0.02	58	0[a]
Pentachlorophenol	6.9	<0.4	>94
Phenol	12	<0.07	>99
Toluene	<0.1	0.4	0[a]
1,2,4-Trichlorobenzene	160	1.8	99
Naphthalene	93	<0.007	>99

[a]Actual data indicate negative removal.

Note: Blanks indicate information was not specified.

TREATMENT TECHNOLOGY: Activated Sludge

Data source: Government report
Point source category: Textile mills
Subcategory:
Plant: N
References: B5, pp. 32-53

Data source status:
Engineering estimate ___
Bench scale ___
Pilot scale ___
Full scale x

Use in system: Secondary
Pretreatment of influent:

DESIGN OR OPERATING PARAMETERS

Aeration method: Surface aeration
Wastewater flow:
Hydraulic aeration detention time:
Volumetric loading:
MLSS:
Volatile fraction of MLSS:
F/M:
Mean cell residence time:
Sludge recycle ratio:
Mixed liquor dissolved oxygen:
Oxygen consumption:
Aerator power requirement:

Secondary clarifier configuration:
Depth:
Hydraulic loading (overflow rate):
Solids loading:
Weir loading:
Sludge underflow:
Percent solids in sludge:

REMOVAL DATA

Sampling period: 1 day

Pollutant/parameter	Concentration		Percent
	Influent	Effluent	removal
Conventional pollutants, mg/L:			
BOD_5	334	36	89
COD	1,140	286	75
TSS	68	77	0[a]
Total phenol	0.156	0.068	56
Total phosphorus	0.43	5.2	0[a]
Toxic pollutants, µg/L:			
Antimony	0.2	2	0[a]
Cadmium	46	<0.5	>99
Chromium	880	1,800	0[a]
Copper	20	8	60
Nickel	<10	30	0[a]
Zinc	7,500	38,000	0[a]
Bis(2-ethylhexyl) phthalate	10	17	0[a]
Diethyl phthalate	5.9	9.4	0[a]
2,4-Dimethylphenol	<0.1	8	0[a]
Phenol	11	<0.07	>99
1,2-Dichlorobenzene	290	6.0	98
1,4-Dichlorobenzene	220	1.5	99
Ethylbenzene	1,800	75	96
Toluene	44	17	62
Naphthalene	17	<0.007	>99
Trichloroethylene	21	<0.5	>98

[a]Actual data indicate negative removal.

Note: Blanks indicate information was not specified.

TREATMENT TECHNOLOGY: Activated Sludge

Data source: Government report
Point source category: Textile mills
Subcategory:
Plant: NN
References: B5, pp. 32-53

Data source status:
Engineering estimate ___
Bench scale ___
Pilot scale ___
Full scale x

Use in system: Secondary
Pretreatment of influent:

DESIGN OR OPERATING PARAMETERS

Aeration method: Surface aeration
Wastewater flow:
Hydraulic aeration detention time:
Volumetric loading:
MLSS:
Volatile fraction of MLSS:
F/M:
Mean cell residence time:
Sludge recycle ratio:
Mixed liquor dissolved oxygen:
Oxygen consumption:
Aerator power requirement:

Secondary clarifier configuration:
Depth:
Hydraulic loading (overflow rate):
Solids loading:
Weir loading
Sludge underflow:
Percent solids in sludge:

REMOVAL DATA

Sampling Period—1 Day

Pollutant/Parameter	Concentration Influent	Concentration Effluent	Percent Removal
Conventional pollutants, mg/ℓ			
COD	938	236	75
Total phenol	0.043	0.014	67
Total phosphorus	48.8	46.8	4
Toxic pollutants, μg/ℓ			
Cadmium	2	4	0*
Chromium	23	170	0*
Copper	47	46	2
Cyanide	40	<4	>90
Lead	33	25	24
Nickel	98	79	19
Silver	42	33	21
Zinc	84	130	0*
Bis(2-ethylhexyl) phthalate	23	27	0*
Phenol	10	<0.07	>99

*Actual data indicate negative removal.

Note: Blanks indicate information was not specified.

TREATMENT TECHNOLOGY: Activated Sludge

Data source: Government report
Point source category: Textile mills
Subcategory:
Plant: OO
References: B5, pp. 32-53

Data source status:
Engineering estimate ___
Bench scale ___
Pilot scale ___
Full scale x

Use in system: Secondary
Pretreatment of influent:

DESIGN OR OPERATING PARAMETERS

Aeration method: Surface aeration
Wastewater flow:
Hydraulic aeration detention time:
Volumetric loading:
MLSS:
Volatile fraction of MLSS:
F/M:
Mean cell residence time:
Sludge recycle ratio:
Mixed liquor dissolved oxygen:
Oxygen consumption:
Aerator power requirement:

Secondary clarifier configuration:
Depth:
Hydraulic loading (overflow rate):
Solids loading:
Weir loading:
Sludge underflow:
Percent solids in sludge:

REMOVAL DATA

Sampling period: 1 day

Pollutant/parameter	Concentration		Percent removal
	Influent	Effluent	
Conventional pollutants, mg/L:			
COD	1,890	635	66
Total phenol	0.082	0.026	68
Total phosphorus	4.6	0.66	86
Toxic pollutants, µg/L:			
Cadmium	4	5	0[a]
Chromium	11	12	0[a]
Copper	39	37	5
Lead	43	84	0[a]
Nickel	110	120	0[a]
Silver	46	50	0[a]
Zinc	120	2,300	0[a]
Bis(2-ethylhexyl) phthalate	26	3.2	88
Di-n-butyl phthalate	61	<0.02	>99
Phenol	23	<0.07	>99
Toluene	<0.1	3	0[a]
Chloroform	48	10	79
Trichloroethylene	42	<0.5	>99

[a]Actual data indicate negative removal.

Note: Blanks indicate information was not specified.

TREATMENT TECHNOLOGY: Activated Sludge

Data source: Government report
Point source category: Textile mills
Subcategory:
Plant: P
References: B5, pp. 32-53

Data source status:
Engineering estimate ___
Bench scale ___
Pilot scale ___
Full scale x

Use in system: Secondary
Pretreatment of influent:

DESIGN OR OPERATING PARAMETERS

Aeration method: Surface aeration
Wastewater flow:
Hydraulic aeration detention time:
Volumetric loading:
MLSS:
Volatile fraction of MLSS:
F/M:
Mean cell residence time:
Sludge recycle ratio:
Mixed liquor dissolved oxygen:
Oxygen consumption:
Aerator power requirement:

Secondary clarifier configuration:
Depth:
Hydraulic loading (overflow rate):
Solids loading:
Weir loading:
Sludge underflow:
Percent solids in sludge:

REMOVAL DATA

Sampling period: 1 day

Pollutant/parameter	Concentration		Percent
	Influent	Effluent	removal
Conventional pollutants, mg/L:			
BOD_5	680	28	96
COD	172	45	74
TSS	6	45	0[a]
Total phenol	0.228	0.032	86
Total phosphorus	5.7	2.2	61
Toxic pollutants, µg/L:			
Chromium	3	<0.2	>93
Cyanide	190	140	26
Lead	13	<1	>92
Nickel	100	40	60
Silver	30	8	73
Zinc	200	140	30
Bis(2-ethylhexyl) phthalate	30	72	0[a]
Di-n-butyl phthalate	9.8	<0.02	>99
Diethyl phthalate	1.7	<0.03	>98
Dimethyl phthalate	12	<0.03	>99
N-nitroso-di-n-propylamine	<0.2	19	0[a]
Phenol	6.6	<0.07	>99
Chlorobenzene	25	<0.2	>99
Ethylbenzene	1,200	280	77
Toluene	36	22	38
Naphthalene	1.9	<0.007	>99
Chloroform	17	6.9	60

[a]Actual data indicate negative removal.

Note: Blanks indicate information was not specified.

TREATMENT TECHNOLOGY: Activated Sludge

Data source: Effluent Guidelines
Point source category: Textile mills
Subcategory: Knit fabric finishing
Plant: Q
References: A6, p. VIII-58

Data source status:
Engineering estimate ___
Bench scale ___
Pilot scale ___
Full scale x

Use in system: Secondary
Pretreatment of influent:

DESIGN OR OPERATING PARAMETERS

Aeration method: Surface aeration
Wastewater flow: 9,500 m^3/d (2.5 mgd)
Hydraulic aeration detention time: 15 hr
Volumetric loading:
MLSS:
Volatile fraction of MLSS:
F/M:
Mean cell residence time:
Sludge recycle ratio:
Mixed liquor dissolved oxygen:
Oxygen consumption:
Aerator power requirement: 29.2 W/m^3 (148 hp/Mgal)

Secondary clarifier configuration:
Depth:
Hydraulic loading (overflow rate):
Solids loading:
Weir loading:
Sludge underflow:
Percent solids in sludge:

REMOVAL DATA

Sampling period: Effluent concentration is an average of two 24-hr composite samples, conventional pollutant influent concentration is a 48-hr composite sample, toxic pollutant influent concentration is an average of two 24-hr composite samples

Pollutant/parameter	Concentration Influent	Concentration Effluent	Percent removal
Conventional pollutants, mg/L:			
COD	782	312	60
TSS	17	28	0[a]
Oil and grease	324	303	6
Toxic pollutants, μg/L:			
Antimony	95	670	0[a]
Chromium	14	32	0[a]
Copper	44	100	0[a]
Cyanide	10	ND[b]	>0
Lead	36	48	0[a]
Nickel	36	ND	<72
Selenium	15	41	0[a]
Silver	12	13	0[a]
Zinc	56	48	14
Bis(2-ethylhexyl) phthalate	41	15	63
Phenol	55	ND	>82
Ethylbenzene	100	ND	>90
1,2,4-Trichlorobenzene	2,700	ND	>99
Naphthalene	45	ND	>78
Tetrachloroethylene	ND	17	0[a]
Trichloroethylene	840	ND	>99

[a]Actual data indicate negative removal.

[b]Not detected.

Note: Blanks indicate information was not specified.

TREATMENT TECHNOLOGY: Activated Sludge

Data source: Government report
Point source category: Textile mills
Subcategory:
Plant: S
References: B5, pp. 32-53

Data source status:
Engineering estimate ___
Bench scale ___
Pilot scale ___
Full scale x

Use in system: Secondary
Pretreatment of influent:

DESIGN OR OPERATING PARAMETERS

Aeration method: Surface aeration
Wastewater flow:
Hydraulic aeration detention time:
Volumetric loading:
MLSS:
Volatile fraction of MLSS:
F/M:
Mean cell residence time:
Sludge recycle ratio:
Mixed liquor dissolved oxygen:
Oxygen consumption:
Aerator power requirement:

Secondary clarifier configuration:
Depth:
Hydraulic loading (overflow rate):
Solids loading:
Weir loading:
Sludge underflow:
Percent solids in sludge:

REMOVAL DATA

Sampling period: 1 day

Pollutant/parameter	Concentration		Percent removal
	Influent	Effluent	
Conventional pollutants, mg/L:			
BOD_5	219	59	73
COD	559	1,040	0[a]
TSS	25	581	0[a]
Total phenol	0.107	0.029	73
Total phosphorus	1.6	5.0	0[a]
Toxic pollutants, μg/L:			
Antimony	57	74	0[a]
Arsenic	5	<5	>0
Chromium	0.7	<0.2	>71
Copper	40	60	0[a]
Cyanide	7	<4	>43
Zinc	120	84	30
Bis(2-ethylhexyl) phthalate	140	41	70
Chlorobenzene	14	<0.2	>99
Ethylbenzene	850	110	87
Toluene	61	21	65
1,2,4-Trichlorobenzene	190	920	0[a]
Naphthalene	140	260	0[a]
Chloroform	71	<5	>93
Tetrachloroethylene	39	0.4	99

[a]Actual data indicate negative removal.

Note: Blanks indicate information was not specified.

TREATMENT TECHNOLOGY: Activated Sludge

Data source: Government report
Point source category: Textile mills
Subcategory:
Plant: T
References: B5, pp. 32-53

Data source status:
Engineering estimate ___
Bench scale ___
Pilot scale ___
Full scale x

Use in system: Secondary
Pretreatment of influent:

DESIGN OR OPERATING PARAMETERS

Aeration method: Surface aeration
Wastewater flow:
Hydraulic aeration detention time:
Volumetric loading:
MLSS:
Volatile fraction of MLSS:
F/M:
Mean cell residence time:
Sludge recycle ratio:
Mixed liquor dissolved oxygen:
Oxygen consumption:
Aerator power requirement:

Secondary clarifier configuration:
Depth:
Hydraulic loading (overflow rate):
Solids loading:
Weir loading:
Sludge underflow:
Percent solids in sludge:

REMOVAL DATA

Sampling period: 1 day

Pollutant/parameter	Concentration Influent	Effluent	Percent removal
Conventional pollutants, mg/L:			
BOD_5	501	32	94
COD	500	414	17
TSS	28	35	0[a]
Total phenol	0.073	0.041	44
Total phosphorus	12	17	0[a]
Toxic pollutants, μg/L:			
Copper	120	60	50
Lead	25	<1	>96
Mercury	0.7	<0.5	>29
Nickel	50	4	92
Zinc	290	80	72
Bis(2-ethylhexyl) phthalate	140	23	83
N-nitrosodiphenylamine	11	<0.07	>99
Ethylbenzene	18	<0.2	>99
Toluene	300	33	89
Tetrachloroethylene	6.4	2.9	55

[a]Actual data indicate negative removal.

Note: Blanks indicate information was not specified.

TREATMENT TECHNOLOGY: Activated Sludge

Data source: Government report
Point source category: Textile mills
Subcategory:
Plant: U
References: B5, pp. 32-53

Data source status:
Engineering estimate ___
Bench scale ___
Pilot scale ___
Full scale x

Use in system: Secondary
Pretreatment of influent:

DESIGN OR OPERATING PARAMETERS

Aeration method: Surface aeration
Wastewater flow:
Hydraulic aeration detention time:
Volumetric loading:
MLSS:
Volatile fraction of MLSS:
F/M:
Mean cell residence time:
Sludge recycle ratio:
Mixed liquor dissolved oxygen:
Oxygen consumption:
Aerator power requirement:

Secondary clarifier configuration:
Depth:
Hydraulic loading (overflow rate):
Solids loading:
Weir loading:
Sludge underflow:
Percent solids in sludge:

REMOVAL DATA

Sampling period: 1 day

Pollutant/parameter	Concentration		Percent
	Influent	Effluent	removal
Conventional pollutants, mg/L:			
BOD_5	400	24	94
COD	1,460	748	49
TSS	111	92	17
Total phenol	0.057	0.007	86
Total phosphorus	3.5	3.7	0[a]
Toxic pollutants, μg/L:			
Antimony	7	1	86
Chromium	27	14	48
Copper	40	23	42
Cyanide	<4	210	0[a]
Zinc	260	190	27
Bis(2-ethylhexyl) phthalate	14	140	0[a]
Diethyl phthalate	6.1	<0.03	>99
Pentachlorophenol	1.6	<0.4	>75
Phenol	0.7	<0.07	>90
1,2-Dichlorobenzene	2.0	<0.05	>97
Toluene	<0.1	13	0[a]
Naphthalene	1.5	22	0[a]
Chloroform	<5.0	18	0[a]
Dichlorobromomethane	<0.9	1.5	0[a]
1,1-Dichloroethane	3.7	<3.0	>18
1,3-Dichloropropene	<0.5	0.89	0[a]
1,1,1-Trichloroethane	310	<2.0	>99

[a]Actual data indicate negative removal.

Note: Blanks indicate information was not specified.

TREATMENT TECHNOLOGY: Activated Sludge

Data source: Government report
Point source category: Textile mills
Subcategory:
Plant: V
References: B5, pp. 32-53

Data source status:
Engineering estimate ___
Bench scale ___
Pilot scale ___
Full scale x

Use in system: Secondary
Pretreatment of influent:

DESIGN OR OPERATING PARAMETERS

Aeration method: Surface aeration
Wastewater flow:
Hydraulic aeration detention time:
Volumetric loading:
MLSS:
Volatile fraction of MLSS:
F/M:
Mean cell residence time:
Sludge recycle ratio:
Mixed liquor dissolved oxygen:
Oxygen consumption:
Aerator power requirement:

Secondary clarifier configuration:
Depth:
Hydraulic loading (overflow rate):
Solids loading:
Weir loading:
Sludge underflow:
Percent solids in sludge:

REMOVAL DATA

Sampling period: 1 day

Pollutant/parameter	Concentration Influent	Effluent	Percent removal
Conventional pollutants, mg/L:			
BOD_5	53	<5	91
TSS	54	26	52
Total phenol	0.018	0.016	11
Total phosphorus	0.75	0.78	0[a]
Toxic pollutants, µg/L:			
Antimony	<0.5	4	0[a]
Cadmium	5	<0.5	>90
Chromium	4	3	25
Copper	230	170	26
Cyanide	6	18	0[a]
Zinc	460	340	26
Bis(2-ethylhexyl) phthalate	5.3	9.5	0[a]
Dimethyl phthalate	13	<0.03	>99
Ethylbenzene	4.9	<0.2	>96
Hexachlorobenzene	2.0	<0.05	>97
Toluene	8.4	1,400	0[a]
1,2,4-Trichlorobenzene	28	<0.09	>99
Acenaphthene	8.7	<0.04	>99

[a]Actual data indicate negative removal.

Note: Blanks indicate information was not specified.

TREATMENT TECHNOLOGY: Activated Sludge

Data source: Government report
Point source category: Textile mills
Subcategory:
Plant: X
References: B5, pp. 32-53

Data source status:
- Engineering estimate ___
- Bench scale ___
- Pilot scale ___
- Full scale x

Use in system: Secondary
Pretreatment of influent:

DESIGN OR OPERATING PARAMETERS

Aeration method: Surface aeration
Wastewater flow:
Hydraulic aeration detention time:
Volumetric loading:
MLSS:
Volatile fraction of MLSS:
F/M:
Mean cell residence time:
Sludge recycle ratio:
Mixed liquor dissolved oxygen:
Oxygen consumption:
Aerator power requirement:

Secondary clarifier configuration:
Depth:
Hydraulic loading (overflow rate):
Solids loading:
Weir loading:
Sludge underflow:
Percent solids in sludge:

REMOVAL DATA

Sampling period: 1 day

Pollutant/parameter	Concentration		Percent removal
	Influent	Effluent	
Conventional pollutants, mg/L:			
BOD_5	237	15	94
COD	786	258	67
TSS	24	18	25
Total phenol	0.940	0.035	96
Total phosphorus	4.6	5.4	0[a]
Toxic pollutants, μg/L:			
Antimony	0.3	0.9	0[a]
Cadmium	5	7	0[a]
Chromium	24	39	0[a]
Copper	84	110	0[a]
Cyanide	<4	100	0[a]
Lead	32	26	19
Mercury	<0.5	0.9	0[a]
Nickel	110	72	35
Silver	17	33	0[a]
Zinc	34	78	0[a]
Bis(2-ethylhexyl) phthalate	1	2.3	0[a]
Diethyl phthalate	<0.03	3.2	0[a]
Phenol	3.8	<0.07	>98
Ethylbenzene	370	<0.2	>99
Hexachlorobenzene	<0.05	0.5	0[a]
Toluene	64	40	38
Acenaphthene	53	-[b]	>99
Naphthalene	1	-[b]	>99
Tetrachloroethylene	410	40	90
1,1,1-Trichloroethane	8.2	<2.0	>76
Trichlorofluoromethane	<2.0	35	0[a]

[a]Actual data indicate negative removal.
[b]No priority pollutant observed.

Note: Blanks indicate information was not specified.

TREATMENT TECHNOLOGY: Activated Sludge

Data source: Government report
Point source category: Textile mills
Subcategory:
Plant: Y-001
References: B5, pp. 32-53

Data source status:
Engineering estimate ___
Bench scale ___
Pilot scale ___
Full scale x

Use in system: Secondary
Pretreatment of influent:

DESIGN OR OPERATING PARAMETERS

Aeration method: Surface aeration
Wastewater flow:
Hydraulic aeration detention time:
Volumetric loading:
MLSS:
Volatile fraction of MLSS:
F/M:
Mean cell residence time:
Sludge recycle ratio:
Mixed liquor dissolved oxygen:
Oxygen consumption:
Aerator power requirement:

Secondary clarifier configuration:
Depth:
Hydraulic loading (overflow rate):
Solids loading:
Weir loading:
Sludge underflow:
Percent solids in sludge:

REMOVAL DATA

Sampling period: 1 day

Pollutant/parameter	Concentration		Percent
	Influent	Effluent	removal
Conventional pollutants, mg/L:			
Total phosphorus	11.7	6.8	42
Toxic pollutants, μg/L:			
Cadmium	6	7	0[a]
Chromium	650	290	55
Copper	41	<0.2	>99
Cyanide	<4	29	0[a]
Lead	160	160	0
Nickel	200	160	20
Silver	68	57	16
Zinc	130	100	23
Bis(2-ethylhexyl) phthalate	3	13	0[a]
Diethyl phthalate	15	12	22
Phenol	19	2.9	85
p-Chloro-*m*-cresol	<0.1	1.6	0[a]
Chlorobenzene	1.6	<0.2	>85
Ethylbenzene	1.9	<0.2	>89
Toluene	12	15	0[a]
Acenaphthene	13	<0.04	>99
Indeno(1,2,3-cd)pyrene	2	<0.02	>99
Naphthalene	4	4.5	0[a]
Chloroform	14	<5	>65

[a]Actual data indicate negative removal.

Note: Blanks indicate information was not specified.

TREATMENT TECHNOLOGY: Activated Sludge

Data source: Government report
Point source category: Textile mills
Subcategory:
Plant: W
References: B5, pp. 32-53

Data source status:
Engineering estimate ___
Bench scale ___
Pilot scale ___
Full scale x

Use in system: Secondary
Pretreatment of influent:

DESIGN OR OPERATING PARAMETERS

Process modification: Oxidation ditch
Wastewater flow:
Hydraulic aeration detention time:
Volumetric loading:
MLSS:
Volatile fraction of MLSS:
F/M:
Mean cell residence time:
Sludge recycle ratio:
Mixed liquor dissolved oxygen:
Oxygen consumption:
Aerator power requirement:

Secondary clarifier configuration:
Depth:
Hydraulic loading (overflow rate):
Solids loading:
Weir loading:
Sludge underflow:
Percent solids in sludge:

REMOVAL DATA

Sampling period: 1 day

Pollutant/parameter	Concentration		Percent removal
	Influent	Effluent	
Conventional pollutants, mg/L:			
BOD_5	1,920	84	96
COD	6,120	837	86
TSS	2,300	300	87
Total phenol	0.670	0.232	65
Total phosphorus	5.1	0.15	97
Toxic pollutants, µg/L:			
Cadmium	9	13	0[a]
Chromium	12	3	75
Copper	23	2	91
Cyanide	15	20	0[a]
Lead	18	57	0[a]
Mercury	<0.5	0.5	0[a]
Nickel	54	60	0[a]
Silver	65	95	0[a]
Zinc	190	90	53
Bis(2-ethylhexyl) phthalate	18	19	0[a]
Phenol	100	<0.07	>99
Benzene	19	<0.2	>99
Ethylbenzene	1.1	<0.2	>82
Hexachlorobenzene	0.5	<0.05	>90
Toluene	62	1.7	97
Trichloroethylene	13	<0.5	>96

[a]Actual data indicate negative removal.

Note: Blanks indicate information was not specified.

TREATMENT TECHNOLOGY: Activated Sludge

Data source: Government report
Point source category: Textile mills
Subcategory:
Plant: Z
References: B5, pp. 32-53

Data source status:
- Engineering estimate ___
- Bench scale ___
- Pilot scale ___
- Full scale x

Use in system: Secondary
Pretreatment of influent:

DESIGN OR OPERATING PARAMETERS

Aeration method: Surface aeration
Wastewater flow:
Hydraulic aeration detention time:
Volumetric loading:
MLSS:
Volatile fraction of MLSS:
F/M:
Mean cell residence time:
Sludge recycle ratio:
Mixed liquor dissolved oxygen:
Oxygen consumption:
Aerator power requirement:

Secondary clarifier configuration:
Depth:
Hydraulic loading (overflow rate):
Solids loading:
Weir loading:
Sludge underflow:
Percent solids in sludge:

REMOVAL DATA

Sampling period: 1 day

Pollutant/parameter	Concentration Influent	Concentration Effluent	Percent removal
Conventional pollutants, mg/L:			
BOD_5	351	<5	>99
COD	812	105	87
TSS	20	13	35
Total phenol	0.56	0.023	96
Total phosphorus	1.1	0.5	55
Toxic pollutants, µg/L:			
Antimony	11	12	0[a]
Copper	97	50	48
Nickel	11	<10	>9
Zinc	110	370	0[a]
Bis(2-ethylhexyl) phthalate	220	2	99
Phenol	34	<0.07	>99
Chlorobenzene	<0.2	3.5	0[a]
Ethylbenzene	0.7	3,000	0[a]
Toluene	5.5	110	0[a]
1,2,4-Trichlorobenzene	45	-[b]	>99
Naphthalene	310	-[b]	>99
Tetrachloroethylene	12.0	<0.9	>92
Trichlorofluoromethane	<2.0	89	0[a]

[a]Actual data indicate negative removal.
[b]No priority pollutant observed.

Note: Blanks indicate information was not specified.

TREATMENT TECHNOLOGY: Activated Sludge

Data source: Effluent Guidelines
Point source category: Textile mills
Subcategory: Stock and yarn finishing
Plant:
References: A6, p. VII-61

Data source status:
Engineering estimate ___
Bench scale ___
Pilot scale ___
Full scale x

Use in system: Secondary
Pretreatment of influent: Screening, neutralization

DESIGN OR OPERATING PARAMETERS

Process modification: One 19,900 m^3 (5.25 Mgal) basin, surface aeration (8 aerators)
Wastewater flow: 3,500 m^3/d (925,000 gpd)
Hydraulic aeration detention time: 120 hr
Volumetric loading:
MLSS:
Volatile fraction of MLSS:
F/M:
Mean cell residence time:
Sludge recycle ratio:
Mixed liquor dissolved oxygen:
Oxygen consumption:
Aerator power requirement: 22.5 W/m^3 (114 hp/Mgal)

Secondary clarifier configuration:
Depth:
Hydraulic loading (overflow rate):
Solids loading:
Weir loading:
Sludge underflow:
Percent solids in sludge:

REMOVAL DATA

Sampling period: 72-hr composite

Pollutant/parameter	Concentration Influent	Concentration Effluent	Percent removal
Conventional pollutants, mg/L:			
COD	226	133[a]	41
TSS	25	135[a]	0[b]
Toxic pollutants, µg/L:			
Arsenic	19	<10	>47
Bis(chloromethyl) ether	59	ND[c]	>83
Di-n-butyl phthalate	25	ND[c]	>60
Dimethyl phthalate	18	ND[c]	>44
2,4-Dichlorophenol	20	ND[c]	>50
2,4-Dimethylphenol	190	ND[c]	>95
2,4,6-Trichlorophenol	16	<10	>37
p-Chloro-*m*-cresol	29	ND[c]	>66
1,2-Dichlorobenzene	56	ND[c]	>82
1,2-Dichloropropane	56	ND[c]	>82
Tetrachloroethylene	310	<10	>96
Trichloroethylene	10	ND[c]	>0

[a]Average of maximum and minimum values
[b]Data indicates negative percent removal.
[c]Not detected; assumed to be <10 µg/L.

Note: Blanks indicate information was not specified.

TREATMENT TECHNOLOGY: Activated Sludge

Data source: Effluent Guidelines
Point source category: Timber products
Subcategory: Plywood, hardwood, and wood preserving unless specified
Plant: See below
References: A24, p. 169 unless specified

Data source status:
Engineering estimate ___
Bench scale ___
Pilot scale ___
Full scale ___

Use in system: Secondary
Pretreatment of influent: See below

DESIGN OR OPERATING PARAMETERS

Process modification:
Wastewater flow:
Hydraulic aeration detention time:
Volumetric loading:
MLSS:
Volatile fraction of MLSS:
F/M:
Mean cell residence time:
Sludge recycle ratio:
Mixed liquor dissolved oxygen:
Oxygen consumption:
Aerator power requirement:

Secondary clarifier configuration:
Depth:
Hydraulic loading (overflow rate):
Solids loading:
Weir loading:
Sludge underflow:
Percent solids in sludge:

REMOVAL DATA

Sampling period:

Plant	Pretreatment of influent	BOD_5 Concentration, mg/L Influent	Effluent	Percent removal	TSS Concentration, mg/L Influent	Effluent	Percent removal
24[a,b,c]	Screening, primary clarification, flow equalization	1,980	436	78	523	157	70
5	Primary settling pond	3,500	175	95	151	388	0[d]
3	Primary clarifier	1,800	54	96	114	295	0[d]
4	Primary settling pond	2,400	552	77	60	360	0[d]

[a]Process modification: two contact stabilization activated sludge systems operating in parallel.

[b]Subcategory: hardboard.

[c]References: A1, pp. 7-103.

[d]Actual data indicate negative removal.

Note: Blanks indicate information was not specified.

TREATMENT TECHNOLOGY: Activated Sludge

Data source: Government report
Point source category:
Subcategory:
Plant:
References: B20, pp. 24, 27, 38, 44-47

Data source status:
Engineering estimate ___
Bench scale ___
Pilot scale ___
Full scale x

Use in system:
Pretreatment of influent:

DESIGN OR OPERATING PARAMETERS

Process modification:
Wastewater flow:
Hydraulic aeration detention time:
Volumetric loading:
MLSS:
Volatile fraction of MLSS:
F/M:
Mean cell residence time:
Sludge recycle ratio:
Mixed liquor dissolved oxygen:
Oxygen consumption:
Aerator power requirement:

Secondary clarifier configuration:
Depth:
Hydraulic loading (overflow rate):
Solids loading:
Weir loading:
Sludge underflow:
Percent solids in sludge:

REMOVAL DATA

Sampling period:

Pollutant/parameter	Concentration, μg/L		Percent removal
	Influent	Effluent	
Toxic pollutants:			
Cadmium	6	1	83
Chromium	290	60	88
Copper	310	80	74
Mercury	7	<1	>86
Nickel	330	270	18
Zinc	360,000	150,000	57
Bis(2-ethylhexyl) phthalate	5,000	1,300	74
Phenol	35,000	300	99
Benzene	170,000	37,000	90

Note: Blanks indicate information was not specified.

TREATMENT TECHNOLOGY: Activated Sludge

Data source: Effluent Guidelines
Point source category:
Subcategory:
Plant: Berwick POTW
References: A34, p. 208

Data source status:
Engineering estimate ___
Bench scale ___
Pilot scale ___
Full scale x

Use in system: Secondary
Pretreatment of influent:

DESIGN OR OPERATING PARAMETERS

Process modification:
Wastewater flow:
Hydraulic aeration detention time:
Volumetric loading:
MLSS:
Volatile fraction of MLSS:
F/M:
Mean cell residence time:
Sludge recycle ratio:
Mixed liquor dissolved oxygen:
Oxygen consumption:
Aerator power requirement:

Secondary clarifier configuration:
Depth:
Hydraulic loading (overflow rate):
Solids loading:
Weir loading:
Sludge underflow:
Percent solids in sludge:

REMOVAL DATA

Sampling Period—3 Days

Pollutant/Parameter	Concentration Influent	Concentration Effluent	Percent Removal
Conventional pollutants, mg/ℓ			
BOD_5	933	77	92
COD	2,600	430	84
TSS	1,150	114	90
Oil and grease	263	20	92
TKN	130	70	46
Toxic pollutants, μg/ℓ			
Chromium	50,000	3,900	92
Copper	350	28	92
Cyanide	30	TR*	>67
Lead	1,500	90	94
Nickel	8	5	38
Zinc	1,700	280	84
Bis(2-ethylhexyl) phthalate	29	4	86
Pentachlorophenol	200	22	89
Phenol	8,500	ND**	>99
2,4,6-Trichlorophenol	330	5	98
Ethylbenzene	>100	TR***	>99
Toluene	>100	TR***	>99
Anthracene/phenanthrene	6.6	0.7	89
Naphthalene	29	ND†	>99
Chloroform	11	10	9

*Trace; <10 μg/ℓ based on reported influent concentration and percent removal.
**Not detected; assumed to be <10 μg/ℓ.
***Trace; <10 μg/ℓ based on reported influent concentration and percent removal.
†Not detected; <0.3 μg/ℓ based on reported influent concentration and percent removal.

Note: Blanks indicate information was not specified.

TREATMENT TECHNOLOGY: Activated Sludge

Data source:
Point source category:
Subcategory:
Plant: Reichhold Chemical, Inc.
References: B4, pp. 23, 25, 28, 29, 31, 32

Data source status:
- Engineering estimate ___
- Bench scale ___
- Pilot scale ___
- Full scale x

Use in system: Secondary
Pretreatment of influent: Clarification

DESIGN OR OPERATING PARAMETERS

Process modification:
Wastewater flow: See below
Hydraulic aeration detention time: See below
Volumetric loading:
MLSS: See below
Volatile fraction of MLSS:
F/M: See below
Mean cell residence time:
Sludge recycle ratio: See below
Mixed liquor dissolved oxygen:
Oxygen consumption: See below
Aerator power requirement:

Secondary clarifier configuration:
Depth:
Hydraulic loading (overflow rate):
Solids loading:
Weir loading:
Sludge underflow:
Percent solids in sludge:

REMOVAL DATA

Sampling period: Average performance data

MLSS, mg/L	F/M	Recycle ratio	Oxygen consumption, mg/L/hr	BOD_5 Concentration, mg/L Influent	BOD_5 Concentration, mg/L Effluent	BOD_5 Percent removal	COD Concentration, mg/L Influent	COD Concentration, mg/L Effluent	COD Percent removal	TSS Concentration, mg/L Influent	TSS Concentration, mg/L Effluent	TSS Percent removal
2,220[a]	0.43	46:54	14	2,010	357	82	5,120	1,120	78	119	84	29
3,020[b]	0.22	100:0	21	3,440	137	84	7,180	657	71	131	85	35
3,920[c]	0.5	100;0	23	1,340	345	74	3,220	1,220	61	134	87	35
5 640[d]	0.24	100;0	34	1,460	403	74	3,080	1,280	59	102	97	5
4,130[e]	0.08	100;0	189	1,270	47	96	3,430	796	77	158	127	20
4,900[f]	0.23		24.1	1,970	43	98	4,020	676	82	158	204	0[g]

[a]Wastewater flow: 1,500 m^3/d (0.4 MGD); hydraulic aeration detention time: 24 hr.
[b]Wastewater flow: 2,080 m^3/d (0.55 MGD); hydraulic aeration detention time: 48 hr.
[c]Wastewater flow: 6,600 m^3/d (1.76 MGD); hydraulic aeration detention time: 24 hr.
[d]Wastewater flow: 5,030 m^3/d (1.33 MDG); hydraulic aeration detention time: 36 hr.
[e]Wastewater flow: 2,080 m^3/d (0.55 MGD); hydraulic aeration detention time: 144-96 hr.
[f]Wastewater flow: 1,970 m^3/d (0.52 MGD); hydraulic aeration detention time: 48 hr.
[g]Actual data indicate negative removal.

Note: Blanks indicate information was not specified.

TREATMENT TECHNOLOGY: Activated Sludge

Data source:
Point source category:[a]
Subcategory:
Plant: Reichhold Chemical, Inc.
References: B4, pp. 23, 25, 28, 29, 31, 32

Data source status:
- Engineering estimate ___
- Bench scale ___
- Pilot scale ___
- Full scale x

Use in system: Secondary
Pretreatment of influent: Clarification

[a]Organic and inorganic wastes.

DESIGN OR OPERATING PARAMETERS

Process modification:
Wastewater flow: 1,500-6,600 m^3/D (0.4-1.75 mgd)
Hydraulic aeration detention time: 24-144 hr
Volumetric loading:
MLSS: 2,200-4,900 mg/L
Volatile fraction of MLSS:
F/M: 0.02-0.5
Mean cell residence time:
Sludge recycle ratio: (recycled:wasted) 100:0-46:54
Mixed liquor dissolved oxygen:
Oxygen consumption: 14-190 mg/L/hr
Aerator power requirement:

Secondary clarifier configuration:
Depth:
Hydraulic loading (overflow rate):
Solids loading:
Weir loading:
Sludge underflow:
Percent solids in sludge:

REMOVAL DATA

Sampling period:

Pollutant/parameter	Concentration, mg/L Influent[a]	Effluent	Percent removal
Conventional pollutants:			
BOD_5	1,920	222	88
COD	4,340	957	78
TSS	134	114	15

[a]Average of six samples.

Note: Blanks indicate information was not specified.

TREATMENT TECHNOLOGY: Activated Sludge

Data source: Effluent Guidelines
Point source category:
Subcategory:
Plant:
References: Al, Appendix D-1

Data source status:
Engineering estimate ___
Bench scale x
Pilot scale ___
Full scale ___

Use in system: Secondary
Pretreatment of influent:

DESIGN OR OPERATING PARAMETERS

Process modification:
Wastewater flow:
Hydraulic aeration detention time: 8-50 hr
Volumetric loading: 144-1,600 kg phenol/100 m^3/d
MLSS:
Volatile fraction of MLSS:
F/M:
Mean cell residence time:
Sludge recycle ratio:
Mixed liquor dissolved oxygen:
Oxygen consumption:
Aerator power requirement:

Secondary clarifier configuration:
Depth:
Hydraulic loading (overflow rate):
Solids loading:
Weir loading:
Sludge underflow:
Percent solids in sludge:

REMOVAL DATA

Sampling period:

Pollutant/parameter	Concentration, mg/L Influent	Effluent[a]	Percent removal
Conventional pollutants:			
Total phenol	281	62	78

[a]Calculated from influent concentration and percent removal.

Note: Blanks indicate information was not specified.

TREATMENT TECHNOLOGY: Activated Sludge

Data source: Government report
Point source category: Unspecified industrial/ domestic wastewater (70:30)
Subcategory:
Plant:
References: B16, p. 260, 262

Data source status:
Engineering estimate ___
Bench scale ___
Pilot scale ___
Full scale x

Use in system: Secondary
Pretreatment of influent:

DESIGN OR OPERATING PARAMETERS

Process modification: Covered basin pure oxygen activated sludge system
Wastewater flow:
Hydraulic aeration detention time:
Volumetric loading:
MLSS: 9,250 mg/L
Volatile fraction of MLSS: 75%
F/M:
Mean cell residence time: Average 9.6 d
Sludge recycle ratio:
Mixed liquor dissolved oxygen:
Oxygen consumption:
Aerator power requirement:

Secondary clarifier configuration:
Depth:
Hydraulic loading (overflow rate):
Solids loading:
Weir loading:
Sludge underflow:
Percent solids in sludge: 2.2

REMOVAL DATA

Sampling period:

Solids retention time (sludge age)	BOD_5 Concentration, mg/L			COD Concentration, mg/L		
	Influent[a]	Effluent	Percent removal	Influent[a]	Effluent	Percent removal
5.9	929	158	83	2,030	1,080	47
7.8	569	91	84	885	425	52
8.0	1,250	212	83	2,250	1,190	47
8.1	653	124	81	902	550	39
10.0	620	62	90	922	249	73
12.7	660	99	85	897	296	67
17.3	420	42	90	681	286	58
17.3	517	62	88	756	257	66
17.3	854	111	87	1,420	397	72
23.9	633	57	91	1,000	200	80
49.7	362	47	87	559	229	59

[a]Calculated from effluent and percent removal.

Note: Blanks indicate information was not specified.

5.2 TRICKLING FILTERS [1, 2]

5.2.1 Function

Trickling filters are used to remove dissolved and collodial biodegradable organics.

5.2.2 Description

Most trickling filters are classified as low rate, using rock media; other types include high rate, using rock media, and plastic media. Trickling filter using wood media is rarely used in wastewater treatment. Therefore, it is not included in the description section.

Low Rate/Rock Media. The process consists of a fixed bed of rock media over which wastewater is applied for aerobic biological treatment. Zoogleal slimes form on the media which assimilate and oxidize substances in the wastewater. The bed is dosed by a distributor system, and the treated wastewater is collected by an underdrain system. Recirculation is usually not used. Primary treatment is normally required to optimize trickling filter performance.

The rotary distributor has become the standard because of its reliability and ease of maintenance. In contrast to the high rate trickling filter which uses continuous recirculation of filter effluent to maintain a constant hydraulic loading to the distributor arms, either a suction-level controlled pump or a dosing siphon is employed for that purpose with a low rate filter. Low-rate filters operate intermittently, dosing and resting. This operation is required because of the low hydraulic load on the filter.

Underdrains are manufactured from specially designed vitrified-clay blocks that support the filter media and pass the treated sewage to a collection sump for transfer to the final clarifier. The filter media consists of 1- to 5-inch stone. Containment structures are normally made of reinforced concrete and installed in the ground to support the weight of the media.

The low rate trickling filter media bed generally is circular in plan, with a depth of 5 to 10 feet. Although filter effluent recirculation is generally not utilized, it can be provided as a standby tool to keep filter media wet during low flow periods.

The organic material present in the wastewater is degraded by a population of microorganisms attached to the filter media. As the microorganisms grow, the thickness of the slime layer increases.

Periodically, wastewater washes the slime off the media, and a new slime layer will start to grow. This phenomenon of losing the slime layer is called sloughing and is primarily a function of the organic and hydraulic loadings on the filter.

Rock Media/High Rate. This process also consists of a fixed bed of rock media over which wastewater is applied for aerobic biological treatment. Zoogleal slimes form on the media which assimilate and oxidize substances in the wastewater. The bed is dosed by a distributor system, and the treated wastewater is collected by an underdrain system. Primary treatment is normally required to optimize trickling filter performance, and post-treatment is often necessary to meet secondary standards or water quality limitations.

The rotary distributor has become the standard because of its reliability and ease of maintenance. It consists of two or more arms that are mounted on a pivot in the center of the filter. Nozzles distribute the wastewater as the arms rotate due to the dynamic action of the incoming primary effluent. Continuous recirculation of filter effluent is used to maintain a constant hydraulic loading to the distributor arms.

Underdrains are manufactured from specially designed vitrified-clay blocks that support the filter media and pass the treated sewage to a collection sump for transfer to the final clarifier.

The filter media consists of 1- to 5-inch stone. The high rate trickling filter media bed generally is circular in plan, with a depth of 3 to 6 feet. Containment structures are normally made of reinforced concrete and installed in the ground to support the weight of the media.

The organic material present in the wastewater is degraded by a population of microorganisms attached to the filter media. As the microorganisms grow, the thickness of the slime layer increases. As the slime layer increases in thickness, either the absorbed organic matter or oxygen is metabolized before it can reach the microorganisms near the media face. As a result, the microorganisms near the media face enter into either an endogenous phase or anaerobic phase of growth depending on whether organic matter or oxygen is limiting. In this phase, the microorganisms lose their ability to cling to the media surface. The liquid then washes the slime off the media, and a new slime layer will start to grow. This phenomenon of losing the slime layer is called sloughing and is primarily a function of the organic and hydraulic loadings on the filter. Filter effluent recirculation is vital with high rate trickling filters to promote the flushing action necessary for effective sloughing control, without which media clogging and anaerobic conditions could develop due to the high organic loading rates employed.

Plastic Media. The process consists of a fixed bed of plastic media over which wastewater is applied for aerobic biological treatment. Zoogleal slimes form on the media which assimilate and oxidize substances in the wastewater. The bed is dosed by a distributor system, and the treated wastewater is collected by an underdrain system. Primary treatment is normally required to optimize trickling filter performance, whereas post-treatment is generally not required to meet secondary standards.

The rotary distributor has become the standard because of its reliability and ease of maintenance, however, fixed nozzles are often used in roughing filters. Plastic media is comparatively light with a specific weight 10 to 30 times less than rock media. Its high void space (approximately 95 percent) promotes better oxygen transfer during passage through the filter than rock media with its approximate 50 percent void space. Because of its light weight, plastic media containment structures are normally constructed as elevated towers 20 to 30 feet high. Excavated containment structures for rock media can sometimes serve as a foundation for elevated towers for converting an existing facility to plastic media.

Plastic media trickling filters can be employed to provide independent secondary treatment or roughing ahead of a second-stage biological process. When used for secondary treatment, the media bed is generally circular in plan and dosed by a rotary distributor. Roughing applications often utilize rectangular media beds with fixed nozzles for distribution.

The slimes formed on the plastic media assimilate and oxidize organic materials in the wastewater. The amount of sloughed solids that are washed off the media depend on the organic and hydraulic loadings on the filter. Filter effluent recirculation is vital with plastic media trickling filters to ensure proper wetting of the media and to promote effective sloughing control compatible with the high organic loadings employed.

Modifications common to all types of trickling filtration include addition of recirculation, multistaging, electrically powered distributors, forced ventilation, filter covers, and use of various methods of pretreatment and post-treatment of wastewater.

5.2.3 Technology Status

Low Rate/Rock Media. The low rate/rock media process is in widespread use in older plants. The process is highly dependable in moderate climates. Use of aftertreatment or multistaging has frequently been found necessary to insure uniform compliance with effluent limitations in colder regions. The process is being superseded by plastic media systems.

High Rate/Rock Media. The high rate/rock media process has been in widespread use since 1936. The process is a modification of the low-rate trickling filter process.

Plastic Media. The plastic media process has been used as a modification of rock media filters for the past 10 to 20 years.

5.2.4 Applications

Low Rate/Rock Media. Treatment of domestic and compatible industrial wastewaters amenable to aerobic biological treatment in conjunction with suitable pretreatment; process is good for removal of suspended or colloidal materials and is somewhat less effective for removal of soluble organics; can be used for nitrification following prior (first-stage) biological treatment or as stand-alone process in warm climates if organic loading is low enough.

High Rate/Rock Media. Treatment of domestic and compatible industrial wastewaters amenable to aerobic biological treatment in conjunction with suitable pre- and post-treatment; industrial and joint wastewater treatment facilities may use process as roughing filter prior to activated sludge or other unit processes; process is effective for removal of suspended or colloidal materials and is less effective for removal of soluble organics.

Plastic Media. Treatment of domestic and compatible industrial wastewaters amenable to aerobic biological treatment; industrial and joint wastewater treatment facilities may use process as roughing filter prior to activated sludge or other unit processes; existing rock filter facilities can be upgraded via elevation of containment structure and conversion to plastic media; can be used for nitrification following prior (first-stage) biological treatment.

5.2.5 Limitations

Low Rate/Rock Media. Vulnerable to below freezing weather; recirculation may be restricted during cold weather due to cooling effects; marginal treatment capability in single-stage operation; less effective in treatment of wastewater containing high concentrations of soluble organics; has limited flexibility and control in comparison with competing processes, and has potential for vector and odor problems, although they are not as prevalent as with low rate trickling filters; long recovery times with upsets; limited to 60-80% BOD_5 removal.

High Rate/Rock Media. Vulnerable to climate changes and low temperatures; filter flies and odors are common, periods of inadequate moisture for slimes can be common; less effective in treatment of wastewater containing high concentrations of soluble organics; limited flexibility and process control in comparison with competing processes; high land and capital cost requirements; recovery times of several weeks with upsets.

Plastic Media. Vulnerable to below freezing weather; recirculation may be restricted during cold weather due to cooling effects; marginal treatment capability in single-stage operation; less effective in treatment of wastewater containing high concentrations of soluble organics; has limited flexibility and control in comparison with competing processes; has potential for vector and odor problems, although they are not as prevalent as with low rate/rock media trickling filters; long recovery times with upsets.

5.2.6 Typical Equipment

Underdrains, distributors, filter covers, plastic media.

5.2.7 Chemicals Required

None.

5.2.8 Residuals Generated

The following data refer to municipal wastewater. For industrial wastewater, greater residuals will be generated for higher strength waste and for wastes containing greater suspended solids concentrations.

Low Rate/Rock Media. Sludge is withdrawn from the secondary clarifier at a rate of 3,000 to 4,000 gal/Mgal of wastewater, containing 500 to 700 lb dry solids.

High Rate/Rock Media. Sludge is withdrawn from the secondary clarifier at a rate of 2,500 to 3,000 gal/Mgal wastewater, containing 400 to 500 lb dry solids.

Plastic Media. Sludge is withdrawn from the secondary clarifier at a rate of 3,000 to 4,000 gal/Mgal of wastewater, containing 500 to 700 lb dry solids.

5.2.9 Reliability

Low Rate/Rock Media. Highly reliable under conditions of moderate climate; mechanical reliability high; process operation requires little skill.

High Rate/Rock Media. Process can be expected to have a high degree of reliability if operating conditions minimize variability, and installation is in a climate where wastewater temperatures do not fall below 13°C for prolonged periods; mechanical reliability is high; process is simple to operate.

Plastic Media. Process can be expected to have a high degree of reliability if operating conditions minimize variability, and installation is in a climate where wastewater temperatures do not fall below 13°C for prolonged periods; mechanical reliability is high; process is simple to operate.

5.2.10. Environmental Impact

Rock Media. Odor problems; high land requirement relative to many alternative processes; filter flies.

Plastic Media. Odor problems if improperly operated.

5.2.11 Design Criteria

Low Rate/Rock Media

Criteria/factor	Unit	Value/range
Hydraulic loading	Mgal/acre/d or	1 - 4
	gal/d/ft²	25 - 90
Organic loading	lb BOD_5/d/acre ft or	200 - 900
	lb BOD_5/d/1,000 ft³	5 - 20
Recirculation ratio	-	0
Bed depth	ft	5 - 10
Under drain minimum slope	-	1
Effluent channel minimum velocity (at average daily flow)	ft/s	2
Media - rock (must meet sodium sulfate soundness test)	in.	1 - 5
Sloughing		Intermittent
Dosing interval		Continuous for majority of daily operating schedule, but become intermittent during low flow periods

High Rate/Rock Media

Criteria/factor	Unit	Value/range
Hydraulic loading (with recirculation)	Mgal/acre/d or	10 - 40
	gal/d/ft²	230 - 900
Organic loading	lb BOD_5/d/acre ft or	900 - 2,600
	lb BOD_5/d/1,000 ft³	20 - 60
Recirculation ratio	-	0.5 - 4
Bed depth	ft	3 - 6
Under drain minimum slope	-	1
Power requirements	hp/Mgal	10 - 50
Dosing interval	s	≥15 (continuous)
Sloughing	-	Continuous
Media - rock (must meet sodium sulfate soundness test)	in.	1 - 5

Plastic Media

Criteria/factor	Unit	Value/range
Hydraulic loading (with recirculation)		
a) Secondary treatment	Mgal/acre/d or	30 - 60
	gal/d/ft²	700 - 1,400
b) Roughing	Mgal/acre/d or	100 - 200
	gal/d/ft²	2,300 - 4,600
Organic loading		
a) Secondary treatment	lb BOD_5/d/acre ft or	450 - 2,200
	lb BOD_5/d/1,000 ft³	10 - 50
b) Roughing	lb BOD_5/d/acre ft or	4,500 - 22,000
	lb BOD_5/d/1,000 ft³	100 - 500
Recirculation ratio	-	0.5 - 5
Dosing interval (continuous)	s	≥15
Sloughing	-	continuous
Bed depth	ft	20 - 30
Power requirement	hp/Mgal	10 - 50
Under drain minimum slope	-	1

5.2.12 Flow Diagrams

Low Rate/Rock Media.

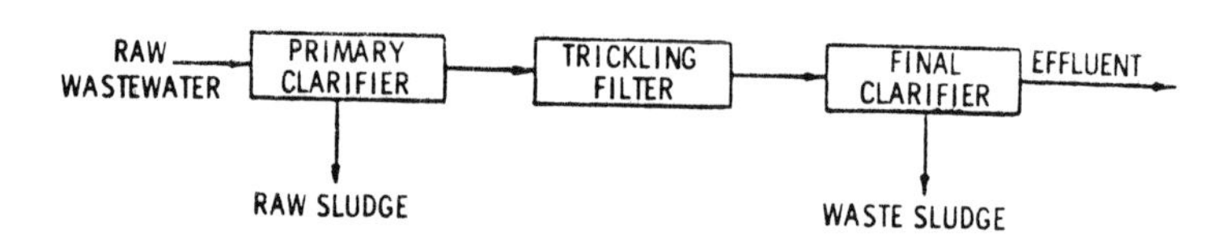

High Rate/Rock Media.

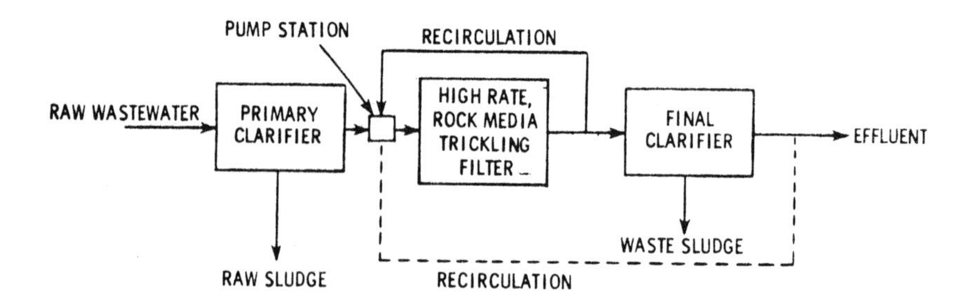

Plastic Media.

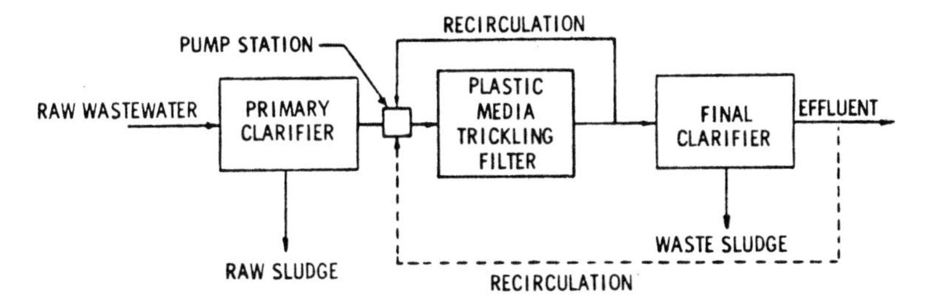

5.2.13 Performance

Subsequent data sheets provide performance data from studies on the following industries and/or wastestreams:

Hospital wastewaters

Leather tanning and finishing
  Chrome tanning

Pulp, paper, and paperboard production
  Wastepaper board

Rubber processing
  Styrene-butadiene rubber

Timber products processing
  Wood preserving (creosote wastewater)

References

1. Innovative and Alternative Technology Assessment Manual. EPA-430/9-78-009 (draft), U.S. Environmental Protection Agency, Cincinnati, Ohio, 1978. 252 pp.

2. Metcalf & Eddy, Wastewater Engineering: Collection, Treatment, Disposal. McGraw-Hill Book Co., New York, 1972. pp. 433-435.

CONTROL TECHNOLOGY SUMMARY FOR TRICKLING FILTER

Pollutant	Number of data points	Effluent concentration				Removal efficiency, %			
		Minimum	Maximum	Median	Mean	Minimum	Maximum	Median	Mean
Conventional pollutants, mg/L:									
BOD_5	11	4	137	27	39	76	98	92	90
COD	3	290	709	623	541	0[a]	77	23	33
TSS	1				45				59
Total phenol	2	<1.0	1.0		<1.0	96	>97		>96.5
Toxic pollutants, μg/L:									
Chromium	1				17				0[a]
Copper	1				42				0[a]
Cyanide	1				16				79
Lead	1				49				0[a]
Bis(2-ethylhexyl) phthalate	1				6				83
Di-n-butyl phthalate	1				6				25
Diethyl phthalate	1				140				0[a]
Pentachlorophenol	1				3				0[a]
Phenol	1				37				0[a]
2,4,6-Trichlorophenol	1				2				0[a]
Naphthalene	1				55				0[a]
Chloroform	1				19				0[a]
Methylene chloride	1				1				0[a]
Trichloroethylene	1				1				0[a]
Other pollutants, μg/L:									
Xylenes	1				2				0[a]

Note: Blanks indicate data not applicable.

[a]Actual data indicate negative removal.

TREATMENT TECHNOLOGY: Trickling Filter

Data source: Effluent Guidelines

Data source status: See below
- Engineering estimate ___
- Bench scale ___
- Pilot scale ___
- Full scale ___

Point source category:[a]
Subcategory:
Plant:
References: A1, Appendix D-7,8

Use in system: Secondary
Pretreatment of influent:

[a]Synthesized wastewater

DESIGN OR OPERATING PARAMETERS

Process modification:
Wastewater flow:
Total hydraulic loading:
Recirculation ratio:
Dosing interval:
Sloughing:
Organic loading:
Bed depth:
Power requirements:

REMOVAL DATA

Sampling Period:

Scale	Total phenol concentration, mg/L		
	Influent	Effluent	Percent removal
Pilot	400	288-308	23-28
Full	25	1	96

Note: Blanks indicate information was not specified.

TREATMENT TECHNOLOGY: Trickling Filter

Data source: Effluent Guidelines
Point source category: Hospital
Subcategory:
Plant: See below
References: A22, p. 52

Data source status:
Engineering estimate ___
Bench scale ___
Pilot scale ___
Full scale x

Use in system: Secondary
Pretreatment of influent:

DESIGN OR OPERATING PARAMETERS

Process modification:
Wastewater flow:
Total hydraulic loading:
Recirculation ratio:
Dosing interval:
Sloughing:
Organic loading:
Bed depth:
Power requirements:

REMOVAL DATA

Sampling period:

	BOD_5[a]		
	Concentration, mg/L		Percent
Plant	Influent[b]	Effluent	removal
94	225	27	88
95	400	32	92
96	183	11	94
97	240	24	90
98	200	4	98
99	275	11	96
100	250	10	96
101	233	56	76

[a]Values based on annual average removal efficiency.

[b]Calculated from effluent and percent removal.

Note: Blanks indicate information was not specified.

TREATMENT TECHNOLOGY: Trickling Filter

Data source: Effluent Guidelines
Point source category: Leather tanning and finishing
Subcategory:
Plant: See below
References: A15, p. 80, 79

Data source status:
Engineering estimate ___
Bench scale ___
Pilot scale ___
Full scale x

Use in system: Secondary unless otherwise specified
Pretreatment of influent: See below

DESIGN OR OPERATING PARAMETERS

Process modification:
Wastewater flow:
Total hydraulic loading:
Recirculation ratio:
Dosing interval:
Sloughing:
Organic loading:
Bed depth:
Power requirements:

REMOVAL DATA

Sampling period:

Plant	Pretreatment of influent	BOD_5 Concentration, mg/L Influent	BOD_5 Concentration, mg/L Effluent	BOD_5 Percent removal	TSS Concentration, mg/L Influent	TSS Concentration, mg/L Effluent	TSS Percent removal
In India	Dilution, primary sedimentation	860	52	94			
3[a]	-[b]	270	62	77	110	45	59
	Primary coagulation, sedimentation	150-400	30-80	80			

[a]Use in system: primary.

[b]Wastewater flow: 3,780 m^3/d, recirculation ratio: 50%.

Note: Blanks indicate information was not specified.

TREATMENT TECHNOLOGY: Trickling Filter

Data source: Effluent Guidelines
Point source category: Pulp, paper and paperboard
Subcategory: Wastepaper board
Plant:
References: A26, pp. A-78-85

Data source status:
- Engineering estimate ___
- Bench scale ___
- Pilot scale ___
- Full scale x

Use in system: Secondary
Pretreatment of influent: Lagooning

DESIGN OR OPERATING PARAMETERS

Process modification:
Wastewater flow:
Total hydraulic loading:
Recirculation ratio:
Dosing interval:
Sloughing:
Organic loading:
Bed depth:
Power requirements:

REMOVAL DATA

Sampling period:

Pollutant/parameter	Concentration[a] Influent	Effluent	Percent removal
Conventional pollutants, mg/L:			
COD	563	623	0[b]
Toxic pollutants, µg/L:			
Chromium	ND[c]	17	0[b]
Copper	ND	42	0[b]
Cyanide	76	16	79
Lead	ND	49	0[b]
Bis(2-ethylhexyl) phthalate	35	6	83
Di-n-butyl phthalate	8	6	25
Diethyl phthalate	ND	140	0[b]
Pentachlorophenol	ND	3	0[b]
Phenol	22	37	0[b]
2,4,6-Trichlorophenol	ND	2	0[b]
Napthalene	34	55	0[b]
Chloroform	ND	19	0[b]
Methylene chloride	ND	1	0[b]
Trichloroethylene	ND	1	0[b]
Other pollutants, µg/L:			
Xylenes	ND	2	0[b]

[a]Average values.

[b]Actual data indicate negative removal.

[c]Not detected.

Note: Blanks indicate information was not specified.

TREATMENT TECHNOLOGY: Trickling Filter

Data source: Government report
Point source category: Rubber manufacturing
Subcategory: Butadiene-styrene synthetic rubber
Plant: General Tire & Rubber Co., (Odessa, Texas)
References: B14, p. 45

Data source status:

Engineering estimate	
Bench scale	
Pilot scale	x
Full scale	

Use in system: Secondary
Pretreatment of influent: Settling

DESIGN OR OPERATING PARAMETERS

Process modification:
Wastewater flow:
Total hydraulic loading:
Recirculation ratio:
Dosing interval:
Sloughing:
Organic loading:
Bed depth:
Power requirements:

REMOVAL DATA

Sampling period:

Pollutant/parameter	Concentration,[a] mg/L Influent	Effluent	Percent removal
COD	379	290	23

[a]Average of six samples.

Note: Blanks indicate information was not specified.

TREATMENT TECHNOLOGY: Trickling Filter

Data source: Effluent Guidelines
Point source category: Timber products
Subcategory: Wood preserving
(creosote wastewater)
Plant:
References: A1, p. D-8

Data source status:
Engineering estimate ___
Bench scale ___
Pilot scale x
Full scale ___

Use in system: Secondary
Pretreatment of influent: Equalization, coagulation/sedimentation, dilution, nitrogen/phosphorus addition

DESIGN OR OPERATING PARAMETERS

Process modification: Plastic media
Wastewater flow: 0.00286 $m^3/min/m^2$ (0.07 gpm/ft^2)
Total hydraulic loading: 0.044 $m^3/min/m^2$ (1.07 gpm/ft^2)
Recirculation ratio: 14.3 (recycle-to-raw wastewater)
Dosing interval:
Sloughing:
Organic loading: 1,060 kg BOD/1,000 m^3/d (66.3 lb BOD/1,000 ft^3/d)
1,940 kg COD/1,000 m^3/d (121 lb COD/1,000 ft^3d)
19.4 kg phenol/1,000 m^3/d (1.2 lb phenol/1,000 ft^3/d)
Bed depth: 6.4 m (21 ft)
Power requirements:
Recycle flow rate: 0.0408 $m^3/min/m^2$ (1.0 gpm/ft^2)

REMOVAL DATA

Sampling period: Seven months

Pollutant/parameter	Concentration, mg/L Influent	Effluent	Percent removal
Conventional pollutants:			
BOD_5	1,700	137	93
COD	3,110	709	77
Total phenol	31	<1.0	>97

Note: Blanks indicate information was not specified.

5.3 LAGOONS (STABILIZATION PONDS) [1,2,3]

5.3.1 Function

Lagoons (stabilization ponds) are used to remove dissolved and colloidal biodegradable organics, and suspended solids.

5.3.2 Description

A shallow body of water contained in an earthen dike and designed for wastewater treatment is termed a stabilization pond or lagoon. Another term which is synonomous and often used is "oxidation pond." The shape of the basin is carefully controlled for optimum performance. Due to their low construction and operating costs, ponds offer a financial advantage over other treatment methods and for this reason have become very popular with small communities. Ponds are also used extensively to treat industrial wastes and mixtures of industrial wastes and domestic sewage that are amenable to biological treatment. Industries which now employ stabilization ponds are dairies, poultry-processing plants, paint manufacturing plants, slaughterhouses, oil refineries, textile mills, and rendering plants. The aerated, anaerobic, facultative, aerobic and tertiary lagoons represent the common types.

Aerated Lagoons. Aerated lagoons are medium-depth basins designed for the biological treatment of wastewater on a continuous basis. In contrast to stabilization ponds, which obtain oxygen from photosynthesis and surface reaeration, aerated lagoons employ aeration devices that supply supplemental oxygen to the system. The aeration systems may be mechanical (i.e., surface aerator) or diffused air types. Surface aerators may be either high speed small diameter impeller or low speed large diameter impeller devices, either fixed mounted on piers or float mounted on pontoons. Diffused aerators may be plastic pipe with regularly spaced holes, static mixers (plastic tubes over the holes extending toward the surface to get better mixing) and others. Because aerated lagoons are normally designed to achieve partial mixing only, aerobic-anaerobic stratification will occur, and a large fraction of the incoming solids and a large fraction of the biological solids produced from waste conversion settle to the bottom of the lagoon cells. As the solids begin to build up, a portion will undergo anaerobic decomposition. Volatile toxics can potentially be removed by the aeration process, similar to an activated sludge system. Several smaller aerated lagoon cells in series are more effective than one large cell. Tapering aeration intensity downward in the direction of flow promotes settling out of solids in the last cell. A nonaerated polishing cell following the last aerated cell is an optional, but recommended, design technique to enhance suspended solids removal prior to discharge.

Lagoons may be lined with concrete or an impervious flexible lining, depending on soil conditions and environmental regulations.

Anaerobic Lagoons. Anaerobic lagoons are relatively deep (up to 20 ft) ponds with steep sidewalls in which anaerobic conditions are maintained by keeping loading so high that complete deoxygenation is prevalent. Although some oxygenation is possible in a shallow surface zone, once greases form an impervious surface layer, complete anaerobic conditions develop. Treatment or stabilization results from thermophilic anaerobic digestion of organic wastes. The treatment process is analogous to that occurring in the single-state anaerobic digestion of sludge in which acid-forming bacteria break down organics. The resultant acids are then converted to carbon dioxide, methane, cells, and other end products.

In the typical anaerobic lagoon, raw wastewater enters near the bottom of the pond (often at the center) and mixes with the active microbial mass in the sludge blanket, which is usually about 6 feet deep. The discharge is located near one of the sides of the pond, submerged below the liquid surface. Excess undigested grease floats to the top, forming a heat-retaining and relatively air-tight cover. Wastewater flow equalization and heating are generally not practiced. Excess sludge is washed out with the effluent. Recirculation of waste sludge is not required.

Anaerobic lagoons are capable of providing treatment of high strength wastewaters and are resistant to shock loads.

Anaerobic lagoons are customarily contained within earthen dikes. Depending on soil characteristics, lining with various impervious materials such as rubber, plastic or clay may be necessary. Pond geometry may vary, but surface-area-to-volume ratios are minimized to enhance heat retention.

Facultative Lagoons. Facultative lagoons are intermediate depth (3 to 8 feet) ponds in which the wastewater is stratified into three zones. These zones consist of an anaerobic bottom layer, an aerobic surface layer, and an intermediate zone. Stratification is a result of solids settling and temperature-water density variations. Oxygen in the surface stabilization zone is provided by reaeration and photosynthesis. This is in contrast to aerated lagoons in which mechanical aeration is used to create aerobic surface conditions. In general, the aerobic surface layer serves to reduce odors while providing treatment of soluble organic by-products of the anaerobic processes operating at the bottom.

Sludge at the bottom of facultative lagoons will undergo anaerobic digestion producing carbon dioxide, methane, and cells. The photosynthetic activity at the lagoon surface produces oxygen diurnally, increasing the dissolved oxygen during daylight hours, while surface oxygen is depleted at night.

Facultative lagoons are often and for optimum performance should be operated in series. When three or more cells are linked, the effluent from either the second or third cell may be recirculated to the first. Recirculation rates of 0.5 to 2.0 times the plant flow have been used to improve overall performance.

Facultative lagoons are customarily contained within earthen dikes. Depending on soil characterstics, lining with various impervious material such as rubber, plastic or clay may be necessary. Use of supplemental top-layer aeration can improve overall treatment capacity, particularly in northerm climates where icing over of facultative lagoons is common in the winter.

Aerobic Lagoons. Aerobic lagoons contain algae and bacteria in suspension, and aerobic conditions prevail throughout the depth. Waste is stabilized as a result of the symbiotic relationship between aerobic bacteria and algae. Bacteria break down waste and generate carbon dioxide and nutrients (primarily nitrogen and phosphorus). Algae, in the presence of sunlight, utilize the nutrients and inorganic carbon; they, in turn, supply oxygen that is utilized by aerobic bacteria. Aerobic lagoons are usually 3 to 5 feet deep and must be periodically mixed to maintain aerobic conditions throughout. In order to achieve effective removals from aerobic lagoons, some means of removing algae (coagulation, filtration, multiple cell design) is necessary. Algae have a high degree of mobility and do not settle well using conventional clarification.

Tertiary Lagoons/Polishing Ponds. Tertiary lagoons serve as a polishing step following other biological treatment processes. They are often called maturation or polishing ponds and primarily serve the purpose of reducing suspended solids. Water depth is generally 3 to 5 feet. Tertiary lagoons are quite popular as a final treatment step for textile wastewater treated with the extended-aeration activated sludge process.

5.3.3 Technology Status

Aerated Lagoons. While not as widely used when compared with the large number of facultative lagoons in common use throughout the United States, aerated lagoons have been fully demonstrated and used for years.

Anaerobic Lagoons. Although anaerobic processes are common for sludge digestion, anaerobic lagoons for wastewater treatment have found only limited application. The process is well demonstrated for the stabilization of highly concentrated organic wastes; such as slaughterhouse wastes.

Facultative Lagoons. Facultative lagoons have been fully demonstrated and are in moderate use especially for treatment of relatively weak municipal wastewater in areas where land costs are not a restricting factor.

5.3.4 Applications

Aerated Lagoons. Used for domestic and industrial wastewater of low and medium strength; commonly used where land is inexpensive, and costs and operational control are to be minimized; existing oxidation ponds, lagoons, and natural bodies of water can be upgraded in a relatively simple manner to this type of treatment; aeration increases the oxidation capacity of the pond and is useful in overloaded ponds that generate odors; useful when supplemental oxygen requirements are high or when the requirements are either seasonal or intermittent.

Anaerobic Lagoons. Typically used in series with aerobic or facultative lagoons; effective as roughing units prior to aerobic treatment of high strength wastes.

Facultative Lagoons. Used for treating raw, screened, or primary settled domestic wastewater and weak biodegradable industrial wastewaters; most applicable when land costs are low, and operation and maintenance costs are to be minimzed.

5.3.5 Limitations

Aerated Lagoons. May experience reduced biological activity and treatment efficiency, and the formation of ice in very cold climates.

Anaerobic Lagoons. May generate odors; require relatively large land area; water temperatures should be maintained above 75°F for efficient operation.

Facultative Lagoons. May experience reduced biological activity and treatment efficiency in very cold climates; ice formation can also hamper operations; odors can be a problem in overloading situations.

5.3.6 Chemicals Required

Aerated Lagoons. None

Anaerobic Lagoons. Nutrients are needed to make deficiencies in raw wastewater; no other chemicals required.

Facultative Lagoons. If wastewater is nutrient deficient, a source of supplemental nitrogen or phosphorus may be needed; no other chemicals required.

5.3.7 Residuals Generated

Aerated Lagoons. Settled solids on pond bottom may require clean-out every 10 to 20 years, or possibly more often if a polishing pond is used behind the aerated pond.

Anaerobic Lagoons. Excess sludge is usually washed out in the effluent; since anaerobic lagoons are often used for preliminary treatment, recirculation or removal of sludge not generally required.

Facultative Lagoons. Settled solids may require clean-out and removal once every 10 to 20 years.

5.3.8 Reliability

Aerated Lagoons. Service life estimated at 30 years or more; reliability of equipment and process is high; little operator expertise required.

Anaerobic Lagoons. Generally resistant to upsets; highly reliable if pH in the relatively narrow optimum range is maintained.

Facultative Lagoons. Service life estimated to be 50 years; little operator expertise required; overall, the system is highly reliable.

5.3.9 Environmental Impact

Aerated Lagoons. Opportunity exists for volatile organic material and pathogens in aerated lagoons to enter the air (as with any aerated wastewater treatment process); opportunity depends on air/water contact afforded by aeration system; potential exists for seepage of wastewater into groundwater unless lagoon is lined; aerated lagoons generate less solid residue, compared to other secondary treatment processes.

Anaerobic Lagoons. May create odors; relatively high land requirement; potential exists for seepage of wastewater into groundwater unless lagoon is lined.

5.3.10 Design Criteria

Criteria/factor	Unit	Aerated lagoon	Anaerobic lagoon	Facultative lagoon
Detention time	d	3 - 10	1 - 50	20 - 180
Depth	ft	6 - 20	8 - 20	3 - 8
pH	-	6.5 - 8	6.8 - 7.2	6.5 - 9.0
Water temperature	°C	0 - 40	6 - 49	2 - 32
Optimum water temperature	°C	20	30	20
Oxygen required	-	0.7 - 1.4 times BOD_5 removed	-	-
Organic loading	lb BOD_5/acre/d	10 - 300	220 - 2,200	10 - 100
Operation	-	One or more cells	Parallel	At least 3 cells in series

5.3.11 Flow Diagrams

Aerated Lagoons.

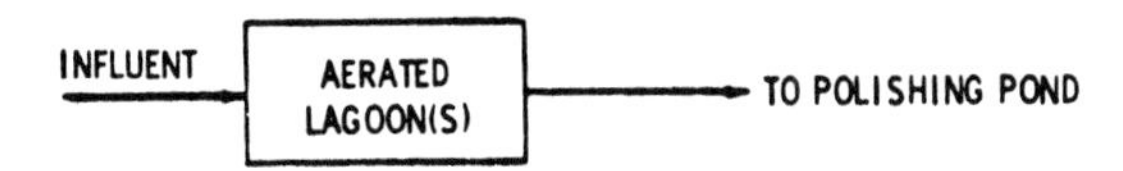

Anaerobic Lagoons.

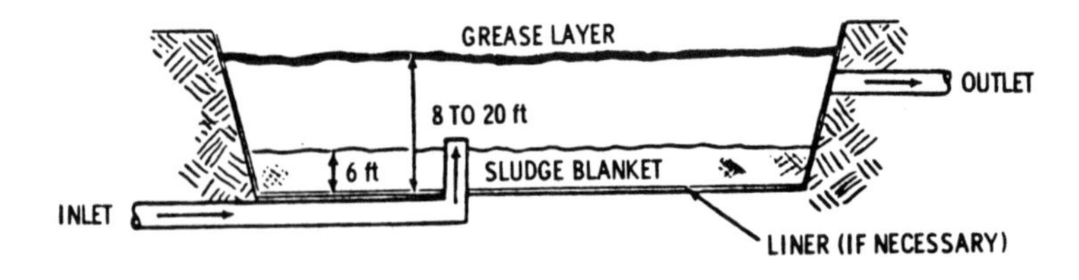

Facultative Lagoons.

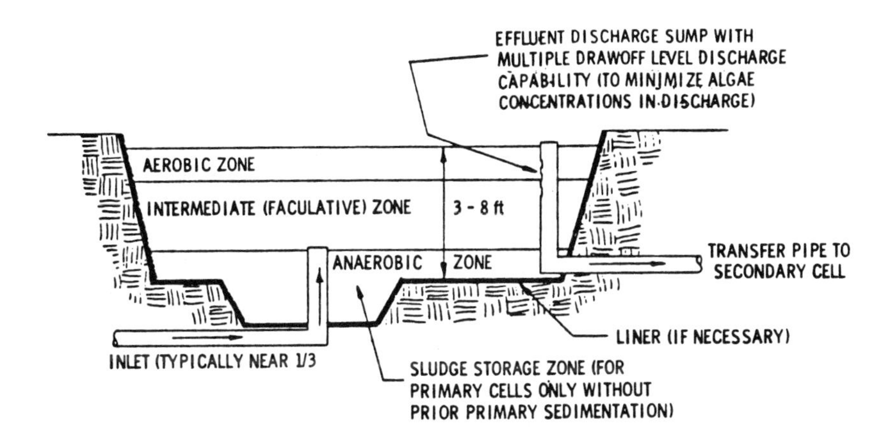

5.2.12 Performance

Subsequent data sheets provide performance data from studies on the following industries and/or wastestreams:

Aerated Lagoons

Leather tanning and finishing
- Hair pulping, chrome tanning, retanning - wet finishing
- Vegetable tanning

Organic chemicals production
- Aqueous liquid-phase reaction systems
- Processes with process water contact as steam diluent or absorbent

Paint manufacturing

Pharmaceuticals production
- Biological and natural extraction products
- Chemical synthesis products, and formulation products

Textile milling
- Knit fabric finishing
- Stock and yarn finishing
- Woven fabric finishing

Timber products processing
Hardwood

Anaerobic Lagoons

Organic chemicals production
Petrochemicals

Facultative Lagoons

Leather tanning and finishing
Cattle - sheep save, chrome tanning
Vegetable tanning

Timber products processing
Hardboard

Tertiary Effluent Polishing Lagoons

Textile milling
Felted fabric processing
Stock and yarn finishing

5.3.13 References

1. Innovative and Alternative Technology Assessment Manual. EPA-430/9-78-009 (draft), U.S. Environmental Protection Agency, Cincinnati, Ohio, 1978. 252 pp.

2. Metcalf & Eddy, Wastewater Engineering: Collecting, Treatment Disposal. McGraw-Hill Book Co., New York, 1972. pp. 551-552.

3. Technical Study Report, BATEA-NSPS-PSES-PSNS, Textile Mills Point Source Category (contractor's draft report). Contracts 68-01-3289 and 68-01-3884, U.S. Environmental Protection Agency, Washington, D. C., November 1978.

CONTROL TECHNOLOGY SUMMARY FOR AERATED LAGOON

Pollutant	Number of data points	Effluent concentration Minimum	Effluent concentration Maximum	Effluent concentration Median	Effluent concentration Mean	Removal efficiency, % Minimum	Removal efficiency, % Maximum	Removal efficiency, % Median	Removal efficiency, % Mean
Conventional pollutants, mg/L:									
BOD_5	16	6	869	90	150	0	>99	77 5	71
COD	10	92	1,610	591	679	3	>99	63	50
TOC	4	47	573	126	220	11	99	46	133
TSS	13	3	1,790	155	410	0[a]	99	24	33
Oil and grease	1				17				98
TKN	2	22	105		64	75	79		77
Total phenol	2	0.003	0.018		0.0105	31	>99		>65
Toxic pollutants, µg/L:									
Antimony	1				30				82
Beryllium	1				<1				>50
Cadmium	1				<2				>97
Chromium	3	9	1,100	16	380	0	99	91	63
Copper	5	5	110	26	40	0[a]	94	36	49
Cyanide	2	52	150		100	0[a]	91		46
Lead	2	<20	80		<50	>83	93		>88
Mercury	1				0.1				82
Nickel	3	30	40	32	34	0[a]	50	0	17
Selenium	1				<200				>50
Thallium	2	13	<20		<33	7	>80		>44
Zinc	4	49	510	<80	210	0[a]	>99	61	55
Bis(2-chloroethoxy)methane	1				<10[b]				>60
Bis(2-chloroisopropyl) ether	1				<10[b]				>0
Bis(2-ethylhexyl) phthalate	5	1	28	<10[c]	<11	26	96	>78	>70
Butyl benzyl phthalate	1				6				0
Di-n-butyl phthalate	1				1				0
Diethyl phthalate	1				4				0[a]
Dimethyl phthalate	1				6				25
Benzidine	1				7				41
1,2-Diphenylhydrazine	1				14				0[a]
N-nitrosodiphenylamine	1				1				67
4-Nitrophenol	1				<10				23
Pentachlorophenol	1				<10[c]				>71
Phenol	3	<10[c]	24	<10[b]	<15	>0	>99	25	>41
2,4,6-Trichlorophenol	1				<10[c]				>99
Benzene	2	<5	<10	<20	<7.5	>74	>95		>84
1,2-Dichlorobenzene	1				<10[c]				>96
1,4-Dichlorobenzene	1				<10[c]				>81
2,4-Dinitrotoluene	1				3				0[a]
2,6-Dinitrotoluene	1				2				83
Ethylbenzene	2	<10[c]	<10[c]		<10[c]	>89	>94		>92
Hexachlorobenzene	1				<10[b]				>0
Nitrobenzene	1				<10[b]				>0
Toluene	3	<10[c]	<10[c]	<10[c]	<10[c]	>90	>95	>95	>93
Acenaphthene	1				4				0
Acenaphthylene	1				5				0[a]
Benzo(a)pyrene	1				2				33
Benzo(b)fluoranthene	1				0 4				97
Fluoranthene	1				<10[b]				>0
Fluorene	1				0.2				99
Anthracene/phenanthrene	1				3				0[a]
Naphthalene	2	<10[b]	<10[b]		<10[b]	>0	>58		>29
Pyrene	1				1				67
2-Chloronaphthalene	1				<10[b]				>47
Chloroform	3	<10[c]	1,000	<10[c]	340	0[a]	>57	>50	>36
Methyl chloride	1				<5				>91
Methylene chloride	3	32	1,000	130	390	0[a]	97	97	65
Tetrachloroethylene	1				<10[c]				>60
1,1,1-Trichloroethane	1				22				96
Isophorone	1				2				33

Note: Blanks indicate data not applicable.

[a]Actual data indicate negative removal.

[b]Below detection limit, assumed to be <10 µg/L.

[c]Not detected, assumed to be <10 µg/L.

CONTROL TECHNOLOGY SUMMARY FOR ANAEROBIC LAGOONS

Pollutant	Number of data points	Effluent concentrations				Removal efficiency, %			
		Minimum	Maximum	Median	Mean	Minimum	Maximum	Median	Mean
Conventional pollutants, mg/L:									
BOD_5	5	80	2,750	488	877	52	90	65	65
COD	4	348	5,910	2,300	2,720	29	47	34.5	36
Toxic pollutants, μg/L:									
Benzene	1				5,000				50
Other pollutants, μg/L:									
Acetaldehyde	3	10	40	35	28	50	67	56	58
Acetic acid	3	220	2,600	2,300	1,700	0[a]	0[a]	0[a]	0[a]
Butyric acid	2	300	330		320	0[a]	0[a]		0[a]
Propionic acid	2	470	500		490	0[a]	0[a]		0[a]

Note: Blanks indicate data not applicable.

[a]Actual data indicate negative removal.

CONTROL TECHNOLOGY SUMMARY FOR FACULTATIVE LAGOONS

Pollutant	Number of data points	Effluent concentration Minimum	Effluent concentration Maximum	Effluent concentration Median	Effluent concentration Mean	Removal efficiency, % Minimum	Removal efficiency, % Maximum	Removal efficiency, % Median	Removal efficiency, % Mean
Conventional pollutants, mg/L:									
BOD_5	3	53	274	152	160	77	92	87	85
COD	2	717	2,110		1,410	55	68		62
TSS	2	48	105		76	74	86		80
TKN	2	35	100		68	33	67		50

Note: Blanks indicate data not applicable.

CONTROL TECHNOLOGY SUMMARY FOR TERTIARY POLISHING LAGOONS

Pollutant	Number of data points	Effluent concentration Minimum	Effluent concentration Maximum	Effluent concentration Median	Effluent concentration Mean	Removal efficiency, % Minimum	Removal efficiency, % Maximum	Removal efficiency, % Median	Removal efficiency, % Mean
Conventional pollutants, mg/L:									
COD	2	142	263		202	0[a]	52		26
TSS	2	22	28		25	24	76		50
Total phenol	2	0.028	0.051		0.04	0[a]	46		23
Toxic pollutants, µg/L:									
Chromium	1				<10[b]				>71
Copper	1				18				0[a]
Lead	1				<10[b]				>72
Selenium	1				18				44
Zinc	2	100	120		110	0[a]	86		43
Bis(2-ethylhexyl) phthalate	2	<10[b]	11		<11	>44	72		>58
Naphthalene	1				<10[b]				>82
Trichlorofluoromethane	1				<10[b]				>79

Note: Blanks indicate data not applicable.

[a]Actual data indicate negative removal.

[b]Reported as not detected; assumed to be <10 µg/L.

TREATMENT TECHNOLOGY: Lagoon, Aerated

Data source: Effluent Guidelines
Point source category: Leather tanning and finishing
Subcategory: Hair pulp, chrome tan, retan-wet finish
Plant: 184
References: A34, p. 208

Data source status:
Engineering estimate ___
Bench scale ___
Pilot scale ___
Full scale x

Use in system: Secondary
Pretreatment of influent:

DESIGN OR OPERATING PARAMETERS

System configuration:
Wastewater flow:
Hydraulic detention time:
Hydraulic loading:
Organic loading:
Oxygen requirement:
Aerator power requirement:
Depth:

REMOVAL DATA

Sampling period: Three days

Pollutant/parameter	Concentration Influent	Concentration Effluent	Percent removal
Conventional pollutants, mg/L:			
BOD_5	1,870	21	99
COD	2,600	430	84
TSS	2,910	155	95
Oil and grease	720	17	
TKN	500	105	79
Toxic pollutants, µg/L:			
Chromium	160,000	1,100	99
Copper	50	5	90
Cyanide	60	150	0[a]
Lead	1,100	80	93
Nickel	60	30	50
Zinc	500	49	90
Bis(2-ethylhexyl) phthalate	51	2	96
Phenol	4,400	ND[b]	>99
2,4,6-Trichlorophenol	880	ND	>99
1,2-Dichlorobenzene	250	ND	>96
1,4-Dichlorobenzene	54	ND	>81
Ethylbenzene	88	ND	>89
Toluene	<100	>10	>90
Naphthalene	24	ND	>58

[a]Actual data indicate negative removal.

[b]Not detected.

Note: Blanks indicate information was not specified.

TREATMENT TECHNOLOGY: Lagoon, Aerated

Data source: Effluent Guidelines
Point source category: Leather tanning and finishing
Subcategory: Vegetable tanning process
Plant: 13
References: A15, p. 82

Data source status:

Engineering estimate	
Bench scale	
Pilot scale	x
Full scale	

Use in system: Secondary
Pretreatment of influent:

DESIGN OR OPERATING PARAMETERS

System configuration: Volume - 2,980 m^3
Wastewater flow:
Hydraulic detention time: 16-35 d
Hydraulic loading:
Organic loading: 16.2-130 kg BOD_5/d/1,000 m^3
Oxygen requirement:
Aerator power requirement: 7.5 kw (10 hp)
Depth:

REMOVAL DATA

Sampling period:

Pollutant/parameter	Concentration, mg/L Influent	Effluent	Percent removal
Conventional pollutants:			
BOD_5	1,040	86	92
COD	4,470	1,610	64
TSS	539	571	0[a]
TKN	88	22	75

[a]Actual data indicate negative removal.

Note: Blanks indicate information was not specified.

TREATMENT TECHNOLOGY: Lagoon, Aerated

Data source: Government report
Point source category: Organic chemicals
Subcategory:
Plant:
References: B16, p. 274

Data source status:
- Engineering estimate ___
- Bench scale ___
- Pilot scale ___
- Full scale x

Use in system: Secondary
Pretreatment of influent: Equalization, limited aeration

DESIGN OR OPERATING PARAMETERS

System configuration: Two lagoons in parallel
Wastewater flow: 49-57 x 10^3 m^3/day
Hydraulic detention time:
Hydraulic loading:
Organic loading:
Oxygen requirement:
Aerator power requirement:
Depth:

REMOVAL DATA

Sampling Period: 4 Days

Pollutant/Parameter	. . Concentration (μg/ℓ) . . Influent	Effluent	Percent Removal
Toxic pollutants			
Bis(2-chloroethoxy)methane	25	BDL*	>60
Bis(2-chloroisopropyl) ether	2	BDL**	>0
Bis(2-ethylhexyl) phthalate	60	1	98
Butyl benzyl phthalate	6	6	0
Di-n-butyl phthalate	1	1	0
Diethyl phthalate	2	4	0***
Dimethyl phthalate	8	6	25
Benzidine	12	7	41
1,2-Diphenylhydrazine	5	14	0***
N-nitrosodiphenylamine	3	1	67
Phenol	1	BDL**	>0
2,4-Dinitrotoluene	2	3	0***
2,6-Dinitrotoluene	12	2	83
Hexachlorobenzene	4	BDL**	>0
Nitrobenzene	3	BDL**	>0
Acenaphthene	4	4	0
Acenaphthylene	2	5	0***
Benzo(a)pyrene	3	2	33
Benzo(b)fluoranthene	12	0.4	97
Fluoranthene	2	BDL**	>0
Fluorene	16	0.2	99
Naphthalene	1	BDL**	>0
Phenanthrene	1	3	>0***
Pyrene	3	1	67
2-Chloronaphthalene	19	BDL*	>47
Isophorone	3	2	33

*Below detectable limits; assumed to be <10 μg/ℓ.
**Below detectable limits; assumed to be less than the corresponding effluent concentration.
***Actual data indicate negative removal.

Note: Blanks indicate information was not specified.

TREATMENT TECHNOLOGY: Lagoon, Aerated

Data source: Effluent Guidelines
Point source category: Organic chemicals
Subcategory: See below
Plant: See below
References: A25, p. 300

Data source status:
- Engineering estimate ___
- Bench scale ___
- Pilot scale ___
- Full scale x

Use in system: Secondary
Pretreatment of influent:

DESIGN OR OPERATING PARAMETERS

System configuration:
Wastewater flow:
Hydraulic detention time:
Hydraulic loading:
Organic loading:
Oxygen requirement:
Aerator power requirement:
Depth:

REMOVAL DATA

Plant	BOD_5 Concentration, mg/L Influent[a]	BOD_5 Concentration, mg/L Effluent	BOD_5 Percent removal	COD Concentration, mg/L Influent[a]	COD Concentration, mg/L Effluent	COD Percent removal
21[b]	123	27	78	1,580	600	62
8[c]	38	6	84	297	92	69
6[d]	870	235	73	2,380	980	66

Plant	TOC Concentration, mg/L Influent[a]	TOC Concentration, mg/L Effluent	TOC Percent removal	TSS Concentration, mg/L Influent[a]	TSS Concentration, mg/L Effluent	TSS Percent removal
21[b]	138	47	66	273	30	89
8[c]	70	52	26	300	3	99
6[d]	644	573	11	<362	362	0[e]

[a]Calculated from effluent and percent removal.

[b]Subcategory: aqueous liquid phase reaction.

[c]Subcategory: process with process water contact as steam diluent or absorbent.

[d]Subcategory: process with process water contact as steam diluent or absorbent and aqueous liquid phase reaction system.

[e]Actual data indicate negative removal.

Note: Blanks indicate information was not specified.

TREATMENT TECHNOLOGY: Lagoon, Aerated

Data source: Effluent Guidelines
Point source category: Paint manufacturing
Subcategory:
Plant:
References: A4, p. VII-18

Data source status:
Engineering estimate ___
Bench scale ___
Pilot scale ___
Full scale x

Use in system: Secondary
Pretreatment of influent: Physical/chemical primary treatment (unspecified)

DESIGN OR OPERATING PARAMETERS

System configuration:
Wastewater flow:
Hydraulic detention time:
Hydraulic loading:
Organic loading:
Oxygen requirement:
Aerator power requirement:
Depth:

REMOVAL DATA

Sampling period:

Pollutant/parameter	Concentration Influent	Concentration Effluent	Percent removal
Conventional pollutants, mg/L:			
BOD_5	23,400	17	>99
COD	260,000	675	>99
TOC	25,000	200	99
TSS	400	42	90
Total phenol	1.1	0.003	>99
Toxic pollutants, µg/L:			
Antimony	170	30	82
Beryllium	2	<1	>50
Cadmium	58	<2	>97
Chromium	100	9	91
Copper	120	7	94
Lead	98	<20	>80
Mercury	140	0.1	>99
Selenium	400	<200	>50
Thallium	100	<20	>80
Zinc	4,200	<60	>99
Benzene	200	<10	>95
Toluene	200	ND[a]	>95
Chloroform	23	ND	>57
Methylene chloride	31,000	1,000	97
Tetrachloroethylene	25	ND	>60
1,1,1,-Trichloroethane	560	22	96

[a]Not detected.

Note: Blanks indicate information was not specified.

TREATMENT TECHNOLOGY: Lagoon, Aerated

Data source: Effluent Guidelines
Point source category: Pharmaceuticals
Subcategory: Biological and natural extraction products, chemical synthesis products, formulation products
Plant: F
References: A32, Supplement 2

Data source status:
- Engineering estimate ___
- Bench scale ___
- Pilot scale ___
- Full scale x

Use in system: Primary
Pretreatment of influent:

DESIGN OR OPERATING PARAMETERS

System configuration:
Wastewater flow: 37.9 m^3/d (0.01 mgd)
Hydraulic detention time:
Hydraulic loading:
Organic loading:
Oxygen requirement:
Aerator power requirement:
Depth:

REMOVAL DATA

Sampling period:

Pollutant/parameter	Concentration, μg/L Influent	Concentration, μg/L Effluent	Percent removal
Toxic pollutants:			
Copper	60	110	0[a]
Zinc	140	510	0[a]
Bis(2-ethylhexyl) phthalate	160	15	57
Methylene chloride	63	130	0[a]

[a]Actual data indicate negative removal.

Note: Blanks indicate information was not specified.

TREATMENT TECHNOLOGY: Lagoon, Aerated

Data source: Effluent Guidelines
Point source category: Pharmaceuticals
Subcategory: Biological and natural extraction products, chemical syntheses products, formulation products
Plant: E
References: A32, Supplement 2

Data source status:
- Engineering estimate ___
- Bench scale ___
- Pilot scale ___
- Full scale x

Use in system: Secondary
Pretreatment of influent: Equalization, neutralization

DESIGN OR OPERATING PARAMETERS

System configuration: Aeration tank with turbine aerators
Wastewater flow: 1,330 m^3/d (0.35 mgd)
Hydraulic detention time:
Hydraulic loading:
Organic loading:
Oxygen requirement:
Aerator power requirement:
Depth:

REMOVAL DATA

Sampling period:

Pollutant/parameter	Concentration[a] Influent	Effluent	Percent removal
Conventional pollutants, mg/L:			
BOD_5	7,100	869	88
TSS	369	1,790	0[b]
Toxic pollutants, μg/L:			
Chromium	16	16	0
Copper	35	26	26
Cyanide	590	52	91
Nickel	20	40	0[b]
Zinc	146	99	32
Bis(2-ethylhexyl) phthalate	38	28	26
Chloroform	860	1,000	0[b]
Methylene chloride	1,100	32	97

[a]Average of three samples.

[b]Actual data indicate negative removal.

Note: Blanks indicate information was not specified.

TREATMENT TECHNOLOGY: Lagoon, Aerated

Data source: Effluent Guidelines
Point source category: Textile mills
Subcategory: See below
Plant:
References: p. VII-22

Data source status:
Engineering estimate ___
Bench scale ___
Pilot scale ___
Full scale x

Use in system:
Pretreatment of influent:

DESIGN OR OPERATING PARAMETERS

System configuration:
Wastewater flow:
Hydraulic detention time: See below
Hydraulic loading:
Organic loading:
Oxygen requirement:
Aerator power requirement: See below
Depth:

REMOVAL DATA

Sampling period:

Subcategory	Hydraulic detention time, hr	BOD_5 Concentration, mg/L Influent	BOD_5 Concentration, mg/L Effluent	BOD_5 Percent removal	COD Concentration, mg/L Influent	COD Concentration, mg/L Effluent	COD Percent removal	TSS Concentration, mg/L Influent	TSS Concentration, mg/L Effluent	TSS Percent removal
Knit fabric finishing[a]	18	388	189	51				1,760	1,220	31
Stock and yarn finishing[b]	0.5	252	249	1	556	429	23			
Stock and yarn finishing[c]	75	108	14	87				21	12	43
Woven fabric finishing[d]	24	69	69	0	644	581	10	54	68	0[e]
Woven fabric finishing[f]	60	366	94	74	835	814	3			
Woven fabric finishing[g]	86	1,740	157	91				556	599	0[e]

[a]Aerator power requirement: 30 W/m³, 150 hp/Mgal.
[b]Aerator power requirement: 197 W/m³, 1,000 hp/Mgal.
[c]Aerator power requirement: 5 W/m³, 25 hp/Mgal.
[d]Aerator power requirement: 79 W/m³, 400 hp/Mgal.
[e]Actual data indicate negative removal.
[f]Aerator power requirement: 9 W/m³, 45 hp/Mgal.
[g]Aerator power requirement: 154 W/m³, 780 hp/Mgal.

Note: Blanks indicate information was not specified.

TREATMENT TECHNOLOGY: Lagoon, Aerated

Data source: Effluent Guidelines
Point source category: Textile mills
Subcategory: Woven fabric finishing
Plant:
References: A6, pp. VII-59-60

Data source status:
- Engineering estimate ___
- Bench scale ___
- Pilot scale ___
- Full scale x

Use in system: Secondary
Pretreatment of influent: Equalization, grit removal, screening, dissolved air flotation with chemical addition

DESIGN OR OPERATING PARAMETERS

System configuration: Two lagoons in series, surface aeration
Wastewater flow: 570 m^3/d (150,000 gpd)
Hydraulic detention time: 170 hr
Hydraulic loading:
Organic loading:
Oxygen requirement:
Aerator power requirement: 3.5 watt/m^3 (18 hp/Mgal)
Depth:

REMOVAL DATA

Sampling period: Two 24-hr composite samples

Pollutant/parameter	Concentration		Percent removal
	Influent	Effluent	
Conventional pollutants, mg/L:			
BOD_5	200	67	66
COD	725	577	20
TSS	32	17	47
Total phenol	0.026	0.018	31
Toxic pollutants, μg/L:			
Copper	81	52	36
Nickel	32	32	0
Thallium	14	13	7
Bis(2-ethylhexyl) phthalate	45	ND[a]	>78
4-Nitrophenol	13	<10	>23
Pentachlorophenol	34	ND	>71
Phenol	32	24	25
Benzene	19	<5	>74
Ethylbenzene	160	ND	>94
Toluene	200	ND	>95
Methylene chloride	56	<5	>91

[a]Not detected.

Note: Blanks indicate information was not specified.

TREATMENT TECHNOLOGY: Lagoon, Aerated

Data source: Effluent Guidelines
Point source category: Timber products
Subcategory: Hardboard
Plant: See below
References: A1, p. 7-10, 7-105

Data source status:
- Engineering estimate ___
- Bench scale ___
- Pilot scale ___
- Full scale x

Use in system: See below
Pretreatment of influent: See below

DESIGN OR OPERATING PARAMETERS

System configuration: See below
Wastewater flow:
Hydraulic detention time: See below
Hydraulic loading:
Organic loading:
Oxygen requirement:
Aerator power requirement:
Depth:

REMOVAL DATA

Sampling period:

Plant	Use in system	BOD_5 Concentration, mg/L Influent	BOD_5 Concentration, mg/L Effluent	BOD_5 Percent removal	TSS Concentration, mg/L Influent	TSS Concentration, mg/L Effluent	TSS Percent removal
24[a]	Tertiary	436	102	77	157	120	24
444[b]	Secondary	686	192	72	148	365	0[c]

[a]Pretreatment of influent: screening, primary clarifier, flow equalization, two contact stabilization activated sludge systems operating in parallel; hydraulic detection time: 6.

[b]Pretreatment of influent: primary settling (2 ponds); system configuration: aerated lagoon plus secondary settling pond.

[c]Actual data indicate negative removal.

Note: Blanks indicate information was not specified.

TREATMENT TECHNOLOGY: Lagoon, Anaerobic

Data source: Government Report
Point source category: Organic chemicals
Subcategory: Petrochemical wastes
Plant:
References: B16, pp. 75-78

Data source status:
Engineering estimate ___
Bench scale ___
Pilot scale ___
Full scale x

Use in system: Primary
Pretreatment of influent:

DESIGN OR OPERATING PARAMETERS

System configuration:
Wastewater flow:
Hydraulic detention time: See below
Hydraulic loading:
Organic loading: See below
Depth:

REMOVAL DATA

Sampling period:

Pollutant/parameter	Concentration[a] Influent	Effluent	Percent removal
Toxic pollutants, μg/L:			
Benzene[b]	10,000	5,000	50
Other pollutants, μg/L:			
Acetaldehyde[b]	30	10	67
-[c]	80	40	50
-[d]	80	35	56
Acetic acid[b]	215	220	0[e]
-[c]	2,100	2,600	0[e]
-[d]	2,100	2,300	0[e]
Butyric acid[c]	ND[f]	300	0[e]
-[d]	ND	330	0[e]
Propionic acid[c]	ND	470	0[e]
-[d]	ND	500	0[e]

[a]Averaged from five to twelve occurrences.

[b]Hydraulic loading: 209 kg/day/1,000 m^3.

[c]Organic loading: 770 kg COD/1,000 m^3/day.

[d]Organic loading: 353 kg COD/1,000 m^3/day.

[e]Actual data indicate negative removal.

Note: Blanks indicate information was not specified.

TREATMENT TECHNOLOGY: Lagoon, Anaerobic

Data source: Government Report
Point source category: Organic chemicals
Subcategory:
Plant: See below
References: B16, p. 61, 79

Data source status: See below
Engineering estimate ___
Bench scale ___
Pilot scale ___
Full scale ___

Use in system: Primary unless otherwise specified
Pretreatment of influent:

DESIGN OR OPERATING PARAMETERS

System configuration:
Wastewater flow:
Hydraulic detention time:
Hydraulic loading:
Organic loading: See below
Depth:

REMOVAL DATA

Sampling period:

		BOD_5			COD		
	Volume,	Concentration, mg/L		Percent	Concentration, mg/L		Percent
Plant	m^3	Influent	Effluent	removal	Influent	Effluent	removal
In Texas City[a]	0.189				1,340	348	47
In Texas City[b]	420,000	2,650	928	65	5,440	3,320	39
_[c]		800	80	90			
In Texas City[d]	98,400	4,820	2,750[e]	43	8,440	5,910[e]	30
In Texas City[f]	55	1,060	488[e]	52	2,090	1,280[e]	39
Seadrift Plant[g]	680,000	570	137	76			

[a] Organic loading: 0.11 kg COD m^2/day; volumetric loading: 139 kg COD/1,000 m^3/day; pilot scale.

[b] Organic loading: 0.02 kg BOD_5/m^2/day, 0.03 kg COD/m^2/day/ volumetric loading: 15.2 kg BOD_5/1,000 m^3/day, 29.9 kg COD/1,000 m^3/day.

[c] Petrochemical diluent; system configuration: several lagoons in series; wastewater flow: 1.9 x 10^3 m^3/d; hydraulic detection time: 20 days (entire system); use in system: secondary.

[d] Organic loading: 0.07 kg BOD_5/m^2/day, 0.13 kg COD/1,000 m^3/day; volumetric loading: 134 kg BOD_5/1,000 m^3/day, 279 kg COD/1,000 m^3/day.

[e] Calculated from influent and percent removal.

[f] Lagoon of irregular prismoid shape; organic loading: 0.104 kg BOD_5/m^2/day, 0.227 kg COD/m^2/day; volumetric loading: 110 kg BOD_5/1,000 m^3/day, 248 kg COD/1,000 m^3/day.

[g] Organic loading: 0.02 kg BOD_5/m^2/day; volumetric loading: 17.5 kg BOD_5/1,000 m^3/day.

Note: Blanks indicate information was not specified.

TREATMENT TECHNOLOGY: Lagoon, Facultative

Data source: Effluent Guidelines
Point source category: Leather tanning and finishing
Subcategory: See below
Plant: See below
References: A15, p. 84, 85, 86

Data source status: See below
Engineering estimate ___
Bench scale ___
Pilot scale ___
Full scale ___

Use in system: Secondary
Pretreatment of influent: See below

DESIGN OR OPERATING PARAMETERS

System configuration:
Wastewater flow: See below
Hydraulic detention time: See below
Hydraulic loading:
Organic loading: See below
Depth:

REMOVAL DATA

Sampling period:

	BOD_5			COD		
	Concentration, mg/L		Percent	Concentration, mg/L		Percent
Subcategory	Influent	Effluent	removal	Influent	Effluent	removal
Cattle-sheep save chrome[a]	673	53	92			
Vegetable tanning process[b]	1,170	274	77	4,730	2,110	55
Vegetable tanning process[c]	1,150	152	87	2,220	717	68

	TSS			TKN		
	Concentration, mg/L		Percent	Concentration, mg/L		Percent
	Influent	Effluent	removal	Influent	Effluent	removal
Cattle-sheep save chrome[a]	339	48	86			
Vegetable tanning process[b]				107	35	67
Vegetable tanning process[c]	408	105	74	150	100	33

[a]Plant: Pownal Tanning Co., North Pownal, Vermont; pretreatment of influent: screening; wastewater flow: 2.27 m^3/d; full scale.

[b]Hydraulic detection time: 4-8 d; organic loading: 142 kg BOD_5/d/1,000 m^3; pilot scale.

[c]Organic loading: 32.4 - 325 kg BOD_5/d/1,000 m^3; full scale.

Note: Blanks indicate information was not specified.

TREATMENT TECHNOLOGY: Lagoon, Tertiary Effluent Polishing

Data source: Effluent Guidelines
Point source category: Textile mills
Subcategory: Felted fabric processing
Plant:
References: A6, p. VII-32
Use in system: Tertiary
Pretreatment of influent: Equalization, activated sludge

Data source status:
- Engineering estimate ___
- Bench scale ___
- Pilot scale ___
- Full scale x

DESIGN OR OPERATING PARAMETERS

System configuration: One basin
Wastewater flow: 380 m^3/d (0.1 mgd)
Hydraulic detention time: 25 d
Hydraulic loading:
Organic loading:
Depth:
Total volume: 9,500 m^3 (2.5 Mgal)

REMOVAL DATA

Sampling period: 24 hr

Pollutant/paremeter	Concentration Influent	Concentration Effluent	Percent removal
Conventional pollutants, mg/L:			
COD	552	263	52
TSS	91	22	76
Total phenol	0.052	0.028	46
Toxic pollutants, µg/L			
Chromium	35	ND[a]	>71
Copper	ND	18	0[b]
Selenium	32	18	44
Zinc	45	100	0[b]
Bis(2-ethylhexyl) phthalate	18	ND	>44
Naphthalene	56	ND	>82

[a]Not detected.

[b]Actual data indicate negative removal.

Note: Blanks indicate information was not specified.

TREATMENT TECHNOLOGY: Lagoon, Tertiary Effluent Polishing

Data source: Effluent Guidelines
Point source category: Textile mills
Subcategory: Stock and yarn finishing
Plant:
References: A6, p. VII-31

Data source status:
- Engineering estimate ___
- Bench scale ___
- Pilot scale ___
- Full scale x

Use in system: Tertiary
Pretreatment of influent: Screening, equalization, activated sludge

DESIGN OR OPERATING PARAMETERS

System configuration: Parallel "primary" and "secondary" oxidation ponds
Wastewater flow: 2,800 m^3/d (0.75 mgd)
Hydraulic detention time: 20 d
Hydraulic loading:
Organic loading:
Depth:
Total volume: 57,000 m^3 (15 Mgal)

REMOVAL DATA

Sampling period: 24-hr composite sample

Pollutant/parameter	Concentration Influent	Concentration Effluent	Percent removal
Conventional pollutants, mg/L:			
COD	78	142	0[a]
TSS	37	28	24
Total phenol	0.036	0.051	0[a]
Toxic pollutants, µg/L:			
Lead	36	ND[b]	>72
Zinc	860	120	86
Bis(2-ethylhexyl) phthalate	40	11	72
Trichlorofluoromethane	48	ND[b]	>79

[a]Actual data indicate negative removal.

[b]Not detected, assumed to be <10 µg/L.

Note: Blanks indicate information was not specified.

5.4 ROTATING BIOLOGICAL CONTACTORS [1]

5.4.1. Function

Rotating biological contactors (RBC) are used to remove dissolved and colloidal biodegradable organics.

5.4.2. Description

The process utilizes a fixed-film biological reactor consisting of plastic media mounted on a horizontal shaft and placed in a tank. Common media forms are a disc-type made of styrofoam and a denser lattice-type made of polyethylene. While wastewater flows through the tank, the media are slowly rotated, about 40% immersed, for contact with the wastewater to remove organic matter by the biological film that develops on the media. Rotation results in exposure of the film to the atmosphere as a means of aeration. Excess biomass on the media is stripped off by rotational shear forces, and the stripped solids are maintained in suspension by the mixing action of the rotating media. Multiple staging of RBC's increases treatment efficiency and could aid in achieving nitrification year round. A complete system could consist of two or more parallel trains, each consisting of multiple stages in series.

5.4.3. Common Modifications

Common modifications of RBC's include the following: multiple staging; use of dense media for latter stages in train; use of molded covers of housing of units; various methods of pretreatment and after treatment of wastewater; use in combination with trickling filter or activated sludge processes; use of air driven system with tapered gas flow in lieu of mechanically driven system; addition of air to the tanks; addition of chemicals for pH control; and sludge recycle to enhance nitrification.

5.4.4. Technology Status

The process has been in use in the United States only since 1969 and is not yet in widespread use. Use of the process is growing, however, because of its characteristic modular construction, low hydraulic head loss, and shallow excavation, which make it adaptable to new or existing treatment facilities.

5.4.5. Applications

Treatment of domestic and compatible industrial wastewater amenable to aerobic biological treatment in conjunction with suitable pretreatment and post-treatment; can be used for nitrification, roughing secondary treatment, and polishing.

5.4.6. Limitations

Can be vulnerable to climatic changes and low temperatures if not housed or covered; performance may diminish significantly at temperature below 55°F; enclosed units can result in considerable wintertime condensation if the heat is not added to enclosure; high organic loadings can result in first-stage septicity and supplemental aeration may be required; use of dense media for early stages can result in media clogging; alkalinity deficit can result from nitrification; supplemental alkalinity source may be required.

5.4.7 Residuals Generated

Sludge in secondary clarifier; 3,000 to 4,000 gal sludge/Mgal wastewater; 500 to 700 lb dry solids/Mgal wastewater. These data are based on municipal wastewater.

5.4.8 Reliability

Moderately reliable in the absence of high organic loading and temperatures below 55°F; mechanical reliability is generally high, provided first stage of system is designed to hold large biomass; dense media in first stage can result in clogging and structural failure.

5.4.9 Environmental Impact

Negative impacts have not been documented; presumably, odor can be a problem if septic conditions develop in first stage.

5.4.10 Design Criteria

Criteria	Units	Range/Value
Organic loading	lb BOD_5 1,000 ft^3 of media	Without nitrification: 30-60 With nitrification: 15-20
Hydraulic loading	gpd/ft^2 of media	Without nitrification: 0.75-1.5 With nitrification: 0.3-0.6
Stages/train	–	At least 4
Parallel trains	–	At least 2
Rotational velocity	fpm (peripheral)	60
Media surface area	ft^2/ft^3	Disc type: 20-25 Lattice type: 30-35
Media submerged	percent	40
Tank volume	gal/ft^2 of disc area	0.12
Detention time	min (based on 0.12 gal/ft^2)	Without nitrification: 40-90 With nitrification: 90-230
Secondary	–	–
Clarifier overflow	gpd/ft^2	500-700
Power	horsepower/25 ft shaft	7.5

5.4.11 Flow Diagram

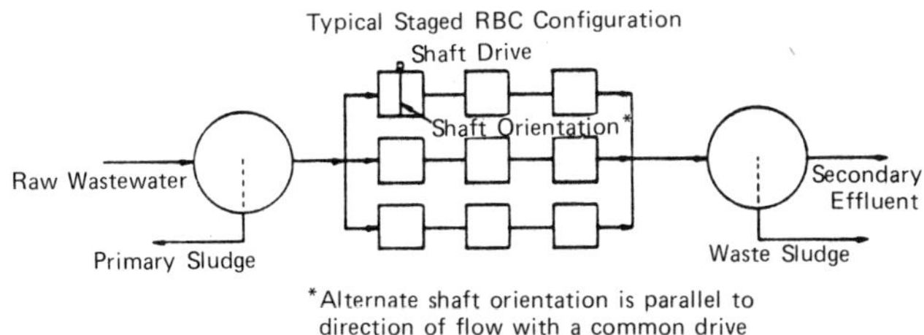

5.4.12 Performance

Subsequent data sheets provide performance data from studies on the following industries and/or wastestreams:

Coal mining

Soap and detergent production
Liquid detergents

5.4.13 References

1. Innovative and Alternative Technology Assessment Manual. EPA-430/9-78-009 (draft), U.S. Environmental Protection Agency, Cincinnati, Ohio, 1978. 252 pp.

CONTROL TECHNOLOGY SUMMARY FOR ROTATING BIOLOGICAL CONTACTORS

Pollutant	Number of data points	Effluent concentration, ng/L				Removal efficiency, %			
		Minimum	Maximum	Median	Mean	Minimum	Maximum	Median	Mean
Conventional pollutants:									
BOD_5	4	18	71	18	31	69	82	72	74
COD	4	340	1,000	750	710	28	54	40.5	41
TSS	8	23	68	62	54	0[a]	35	0[a]	8
Oil and grease	5	13	47	29	27.8	0[a]	21	6	9
Phosphorous	5	3.0	5	3.4	3.62	0[a]	21	11	11
TKN	5	6	38	15	17.4	5	57	33	36

[a]Actual data indicate negative removal.

TREATMENT TECHNOLOGY: Rotating Biological Contactors

Data source: Government report
Point source category: Coal mining
Subcategory:
Plant:
References: B22, pp. 20, 33, 42

Data source status:
Engineering estimate ___
Bench scale ___
Pilot scale x
Full scale ___

Use in system: Primary
Pretreatment of influent:

DESIGN OR OPERATING PARAMETERS

Wastewater flow: See below
Organic loading:
Hydraulic loading: See below
Rotational velocity: 19 m/min (63 fpm)
Percent media submerged:
Number of trains:
Secondary clarifier overflow rate:
Theoretical retention time: See below

REMOVAL DATA

Sampling period: See below

Wastewater flow, m^3/d (gpd)	Hydraulic loading time, $m^3/d/m^2$ (gpd/ft^2)	Theoretical retention time, min	TTS Concentration, mg/L Influent	Effluent	Percent removal	Grab samples taken, weeks
6.3(1,800)	0.31(7.5)	29	3	23	0[a]	9
9.8(2,600)	0.44(10.8)	20	4	26	0[a]	10
4.92(1,300)	0.22(5.4	40	20	68	0[a]	8

[a]Actual data indicate negative removal

Note: Blanks indicate information was not specified.

TREATMENT TECHNOLOGY: Rotating Biological Contactors

Data source: Government report
Point source category:
Subcategory: Liquid detergent
Plant: Texize Chemical Co. (Greenville, SC)
References: B21

Data source status:
- Engineering estimate ___
- Bench scale ___
- Pilot scale x
- Full scale ___

Use in system: Tertiary
Pretreatment of influent: Equalization

DESIGN OR OPERATING PARAMETERS

Wastewater flow: See below
Organic loading: 0.0146-0.0175 kg $BOD_5/m^2/d$
Hydraulic loading:
Rotational velocity: 10 rpm
Percent media submerged:
Number of trains:
Secondary clarifier overflow rate:
Temperature: See below

REMOVAL DATA

Sampling period:

Wastewater flow, L/min	Temperature °C	BOD_5 Concentration, mg/L Influent	BOD_5 Concentration, mg/L Effluent	BOD_5 Percent removal	COD Concentration, mg/L Influent	COD Concentration, mg/L Effluent	COD Percent removal	TSS Concentration, mg/L Influent	TSS Concentration, mg/L Effluent	TSS Percent removal
1.9	7-28	228	71	69	1,400	1,000	29	75	68	9
2.85	16-22	100	18	82	1,240	570	54	54	56	0[a]
3.8	7-23	64	18	72	710	340	52	97	63	35
7.6	9-14	65	18	72	1,290	930	28	60	61	0[a]
0.95	9-25							82	67	18

Wastewater flow, L/min	Temperature °C	Oil and grease Concentration, mg/L Influent	Oil and grease Concentration, mg/L Effluent	Oil and grease Percent removal	Phosphorus Concentration, mg/L Influent	Phosphorus Concentration, mg/L Effluent	Phosphorus Percent removal	TKN Concentration, mg/L Influent	TKN Concentration, mg/L Effluent	TKN Percent removal
1.9	7-28	26	29	0[a]	3.6	3.4	6	35	15	57
2.85	16-22	22	47	0[a]	6.3	5	21	40	38	5
3.8	7-23	16	13	19	3.6	3.0	17	9	6	33
7.6	9-14	33	31	6	3.25	3.50	0[a]	22	15	32
0.95	9-25	24	19	21	3.6	3.2	11	29	13	55

[a]Actual data indicate negative removal.

Note: Blanks indicate information was not specified.

5.5 STEAM STRIPPING [1]

5.5.1 Function

Steam stripping is used to remove gases or volatile organics from dilute wastewater streams.

5.5.2 Description

Steam stripping is essentially a fractional distillation of volatile compounds from a wastewater stream. The volatile component may be a gas or volatile organic compound with solubility in the wastewater stream. In most instances, the volatile component, such as methanol or ammonia, is quite water soluble.

Steam stripping is usually conducted as a continuous operation in a packed tower or conventional fractionating distillation column (bubble cap or sieve tray) with more than one stage of vapor/liquid contact. The preheated wastewater from the heat exchanger enters near the top of the distillation column and then flows by gravity countercurrent to the steam and organic vapors (or gas) rising up from the bottom of the column. As the wastewater passes down through the column, it contacts the vapors rising from the bottom of the column that contain progressively less volatile organic compound or gas until it reaches the bottom of the column where the wastewater is finally heated by the incoming steam to reduce the concentration of volatile component(s) to their final concentration. Much of the heat in the wastewater discharged from the bottom of the column is recovered in preheating the feed to the column.

Reflux (condensing a portion of the vapors from the top of the column and returning it to the column) may or may not be practiced depending on the composition of the vapor stream that is desired. Although many of the steam strippers in industrial use introduce the wastewater at the top of the stripper, there are advantages to introducing the feed to a tray below the top tray when reflux is used.

Introducing the feed at a lower tray (while still using the same number of trays in the stripper) will have the effect of either reducing steam requirements (due to the need for less reflux) or yielding a vapor stream richer in volatile component). The combination of using reflux and introducing the feed at a lower tray will increase the concentration of the volatile organic component beyond that obtainable by reflux alone.

5.5.3 Technology Status

Steam stripping has been used for many years for the recovery of ammonia from coke oven gas. Recently, as water effluent regulations have become more stringent, other aqueous waste streams are being treated by this unit operation for removal of volatile organic components (i.e., methanol from pulp mill condensate).

5.5.4 Applications

Used in both industrial chemical production (for recovery and/or recycle of product) and in industrial waste treatment; three common examples of product recovery by steam stripping are ammonia recover for sale as ammonia or ammonium sulfate from coke oven gas scrubber water, sulfur from refinery sour water, and phenol from water solution in the production of phenol; has been recently applied to wastewater treatment; newer applications include removal of phenols, mercaptans, and chlorinated hydrocarbons from wastewater.

5.5.5 Limitations

May be designed for pure nonreactive volatile components in the wastewater by using tray-by-tray calculations and vapor/liquid equilibrium data reported in the literature although a "wastewater stream" rarely contains only nonreactive components; if volatile components react with each other, as in refinery sour water containing H_2S and ammonia, the vapor pressure exerted by each component in water solution no longer follows Raoult's Law; thus, where vapor/liquid equilibrium data do not exist for a specific combination of water soluble components, these data must be experimentally developed.

5.5.6 Typical Equipment

Equipment is nearly the same as that required for conventional fractional distillation (i.e., packed column or tray tower, reboiler, reflux condenser and feed tanks, and pumps); however, heat exchanger is used for heating feed entering column and cooling stripped wastewater leaving column; reboiler is often an integral part of tower body rather than a separate vessel; materials of construction depend on operating pH and presence (or absence) of corrosive ions (i.e., sulfides, chlorides); in a single-column sour-water steam stripper, the high pH (from the presence of ammonia) allows use of mild steel; if sour water is stripped in two columns (H_2S removed in one and NH_3 removed in other) alloy steel or alloy clad steel should be used in unit in which H_2S is removed.

5.5.7 Residuals Generated/Environmental Impacts

Steam stripped volatiles are usually processed further for recovery or incinerated; if stripped volatiles contain sulfur and are incinerated, the impact of SO_2 emissions must be considered; impact of the stripped wastewater depends on the quantity and type of residual volatile organics remaining in the stripped wastewater; land requirements are small; there are generally no discharges except for the treated wastewater.

5.5.8 Reliability

Dependent on specific wastewater application; in refinery operations, steam stripping has proven to be highly dependable.

5.5.9 Design Criteria

Criteria	Units	Value/range
Column height	ft	20-60
Column diameter	ft	3-6
Steam requirement	lb/gal	0.6-2.0
Typical wastewater flow	gpm	200

5.5.10 Flow Diagram

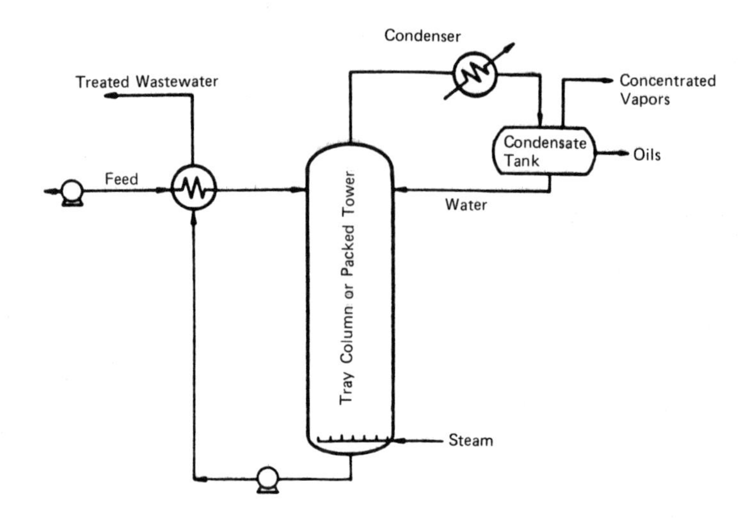

5.5.11 Performance

Subsequent data sheets provide performance data from studies on the following industries and/or wastestreams:

Organic Chemicals

References

1. Physical, Chemical, and Biological Treatment Techniques for Industrial Wastes, PB 275 287, U.S. Environmental Protection Agency, Washington, D.C., November 1976. pp. 42-2 - 42-16.

CONTROL TECHNOLOGY SUMMARY FOR STEAM STRIPPING

Pollutant	Number of data points	Effluent concentration				Removal efficiency, %			
		Minimum	Maximum	Median	Mean	Minimum	Maximum	Median	Mean
Conventional pollutants, mg/L:									
COD	6	118	233	173	170	44	72	62	59
TOC	40	14	593	110	118	0[a]	94	72	57
Toxic pollutants, μg/L:									
Chloroform	5	<10[b]	65,000	<10[b]	13,000	49	>99	>99	89
1,2-Dichloroethane	45	300	440,000	7,000	33,000	70	>99	>99	97
1,2-Trans-dichloroethylene	5	<10[b]	1,300,000	16,000	340,000	9	>99	99	76
Methylene chloride	5	90,000	300,000	130,000	160,000	54	87	81	75
1,1,2,2-Tetrachloroethane	5	<10[b]	78,000	33,000	32,000	0[a]	>99	0[a]	40
Tetrachloroethylene	3	<10[b]	6,800	<10[b]	2,300	37	>99	>99	78
1,1,1-Trichloroethane	1				42,000				9
1,1,2-Trichloroethane	5	<10[b]	200	<10[b]	<48	98	>99	>99	>99
Trichloroethylene		<10[b]	34,000	<10[b]	11,000	24	>99	>99	74

[a] Actual data indicate negative removal.

[b] Reported as not detected; assumed to be <10 μg/L.

TREATMENT TECHNOLOGY: Steam Stripping

Data source: Government report
Point source category:[a] Organic chemicals
Subcategory:
Plant:
References: B2, p. 134

Data source status:
- Engineering estimate ___
- Bench scale ___
- Pilot scale x
- Full scale ___

Use in system: Primary
Pretreatment of influent:

[a]Halogenated hydrocarbons wastewater.

DESIGN OR OPERATING PARAMETERS

Unit configuration:
Flow--wastewater feed: See below, 3.8 L/min, design
steam feed: See below
overhead: See below
bottoms: See below
Temperature--feed:
overhead:
bottoms:
Steam pressure:
Pressure drop:
Reflux ratio (if applicable):
Cooling water requirement:
Column height: 3.67 m
Column diameter: 5.08 cm
Plate/packing characteristics:
Plate/packing spacing: Pall rings made from polypropylene
Number of plates (if applicable):
Distillate, percent of feed:

REMOVAL DATA

Sampling period:

Flow wastewater feed, mL/min	Steam feed, mL/min	Overhead, mL/min	Bottoms, mL/min	TOC Concentration, mg/L			Percent removal[a]
				Feed	Overhead	Bottoms	
243[b]		9.4	272	99	132	76	14[b]
276	54	7.8	388	150	64	142	0[c]
255	50	13.5	321	158	115	139	0[c]
245	51	5.3	290	16	84	15	0[c]
235	65	11.4	340	24	88	16	4

[a]Percent removal calculated on a volumetric basis.
[b]Reflux ratio is 1.6:1 (reflux: overhead).
[c]Actual data indicate negative removal.

Note: Blanks indicate information was not specified.

TREATMENT TECHNOLOGY: Steam Stripping

Data source: Government report
Point source category:[a] Organic chemicals
Subcategory:
Plant:
References: B2, p. 127, 129

Data source status:
Engineering estimate ___
Bench scale ___
Pilot scale x
Full scale ___

Use in system: Primary
Pretreatment of influent:

DESIGN OR OPERATING PARAMETERS

Unit configuration:
Flow--wastewater feed: 250 mL/min, 3.8 L/min, design
steam feed: 53.1 mL/min
overhead: 7.1 mL/min
bottoms: 281 mL/min
Temperature--feed:
overhead: 102°C
bottoms: 103°C
Steam pressure:
Pressure drop:
Reflux ratio (if applicable):
Cooling water requirement:
Column height: 3.67 m
Column diameter: 5.08 cm
Plate/packing characteristics: Pall rings made from polypropylene
Plate/packing spacing:
Number of plates (if applicable):
Distillate, percent of feed: 2.8

REMOVAL DATA

Sampling period:

Pollutant/parameter	Concentration Feed	Concentration Overhead	Concentration Bottoms	Percent removal[a]
Conventional pollutants, mg/L:				
TOC	668	10,500	292	51
Toxic pollutants, μg/L:				
Chloroform	140,000	880,000	ND[b]	>99
1,2-Dichloroethane	1,600,000	4,100,000	65,000	95
1,2-*Trans*-dichloroethylene	1,600,000	270,000	1,300,000	9
Methylene chloride	800,000	3,300,000	90,000	87
1,1,2,2-Tetrachloroethane	15,000	120,000	50,000	0[c]
Tetrachloroethylene	15,000	50,000	ND[b]	>99
1,1,2-Trichloroethane	14,000	34,000	ND	>99
Trichloroethylene	16,000[d]	570,000	ND	>99

[a]Percent removal calculated on a volumetric basis.
[b]Note detected; assumed to be <10 μg/L.
[c]Actual data indicate negative removal.
[d]Based on mass balance.

Note: Blanks indicate information was not specified.

TREATMENT TECHNOLOGY: Steam Stripping

Data source: Government report
Point source category:[a] Organic chemicals
Subcategory:
Plant:
References: B2, pp. 127, 129

Data source status:
Engineering estimate ___
Bench scale ___
Pilot scale x
Full scale ___

Use in system: Primary
Pretreatment of influent:

[a]Halogenated hydrocarbons wastewater.

DESIGN OR OPERATING PARAMETERS

Unit configuration:
Flow--wastewater feed: 250 mL/min, 3.8 L/min, design
steam feed: 45 mL/min
overhead: 5.8 mL/min
bottoms: 350 mL/min
Temperature--feed:
overhead: 103°C
bottoms: 104°C
Steam pressure:
Pressure drop:
Reflux ratio (if applicable):
Cooling water requirement:
Column height: 3.67 m
Column diameter: 5.08 cm
Plate/packing characteristics: Pall rings made from polypropylene
Plate/packing spacing:
Number of plates (if applicable):
Distillate, percent of feed: 2.3

REMOVAL DATA

Sampling period:

Pollutant/parameter	Concentration			Percent removal[a]
	Feed	Overhead	Bottoms	
Conventional pollutants, mg/L:				
TOC	645	10,400	256	44
Toxic pollutants, μg/L:				
Chloroform	140,000	1,200,000	ND[b]	>99
1,2-Dichloroethane	1,600,000	4,400,000	42,000	96
1,2-*Trans*-dichloroethylene	1,600,000	350,000	370,000	75
Methylene chloride	800,000	3,500,000	110,000	81
1,1,2,2-Tetrachloroethane	15,000	15,000	33,000	0[c]
Tetrachloroethylene	15,000	~240,000[d]	6,800	37
1,1,2-Trichloroethane	14,000	250,000	200	98
Trichloroethylene	~62,000[d]	640,000	34,000	24[d]

[a]Percent removal calculated on a volumetric basis.
[b]Not detected; assumed to be <10 μg/L.
[c]Actual data indicate negative removal.
[d]Based on mass balance calculation.

Note: Blanks indicate information was not specified.

TREATMENT TECHNOLOGY: Steam Stripping

Data source: Government report
Point source category:[a] Organic chemicals
Subcategory:
Plant:
References: B2, pp. 127, 129

Data source status:
Engineering estimate ___
Bench scale ___
Pilot scale x
Full scale ___

Use in system: Primary
Pretreatment of influent:

[a]Halogenated hydrocarbons wastewater.

DESIGN OR OPERATING PARAMETERS

Unit configuration:
Flow--wastewater feed: 250 mL/min, 3.8 L/min, design
steam feed: 39.7 mL/min
overhead: 13.5 mL/min
bottoms: 275 mL/min
Temperature--feed:
overhead:
bottoms:
Steam pressure:
Pressure drop:
Reflux ratio (if applicable): 0.9:1 (reflux: overhead)
Cooling water requirement:
Column height: 3.67 m
Column diameter: 508 mm
Plate/packing characteristics: Pall rings made from polypropylene
Plate/packing spacing:
Number of plates (if applicable):
Distillate, percent of feed: 2.5

REMOVAL DATA

Sampling period:

Pollutant/parameter	Concentration			Percent removal[a]
	Feed	Overhead	Bottoms	
Conventional pollutants, mg/L				
TOC	636	9,810	243	58
Toxic pollutants, µg/L:				
Chloroform	140,000	1,100,000	65,000	49
1,2-Dichloroethane	1,600,000	5,500,000	440,000	70
1,2-*Trans*-dichloroethylene	1,600,000	1,300,000	ND[b]	>99
Methylene chloride	800,000	5,200,000	130,000	82
1,1,2,2-Tetrachloroethane	15,000	24,000	100	99
Tetrachloroethylene	15,000	9,600	ND	>99
1,1,1-Trichloroethane	51,000	170,000	42,000	9
1,1,2-Trichloroethane	14,000	66,000	ND	>99
Trichloroethylene		640,000	ND	>99

[a]Percent removal calculated on a volume basis.

[b]Not detected; assumed to be <10 µg/L.

Note: Blanks indicate information was not specified.

TREATMENT TECHNOLOGY: Steam Stripping

Data source: Government report
Point source category:[a] Organic chemicals
Subcategory:
Plant:
References: B2, pp. 127, 129

Data source status:
Engineering estimate ___
Bench scale ___
Pilot scale x
Full scale ___

Use in system: Primary
Pretreatment of influent:

[a]Halogenated hydrocarbons wastewater.

DESIGN OR OPERATING PARAMETERS

Unit configuration:
Flow--wastewater feed: 250 mL/min, 3.8 L/min, design
steam feed: 59.7 mL/min
overhead: 4.3 mL/min
bottoms: 305 mL/min
Temperature--feed:
overhead: 104°C
bottoms: 104°C
Steam pressure:
Pressure drop:
Reflux ratio (if applicable): 1.4:1 (reflux: overhead)
Cooling water requirement:
Column height: 3.67 m
Column diameter: 5.08 cm
Plate/packing characteristics: Pall rings made from polypropylene
Plate/packing spacing:
Number of plates (if applicable):
Distillate, percent of feed: 2.3

REMOVAL DATA

Sampling period:

Pollutant/parameter	Concentration			Percent removal[a]
	Feed	Overhead	Bottoms	
Conventional pollutants, mg/L:				
TOC	785	4,520	241	63
Toxic pollutants, μg/L:				
Chloroform	140,000	400,000	ND[b]	>99
1,2-Dichloroethane	1,600,000	3,700,000	39,000	97
1,2-*Trans*-dichloroethylene	1,600,000	1,300,000	16,000	99
Methylene chloride	800,000	1,000,000	300,000	54
1,1,2,2-Tetrachloroethane	14,000	8,000	ND	>99
1,1,2-Trichloroethane	14,000	42,000	ND	>99

[a]Percent removal calculated on a volumetric basis.
[b]Not detected; assumed to be <10 μg/L.

Note: Blanks indicate information was not specified.

TREATMENT TECHNOLOGY: Steam Stripping

Data source: Government report
Point source category:[a] Organic chemicals
Subcategory:
Plant:
References: B2, pp. 127, 129

Data source status:
Engineering estimate ___
Bench scale ___
Pilot scale x
Full scale ___

Use in system: Primary
Pretreatment of influent:

[a]Halogenated hydrocarbons wastewater.

DESIGN OR OPERATING PARAMETERS

Unit configuration:
Flow--wastewater feed: 250 mL/min, 3.8 L/min, design
steam feed: 50.8 mL/min
overhead: 12.75 mL/min
bottoms: 302.5 mL/min
Temperature--feed:
overhead: 104°C
bottoms: 104°C
Steam pressure:
Pressure drop:
Reflux ratio (if applicable): 5.1:94.9
Cooling water requirement:
Column height: 3.67 m
Column diameter: 5.08 cm
Plate/packing characteristics: Pall rings made from polypropylene
Plate/packing spacing:
Number of plates (if applicable):
Distillate, percent of feed: 5.1

REMOVAL DATA

Sampling period:

Pollutant/parameter	Concentration Feed	Overhead	Bottoms	Percent removal[a]
Conventional pollutants, mg/L:				
TOC	645	4,770	593	0[b]
Toxic pollutants, µg/L:				
Chloroform	140,000	840,000	ND[c]	>99
1,2-Dichloroethane	1,600,000	4,800,000	43,000	97
1,2-*Trans*-dichloroethylene	1,600,000	480,000	15,000	99
Methylene chloride	800,000	2,800,000	180,000	73
1,1,2,2-Tetrachloroethane	14,000	440,000	78,000	0[b]
1,1,2-Trichloroethane	14,000	76,000	ND	>99

[a]Percent removal calculated on a volumetric basis.

[b]Actual data indicate negative removal.

[c]Not detected; assumed to be <10 µg/L.

Note: Blanks indicate information was not specified.

TREATMENT TECHNOLOGY: Steam Stripping

Data source: Government report
Point source category:[a] Organic chemicals
Subcategory:
Plant:
References: B2, p. 130

Data source status:
Engineering estimate ___
Bench scale ___
Pilot scale x
Full scale ___

Use in system: Primary
Pretreatment of influent:

[a]Halogenated hydrocarbons wastewater.

DESIGN OR OPERATING PARAMETERS

Unit configuration:
Flow--wastewater feed: 245 to 500 mL/min
steam feed: 46 to 102 mL/min
overhead:
bottoms:
Temperature--feed:
overhead:
bottoms:
Steam pressure:
Pressure drop:
Reflux ratio (if applicable):
Cooling water requirement:
Column height: 3.67 m
Column diameter: 5.08 cm
Plate/packing characteristics: Pall rings made from polypropylene
Plate/packing spacing:
Number of plates (if applicable):
Distillate, percent of feed:

REMOVAL DATA

See table on the next page

Note: Blanks indicate information was not specified.

REMOVAL DATA

			Reflux ratio		COD			TOC			1,2-Dichloroethane			
Feed, mL/min	Overhead, mL/min	Bottoms, mL/min	(reflux: overhead)	Steam, mL/min	Feed, mg/L	Bottoms, mg/L	Percent[a] removal	Feed, mg/L	Bottoms, mg/L	Percent[a] removal	Feed, mg/L	Overhead, mg/L	Bottoms, mg/L	Percent[a] removal
395	11.4	410		69							1,700	8,800	65	96
300	16.0	340		69							1,600	10,000	65	>99
300	10.0	335		59							1,600	9,900	65	95
300	5.2	330		52							1,600	7,800	65	96
390	16.0	420		75							2,000	10,000	65	96
390	11.0	410		69							2,000	11,000	65	97
335	6.4	400		58							1,200	7,000	65	94
335	12.1	425		65							1,200	16,000	65	93
335	20.0	350		72							1,200	11,000	65	94
370	9.8	420		65							2,100	8,100	65	97
370	5.0	410		55							2,100	22,000	65	97
245	55.0	290		87							11,000	16,000	0.3	>99
245	34.0	270		65							11,000	16,000	1.0	>99
245	19.0	255		46							13,000	17,000	1.0	>99
245	52.0	290	0.13	87							13,000	15,000	1.0	>99
290	30.0	340		80							8,000	14,000	1.0	>99
290	11.0	345		65							8,000	13,000	1.0	>99
290	36.0	380	0.086	83							8,000	15,000	1.0	>99
370	18.5	410		71							7,000	14,000	1.0	>99
400	31.5	425		76							7,000	12,000	1.0	>99
400	16.6	410		69							7,000	13,000	1.0	>99
400	6.0	410		69							7,000	13,000	1.0	>99
400	13.8	475		77				801	129	81				
500	12.5	530		84				801	102	87				
280	10.9	350		50				801	109	83				
380	19.4	450		77				739	115	82	5,600	15,000	1.0	>99
380	12.3	390		72				739	120	83	5,600	14,000	1.0	>99
390	21.7	460		84				710	185	69	5,600	20,000	43	99
390	10.8	490		80				740	191	68	5,600	21,000	3.6	>99
390	33.0	470	0.056	87				740	166	73	5,600	16,000	4.0	>99
260	15.7	300		56				732	160	75	5,200	16,000	3.7	>99
290	9.1	350		65				765	69	89	5,800	18,000	5.7	>99
290	35.1	391	0.093	84	519	125	68	531	67	83	5,800	19,000	1.0	>99
335	10.3	394		60	519	146	67	531	30	93	5,100	17,000	1.0	>99
335	42.0	418	0.096	84	519	118	72	531	57	87	5,100	17,000	1.4	>99
350	18.3	400		80	519	200	56	531	121	74	1,500	6,600	6.0	>99
225	12.0	279		55	519	233	44	531	111	74	1,200	6,800	9.2	99
225	47.8	324	0.161	77	519	200	45	531	111	70	1,200	5,200	9.5	99
240	14.1	309		56	519			531	125	70	1,400	6,000	7.0	99
250	10.8	271		56	409			512	125	74	1,400	5,900	9.3	99
250	42.3	285	0.133	77	409			512	106	76	1,500	4,900	6.1	>99
280	14.2	301		65	409			512	91	81	1,500	7,100	8.5	99
280	57.1	366	0.153	102	409			512	69	82	1,500	5,900	19.4	98

[a]Percent removal calculated on a volumetric basis.

TREATMENT TECHNOLOGY: Steam Stripping

Data source: Government report
Point source category:
Subcategory:
Plant:
References: B3

Data source status:
Engineering estimate ___
Bench scale ___
Pilot scale ___
Full scale ___

Use in system:
Pretreatment of influent:

DESIGN OR OPERATING PARAMETERS

	Run number								
	1	2	3	4	5	6	7	8	9
Steam feed rate, mL/min	54.7	63.9	59.7	53.1	36.7	-	-	-	-
Volumetric flow rate, mL/min:									
Overhead:	10	13.8	3.0	11.5	13.4	6.5	8.2	7.5	14.0
Bottoms:	207	290	317	312	344	338	342	452	380
Temperature, °C									
Overhead:									
Bottoms:									
Column pressure, BTM/TOP:									
Reflux ratio:									
Feed to column, mL/min:	250	250	250	258	255	252	250	255	261

REMOVAL DATA

Pollutant/parameter	Concentration, mg/L			Percent removal[a]
	Feed	Overhead	Bottoms	
Conventional pollutants:				
TOC Run number				
1	315	65	24	94
2	2,420	98	118	94
3	20	193	15	5
4	67	83	45	19
5	26	94	21	0[b]
6	90	147	40	40
7	80	280	46	21
8	58	209	37	0[b]
9	155	737	14	87

[a]Calculated on a volumetric basis.

[b]Actual data indicate negative removal.

Note: Blanks indicate information was not specified.

5.6 SOLVENT EXTRACTION [1]

5.6.1 Function

Liquid-liguid solvent extraction, hereinafter referred to as solvent extraction, is the separation of the constituents of a liquid solution by contact with another immiscible liquid. If the substances comprising the original solution distribute themselves differently between the two liquid phases, a certain degree of separation will result, and this may be enhanced by use of multiple contacts.

5.6.2 Description

The solvent extration process is shown schematically in the Flow Diagram section. The diagram shows a single solvent extraction unit operating on an aqueous stream; in practice this unit might consist of (1) a single-stage mixing and settling unity, (2) several mixers and settlers (single-stage unit) in series, or (3) a multi-stage unit operating by countercurrent flows in one device (e.g., a column or differential centrifuge).

As the Flow Diagram indicates, reuse of the extracting solvent (following solute removal) and recovery of that portion of the extracting solvent that dissolves in the extracted phase are usually necessary aspects of the solvent extraction process. Solvent reuse is necessary for economic reasons; the cost of the solvent is generally too high to consider disposal after use. Only in a very few cases may solvent reuse be eliminated; these cases arise where an industrial chemical feed stream can be used as the solvent and then sent on for normal processing, or where water is the solvent. Solvent recovery from extracted water may be eliminated in cases where the concentration in the water to be discharged is not harmful, and where the solvent loss does not incur a high cost.

The end result of solvent extraction is to separate the original solution into two streams: a treated stream (the raffinate), and a recovered solute stream (which may contain small amounts of water and solvent). Solvent extraction may thus be considered a recovery process since the solute chemicals are generally recovered for reuse, resale, or further treatment and disposal.

A process for solvent extracting a solution will typically include three basic steps: the actual extraction, solute removal from the extracting solvent, and solvent recovery from the raffinate (treated stream). The process may be operated continuously.

The first step, extraction, brings the two liquid phases (feed and solvent) into intimate contact to allow solute transfer either by forced mixing or by countercurrent flow caused by density differences. The extractor will also have provisions to allow separation of the two phases after mixing. One output stream from the extractor is the solute-laden solvent; some water may also be present. Solute removal may be via a second solvent extraction step, distillation, or some other process. For example, a second extraction, with caustic, is sometimes used to extract phenol from light oil, which is used as the primary solvent in dephenolizing coke plant wastewaters. Distillation will usually be more common, except where problems with azeotropes are present. In certain cases, it may be possible to use the solute-laden solvent as a feed stream in some industrial process, thus eliminating solute recovery. This is apparently the case at some refineries where crude or light oil can be used as a solvent (for phenol removal from water) and later processed with the solute in it. Other similar applications probably exist and are particularly attractive since they eliminate one costly step. Solvent recovery from the treated stream may be required if solvent losses would otherwise add significantly to the cost of the process, or cause a problem with the discharge of the raffinate. Solvent recovery may be accomplished by stripping, distillation, adsorption, or other suitable process.

5.6.3 Technology Status

Solvent extraction should be regarded as a process for treating concentrated, selected, and segregated waste water streams primarily where material recovery is possibe to offset process costs. Solvent extraction, when carried out on the more concentrated waste streams, will seldom produce a treated effluent (the raffinate) that can be directly discharged to surface waters; some form of final polishing will usually be needed. Solvent extraction cannot compete economically with biological oxidation or adsorption in the treatment of large quantities of very dilute wastes, and it will have trouble competing with stream stripping in the recovery of volatile solutes present in moderate to low concentrations. Nevertheless, solvent extraction is a proven method for the recovery of organics from liquid solutions and will be the process of choice in some cases.

5.6.4 Applications

Removal of phenol and related compounds from wastewaters is the principal application; applications are to petroleum refinery wastes, coke-oven liquors and phenol resin plant effluents. Extraction reduces phenol concentrations from levels of several percent down to levels of a few parts per million. Removal efficiencies of 90 to 98% are possible in most applications, and with special equipment (e.g., centrifugal and rotating disc contactors) removal efficiencies around 99% have been achieved.

Commonly used solvents are crude oil, light oil, benzene, toluene, and "benzol;" less common, but more selective solvents are isopropyl ether, tricresyl phosphate, methyl isobutyl ketone, methylene chloride, and butyl acetata. When crude or light oil is used, the phenol is not always covered (i.e., the solvent is not recycled); the phenol is destroyed in downstream operations. Alternatively, extraction with light oil may be followed by phenol recovery via extraction of the oil with caustic; in this case, the phenol is recovered as sodium phenolate.

Solvent recovery via solvent extraction is carried out in at least one hazardous waste management facility (Silresin Chemical Corporation, Lowell, Massachusetts). In one case, waste solvent containing typically 85% methylene chloride (MC) and 15% isopropyl alchohol (IPA) is extracted with water to remove the IPA. Extraction has been carried out in a counter-current column (1 ft diameter, 40 ft high, packed with Berl saddles), which accepts a feed of 1 gpm and produces a purified MC product at around 0.7 to 0.8 gpm. The water/feed ration used in this device was about 3:1. More recently, a single tank has been used as a combination mixer-settler to handle larger flows. The partially purified MC is then further processed through a flash evaporator and calcium chloride absorption bed (for drying) to obtain salable quality MC (98% to 99%) pure. A second example involves the reclamation of Freon solvents. The waste material arrives as a mixture of oil, Freon, and other solvents (e.g., acetone or alcohol); distillation separates out the oil (for use as a fuel), but leaves a Freon/acetone (or alcohol) mixture which is then extracted with water to recover Freon. The material is sold for about half the price (per gallon) for new Freon solvent. A third example involves the removal of water-soluble solvents (e.g., alcohols) from a waste of mixed chlorinated hydrocarbon solvents via extraction with water. Simple mixer-settlers are commonly used, and the process yields a salable quality (mixed) chlorinated hydrocarbon solvent.

Other applications of solvent extraction are briefly described below:

- Extraction of thiazole-based chemicals from rubber processing effluent with benzene.

- Extraction of salicylic and other hydroxy-aromatic acids from wastewaters using methyl isobutyl ketone as solvent.

- Deoiling of quench waters from petroleum operations via solvent extraction has been developed by Gulf Oil Corporation. Quench water containing about 6,000 ppm of dissolved and emulsified oil is extracted with a light aromatic oil solvent and the extract recycled for refinery processing. Additional treatment of the water (e.g., via coalescence) is necessary for water reuse. It is not known if this process is in current use.

- Recovery of acetic acid from industrial wastewater is being studied by Hydroscience. A novel extraction is proposed to handle wastewaters that may contain acetic acid levels of 0.5% to over 5%. The extractant is a solution of trioctylphosphine oxide in a carrier solvent. This process is currently in the developmental stage, but has been demonstrated to be practical.

- A novel process employing solvent extraction is currently being developed by Resources Conservation Co. (Renton, Wash.) to remove essentially all of the water and oils from inorganic and organic sludges. The process, called Basic Extractive Sludge Treatment (B.E.S.T.), converts sludges with 0.05% to 60% solids to output streams of (1) very dry solids (4.5% moisture), (2) a clear water effluent, and (3) recovered oils, if present in the original sludge. The process train includes: (1) extraction of water (and oils) from the sludge with an aliphatic amine at low temperatures (∿50°F), (2) removal of solids with a centrifuge followed by solids drying (and solvent recovery), (3) heating the solvent/water/ oil mixture (to ∿120°F) to force phase separation, (4) steam stripping of the water phase for solvent recovery, and (5) distillation of the solvent phase for oil recovery. The company claims the process is economical; it requires, for example, only 6,400 Btu's per pound to reduce a 7 percent sludge to dry solids versus 15,000 Btu's per pound for conventional high-temperature "brute force" drying methods. A mobile test and demonstration facility has been constructed which can treat 1,500 gpd. Several different types of sludges have been successfully processed.

5.6.5 Limitations

There are relatively few insurmountable technical problems with solvent extraction. The most difficult problem is usually finding a solvent that best meets a long list of desired qualities including low cost, high extraction efficiency, low solubility in the raffinate, easy separation from the solute, adequate density difference with raffinate, no tendency for emulsion formation, nonreactive, and nonhazardous. No one solvent will meet all the desired criteria and, thus, compromise is necessary. There is a wide range of extraction equipment available today, and space requirements are not a problem.

Process costs are always a determining factor with solvent extractions, and they have thus far limited actual applications to situations where a valuable product is recovered in sufficient quantity to offset extraction costs. These costs will be relatively small when a single-stage extraction unit can be used (e.g., simple mixer-settler) and where solvent and solute recovery can be carried out efficiently. In certain cases, the process may yield a profit when credit for recovered material is taken. Any extraction requiring more than the equivalent of about ten theoretical stages may require custom-designed equipment and will, thus, be quite expensive.

5.6.6 Residuals Generated/Environmental Impact

There are no major environmental impacts associated with the proper use of solvent extraction. Solvent extraction will almost always be used for material recovery (for resale or reuse) and, thus, will be of some benefit.

When one or more solutes are recovered from aqueous wastes, minor impacts will result from small losses of the solvent (to the air and/or water), and head (e.g., from stripping or distillation). In addition, solvent extraction systems seldom produce a raffinate that is suitable for direct discharge to surface waters and thus, a polishing treatment is generally required; biological treatment may suffice in many cases.

When mixed organic liquids are treated principally for the recovery of just one component (e.g., the more valuable halogenated hydrocarbons), current economic forces may make the purification of the other components (as required for resale or reuse) impractical and, thus, results in a waste for disposal.

5.6.7 Reliability

Process is highly reliable for proven applications, if properly operated.

5.6.8 Typical Equipment

There are two major categories of equipment for liquid extraction: simple-stage and multi-stage equipment.

In single-stage equipment, the fluids are mixed, extraction occurs, and the insoluble liquids are settled and separated. A cascade of such stages may then be arranged. A single-stage must provide facilities for mixing the insoluble liquids and for settling and decanting the emulsion or dispersion which results. In batch operation, mixing together with settling and decanting may take place in the same or in separate vessels. In continuous operation, different vessels are required.

In multi-stage equipment, the equivalent of many stages may be incorporated into a single device or apparatus. Countercurrent flow is produced by virtue of the difference in densities of the liquids, and with few exceptions the equipment takes the form of a vertical tower which may or may not contain internal devices to influence the flow pattern. Other forms include centrifuges, rotating discs, and rotating buckets. Depending upon the nature of the internal structure, the equipment may be of the stagewise or continuous-contact type.

5.6.9 Flow Diagram

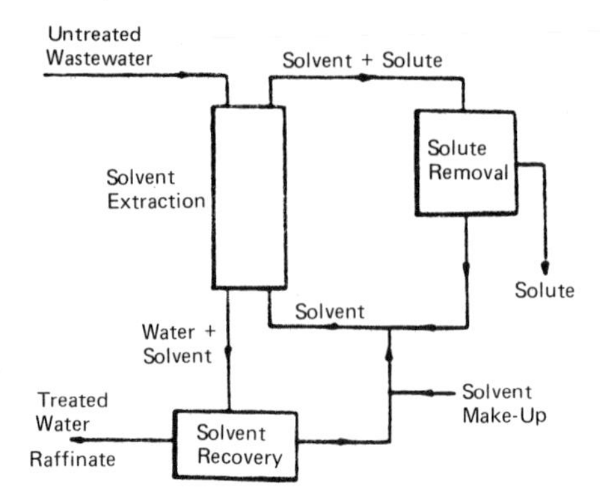

5.6.10 Performance

Subsequent data sheets provide performance data from studies on the following industries and/or wastestreams:

- Organic chemicals production
  - Cresylic acid recovery
  - Ethylene oxychlorination
  - Ethylene quenching
  - Styrene production
- Petroleum refining
  - Lube oil refining
- Phenolic resin production

5.6.11 Reference

1. Physical, Chemical, and Biological Treatment Techniques for Industrial Wastes, PB 275-287, U.S. Environmental Protection Agency, Washington, D.C. November 1976. pp. 32-1 through 32-25.

CONTROL TECHNOLOGY SUMMARY FOR SOLVENT EXTRACTION

Pollutant	Number of data points	Effluent concentration				Removal efficiency, %			
		Minimum	Maximum	Median	Mean	Minimum	Maximum	Median	Mean
Conventional pollutants, mg/L:									
COD	4	699	18,600	1,140	5,390	0[a]	74	50	>43
TOC	6	37	86.5	43.5	53.9	0[a]	49	35	31
Toxic pollutants, μg/L:									
Phenol	15	<1,000	10,000,000	190,000	2,200,000	3	99	80	65
Benzene	6	2,400	35,000	8,100	11,000	75	97	96	90
Ethylbenzene	1				4,000				97
Toluene	2	1,600	2,300	32.0	2,000	94	96		95
1,2-Dichloroethane	6	<20,000	350,000	32,000	84,000	62	>99	89	87
1,1,2,2-Tetrachloroethane	5	1,000	11,000	2,000	4,200	73	99	98	91
1,1,2-Trichloroethane	5	5,400	30,000	16,000	16,000	85	95	92	90
Other pollutants, μg/L:									
Acetone	3	12,000	22,000	16,000	16,700	41	57	52	50
MEK	7	12,000	5,900,000	2,500,000	2,000,000	32	95	50	60
m,p-cresol	1				25,000				91
o-cresol	9	2,800	4,000,000	31,000	500,000	0[a]	>99	90	82
Styrene	1				<1,000				>93
Total chlorine	11	1,800	510,000	81,000	98,000	68	99	94	90
Xylenes	3	<1,000	10,000	<1,000	<4,000	96	>98	>97	>97

Note: Blanks indicate data not available.

[a]Actual data indicate negative removal.

TREATMENT TECHNOLOGY: Solvent Extraction

Data source: Effluent Guidelines
Point source category: Organic chemicals
Subcategory: Petrochemicals
Plant:
References: A25, p. 292

Data source status:
- Engineering estimate ___
- Bench scale ___
- Pilot scale ___
- Full scale x

Use in system: Tertiary
Pretreatment of influent:

DESIGN OR OPERATING PARAMETERS

None given.

REMOVAL DATA

Sampling period:

Pollutant/parameter	Total phenol Concentration, mg/L Influent	Effluent	Percent removal
Solvent used:			
Aromatics, 75% Paraffins, 25%	200	0.2	>99
Aliphatic esters	4,000	60	98
Benzene	750	34	95
Light cycle oil	7,300	30	>99
Light oil	3,000	35	99
Tri-cresyl phosphates	3,000	300-150	90-95

Note: Blanks indicate information was not specified.

TREATMENT TECHNOLOGY: Solvent Extraction

Data source: Government report
Point source category: Organic chemicals
Subcategory: Ethylene oxychlorination process
Plant:
References: B2, pp. 102-117, Appendix

Data source status:
Engineering estimate ___
Bench scale ___
Pilot scale x
Full scale ___

Use in system: Primary
Pretreatment of influent:

DESIGN OR OPERATING PARAMETERS

Unit configuration:
Column specifications: Extractor: 0.10 m diameter, 3.0 m tall
Stripper: 0.05 m diameter, 2.25 m tall
Solvent used: Kerosene-diesel oil mix
Solvent flow rate: 0.205 L/min
Wastewater flow rate: 0.76 to 3.76 L/min

REMOVAL DATA

Sampling period: One-day composites

Pollutant/parameter	Concentration		Percent removal	H_2O to solvent ratio
	Influent	Effluent		
Toxic pollutants, μg/L:				
1,2,-Dichloroethane	920,000	350,000	62	18.3:1
	190,000	20,000	89	13.7:1
	210,000[a]	36,000[a]	83	9.1:1
	460,000[b]	51,000[b]	89	5.5:1
	1,100,000[c]	27,000[c]	98	3.7:1
1,1,2,2-Tetrachloroethane	22,000	6,000	73	18.3:1
	200,000	2,000	99	13.7:1
	85,000[c]	11,000[c]	87	9.1:1
	51,000[d]	1,000[d]	98	5.5:1
	91,000[a]	1,000	99	3.7:1
1,1,2-Trichloroethane	110,000	16,000	85	18.3:1
	360,000	30,000	92	13.7:1
	150,000[a]	22,000[a]	85	9.1:1
	110,000[e]	5,400[e]	95	5.5:1
	110,000[a]	8,700[a]	92	3.7:1
Other pollutants, mg/L:				
Total chlorine	1,590	514	68	18.3:1
	907	81	91	13.7:1
	553[a]	85[a]	85	9.1:1
	1,810[a]	110[a]	94	5.5:1
	1,830[c]	84[c]	95	3.7:1

[a]Average of three one-day composites.
[b]Average of four one-day composites.
[c]Average of two one-day composites.
[d]Average of six one-day composites.
[e]Average of five one-day composites.

Note: Blanks indicate information was not specified.

TREATMENT TECHNOLOGY: Solvent Extraction

Data source: Government report
Point source category: Organic chemicals
Subcategory: Ethylene oxychlorination process
Plant:
References: B2, pp. 102-117

Data source status:
- Engineering estimate ___
- Bench scale ___
- Pilot scale x
- Full scale ___

Use in system: Primary
Pretreatment of influent:

DESIGN OR OPERATING PARAMETERS

Solvent used: C_{10}-C_{12} paraffin
Solvent flow rate: 0.27 L/min
Wastewater flow rate: 1.23 to 5.32 L/min
Column specifications: Extractor: 0.10 m diameter x 3.0 m
Stripper: 0.05 m diameter x 2.25 m

REMOVAL DATA

Sampling period: One-day composites

Pollutant/parameter	Concentration, mg/L		Percent	H_2O to
	Influent	Effluent	removal	solvent ratio
Conventional pollutants:				
TOC	58	37	36	5:1
	73	48	34	6.5:1
	59	38	36	8:1
	76	39	49	10:1
	54	75	0[a]	16.5:1
	124[b]	86.5[b]	30[b]	20:1
Other pollutants:				
Total chlorine	148	3.2	98	5:1
	185	3.0	98	6.5:1
	165	1.8	99	8:1
	297	6.6	98	10:1
	267	16.5	94	16.5:1
	693[b]	178[b]	74	20:1

[a] Actual data indicate negative removal.
[b] Average of 2 1-day composites.

Note: Blanks indicate information was not specified.

TREATMENT TECHNOLOGY: Solvent Extraction

Data source: Government report
Point source category: Organic chemicals
Subcategory: Styrene production process
Plant:
References: B18, pp. 102-109, 241-243, 501

Data source status:
- Engineering estimate ___
- Bench scale ___
- Pilot scale x
- Full scale ___

Use in system: Primary
Pretreatment of influent:

DESIGN OR OPERATING PARAMETERS

Unit configuration: Rotating disc contactor and stripping column
Column specifications: 0.0762 m (3 in.) diameter x 1.22 m (48 in.) glass pipe
Solvent used: Isobutylene
Solvent flow rate: 0.451 m/hr (1.48 ft/hr)
Wastewater flow rate: 2.49 m/hr (8.17 ft/hr)

REMOVAL DATA

Sampling period:

Pollutant/parameter	Concentration, μg/L Influent	Concentration, μg/L Effluent	Percent removal
Toxic pollutants:			
Benzene	290,000	10,000	97
Ethylbenzene	120,000	4,000	97
Other pollutants:			
Styrene	15,000	<1,000	>93

Note: Blanks indicate information was not specified.

TREATMENT TECHNOLOGY: Solvent Extraction

Data source: Government report
Point source category: Organic chemicals
Subcategory: Ethylene quench process
Plant:
References: B18, pp. 102-109, 223-227, 496

Use in system: Primary
Pretreatment of influent:

Data source status:
- Engineering estimate ___
- Bench scale ___
- Pilot scale x
- Full scale ___

DESIGN OR OPERATING PARAMETERS

Unit configuration: Rotating disc contactor and stripping column
Column specifications: 0.0762 m (3 in.) diameter x 1.22 m (48 in.) glass pipe
Solvent used: Isobutane
Solvent flow rate: 0.668 m/hr (2.19 ft/hr)
Wastewater flow rate: 3.81 m/hr (12.5 ft/hr)

REMOVAL DATA

Sampling period:

Pollutant/parameter	Concentration		Percent
	Influent	Effluent	removal
Conventional pollutants, mg/L:			
COD	1,880	699	63
Toxic pollutants, µg/L:			
Phenol	68,000	66,000	3
Benzene	81,000	2,400	97
Toluene	44,000	1,600	96
Other pollutants, µg/L:			
Xylenes	34,000	<1,000	>97

Note: Blanks indicate information was not specified.

TREATMENT TECHNOLOGY: Solvent Extraction

Data source: Government report
Point source category: Organic chemicals
Subcategory: Ethylene quench process
Plant:
References: B18, pp. 102-109, 223-227, 495

Data source status:
Engineering estimate ___
Bench scale ___
Pilot scale x
Full scale ___

Use in system: Primary
Pretreatment of influent:

DESIGN OR OPERATING PARAMETERS

Unit configuration: Rotating disc contactor and stripping column
Column specifications: 0.0762 m (3 in.) diameter x 1.22 m (48 in.) glass pipe
Solvent used: Isobutylene
Solvent flow rate: 0.652 m/hr (2.14 ft/hr)
Wastewater flow rate: 3.84 m/hr (12.6 ft/hr)

REMOVAL DATA

Sampling period:

Pollutant/parameter	Concentration		Percent removal
	Influent	Effluent	
Conventional pollutants, mg/L:			
COD	1,880	1,210	36
Toxic pollutants, μg/L:			
Phenol	67,000	63,000	6
Benzene	71,000	2,900	96
Toluene	41,000	2,300	94
Other pollutants, μg/L:			
Xylenes	41,000	<1,000	>98

Note: Blanks indicate information was not specified.

TREATMENT TECHNOLOGY: Solvent Extraction

Data source: Government report
Point source category: Organic chemicals
Subcategory: Cresylic acid recovery process
Plant:
References: B18, pp. 98-102, 159-165, 465

Data source status:
Engineering estimate ___
Bench scale ___
Pilot scale x
Full scale ___

Use in system: Primary
Pretreatment of influent:

DESIGN OR OPERATING PARAMETERS

Unit configuration: Spray column contactor and stripping column
Column specifications: 0.0254 m (1 in.) diameter x 0.914 m (36 in.) glass pipe
Solvent used: Isobutylene
Solvent flow rate: 18.5 m/hr (60.6 ft/hr)
Wastewater flow rate: 6.14 m/hr (20.1 ft/hr)

REMOVAL DATA

Sampling period:

Pollutant/parameter	Concentration		Percent
	Influent	Effluent	removal
Conventional pollutants, mg/L:			
COD	4,050	1,070	74
Toxic pollutants, μg/L:			
Phenol	580,000	160,000	72
Other pollutants, μg/L:			
o-Cresol	310,000	31,000	90
m, *p*-Cresol	290,000	25,000	91
Xylenes	230,000	10,000	96

Note: Blanks indicate information was not specified.

TREATMENT TECHNOLOGY: Solvent Extraction

Data source: Government report
Point source category: Petroleum refining
Subcategory: Lube oil refining
Plant:
References: B18, pp. 98-109, 159-165, 198-212, 453, 491-493

Data source status:
- Engineering estimate ___
- Bench scale ___
- Pilot scale x
- Full scale ___

Use in system: Primary unless otherwise specified
Pretreatment of influent:

DESIGN OR OPERATING PARAMETERS

Unit configuration: Rotating disc contactor and stripping column
Column specifications: 0.0762 m (3 in.) diameter x 1.22 m (48 in.) glass pipe
Solvent used: Isobutylene unless otherwise specified
Solvent flow rate: See below
Wastewater flow rate: 2.67 m/hr (8.74 ft/hr)

REMOVAL DATA

Sampling period:

Solvent flow rate, m/hr (ft/hr)	Phenol Concentration, µg/L Influent	Phenol Concentration, µg/L Effluent	Phenol Percent removal	MEK Concentration, µg/L Influent	MEK Concentration, µg/L Effluent	MEK Percent removal	o-Cresol Concentration, µg/L Influent	o-Cresol Concentration, µg/L Effluent	o-Cresol Percent removal
0.459 (1.51)	310,000[a]	230,000	26	5,600,000	3,600,000	36	24,000	2,300	90
0.459 (1.51)	230,000[a]	190,000	17	2,800,000	1,900,000	32	18,000	2,800	84
0.306 (1.00)	8,800,000[b]	100,000	99	12,000,000	5,900,000	51	890,000	6,500	99
0.921 (3.02)	8,800,000[b]	77,000	99	12,000,000	2,500,000	79	990,000	4,300	>99
21.8 (71.6)	17,000,000[c]	10,000,000					1,200,000	4,000,000	41

[a] Pretreatment of influent: N-butyl acetate extraction; use in system: secondary.
[b] Solvent used: N-butyl acetate.
[c] Wastewater flow rate: 6.77 m/hr (22.2 ft/hr).

Note: Blanks indicate information was not specified.

TREATMENT TECHNOLOGY: Solvent Extraction

Data source: Government report
Point source category: Petroleum refining
Subcategory: Lube oil refining
Plant:
References: B18, pp. 102-109, 212-216, 494

Data source status:
- Engineering estimate ___
- Bench scale ___
- Pilot scale x
- Full scale ___

Use in system: Primary
Pretreatment of influent:

DESIGN OR OPERATING PARAMETERS

Unit configuration: Rotating disc contactor and stripping column
Column specifications: 0.0762 m (3 in.) diameter x 1.22 m (48 in.) glass pipe
Solvent used: 48.7 wt % n-butyl acetate, 51.3 wt % isobutylene
Solvent flow rate: 0.936 m/hr (3.07 ft/hr)
Wastewater flow rate: 3.35 m/hr (11.0 ft/hr)

REMOVAL DATA

Sampling period:

Pollutant/parameter	Concentration, µg/L		Percent removal
	Influent	Effluent	
Toxic pollutants:			
Phenol	17,000,000	1,900,000	89
Benzene	37,000	9,200	75
Other pollutants:			
Acetone	25,000	12,000	52
MEK	110,000	55,000	50
o-Cresol	2,700,000	120,000	96

Note: Blanks indicate information was not specified.

TREATMENT TECHNOLOGY: Solvent Extraction

Data source: Government report
Point source category: Petroleum refining
Subcategory: Lube oil refining
Plant:
References: B18, pp. 98-102, 159-165, 456

Data source status:
Engineering estimate ___
Bench scale ___
Pilot scale x
Full scale ___

Use in system: Primary
Pretreatment of influent:

DESIGN OR OPERATING PARAMETERS

Unit configuration: Spray column contactor and stripping column
Column specifications: 0.0254 m (1 in.) diameter x
0.914 m (36 in.) glass pipe
Solvent used: Isobutylene
Solvent flow rate: 15.9 m/hr (52.00 ft/hr)
Wastewater flow rate: 6.57 m/hr (21.6 ft/hr)

REMOVAL DATA

Sampling period:

Pollutant/parameter	Concentration, μg/L		Percent removal
	Influent	Effluent	
Toxic pollutants:			
Phenol	23,000,000	9,600,000	58
Benzene	170,000	35,000	79
Other pollutants:			
o-Cresol	2,000,000	330,000	83
Acetone	37,000	22,000	41
MEK	230,000	55,000	76

Note: Blanks indicate information was not specified.

TREATMENT TECHNOLOGY: Solvent Extraction

Data source: Government report
Point source category: Petroleum refining
Subcategory: Lube oil refining
Plant:
References: B18, pp. 98-102, 159-165, 455

Data source status:
- Engineering estimate ___
- Bench scale ___
- Pilot scale x
- Full scale ___

Use in system: Primary
Pretreatment of influent:

DESIGN OR OPERATING PARAMETERS

Unit configuration: Spray column contactor and stripping column
Column specifications: 0.0254 m (1 in.) diameter x 0.914 m (36 in.) glass pipe
Solvent used: Isobutylene
Solvent flow rate: 28.1 m/hr (92.2 ft/hr)
Wastewater flow rate: 6.57 m/hr (21.6 ft/hr)

REMOVAL DATA

Sampling period:

Pollutant/parameter	Concentration, µg/L		Percent
	Influent	Effluent	removal
Toxic pollutants:			
Phenol	23,000,000	4,600,000	80
Benzene	170,000	7,000	96
Other pollutants:			
o-Cresol	2,000,000	50,000	97
Acetone	37,000	16,000	57
MEK	230,000	12,000	95

Note: Blanks indicate information was not specified.

TREATMENT TECHNOLOGY: Solvent Extraction

Data source: Government report
Point source category:
Subcategory: Hydrofiner or phenolic resin plant
Plant:
References: B18, pp. 102-109, 233-241, 499-501

Data source status:
- Engineering estimate ___
- Bench scale ___
- Pilot scale x
- Full scale ___

Use in system: Primary
Pretreatment of influent:

DESIGN OR OPERATING PARAMETERS

Unit configuration: Rotating disc contactor and stripping column
Column specifications: 0.0762 m (3 in.) diameter x 1.22 m (48 in.) glass pipe
Solvent used: See below
Solvent flow rate: See below
Wastewater flow rate: See below

REMOVAL DATA

Solvent flow rate, m/hr (ft/hr)	Wastewater flow rate, m/hr (ft/hr)	COD			Phenol		
		Concentration, mg/L		Percent	Concentration, μg/L		Percent
		Influent	Effluent	removal	Influent	Effluent	removal
0.512(1.68)[a]	3.26(10.7)				400,000	<1,000	>99
0.625(2.05)[b]	2.08(6.81)	17,500	18,600	0[c]	400,000	<1,000	>99
0.561(1.84)[d]	2.01(6.58)				48,000,000	480,000	99
0.245(0.804)[e]	1.81(5.94)				48,000,000	6,100,000	87

[a] Solvent used: methyl isobutyl ketone.

[b] Solvent used: 49.5 wt % methyl isobutyl ketone, 50.5 wt % isobutylene.

[c] Actual data indicate negative removal.

[d] Solvent used: 48.2% n-butyl acetate, 51.8% isobutylene.

[e] Solvent used: N-butyl acetate.

Note: Blanks indicate information was not specified.

TREATMENT TECHNOLOGY: Solvent Extraction

Data source: Government report
Point source category:
Subcategory: Oxychlorination
Plant:
References: B18, pp. 102-109, 227-232, 497

Data source status:
Engineering estimate ___
Bench scale ___
Pilot scale x
Full scale ___

Use in system: Secondary
Pretreatment of influent: Neutralization

DESIGN OR OPERATING PARAMETERS

Unit configuration: Rotating disc contactor and stripping column
Column specifications: 0.0762 m (3 in.) diameter x 1.22 m (48 in.) glass pipe
Solvent used: 2-ethyl hexanol
Solvent flow rate: 0.457 m/hr (1.50 ft/hr)
Wastewater flow rate: 3.60 m/hr (11.8 ft/hr)

REMOVAL DATA

Sampling period:

Pollutant/parameter	Concentration, µg/L Influent	Effluent	Percent removal
Toxic pollutants:			
1,2-Dichloroethane	1,500,000	<20,000	>99

Note: Blanks indicate information was not specified.

6. Tertiary Wastewater Treatment

6.1 GRANULAR ACTIVATED CARBON ADSORPTION

Function. Activated carbon adsorption is used for the removal of dissolved organics and control of such wastewater parameters as COD, TOC, BOD, TOD, and specific soluble organic materials.

Treatability Factor. Adsorbability, kg removed/kg of carbon.

Description. In most cases, activated carbon is used as an individual-stream pretreatment process; however, in other cases activated carbon treatment is used as a final treatment process following biological treatment.

Granular carbon systems generally consist of vessels in which the carbon is placed, forming a "filter" bed. These systems can also include carbon storage vessels and thermal regeneration facilities. Vessels are usually circular for pressure systems or rectangular for gravity flow systems. Once the carbon adsorptive capacity has been fully utilized, the carbon must be disposed of or regenerated. Usually multiple carbon vessels are used to allow continuous operation. Columns can be operated in a series or parallel modes. All vessels must be equipped with carbon removal and loading mechanisms to allow for the removal of spent carbon and the addition of new material. Flow can be either upward or downward through the carbon bed. Vessels are backwashed periodically. Surface wash and air scour systems can also be used as part of the backwash cycle.

Small systems usually dispose of spent carbon or regenerate it offsite. Systems above about 3 to 5 Mgal/d usually provide onsite regeneration of carbon for economic reasons, as do systems where carbon usage exceeds 1,000 lb/d. Activated carbon regeneration is described separately in the table on the following page.

Technology Status. Granular activated carbon has been widely used in water treatment systems for many years. Carbon has been used in waste treatment for 10 to 20 years.

ACTIVATED CARBON REGENERATION

Function:

Remove and thermally oxidize adsorbed organics from spent activated carbon, for reuse of the carbon

Parameters affected:

Carbon adsorption capacity

Effectiveness:

Complete combustion of offgases

Application limits:

None

Design basis:

Multiple-hearth furnace with afterburner on top hearth; carbon loading: 40 to 120 lb/d per ft^2 of hearth surface area; temperature: 1,700°F to 1,800°F; surface area required: design plus 20% for downtime; regeneration fuel: 8,000 Btu/lb of carbon; carbon loss: 10% per cycle

Residues:

Clean offgas and ash, representing the carbon losses

Major equipment:

Regeneration furnace (multiple hearth) with stacks and afterburner; quench chamber; venturi scrubber; separator; venturi recirculation tank and pumps; caustic storage and feed system; combustion and shaft cooling air blowers; fuel oil storage and feed system; carbon transfer pumps; feed slurry tank; dewatering screw conveyor

Applications. Used directly following secondary clarifier, primarily when nitrification obtained in secondary treatment. Often preceded by chemical clarification of secondary effluent. In either case, a high quality effluent is sought.

Limitations. Wastewater should be filtered prior to treatment to remove suspended solids. Requires more sophisticated operation than standard secondary treatment systems. Under certain conditions, granular carbon beds provide favorable conditions for the production of hydrogen sulfide, creating odors and corrosion problems. More mechanical operations, difficult corrosion control and materials handling. Most applicable to low strength or toxic wastewaters. Influent limits: ≤25 mg/L on suspended solids, ≤ mg/L on free oil.

Typical Equipment. Adsorbers [fixed-bed, pressurized, downflow contactors (minimum of two in series, plus a spare), minimum depth:diameter ratio = 1.1]; regenerated-carbon storage tank; spent-carbon holding tank; effluent holding tank; backwash pumps.

Design Criteria. Size: vessels 2 to 12 ft diameter commonly used; area loading: 2 to 10 $gal/min/ft^2$; organic loading: 0.1 to 0.3 lb BOD_5 or COD/lb carbon; backwash: 12 to 20 $gal/min/ft^2$; air scour: 3 to 5 $ft^3/min/ft^2$; bed depth: 5 to 30 ft; contact time: 10 to 50 min; land area: minimal; side stream: spent carbon, 3 to 10 lb/lb of COD removed for tertiary treatment; backwash water, 1% to 5% of wastewater throughout, TSS 100 to 250 mg/L.

Chemicals Required. $NaHSO_3$ for H_2S control. Cl_2 or hypochlorite for biological growth control.

Reliability. Moderately reliable both mechanically and operationally depending on design, construction and manufactured equipment quality.

Toxics Management. Removes many, but not all, nondegradable organic compounds. Most effective for nonpolar, high molecular weight, slightly soluble compounds.

EPA has developed activated carbon adsorption isotherms for 60 toxic organic materials. The isotherms demonstrate removal of 51 of these organic compounds by activated carbon technology. Another study demonstrated that PCB levels can be reduced from 50 μg/L to less than 1 μg/L, and other work showed that aldrin, dieldrin, endrin, DDE, DDT, DDD, Toxaphene, and Aroclors 1242 and 1254 can be removed to values less than 1 μg/L.

Environmental Impact. Very little use of land. There is air pollution generated as a result of regeneration. Sulfide odors sometimes occur from contractors. Spent carbon may be a land disposal problem, unless regenerated.

Improved Joint Treatment Potential. Will remove pollutants discharged by industrial sources that are generally not treated by normal secondary systems such as refractory organic materials and some metals.

Flow Diagram

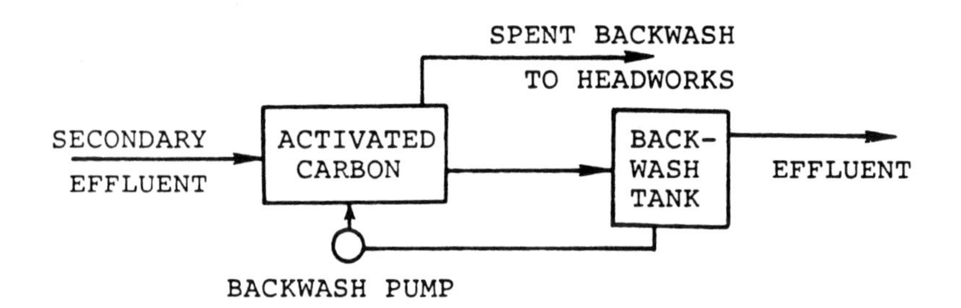

Performances

Subsequent data sheets provide performance data from studies on the following industries and/or wastestreams:

Auto and other laundries industry
  Industrial laundries
  Power laundries

Gum and wood chemicals production

Ore mining and dressing
  Base and precious metals

Organic chemicals production
  Fumaric acid
  Plasticizers
  Vinyl chloride
  Halogenated hydrocarbon wastewaters

Pesticides chemicals production
  Halogenated organic pesticides
  Metallo-organic pesticides
  Noncategorized pesticides
  Organic-nitrogen pesticides

Petroleum refining

Pulp, paper, and paperboard production
  Unbleached kraft mill wastewaters

Textile milling
  Knit fabric finishing
  Stock and yarn finishing
  Wool finishing
  Wool scouring
  Woven fabric finishing

TREATMENT TECHNOLOGY: Granular Activated Carbon Adsorption

Data source: Government report
Point source category: Auto and other laundries
Subcategory: Industrial laundries
Plant:
References: B9, p. 50, 66

Data source status:
Engineering estimate ___
Bench scale ___
Pilot scale x
Full scale ___

Use in system: Tertiary
Pretreatment of influent: Foam filtration, ultrafiltration

DESIGN OR OPERATING PARAMETERS

Unit configuration: 5.08 cm (2 in.) diameter column
Wastewater flow: See below
Contact time: See below
Hydraulic loading:
Organic loading:
Bed depth:
Total carbon inventory:
Carbon exhaustion rate:
Backwash rate:
Air scour rate:

Regeneration technique:
Carbon makeup rate:
Carbon type/
characteristics: See below

REMOVAL DATA

Sampling period:

Wastewater flow, $m^3/min/m^2$ (gpm/ft^2)	Contact time, min	Carbon type/ characteristics	BOD_5 Concentration, mg/L Influent	BOD_5 Concentration, mg/L Effluent	BOD_5 Percent removal	COD Concentration, mg/L Influent	COD Concentration, mg/L Effluent	COD Percent removal
0.27(6.7)	11.3	Filtrasorb 400[a]	330	132	60	520	159	69
			302	190	37	634	353	44

Wastewater flow, $m^3/min/m^2$ (gpm/ft^2)	Contact time, min	Carbon type/ characteristics	TOC Concentration, mg/L Influent	TOC Concentration, mg/L Effluent	TOC Percent removal	Oil and Grease Concentration, mg/L Influent	Oil and Grease Concentration, mg/L Effluent	Oil and Grease Percent removal
0.27(6.7)	11.3	Filtrasorb 400[a]	148	55	63			
			194	123	37	20.4	<9	>56

[a]Total carbon inventory: 2,400 g.

Note: Blanks indicate information was not specified.

TREATMENT TECHNOLOGY: Granular Activated Carbon Adsorption

Data source: Effluent Guidelines
Point source category: Auto and other laundries
Subcategory: Power laundries
Plant: N
References: A28, Appendix C

Data source status:
Engineering estimate ___
Bench scale ___
Pilot scale ___
Full scale x

Use in system: Secondary
Pretreatment of influent: Screening, equalization, sedimentation with alum and polymer addition

DESIGN OR OPERATING PARAMETERS

Unit configuration:
Wastewater flow: 15.2 m^3/d (4,000 gpd)
Contact time:
Hydraulic loading:
Organic loading:
Bed depth:
Total carbon inventory:
Carbon exhaustion rate:
Backwash rate:
Air scour rate:

Regeneration technique:
Carbon makeup rate:
Carbon type/characteristics:

REMOVAL DATA

Sampling period: 2 days

Pollutant/parameter	Concentration Influent	Concentration Effluent	Percent removal
Conventional pollutants, mg/L:			
BOD_5	57	35.5	38
COD	125	136	0[a]
TOC	40	38	5
TSS	46	78	0[a]
Oil and grease	4	8	0[a]
Total phenol	0.028	0.029	0[a]
Total phosphorus	1.6	2.0	0[a]
Toxic pollutants, µg/L:			
Antimony	55	44	20
Cadmium	12	15	0[a]
Chromium	34	36	0[a]
Copper	31	42	0[a]
Lead	66	65	2
Nickel	50	<36	>28
Silver	11	7	36
Zinc	240	210	12
Bis(2-ethylhexyl) phthalate	67	23	66
Butyl benzyl phthalate	36	17	53
Di-n-butyl phthalate	7	5	29
Diethyl phthalate	<0.03	3	0[a]
Di-n-octyl phthalate	5	4	20
Pentachlorophenol	<0.4	3	0[a]
Phenol	2	1	50
Toluene	3	4	0[a]
Chloroform	70	18	74
Methylene chloride	38	3	92
Tetrachloroethylene	100	32	68
Trichloroethylene	12	5	58

[a]Actual data indicate negative removal.

Note: Blanks indicate information was not specified.

TREATMENT TECHNOLOGY: Granular Activated Carbon Adsorption

Data source: Effluent Guidelines
Point source category: Gum and wood chemicals
Subcategory:
Plant: 102
References: A7, pp. 7-10

Data source status:
- Engineering estimate ___
- Bench scale ___
- Pilot scale ___
- Full scale x

Use in system: Tertiary
Pretreatment of influent: Oil-water separation, neutralization, dissolved air flotation, filtration

DESIGN OR OPERATING PARAMETERS

Unit configuration:
Wastewater flow: 1.23×10^4 m^3/d (3.24 mgd) (design)
9,820 m^3/d (2.59 mgd) (actual)
Contact time:
Hydraulic loading:
Organic loading:
Bed depth:
Total carbon inventory:
Carbon exhaustion rate: 1.2 kg COD/kg carbon; 0.44 kg TOC/kg carbon
Backwash rate:
Air scour rate:
Regeneration technique:
Carbon makeup rate:
Carbon type/characteristics:

REMOVAL DATA

Sampling period:

Pollutant/parameter	Concentration Influent	Concentration Effluent	Percent removal
Conventional pollutants, mg/L:			
BOD_5	300	82	73
COD	752	160	79
TOC	203	42	79
TSS	81	13	84
Oil and grease	28	2.2	92
Total phenol	4.66	0.58	88
Toxic pollutants, µg/L:			
Cadmium	91	22	76
Chromium	1,100	260	77
Copper	1,300	360	72
Nickel	1,000	330	68
Zinc	1,100	290	74
Pentachlorophenol	120	49	59
Benzene	590	210	64
Toluene	2,500	630	75

Note: Blanks indicate information was not specified.

TREATMENT TECHNOLOGY: Granular Activated Carbon Adsorption

Data source: Effluent Guidelines
Point source category: Ore mining and dressing
Subcategory: Gold mill
Plant: 4105
References: A2, p. VI-60

Data source status:
- Engineering estimate ___
- Bench scale ___
- Pilot scale x
- Full scale ___

Use in system: Secondary
Pretreatment of influent: Sedimentation

DESIGN OR OPERATING PARAMETERS

Unit configuration:
Wastewater flow:
Contact time:
Hydraulic loading:
Organic loading:
Bed depth:
Total carbon inventory:
Carbon exhaustion rate:
Backwash rate:
Air scour rate:

Regeneration technique:
Carbon makeup rate:
Carbon type/
characteristics:

REMOVAL DATA

Sampling period:

Pollutant/parameter	Concentration, µg/L Influent	Effluent	Percent removal
Toxic pollutants:			
Copper	140	<50	>64
Zinc	40	10	75

Note: Blanks indicate information was not specified.

TREATMENT TECHNOLOGY: Granular Activated Carbon Adsorption

Data source: Government report
Point source category: Organic chemicals
Subcategory: Fumaric acid wastewater
Plant:
References: B15, pp. H-2-H-4

Data source status:
Engineering estimate ___
Bench scale ___
Pilot scale x
Full scale ___

Use in system: Tertiary
Pretreatment of influent: Sedimentation, filtration

DESIGN OR OPERATING PARAMETERS

Unit configuration:
Wastewater flow:
Contact time:
Hydraulic loading: 0.035 $m^3/min/m^2$ (0.85 gpm/ft^2)
Organic loading:
Bed depth:
Total carbon inventory: 4.5 kg/column (10 lb/column)
Carbon exhaustion rate:
Backwash rate:
Air scour rate:

Regeneration technique:
Carbon makeup rate:
Carbon type/ characteristics:

REMOVAL DATA

Sampling period:

Pollutant/parameter	Concentration, mg/L Influent	Effluent	Percent removal	Breakthrough period, hr
Conventional pollutants:				
TOC (1st column)	2,900	78[a]	97	-
TOC (2nd column)	2,430	91[b]	96	12

[a]Average of three samples.

[b]Average of six samples.

Note: Blanks indicate information was not specified.

TREATMENT TECHNOLOGY: Granular Activated Carbon Adsorption

Data source: Government report
Point source category: Organic chemicals
Subcategory: Plasticizer wastestream
Plant:
References: B15, p. 31

Data source status:
- Engineering estimate ___
- Bench scale ___
- Pilot scale x
- Full scale ___

Use in system: Tertiary
Pretreatment of influent: Filtration

DESIGN OR OPERATING PARAMETERS

Unit configuration: Four columns in series
Wastewater flow:
Contact time: Varies, see removal data
Hydraulic loading:
Organic loading:
Bed depth:
Total carbon inventory: 4.5 kg/column (10 lb/column)
Carbon exhaustion rate:
Backwash rate:
Air scour rate:

Regeneration technique:
Carbon makeup rate:
Carbon type/ characteristics:

REMOVAL DATA

Sampling period:

Pollutant/parameter	Concentration[a] Influent	Effluent 1st column contact time, 30 min	2nd column contact time, 60 min	3rd column contact time, 90 min	4th column contact time, 120 min	Percent[b] removal 1[c]	2[c]	3[c]	4[c]
Toxic pollutants, µg/L:									
Di-n-octyl phthalate	1,340	340	120	55	48	75	91	96	96

[a]Mean average.
[b]Calculated from influent and respective effluent columns.
[c]Column number.

Note: Blanks indicate information was not specified.

TREATMENT TECHNOLOGY: Granular Activated Carbon Adsorption

Data source: Journal article
Point source category: Organic chemicals
Subcategory:
Plant: Stepan Chemical Co.
References: C1, pp. 81-84

Data source status:
- Engineering estimate ___
- Bench scale ___
- Pilot scale ___
- Full scale x

Use in system: Primary
Pretreatment of influent:

DESIGN OR OPERATING PARAMETERS

Unit configuration: Three - 1.83 m (6 ft) diameter by 3.05 m (10 ft) carbon columns in series
Wastewater flow: 57 m^3/d (15,000 gal/day)
Contact time: 180 min/column
Hydraulic loading:
Organic loading:
Bed depth: 3.05 m (10 ft), each column
Total carbon inventory: 2,950 kg/column (6,500 lb/column)
Carbon exhaustion rate:
Backwash rate:
Air scour rate:
Regeneration technique:
Carbon makeup rate:
Carbon type/characteristics: Filtrasorb 300

REMOVAL DATA

Sampling period:

Pollutant/parameter	Concentration, mg/L Influent	Concentration, mg/L Effluent	Percent removal
Conventional pollutants:			
TOC[a]	6,310	289	95
COD	29,000	1,300	95

[a]Calculated based on COD/TOC relationship.

Note: Blanks indicate information was not specified.

TREATMENT TECHNOLOGY: Granular Activated Carbon Adsorption

Data source: Government report
Point source category:[a] Organic chemicals
Subcategory:
Plant:
References: B2, p. 43

Data source status:
- Engineering estimate ___
- Bench scale ___
- Pilot scale x
- Full scale ___

Use in system: Primary
Pretreatment of influent:

[a]Chlorinated hydrocarbons contaminated wastewater.

DESIGN OR OPERATING PARAMETERS

Unit configuration: Two columns in series. Columns have a double layer of fiberglass windowscreen and 10-15 cm of pea gravel at the bottom.
Wastewater flow:
Contact time:
Hydraulic loading:
Organic loading:
Bed depth:
Total carbon inventory:
Carbon exhaustion rate:
Backwash rate:
Air scour rate:

Regeneration technique: Thermal
Carbon makeup rate:
Carbon type/characteristics: Westvaco-WVG
pH: 1

REMOVAL DATA

Sampling period:

Pollutant/parameter	Concentration, mg/L Influent	Effluent	Percent removal
Conventional pollutants:			
COD	1,190	225	81
TOC	724	40	94

Note: Blanks indicate information was not specified.

TREATMENT TECHNOLOGY: Granular Activated Carbon Adsorption

Data source: Government report
Point source category:[a] Organic chemicals
Subcategory:
Plant:
References: B2, p. 43

Data source status:
- Engineering estimate ___
- Bench scale ___
- Pilot scale x
- Full scale ___

Use in system: Primary
Pretreatment of influent:

---

[a]Chlorinated hydrocarbons contaminated wastewater.

DESIGN OR OPERATING PARAMETERS

Unit configuration: Two columns in series. Columns have a double layer of fiberglass windowscreen and 10-15 cm of pea gravel in bottom.
Wastewater flow:
Contact time:
Hydraulic loading:
Organic loading:
Bed depth:
Total carbon inventory:
Carbon exhaustion rate:
Backwash rate:
Air scour rate:
pH: 1

Regeneration technique: Thermal
Carbon makeup rate:
Carbon type/characteristics: Regenerated Westvaco-WVG
Adsorption capacity: 0.34 kg organics/kgC

REMOVAL DATA

Sampling period:

Pollutant/parameter	Concentration, mg/L Influent	Effluent	Percent removal
Conventional pollutants:			
COD	995	512	9
TOC	627	347	45

Note: Blanks indicate information was not specified.

TREATMENT TECHNOLOGY: Granular Activated Carbon Adsorption

Data source: Government report
Point source category:[a] Organic chemicals
Subcategory:
Plant:
References: B2, p. 43

Data source status:
Engineering estimate ___
Bench scale ___
Pilot scale x
Full scale ___

Use in system: Primary
Pretreatment of influent:

[a]Chlorinated hydrocarbons contaminated wastewater.

DESIGN OR OPERATING PARAMETERS

Unit configuration: One column with a double layer of fiberglass window-screen and 10-15 cm of pea gravel in bottom.
Wastewater flow:
Contact time:
Hydraulic loading:
Organic loading:
Bed depth:
Total carbon inventory:
Carbon exhaustion rate:
Backwash rate:
Air scour rate:

Regeneration technique: Thermal
Carbon makeup rate:
Carbon type/characteristics: Westvaco-WVG
pH: 1

REMOVAL DATA

Sampling period:

Pollutant/parameter	Concentration Influent	Concentration Effluent	Percent removal
Conventional pollutants, mg/L:			
COD	1,570	1,230	19
TOC	640	394	38
Toxic pollutants, µg/L:			
Chloroethane	170,000	ND[a]	>99
1,1-Dichloroethane	190,000	ND	>99
1,2-Dichloroethane	1,300,000	ND	>99

[a]Not detected; assumed to be <10 µg/L.

Note: Blanks indicate information was not specified.

TREATMENT TECHNOLOGY: Granular Activated Carbon Adsorption

Data source: Government report
Point source category:[a] Organic chemicals
Subcategory:
Plant:
References: B2, p. 43

Data source status:
Engineering estimate ___
Bench scale ___
Pilot scale x
Full scale ___

Use in system: Primary
Pretreatment of influent:

[a]Chlorinated hydrocarbons contaminated wastewater.

DESIGN OR OPERATING PARAMETERS

Unit configuration: Two columns in series. Columns have a double layer of fiberglass windowscreen and 10-15 cm of pea gravel in the bottom.
Wastewater flow:
Contact time:
Hydraulic loading:
Organic loading:
Bed depth:
Total carbon inventory:
Carbon exhaustion rate:
Backwash rate:
Air scour rate:

Regeneration technique: Thermal
Carbon makeup rate:
Carbon type/characteristics: Westvaco-WVG
pH: 1

REMOVAL DATA

Sampling period:

Pollutant/parameter	Concentration Influent	Concentration Effluent	Percent removal
Conventional pollutants, mg/L:			
COD	1,570	898	43
TOC	650	271	58
Toxic pollutants, μg/L:			
Chloroethane	170,000	63	>99
1,2-Dichloroethane	1,300,000	78	>99

Note: Blanks indicate information was not specified.

TREATMENT TECHNOLOGY: Granular Activated Carbon Adsorption

Data source: Government report
Point source category:[a] Organic chemicals
Subcategory:
Plant:
References: B2, p. 43

Data source status:
- Engineering estimate ___
- Bench scale ___
- Pilot scale _x_
- Full scale ___

Use in system: Primary
Pretreatment of influent:

[a]Chlorinated hydrocarbons contaminated wastewater.

DESIGN OR OPERATING PARAMETERS

Unit configuration: Two columns in series. Columns have a double layer of fiberglass windowscreen and 10-15 cm of pea gravel in the bottom.
Wastewater flow:
Contact time:
Hydraulic loading:
Organic loading:
Bed depth:
Total carbon inventory:
Carbon exhaustion rate:
Backwash rate:
Air scour rate:

Regeneration technique: Thermal
Carbon makeup rate:
Carbon type/characteristics: Westvaco-WVG
pH: 1

REMOVAL DATA

Sampling period:

Pollutant/parameter	Concentration Influent	Concentration Effluent	Percent removal
Conventional pollutants, mg/L:			
COD	1,110	1,550	0[a]
TOC	663	588	11
Toxic pollutants, μg/L:			
Chloroethane	330,000	240,000	27
1,2-Dichloroethane	3,000,000	180,000	94

[a]Actual data indicate negative removal.

Note: Blanks indicate information was not specified.

TREATMENT TECHNOLOGY: Granular Activated Carbon Adsorption

Data source: Government report
Point source category:[a] Organic chemicals
Subcategory:
Plant:
References: B2, p. 43

Data source status:
Engineering estimate ___
Bench scale ___
Pilot scale x
Full scale ___

Use in system: Primary
Pretreatment of influent:

[a]Chlorinated hydrocarbons contaminated wastewater.

DESIGN OR OPERATING PARAMETERS

Unit configuration: One column with a double layer of fiberglass window-screen and 10-15 cm of pea gravel in bottom.

Wastewater flow:
Contact time:
Hydraulic loading:
Organic loading:
Bed depth:
Total carbon inventory:
Carbon exhaustion rate:
Backwash rate:
Air scour rate:

Regeneration technique: Thermal
Carbon makeup rate:
Carbon type/ characteristics: Westvaco-WVG
pH: 1

REMOVAL DATA

Sampling period:

Pollutant/parameter	Concentration Influent	Concentration Effluent	Percent removal
Conventional pollutants, mg/L:			
COD	1,110	1,140	0[a]
TOC	663	297	55
Toxic pollutants, μg/L:			
Chloroethane	330,000	ND[b]	>99
1,1-Dichloroethane	310,000	ND	>99
1,2-Dichloroethane	3,000,000	ND	>99

[a]Actual data indicate negative removal.

[b]Not detected; assumed to be <10 μg/L.

Note: Blanks indicate information was not specified.

TREATMENT TECHNOLOGY: Granular Activated Carbon Adsorption

Data source: Government report
Point source category:[a] Organic chemicals
Subcategory:
Plant:
References: B2, p. 43

Data source status:
Engineering estimate ___
Bench scale ___
Pilot scale x
Full scale ___

Use in system: Primary
Pretreatment of influent:

[a]Chlorinated hydrocarbons contaminated wastewater.

DESIGN OR OPERATING PARAMETERS

Unit configuration: Two columns in series. Columns have a double layer of fiberglass windowscreen and 10-15 cm of pea gravel in bottom.

Wastewater flow:
Contact time:
Hydraulic loading:
Organic loading:
Bed depth:
Total carbon inventory:
Carbon exhaustion rate:
Backwash rate:
Air scour rate:

Regeneration technique: Thermal
Carbon makeup rate:
Carbon type/
characteristics: Westvaco-WVG
pH: 1

REMOVAL DATA

Sampling period:

Pollutant/parameter	Concentration		Percent
	Influent	Effluent	removal
Conventional pollutants, mg/L:			
COD	1,550	1,120	28
TOC	567	962	0[a]
Toxic pollutants, μg/L:			
Chloroethane	59,000	190,000	0[a]
1,1-Dichloroethane	78,000	8,000	90
1,2-Dichloroethane	960,000	130,000	86

[a]Actual data indicate negative removal.

Note: Blanks indicate information was not specified.

TREATMENT TECHNOLOGY: Granular Activated Carbon Adsorption

Data source: Government report
Point source category:[a] Organic chemicals
Subcategory:
Plant:
References: B2, p. 43

Data source status:
- Engineering estimate ___
- Bench scale ___
- Pilot scale x
- Full scale ___

Use in system: Primary
Pretreatment of influent:

[a]Chlorinated hydrocarbons contaminated wastewater.

DESIGN OR OPERATING PARAMETERS

Unit configuration: One column with a double layer of fiberglass window-screen and 10-15 cm of pea gravel in bottom.
Wastewater flow:
Contact time:
Hydraulic loading:
Organic loading:
Bed depth:
Total carbon inventory:
Carbon exhaustion rate:
Backwash rate:
Air scour rate:

Regeneration technique: Thermal
Carbon makeup rate:
Carbon type/characteristics: Westvaco-WVG
pH: 1

REMOVAL DATA

Sampling period:

Pollutant/parameter	Concentration Influent	Concentration Effluent	Percent removal
Conventional pollutants, mg/L:			
COD	1,550	1,390	10
TOC	567	614	0[a]
Toxic pollutants, μg/L:			
Chloroethane	59,000	150,000	0[a]
1,1-Dichloroehtane	78,000	45,000	42
1,2-Dichloroethane	960,000	750,000	21

[a]Actual data indicate negative removal.

Note: Blanks indicate information was not specified.

TREATMENT TECHNOLOGY: Granular Activated Carbon Adsorption

Data source: Government report
Point source category:[a] Organic chemicals
Subcategory:
Plant:
References: B2, p. 43

Data source status:
Engineering estimate ___
Bench scale ___
Pilot scale x
Full scale ___

Use in system: Primary
Pretreatment of influent:

[a]Chlorinated hydrocarbons contaminated wastewater.

DESIGN OR OPERATING PARAMETERS

Unit configuration: One column with a double layer of fiberglass windowscreen and 10-15 cm of pea gravel at the bottom.
Wastewater flow:
Contact time:
Hydraulic loading:
Organic loading:
Bed depth:
Total carbon inventory:
Carbon exhaustion rate:
Backwash rate: 15.2 L/min
Air scour rate:

Regeneration technique: Thermal
Carbon makeup rate:
Carbon type/characteristics: Westvaco-WVG
pH: 1
Adsorption capacity: 0.34 kg organics/kg C

REMOVAL DATA

Sampling period:

Pollutant/paremeter	Concentration		Percent removal
	Influent	Effluent	
Conventional pollutants, mg/L:			
COD	995	562	44
TOC	627	437	30
Toxic pollutants, μg/L:			
Chloroethane	110,000	ND[a]	>99
1,1-Dichloroethane	79,000	ND	>99
1,2-Dichloroethane	920,000	ND	>99

[a]Not detected; assumed to be <10 μg/L.

Note : Blanks indicate information was not specified.

TREATMENT TECHNOLOGY: Granular Activated Carbon Adsorption

Data source: Government report
Point source category:[a] Organic chemicals
Subcategory:
Plant:
References: B2, p. 43

Data source status:
Engineering estimate ___
Bench scale ___
Pilot scale x
Full scale ___

Use in system: Primary
Pretreatment of influent:

[a]Chlorinated hydrocarbons contaminated wastewater.

DESIGN OR OPERATING PARAMETERS

Unit configuration: One column with a double layer of fiberglass windowscreen and 10-15 cm of pea gravel at the bottom.
Wastewater flow:
Contact time:
Hydraulic loading:
Organic loading:
Bed depth:
Total carbon inventory:
Carbon exhaustion rate:
Backwash rate:
Air scour rate:

Regeneration technique: Thermal
Carbon makeup rate:
Carbon type/ characteristics: Westvaco-WVG
pH: 1

REMOVAL DATA

Sampling period:

Pollutant/parameter	Concentration Influent	Concentration Effluent	Percent removal
Conventional pollutants, mg/L:			
COD	1,190	446	63
TOC	724	76	90
Toxic pollutants, µg/L:			
Chloroethane	390,000	ND[a]	>99
1,1-Dichloroethane	40,000	ND	>99
1,2-Dichloroethane	950,000	ND	>99

[a]Not detected; assumed to be <10 µg/L.

Note: Blanks indicate information was not specified.

TREATMENT TECHNOLOGY: Granular Activated Carbon Adsorption

Data source: Government report
Point source category:[a] Organic chemicals
Subcategory:
Plant:
References: B2, p. 45, 51

Data source status:
- Engineering estimate ___
- Bench scale ___
- Pilot scale x
- Full scale ___

Use in system: Primary
Pretreatment of influent:

[a]Chlorinated hydrocarbons contaminated wastewater.

DESIGN OR OPERATING PARAMETERS

Unit configuration: Columns have a double layer of fiberglass windowscreen and 10-15 cm of pea gravel at bottom unless otherwise specified.
Wastewater flow: See below
Contact time:
Hydraulic loading:
Organic loading:
Bed depth: See below
Total carbon inventory: See below
Carbon exhaustion rate: See below
Backwash rate:
Air scour rate:

Regeneration technique:
Carbon makeup rate:
Carbon type/characteristics: See below
Total run time: See below

REMOVAL DATA

Sampling period:

Wastewater flow, L/min	Organic loading, kg EDC[a]/kg C	Bed depth, m	Total carbon inventory	Carbon type/ characteristics	Total run, hr	1,2-Dichloroethane Concentration, μg/L Influent	Effluent	Percent removal
0.76	0.33	3.95[b]	16.6 kg (39.1 L)	Filtrasorb 400	3	3,700,000	ND[c]	>99
0.95	0.35	6.87[d]	61.4 kg (125 L)	Witco 718	100	3,500,000	<14,000	>99
0.76-0.95	0.25-0.29	0.14-5.56[e]	9.71-44.4 kg (22-104 L)	WVG	17.5-120	1,700,000	100,000	94
0.84	0.16	1.4	10.2 kg (0.025 m^3)	Monochem/activated soot carbon	23.5	2,500,000	1,100,000	55
0.8	0.13	1.63	11.9 kg (0.03 m^3)	Filtrasorb 400	19	1,800,000	37,000	98
0.84	0.2	1.4	10.2 kg (0.025 m^3)	Monochem/activated soot carbon	19.5	3,100,000	970,000	69

[a]1,2-Dichloroethane.
[b]Unit configuration used three columns in parallel.
[c]Not detected, assumed to be <10 μg/L.
[d]Unit configuration used was five columns in parallel.
[e]Unit configuration used was one to five columns in parallel.

Note: Blanks indicate information was not specified.

TREATMENT TECHNOLOGY: Granular Activated Carbon Adsorption

Data source: Government Report
Point source category:[a] Organic chemicals
Subcategory:
Plant:
References: B2, Appendix

Data source status:
- Engineering estimate ___
- Bench scale ___
- Pilot scale x
- Full scale ___

Use in system:
Pretreatment of influent:

[a]Halogenated hydrocarbons contaminated wastewater.

DESIGN OR OPERATING PARAMETERS (also see removal data)

Unit configuration: Plexiglass columns 1.27 cm in diameter and 1.83 in high. Columns have a double layer of fiberglass window screen and 10-15 cm of pea gravel at the bottom
Wastewater flow: 3.79 L/min
Contact time:
Hydraulic loading: 41.6 L/min-m²
Organic loading:
Bed depth:
Total carbon inventory: 28.3 L, 11.3 kg
Carbon exhaustion rate:
Backwash rate:
Air scour rate:

Regeneration technique:
Carbon makeup rate:
Carbon type/ characteristics: Westvaco WVG
Adsorption capacity: 0.15 kg organics/ kg carbon

REMOVAL DATA

Sampling period:

Running Time, hr	1,2-Dichloroethane Concentration, μg/L		Percent removal	1,2-*Trans*-dichloroethylene Concentration, μg/L		Percent removal	Methylene chloride Concentration, μg/L		Percent removal	1,1,2,2-Tetrachloroethane Concentration, μg/L		Percent removal
	Influent	Effluent		Influent	Effluent		Influent	Effluent		Influent	Effluent	
3	80,000	120	>99	140,000	20	>99	150	20	87	320,000	64,000	80
6	46,000	2,600	94	3,700	100	97	130	50	62	330,000	6,300	98
9	150,000	90	>99	-	-	-	180	40	78	190,000	7,000	96
12	76,000	25,000	67	940	500	47	340	60	82	11,000	24,000	0[a]
15	250,000	42,000	83	2,400	750	69	1,300	170	87	110,000	25,000	77
18	11,000	480	>99	7,000	1,100	84	320	70	78	140,000	680	>99
21	170,000	160,000	6	12,000	2,600	79	200	110	45	18,000	36,000	0[a]
24	170,000	260,000	0[a]	4,400	8,200	0[a]	360	70	81	18,000	2,700	85
27	5,000	140,000	0[a]	320	620	0[a]	70	60	14	50,000	10,000	80
30	400	160,000	0[a]	60	8,600	0[a]	540	10	98	30,000	8,500	71
33	190,000	140,000	24	7,800	12,000	0[a]	320	840	0[a]	9,500	20,000	0[a]
36	160,000	94,000	42	11,000	17,000	0[a]	130	90	31	10,000	3,200	69
39	42,000	130,000	0[a]	1,800	19,000	0[a]	240	100	58	60,000	4,000	93
42	42,000	34,000	19	750	30,000	0[a]	-	-	-	36,000	3,000	92
45	24,000	63,000	0[a]	20	30	0[a]	70	220	0[a]	-	-	-
48	6,200	85,000	0[a]	30	30	0	620	340	45	-	-	-
51	5,400	37,000	0[a]	220	5,400	0[a]	400	230	42	43,000	4,000	91
54	57,000	33,000	42	18,000	7,200	59	-	-	-	50,000	2,600	95
57	6,500	50,000	0[a]	110	6,800	0[a]	120	420	0[a]	50,000	3,800	92
60	2,100	60	97	170	1,220	0[a]	320	56,000	0[a]	20,000	1,600	92

[a]Actual data indicate negative removal.

Note: Blanks indicate information was not specified.

TREATMENT TECHNOLOGY: Granular Activated Carbon Adsorption

Data source: Government report		Data source status:	
Point source category:[a] Organic chemicals		Engineering estimate	___
Subcategory:		Bench scale	___
Plant:		Pilot scale	x
References: B2, Appendix		Full scale	___

Use in system: Primary
Pretreatment of influent:

[a]Halogenated hydrocarbons contaminated wastewater.

DESIGN OR OPERATING PARAMETERS (Also see removal data)

Unit configuration: Plexiglass column 1.27 cm in diameter and 1.83 in high. Columns have a double layer of fiberglass window screen, 10-15 cm of pea gravel at the bottom
Wastewater flow: 3.79 L/min
Contact time:
Hydraulic loading: 41.6 L/min-m^2
Organic loading:
Bed depth:
Total carbon inventory: 28.3 L, 11.3 kg
Carbon exhaustion rate:
Backwash rate: 15.2 L/min
Air scour rate:

Regeneration technique:
Carbon makeup rate:
Carbon type/characteristics: Westvaco WVG
Adsorption capacity: 0.15 kg organics/kg carbon

REMOVAL DATA

Sampling period:

Running time, hr	1,2-Dichloroethane: Concentration, µg/L Influent	Effluent	Percent removal	1,2-*Trans*-dichloroethylene: Concentration, µg/L Influent	Effluent	Percent removal	Methylene chloride: Concentration, µg/L Influent	Effluent	Percent removal	1,1,2,2-Tetrachloroethane: Concentration, µg/L Influent	Effluent	Percent removal
3	2,400	720	71	8,800	250	97	27,000	650	98	-	-	-
6	730	20	97	6,500	230	96	13,000	190	98	4,600	20	>99
9	1,100	180	84	15,000	90	99	19,000	150	99	2,300	2,800	0[a]
12	880	230	74	3,500	140	96	3,200	330	90	-	-	-
15	560	550	2	2,800	200	93	1,300	180	87	-	-	-
18	600	560	7	2,500	20	99	1,400	60	96	-	-	-
21	1,600	1,800	0[a]	3,900	170	96	3,200	230	93	-	-	-
24	1,800	2,300	0[a]	3,900	90	98	2,200	240	89	-	-	-
27	230	4,200	0[a]	2,000	140	93	1,300	390	70	-	-	-
30	2,000	8,400	0[a]	1,400	210	85	230	310	0[a]	-	-	-
33	3,300	4,200	0[a]	3,500	240	93	2,100	380	82	-	-	-
36	2,000	4,200	0[a]	3,000	220	93	2,100	380	82	-	-	-
39	2,000	2,400	0[a]	7,500	410	95	11,000	390	97	-	-	-
42	2,600	4,500	0[a]	12,000	240	98	20,000	280	99	-	-	-
45	50	50	0	230	240	0[a]	22,000	23,000	0[a]	2,800	11,000	0[a]
48	450	110	76	15,000	180	99	22,000	25,000	0[a]	2,000	7,900	0[a]
51	860	3,200	0[a]	1,300	90	93	260	240	8	150	130	13
54	1,100	3,100	0[a]	1,400	90	93	3,900	240	94	920	20	98
57	520	6,800	0[a]	3,200	390	88	1,500	300	80	2,200	20	99
60	250	12,000	0[a]	2,600	140	95	2,100	280	87	-	-	-

[a]Actual data indicate negative removal.

Note: Blanks indicate information was not specified.

TREATMENT TECHNOLOGY: Granular Activated Carbon Adsorption

Data source: Government report
Point source category:[a] Organic chemicals
Subcategory:
Plant:
References: B2, Appendix

Use in system: Primary
Pretreatment of influent:

Data source status:
Engineering estimate ___
Bench scale ___
Pilot scale X
Full scale ___

[a]Halogenated hydrocarbons wastewater.

DESIGN OR OPERATING PARAMETERS

Unit configuration: Columns have a double layer of fiberglass windowscreen and 10-15 cm of pea gravel at bottom
Wastewater flow: See below
Contact time:
Hydraulic loading: See below
Organic loading:
Bed depth:
Total carbon inventory:
Carbon exhaustion rate:
Backwash rate:
Air scour rate:

Regeneration technique:
Carbon makeup rate:
Carbon type characteristics:

REMOVAL DATA

Sampling period:

Wastewater flow, L/min	Hydraulic loading $L/cm^2/min$	Number of samples	Toxic pollutants	Concentration, µg/L Influent	Effluent	Percent removal
1.1	0.74	4	Chloroethane	42,000	4,600	89
1.9	1.23	5		60,000	67,000	0[a]
2.85	1.84	6		35,000	99,000	0[a]
1.1	0.74	4	Chloroform	11,000	ND[b]	>99
1.9	1.23	4		92,000	ND	>99
2.85	1.84	6		12,000	ND	>99
1.1	0.74	4	1,1-Dichloroethane	26,000	ND	>99
1.9	1.23	3		91,000	ND	>99
2.85	1.84	6		64,000	9,000	86
1.1	0.74	4	1,2-Dichloroethane	510,000	ND	>99
1.9	1.23	4		1,170,000	150,000	87
2.85	1.84	6		1,220,000	330,000	73
1.1	0.74		1,2-Dichloropropane			
1.9	1.23	4		28,000	ND	>99
2.85	1.84	6		6,900	ND	>99
1.1	0.74	4	1,1,1-Trichloroethane	5,000	ND	>99
1.9	1.23	4		12,000	ND	>99
2.85	1.84					
1.1	0.74	4	1,1,2-Trichloroethane	23,000	ND	>99
1.9	1.23	4		15,000	ND	>99
2.85	1.84	5		20,000	ND	>99
1.1	0.74	3	Vinly chloride	2,300	1,100	52
1.9	1.23	4		4,400	9,600	0[a]
2.85	1.84	6		4,800	8,600	0[a]

[a]Actual data indicate negative removal.
[b]Not detected, assumed to be <10 µg/L.

Note: Blanks indicate information was not specified.

TREATMENT TECHNOLOGY: Granular Activated Carbon Adsorption

Data source: Effluent Guidelines	Data source status:		
Point source category: Pesticide chemicals	Engineering estimate	___	
Subcategory: See below	Bench scale	___	
Plant: See below	Pilot scale	___	
References: A16, pp. 111, 113	Full scale	x	

Use in system: Primary
Pretreatment of influent: none, unless otherwise specified

DESIGN OR OPERATING PARAMETERS

Unit configuration: Downflow, unless otherwise specified
Wastewater flow:
Contact time: See below
Hydraulic loading: See below
Organic loading:
Bed depth:
Total carbon inventory:
Carbon exhaustion rate:
Backwash rate:
Air scour rate:
Regeneration technique: Thermal
Carbon makeup rate:
Carbon type/characteristics:

REMOVAL DATA

Sampling period:

Plant	Contact time, min	Hydraulic loading, $m^3/min/m^2$ (gpm/ft^2)	BOD_5 Concentration, mg/L Influent	BOD_5 Concentration, mg/L Effluent	BOD_5 Percent removal	COD Concentration, mg/L Influent	COD Concentration, mg/L Effluent	COD Percent removal	TOC Concentration, mg/L Influent	TOC Concentration, mg/L Effluent	TOC Percent removal
8[a]	479	0.013(0.32)				5,770	320	94	696	85.7	98
6[b]	760	0.02(0.60)	1,630	780	52	5,780	2,120	63	2,220	534	76
20[c]	35	0.0857(2.10)	45,200	37,400	17	148,000	109,000	27	79,800	66,700	16
39[d]	230	0.027(0.66)	995	1,100	0[e]	8,310	6,380	23	966	1,950	0[e]
46[f]	120	0.053(1.3)									
50[g]	292	0.021(0.51)	19[illegible]	9.2	95	4,880	31	99	2,170	15.4	99
45[h]	456	0.015(0.36)				4,750	808	83	1,650	153	91

Plant	Contact time, min	Hydraulic loading, $m^3/min/m^2$ (gpm/ft^2)	TSS Concentration, mg/L Influent	TSS Concentration, mg/L Effluent	TSS Percent removal	Total phenol Concentration, mg/L Influent	Total phenol Concentration, mg/L Effluent	Total phenol Percent removal
8[a]	479	0.013(0.32)	1,510	255	83			
6[b]	760	0.02(0.60)	69	109	0[e]	77.9	2.32	97
20[c]	35	0.0857(2.10)	1,460	2,600	0[e]			
39[d]	230	0.027(0.66)	168	165	2			
46[f]	120	0.053(1.3)	29.5	8.78	70			
50[g]	292	0.021(0.51)	674	6.6	99	2.8	<0.7	>75
45[h]	456	0.015(0.36)	68.6	46.6	32	129	4.26	97

[a] Subcategory: halogenated organics.
[b] Subcategory: halogenated organics; unit configuration: upflow.
[c] Subcategory: halogenated organics, organo nitrogen, metallo-organic; pretreatment of influent: two multimedia filters in parallel.
[d] Subcategory: organo nitrogen.
[e] Actual data indicate negative removal.
[f] Subcategory: organo nitrogen.
[g] Subcategory: organo nitrogen, metallo-organic; unit configuration: downflow, two carbon columns in series.
[h] Subcategory: organo nitrogen, noncategorized pesticides; pretreatment of influent: neutralization, dual media filtration, equalization.

Note: Blanks indicate information was not specified.

TREATMENT TECHNOLOGY: Granular Activated Carbon Adsorption

Data source: Conference paper
Point source category: Petroleum refining
Subcategory:
Plant: East coast oil refinery
References: D1, p. 207; D2, p. 217

Data source status:
- Engineering estimate ___
- Bench scale x
- Pilot scale ___
- Full scale ___

Use in system: Tertiary
Pretreatment of influent: Sand filter, API separator

DESIGN OR OPERATING PARAMETERS

Unit configuration: 2 sets of 3 - 0.0338 m (1-1/2 in.) I.D. carbon columns in parallel and in series upflow
Wastewater flow: 0.0816 $m^3/min/m^2$ (2 gpm/ft^2)
Contact time: 18 min (empty bed)
Hydraulic loading:
Organic loading: 0.65 kg COD removed[a]/kg of carbon
0.46 kg COD removed[b]/kg of carbon
Bed depth:
Carbon dosage: 0.111 kg/m^3[a] (0.93 lb/1,000 gal)
0.157 kg/m^3[b] (1.31 lb/1,000 gal)
Carbon exhaustion rate:
Backwash rate:
Air scour rate:
Regeneration technique:
Carbon makeup rate:
Carbon type/characteristics: 12 x 40 mesh lignite[a], 12 x 40 mesh bituminous

[a]First set of columns.
[b]Second set of columns.

REMOVAL DATA

Sampling period:

Pollutant/parameter	Concentration, mg/L Influent	Effluent	Percent removal
Conventional pollutants:			
COD	104[a]	31[b]	70
	104[c]	31[b]	70

[a]First set of columns (lignite carbon).
[b]Breakthrough at 70% removal.
[c]Second set of columns (bituminous carbon).

Note: Blanks indicate information was not specified.

TREATMENT TECHNOLOGY: Granular Activated Carbon Adsorption

Data source: Conference paper
Point source category: Petroleum refining
Subcategory:
Plant: East coast oil refinery
References: D1, p. 207; D2, p. 217

Data source status:
- Engineering estimate ___
- Bench scale x
- Pilot scale ___
- Full scale ___

Use in system: Tertiary
Pretreatment of influent: Sand filtered, API separator

DESIGN OR OPERATING PARAMETERS

Unit configuration: 2 sets of 4 - 0.0338 m (1-1/2 in.) carbon columns in parallel and in series downflow
Wastewater flow: 0.0204 $m^3/min/m^2$ (0.5 gpm/ft^2)
Contact time: 88 min
Hydraulic loading:
Organic loading: 0.21 kg COD removed[a]/kg of carbon
0.16 kg COD removed[b]/kg of carbon
Bed depth:
Carbon dosage: 0.228 kg/m^3[a] (1.91 lb/1,000 gal)
0.297 kg/m^3[b] (2.49 lb/1,000 gal)
Carbon exhaustion rate:
Backwash rate:
Air scour rate:
Regeneration technique:
Carbon makeup rate:
Carbon type/characteristics: 8 x 30 mesh lignite[a], 8 x 30 mesh bituminous[b]

[a]First set of columns.

[b]Second set of columns.

REMOVAL DATA

Sampling period:

Pollutant/parameter	Concentration, mg/L Influent	Effluent	Percent removal
Conventional pollutants:			
COD	70[a]	21[b]	70
	70[c]	21[b]	70

[a]First set of columns (lignite carbon).

[b]Breakthrough at 70% removal.

[c]Second set of columns (bituminous carbon).

Note: Blanks indicate information was not specified.

TREATMENT TECHNOLOGY: Granular Activated Carbon Adsorption

Data source: Effluent Guidelines
Point source category: Petroleum refining
Subcategory:
Plant: B
References: A3, pp. VI-36-42

Data source status:
Engineering estimate ___
Bench scale ___
Pilot scale x
Full scale ___

Use in system: Tertiary
Pretreatment of influent: Dissolved air flotation, multimedia filtration

DESIGN OR OPERATING PARAMETERS

Unit configuration:
Wastewater flow:
Contact time:
Hydraulic loading:
Organic loading:
Bed depth:
Total carbon inventory:
Carbon exhaustion rate:
Backwash rate:
Air scour rate:

Regeneration technique:
Carbon makeup rate:
Carbon type/ characteristics:

REMOVAL DATA

Sampling period: Four days

Pollutant/parameter	Concentration Influent	Effluent	Percent removal
Conventional pollutants, mg/L:			
BOD_5	15	<12	>20
COD	101	22	79
TOC	40	13	67
TSS	21	4.4	79
Oil and grease	8.5	7.5	12
Total phenol	0.022	<0.010	>55
Toxic pollutants, μg/L:			
Beryllium	2	2	0
Chromium	30	19	39
Cyanide	50	<18	>63
Selenium	56	50	11
Zinc	65	30	54

Note: Blanks indicate information was not specified.

TREATMENT TECHNOLOGY: Granular Activated Carbon Adsorption

Data source: Effluent Guidelines
Point source category: Petroleum refining
Subcategory:
Plant: H
References: A3, pp. VI-36-42

Data source status:
Engineering estimate ___
Bench scale ___
Pilot scale x
Full scale ___

Use in system: Tertiary
Pretreatment of influent: API design gravity oil separator, filtration

DESIGN OR OPERATING PARAMETERS

Unit configuration:
Wastewater flow:
Contact time:
Hydraulic loading:
Organic loading:
Bed depth:
Total carbon inventory:
Carbon exhaustion rate:
Backwash rate:
Air scour rate:

Regeneration technique:
Carbon makeup rate:
Carbon type/
characteristics:

REMOVAL DATA

Sampling period: Four days

Pollutant/parameter	Concentration Influent	Concentration Effluent	Percent removal
Conventional pollutants, mg/L:			
COD	29	12	59
TOC	19	9.3	51
TSS	3.8	3.9	0[a]
Oil and grease	8.3	7.1	14
Toxic pollutants, μg/L:			
Copper	12	<6	>50
Zinc	20	20	0

[a]Actual data indicate negative removal.

Note: Blanks indicate information was not specified.

TREATMENT TECHNOLOGY: Granular Activated Carbon Adsorption

Data source: Effluent Guidelines
Point source category: Petroleum refining
Subcategory:
Plant: K
References: A3, pp. VI-36-42

Data source status:
Engineering estimate ___
Bench scale ___
Pilot scale x
Full scale ___

Use in system: Tertiary
Pretreatment of influent: Dissolved air flotation, filtration

DESIGN OR OPERATING PARAMETERS

Unit configuration:
Wastewater flow:
Contact time:
Hydraulic loading:
Organic loading:
Bed depth:
Total carbon inventory:
Carbon exhaustion rate:
Backwash rate:
Air scour rate:

Regeneration technique:
Carbon makeup rate:
Carbon type/ characteristics:

REMOVAL DATA

Sampling period: Four days

Pollutant/paremeter	Concentration Influent	Effluent	Percent removal
Conventional pollutants, mg/L:			
COD	56	12	79
TOC	22	8.8	60
TSS	4	<1.5	>62
Oil and grease	6.3	6	5
Total phenol	0.023	<0.0115	>50
Toxic pollutants, μg/L:			
Chromium	34	12	63
Zinc	92	32	65

Note: Blanks indicate information was not specified.

TREATMENT TECHNOLOGY: Granular Activated Carbon Adsorption

Data source: Effluent Guidelines
Point source category: Petroleum refining
Subcategory:
Plant: M
References: A3, pp. VI-36-42

Data source status:
Engineering estimate ___
Bench scale ___
Pilot scale x
Full scale ___

Use in system: Tertiary
Pretreatment of influent: Dissolved air flotation, filtration

DESIGN OR OPERATING PARAMETERS

Unit configuration:
Wastewater flow:
Contact time:
Hydraulic loading:
Organic loading:
Bed depth:
Total carbon inventory:
Carbon exhaustion rate:
Backwash rate:
Air scour rate:

Regeneration technique:
Carbon makeup rate:
Carbon type/
characteristics:

REMOVAL DATA

Sampling period: Four days

Pollutant/parameter	Concentration		Percent
	Influent	Effluent	removal
Conventional pollutants, mg/L:			
COD	55	11	80
TOC	17	6.2	64
TSS	3	<1.3	>56
Oil and grease	12	8.7	28
Toxic pollutants,[b] μg/L:			
Chromium	48	32	34
Cyanide	65	<22	>65
Selenium	26	24	6
Zinc	200	100	50

[a]Actual data indicate negative removal.

[b]Results from two labs averaged for metals.

Note: Blanks indicate information was not specified.

TREATMENT TECHNOLOGY: Granular Activated Carbon Adsorption

Data source: Effluent Guidelines
Point source category: Petroleum refining
Subcategory:
Plant: O
References: A3, pp. VI-36-42

Data source status:
Engineering estimate ___
Bench scale ___
Pilot scale x
Full scale ___

Use in system: Tertiary
Pretreatment of influent: Dissolved air flotation, filtration

DESIGN OR OPERATING PARAMETERS

Unit configuration:
Wastewater flow:
Contact time:
Hydraulic loading:
Organic loading:
Bed depth:
Total carbon inventory:
Carbon exhaustion rate:
Backwash rate:
Air scour rate:

Regeneration technique:
Carbon makeup rate:
Carbon type/ characteristics:

REMOVAL DATA

Sampling period: Four days

Pollutant/parameter	Concentration Influent	Concentration Effluent	Percent removal
Conventional pollutants, mg/L:			
BOD_5	20	9	55
COD	120	79	34
TOC	44	30	32
TSS	18	22	0[a]
Oil and grease	11	14	0[a]
Total phenol	0.032	<0.005	>84
Toxic pollutants, μg/L:			
Chromium	60	54	10
Chromium (+6)	30	<20	>33
Zinc	<10	14	0[a]

[a]Actual data indicate negative removal.

Note: Blanks indicate information was not specified.

TREATMENT TECHNOLOGY: Granular Activated Carbon Adsorption

Data source: Effluent Guidelines
Point source category: Petroleum refining
Subcategory:
Plant: P
References: A3, pp. VI-36-42

Data source status:
- Engineering estimate ___
- Bench scale ___
- Pilot scale x
- Full scale ___

Use in system: Tertiary
Pretreatment of influent: API design gravity oil separator, filtration

DESIGN OR OPERATING PARAMETERS

Unit configuration:
Wastewater flow:
Contact time:
Hydraulic loading:
Organic loading:
Bed depth:
Total carbon inventory:
Carbon exhaustion rate:
Backwash rate:
Air scour rate:

Regeneration technique:
Carbon makeup rate:
Carbon type/
characteristics:

REMOVAL DATA

Sampling period: Four days

Pollutant/parameter	Concentration Influent	Concentration Effluent	Percent removal
Conventional pollutants, mg/L:			
BOD_5	13	<8.3	>36
COD	127	64	50
TOC	45	33	27
TSS	14	7.7	45
Oil and grease	17	13	24
Total phenol	0.051	<0.005	>90
Toxic pollutants, μg/L:			
Antimony	430	430	0
Cyanide	42	52	0[a]
Zinc	30	28	5

[a]Actual data indicate negative removal.

Note: Blanks indicate information was not specified.

TREATMENT TECHNOLOGY: Granular Activated Carbon Adsorption

Data source: Government report
Point source category: Petroleum refining
Subcategory: Class B refinery
Plant: Marcus Hook Refinery
References: B3

Data source status:
Engineering estimate ___
Bench scale ___
Pilot scale x
Full scale ___

Use in system: Tertiary
Pretreatment of influent: API separator, filtration

DESIGN OR OPERATING PARAMETERS

Unit configuration: Upflow; 4 columns in series
Total flow: 0.002 m^3/min (0.5 gpm)
Hydraulic loading: 0.15 $m^3/min/m^2$ (3.6 gpm/ft^2)
Contact time: 36 min
Total carbon inventory: 0.071 m^3 (2.5 ft^3)
Carbon exhaustion rate: 100 kg/1,000 m^3 (0.86 lb/1,000 gal)
Carbon type: Filtrasorb 300, 8 x 30 mesh

REMOVAL DATA

Pollutant/parameter	Concentration, mg/L		Percent removal
	Influent	Effluent	
Conventional pollutants:			
BOD_5	57	9	83
TOC	37	13	65
TSS	8	3	62
Oil and grease	12.3	1.8	85
Total phenol	2.7	0.02	99

Note: Blanks indicate information was not specified.

TREATMENT TECHNOLOGY: Granular Activated Carbon Adsorption

Data source: Effluent Guidelines
Point source category: Pulp, paper, and paper board
Subcategory: See below
Plant:
References: A26, p. VII-27

Data source status:
- Engineering estimate ___
- Bench scale ___
- Pilot scale _x_
- Full scale ___

Use in system: Secondary, unless otherwise specified
Pretreatment of influent: See below

DESIGN OR OPERATING PARAMETERS

Unit configuration:
Wastewater flow:
Contact time: See below
Hydraulic loading: See below
Organic loading:
Bed depth:
Total carbon inventory:
Carbon exhaustion rate:
Backwash rate:
Air scour rate:
Regeneration technique:
Carbon makeup rate: See below
Carbon type/characteristics:
pH: 11.3 (specified for one sample only)

REMOVAL DATA

Sampling period:

Pretreatment of influent	Contact time, min	Hydraulic loading, $m^3/min/m^2$ (gpm/ft^2)	Carbon makeup rate, kg C/m^3 (lb C/1,000 gal)	BOD_5 Concentration, mg/L Influent	BOD_5 Concentration, mg/L Effluent	BOD_5 Percent removal	COD Concentration, mg/L Influent	COD Concentration, mg/L Effluent	COD Percent removal
Lime, treatment and clarification	108	0.06(1.42)	0.03(2.5)						
Primary clarification		0.029(0.71)	3.36(28)						
Primary clarification		0.06(1.42)	2.46(20.5)						
Lime precipitation[a]		0.15-016(3.6-4.0)		92	22	76	302	209	35
Biological oxidation and clarification[a]	140	0.087(2.13)	0.96(81)						
Lime precipitation and biological oxidation[a]		0.15-0.16(3.6-4.0)		48	23	52			

Pretreatment of influent	TOC Concentration, mg/L Influent	TOC Concentration, mg/L Effluent	TOC Percent removal	TSS Concentration, mg/L Influent	TSS Concentration, mg/L Effluent	TSS Percent removal
Lime, treatment and clarification	177	100	44			
Primary clarification	1,160	202	83			
Primary clarification	220	83	62			
Lime precipitation[a]				1,280	1,200	6
Biological oxidation and clarification	148	57	61			
Lime precipitation and biological oxidation[a]						

[a] Subcategory: unbleached kraft mill waste.

Note: Blanks indicate information was not specified.

TREATMENT TECHNOLOGY: Granular Activated Carbon Adsorption

Data source: Effluent Guidelines, Government report
Point source category: Textile mills
Subcategory: Knit fabric finishing
Plant: W, S (different references)
References: A6, pp. VII-91, 92; B3, pp. 55-59

Data source status:
Engineering estimate ___
Bench scale ___
Pilot scale x
Full scale ___

Use in system: Tertiary
Pretreatment of influent: Screening, primary sedimentation, equalization, nitrogen addition, activated sludge, multimedia filtration

DESIGN OR OPERATING PARAMETERS

Unit configuration: Downflow; 3 columns in series
Wastewater flow: 0.0018 m^3/min
Contact time (empty bed): 45 min
Hydraulic loading: 0.062 $m^3/min/m^2$ (1.5 gpm/ft^2)
Organic loading:
Bed Depth (total): 7.09 m (23.2 ft)
Total carbon inventory: 54 kg (120 lb)
Carbon exhaustion rate:
Backwash rate:
Air scour rate:
Regeneration technique:
Carbon makeup rate:
Carbon type/characteristics: Westvaco WV-L

REMOVAL DATA

Sampling period: 24-hr

Pollutant/parameter	Concentration Influent	Effluent	Percent removal
Conventional pollutants, mg/L:			
BOD_5	3.4	1.5	56
COD	55	19	65
TOC	11	2.9	74
TSS	9.5	2.0	79
Total phenol	0.009	<0.0075	>17
Toxic pollutants, µg/L:			
Antimony	620	590	5
Arsenic	<10	11	0[a]
Cadmium	5	6	0[a]
Copper	27	<4	>85
Lead	81	79	2
Mercury	0.4	0.4	0
Nickel	81	96	0[a]
Zinc	75	31	59
Bis(2-ethylhexyl) phthalate	42	410	0[a]
Di-n-butyl phthalate	6.0	<0.02	>99
Phenol	0.4	<0.07	>82
Toluene	0.4	1.6	0[a]
Acenaphthene	0.6	<0.04	>93
Chloroform	7.0	<5.0	>29
Methylene chloride[b]	4.6	940	0[a]
Trichlorofluoromethane	<2.0	69	0[a]

[a]Actual data indicate negative removal.

[b]Presence may be due to sample contamination.

Note: Blanks indicate information was not specified.

TREATMENT TECHNOLOGY: Granular Activated Carbon Adsorption

Data source: Effluent Guidelines, Government report

Point source category: Textile mills
Subcategory: Knit fabric finishing
Plant: E, P (different references)
References: A6, p. VII-93; B3, pp. 60-64

Data source status:

Engineering estimate	
Bench scale	
Pilot scale	x
Full scale	

Use in system: Tertiary
Pretreatment of influent: Screening, activated sludge, multimedia filtration

DESIGN OR OPERATING PARAMETERS

Unit configuration: Downflow; 3 columns in series
Wastewater flow:
Contact time:
Hydraulic loading:
Organic loading:
Bed depth (total): 7.09 m (23.2 ft)
Total carbon inventory: 54 kg (120 lb)
Carbon exhaustion rate:
Backwash rate:
Air scour rate:

Regeneration technique:
Carbon makeup rate:
Carbon type/characteristics:

REMOVAL DATA

Sampling period:

Pollutant/parameter	Concentration		Percent removal
	Influent	Effluent	
Conventional pollutants, mg/L:			
Total phenol	0.068	0.018	74
Toxic pollutants, μg/L:			
Antimony	48	36	25
Arsenic	<2	12	0[a]
Mercury	0.3	0.4	0[a]
Nickel	58	50	14
Silver	5	<5	>0
Zinc	150	<1	>99
Bis(2-ethylhexyl) phthalate	3.9	3.9	0
Di-n-butyl phthalate	1.6	<0.02	>99
Diethyl phthalate	0.8	1.4	0[a]
Phenol	1.8	<0.07	>96
Benzene	1.0	<0.2	>80
Toluene	2.7	3.6	0[a]
Anthracene/phenanthrene	0.5	0.1	80
Methylene chloride[b]	4.1	7.3	0[a]

[a]Actual data indicate negative removal.
[b]Presence may be due to sample contamination.

Note: Blanks indicate information was not specified.

TREATMENT TECHNOLOGY: Granular Activated Carbon Adsorption

Data source: Effluent Guidelines
Point source category: Textile mills
Subcategory: Knit fabric finishing
Plant: Q
References: A6, p. VII-89

Data source status:
Engineering estimate ___
Bench scale ___
Pilot scale x
Full scale ___

Use in system: Tertiary
Pretreatment of influent: Screening, equalization, activated sludge, sedimentation with chemical addition, multimedia filtration

DESIGN OR OPERATING PARAMETERS

Unit configuration: Downflow; 3 columns in series
Wastewater flow: 0.00084-0.0012 m^3/min (0.22-0.31 gpm)
Contact time (empty bed): 22-30 min
Hydraulic loading: 0.03-0.041 $m^3/min/m^2$ (0.73-1.0 gpm/ft^2)
Organic loading:
Bed depth (total): 7.09 m (23.2 ft)
Total carbon inventory: 54 kg (120 lb)
Carbon exhaustion rate:
Backwash rate:
Air scour rate:
Regeneration technique:
Carbon makeup rate:
Carbon type/characteristics: Westvaco WV-L

REMOVAL DATA

Sampling period: 24-hr composite samples

Pollutant/parameter	Concentration, mg/L Influent	Concentration, mg/L Effluent	Percent removal
Conventional pollutants:			
BOD_5	4	2	50
COD	206	71	66
TOC	22	14	36
TSS	4	2	50

Note: Blanks indicate information was not specified.

TREATMENT TECHNOLOGY: Granular Activated Carbon Adsorption

Data source: Effluent Guidelines, Government report
Point source category: Textile mills
Subcategory: Wool finishing
Plant: B, A (different references)
References: A6, pp. VII-85-87; B3, pp. 39-44

Data source status:

Engineering estimate	
Bench scale	
Pilot scale	x
Full scale	

Use in system: Tertiary
Pretreatment of influent: Screening, equalization, activated sludge, sedimentation with chemical addition (alum, lime), multimedia filtration

DESIGN OR OPERATING PARAMETERS

Unit configuration: Downflow; 3 columns in series
Wastewater flow: 0.001-0.0012 m^3/min (0.26-0.31 gpm)
Contact time (empty bed): 25-30 min
Hydraulic loading: 0.0032-0.0038 m^3/min (0.83-1.0 gpm)
Organic loading:
Bed depth (total): 7.09 m (23.2 ft)
Total carbon inventory: 54 kg (120 lb)
Carbon exhaustion rate:
Backwash rate:
Air scour rate:
Regeneration technique:
Carbon makeup rate:
Carbon type/characteristics: ICI Hydrodarcc

REMOVAL DATA

Sampling period: 24-hr for priority pollutants

Pollutant/parameter	Concentration		Percent
	Influent	Effluent	removal
Conventional pollutants, mg/L:			
BOD_5	25	12	52
COD	184	31	83
TOC	60	16	73
TSS	12	2	83
Total phenol	0.055	0.017	69
Toxic pollutants, μg/L:			
Antimony	<10	24	0[a]
Arsenic	100	<1	>99
Beryllium	1.2	5.4	0[a]
Cadmium	97	5.2	95
Chromium	34	19	44
Copper	110	47	57
Cyanide	10	<4	>60
Lead	79	<22	>72
Zinc	5,900	6,000	0[a]
Bis(2-ethylhexyl) phthalate	14	4.7	66
N-nitrosodiphenylamine	0.4	<0.07	>82
2,4-Dimethylphenol	0.9	<0.1	>89
Pentachlorophenol	10	<0.4	>96
Phenol	3.0	1.5	50
1,2-Dichlorobenzene	5.4	<0.05	>99
Toluene	12	<0.1	>99
1,2,4-Trichlorobenzene	94	<0.09	>99
Benzo(a)pyrene	0.8	<0.02	>97
α-BHC	1.9	<1.0	>47

[a] Actual data indicate negative removal.

Note: Blanks indicate information was not specified.

TREATMENT TECHNOLOGY: Granular Activated Carbon Adsorption

Data source: Effluent Guidelines, Government report
Point source category: Textile mills
Subcategory: Wool finishing
Plant: O, N (different references)
References: A6, pp. VII-94, 95; B3, pp. 65-69

Data source status:

Engineering estimate	___
Bench scale	___
Pilot scale	x
Full scale	___

Use in system: Tertiary
Pretreatment of influent: Neutralization, activated sludge, multimedia filtration

DESIGN OR OPERATING PARAMETERS

Unit configuration: Downflow; 3 columns in series
Wastewater flow:
Contact time:
Hydraulic loading:
Organic loading:
Bed depth (total): 7.09 m (23.2 ft)
Total carbon inventory: 54 kg (120 lb)
Carbon exhaustion rate:
Backwash rate:
Air scour rate:
Regeneration technique:
Carbon makeup rate:
Carbon type/ characteristics:

REMOVAL DATA

Sampling period: 72-hr for conventional pollutants, 24-hr composite samples for toxic pollutants, and grab samples for volatile organics

Pollutant/parameter	Concentration Influent	Concentration Effluent	Percent removal
Conventional pollutants, mg/L:			
COD	210	44	79
TSS	<1	12	0[a]
Total phenol	0.017	0.011	35
Total phosphorous	2.3	1.0	57
Toxic pollutants, μg/L:			
Arsenic	3	3	0
Chromium	95	5.2	95
Copper	130	24	82
Zinc	590	430	27
Bis(2-ethylhexyl) phthalate	29	78	0[a]
Di-n-butyl phthalate	1.1	1.8	0[a]
Diethyl phthalate	0.4	1.2	0[a]
Toluene	0.6	<0.1	>83
Anthracene/phenanthrene	0.5	0.4	20
Fluoranthene	0.08	<0.02	>75
Pyrene	0.1	<0.01	>90
1,2-Dichloropropane	1.0	<0.7	>30
Methylene chloride[b]	47	27	43

[a]Actual data indicate negative removal.
[b]Presence may be due to sample contamination.

Note: Blanks indicate information was not specified.

TREATMENT TECHNOLOGY: Granular Activated Carbon Adsorption

Data source: Government report
Point source category: Textile mills
Subcategory: Woven fabric finishing
Plant: T
References: B3, pp. 76-82

Data source status:
Engineering estimate ___
Bench scale ___
Pilot scale x
Full scale ___

Use in system: Tertiary
Pretreatment of influent: Equalization, aeration, multimedia filtration

DESIGN OR OPERATING PARAMETERS

Unit configuration: Downflow; 3 columns in series
Wastewater flow:
Contact time:
Hydraulic loading:
Organic loading:
Bed depth (total): 7.09 m (23.2 ft)
Total carbon inventory: 54 kg (120 lb)
Carbon exhaustion rate:
Backwash rate:
Air scour rate:

Regeneration technique:
Carbon makeup rate:
Carbon type/
characteristics:

REMOVAL DATA

Sampling period: 24-hr composite, volatile organics were grab sampled

Pollutant/parameter	Concentration Influent	Effluent	Percent removal
Conventional pollutants, mg/L:			
COD	160	340	0[a]
TSS	14	12	14
Total phenol	0.16	0.12	25
Total phosphorous	13	14	0[a]
Toxic pollutants, μg/L:			
Antimony	58	39	33
Arsenic	3	3	0
Chromium	95	84	12
Copper	100	87	13
Cyanide	20	<2	>90
Lead	26	29	0[a]
Nickel	100	90	10
Selenium	2	<1	>50
Silver	32	28	12
Zinc	97	110	0[a]
Bis(2-ethylhexyl) phthalate	19	14	26
Butyl benzyl phthalate	2.5	<0.03	>99
Di-n-butyl phthalate	7.0	1.7	76
Phenol	1.1	0.9	18
p-Chloro-m-cresol	0.6	<0.1	>83
Benzene	6.9	9.8	0[a]
Chlorobenzene	4.8	<0.2	>96
Ethylbenzene	0.2	<0.2	>0
Toluene	0.8	0.6	25
Methylene chloride[b]	19	19	0

[a]Actual data indicate negative removal.
[b]Presence may be due to sample contamination.

Note: Blanks indicate information was not specified.

TREATMENT TECHNOLOGY: Granular Activated Carbon Adsorption

Data source: Effluent Guidelines, Government Report

Point source category: Textile mills

Subcategory: Woven fabric finishing

Plant: V, C (different references)

References: A6, p. VII-91; B3, pp. 45-49

Data source status:

Engineering estimate	___
Bench scale	___
Pilot scale	x
Full scale	___

Use in system: Tertiary

Pretreatment of influent: Screening, neutralization, activated sludge, sedimentation with chemical addition (alum), multimedia filtration

DESIGN OR OPERATING PARAMETERS

Unit configuration: Downflow; 3 columns in series

Wastewater flow: 0.002 m^3/min (0.46 gpm)

Contact time (empty bed): 45 min

Hydraulic loading: 0.061 m^3/min /m^2 (1.5 gpm/ft^2)

Organic loading:

Bed depth (total): 7.09 m (23.2 ft)

Total carbon inventory: 54 kg (120 lb)

Carbon exhaustion rate:

Backwash rate:

Air scour rate:

Regeneration technique:

Carbon makeup rate:

Carbon type/characteristics: Westvaco WV-L

REMOVAL DATA

Sampling period: 24-hr composite samples for toxic pollutants, grab samples for volatile organics

Pollutant/parameter	Concentration		Percent removal
	Influent	Effluent	
Conventional pollutants, mg/L:			
BOD_5	2.5	1.2	52
COD	331	176	47
TOC	62	36	42
TSS	20	20	0
Total phenol	0.019	<0.002	>89
Total phosphorous	2.0	1.9	5
Toxic pollutants, μg/L:			
Antimony	140	120	14
Beryllium	1.2	2.7	0[a]
Cadmium	2.7	9.8	0[a]
Chromium	14	15	0[a]
Copper	25	35	0[a]
Lead	64	64	0
Silver	77	91	0[a]
Zinc	230	83	64
Bis(2-ethylhexyl) phthalate	5.3	11	0[a]
Di-n-butyl phthalate	0.6	0.4	33
Pentachlorophenol	12	<0.4	>97
1,2-Dichlorobenzene	5.8	<0.05	>99
Anthracene/phenanthrene	0.03	0.01	67
Methylene chloride[b]	210	110	48

[a]Actual data indicate negative removal.

[b]Presence may be due to sample contamination.

Note: Blanks indicate information was not specified.

TREATMENT TECHNOLOGY: Granular Activated Carbon Adsorption

Data source: Government report
Point source category: Textile mills
Subcategory: Woven fabric finishing
Plant: V
References: B3, pp. 70-75

Data source status:
- Engineering estimate ___
- Bench scale ___
- Pilot scale x
- Full scale ___

Use in system: Tertiary
Pretreatment of influent: Screening, activated sludge, multimedia filtration

DESIGN OR OPERATING PARAMETERS

Unit configuration: Downflow; 3 columns in series
Wastewater flow:
Contact time:
Hydraulic loading:
Organic loading:
Bed depth (total): 7.09 m (23.2 ft)
Total carbon inventory: 54 kg (120 lb)
Carbon exhaustion rate:
Backwash rate:
Air scour rate:

Regeneration technique:
Carbon makeup rate:
Carbon type/ characteristics:

REMOVAL DATA

Sampling period: 24-hr composite sample, volatile organics were grab sampled

Pollutant/parameter	Concentration		Percent removal
	Influent	Effluent	
Conventional pollutants, mg/L:			
COD	72	22	69
TSS	4	6	0[a]
Total phenol	0.013	0.008	38
Total phosphorus	1.1	1.1	0
Toxic pollutants, μg/L:			
Antimony	<10	24	0[a]
Arsenic	4	5	0[a]
Copper	75	16	79
Cyanide	3	<2	>33
Lead	31	26	16
Nickel	<36	67	0[a]
Selenium	<1	2	0[a]
Silver	<5	15	0[a]
Zinc	190	69	64
Bis(2-ethylhexyl) phthalate	16	17	0[a]
Butyl benzyl phthalate	0.9	<0.03	>97
Di-n-butyl phthalate	12	<0.02	>99
Toluene	1.3	1.0	23
Anthracene/phenanthrene	0.3	<0.01	>97
Methylene chloride[b]	13	17	0[a]
Trichloroethylene	<0.5	0.6	0[a]

[a]Actual data indicate negative removal.

[b]Presence may be due to sample contamination.

Note: Blanks indicate information was not specified.

TREATMENT TECHNOLOGY: Granular Activated Carbon Adsorption

Data source: Effluent Guidelines
Point source category: Textile mills
Subcategory: Woven fabric/stock and yarn finishing
Plant: DD
References: A6, p. VII-85

Data source status:
- Engineering estimate ___
- Bench scale ___
- Pilot scale x
- Full scale ___

Use in system: Tertiary
Pretreatment of influent: Screening, neutralization, activated sludge, multimedia filtration

DESIGN OR OPERATING PARAMETERS

Unit configuration: Downflow; 3 columns in series
Wastewater flow: 0.0018 m^3/min (0.46 gpm)
Contact time (empty bed): 45 min
Hydraulic loading: 0.062 $m^3/min/m^2$ (1.5 gpm/ft^2)
Organic loading:
Bed depth (total): 7.09 m (23.2 ft)
Total carbon inventory: 54 kg (120 lb)
Carbon exhaustion rate:
Backwash rate:
Air scour rate:
Regeneration technique:
Carbon makeup rate:
Carbon type/ characteristics:

REMOVAL DATA

Sampling period: 8 hr

Pollutant/parameter	Concentration, µg/L Influent	Effluent	Percent removal
Toxic pollutants:			
Chromium	58	130	0[a]
Copper	59	42	29
Lead	37	35	5
Nickel	72	81	0[a]
Silver	25	32	0[a]
Zinc	190	370	0[a]

[a]Actual data indicate negative removal.

Note: Blanks indicate information was not specified.

TREATMENT TECHNOLOGY: Granular Activated Carbon Adsorption

Data source: Effluent Guidelines
Point source category: Textile mills
Subcategory: Woven fabric finishing
Plant: D
References: A6, p. VII-84

Data source status:
- Engineering estimate ___
- Bench scale ___
- Pilot scale x
- Full scale ___

Use in system: Tertiary
Pretreatment of influent: Screening, neutralization, activated sludge, multimedia filtration

DESIGN OR OPERATING PARAMETERS

Unit configuration: Downflow; 3 columns in series
Wastewater flow: 0.0018 m^3/min (0.46 gpm)
Contact time (empty bed): 45 min
Hydraulic loading: 0.062 $m^3/min/m^2$ (1.5 gpm/ft^2)
Organic loading:
Bed depth (total): 7.09 m (23.2 ft)
Total carbon inventory: 54 kg (120 lb)
Carbon exhaustion rate:
Backwash rate:
Air scour rate:
Regeneration technique:
Carbon makeup rate:
Carbon type/characteristics: Westvaco WV-L

REMOVAL DATA

Sampling period:

Pollutant/parameter	Concentration, mg/L Influent	Concentration, mg/L Effluent	Percent removal
Conventional pollutants:			
BOD_5	19	13	32
COD	630	422	33
TOC	157	101	36
TSS	85	23	73

Note: Blanks indicate information was not specified.

TREATMENT TECHNOLOGY: Granular Activated Carbon Adsorption

Data source: Effluent Guidelines
Point source category: Textile mills
Subcategory: Woven fabric finishing
Plant: P
References: A6, p. VII-88

Data source status:
Engineering estimate ___
Bench scale ___
Pilot scale x
Full scale ___

Use in system: Tertiary
Pretreatment of: Screening, neutralization, equalization, activated sludge, multimedia filtration with precoagulation

DESIGN OR OPERATING PARAMETERS

Unit configuration: Downflow; 3 columns in series
Wastewater flow: 0.00092-0.0018 m^3/min (0.24-0.46 gpm)
Contact time (empty bed): 23-45 min
Hydraulic loading: 0.032-0.062 $m^3/min/m^2$ (0.77-1.5 gpm/ft^2)
Organic loading:
Bed depth (total): 7.09 m (23.2 ft)
Total carbon inventory: 54 kg (120 lb)
Carbon exhaustion rate:
Backwash rate:
Air scour rate:
Regeneration technique:
Carbon makeup rate:
Carbon type/characteristics: Westvaco WV-L

REMOVAL DATA

Sampling period:

Pollutant/parameter	Concentration, mg/L Influent	Effluent	Percent removal
Conventional pollutants:			
BOD_5	14	8	43
COD	107	81	24
TOC	24	11	54
TSS	19	19	0

Note: Blanks indicate information was not specified.

TREATMENT TECHNOLOGY: Granular Activated Carbon Adsorption

Data source: Effluent Guidelines, Government report
Point source category: Textile mills
Subcategory: Wool scouring
Plant: A, W (different references)
References: A6, p. VII-94; B3, pp. 50-54

Data source status:
Engineering estimate ___
Bench scale ___
Pilot scale x
Full scale ___

Use in system: Tertiary
Pretreatment of influent: Grit removal, activated sludge, tertiary sedimentation, multimedia filtration

DESIGN OR OPERATING PARAMETERS

Unit configuration: Downflow; 3 columns in series
Wastewater flow:
Contact time:
Hydraulic loading:
Organic loading:
Bed depth (total): 7.09 m (23.2 ft)
Total carbon inventory: 54 kg (120 lb)
Carbon exhaustion rate:
Backwash rate:
Air scour rate:
Regeneration technique:
Carbon makeup rate:
Carbon type/characteristics:

REMOVAL DATA

Sampling period: 24-hr composite, volatile organics were grab sampled

Pollutant/parameter	Concentration Influent	Concentration Effluent	Percent removal
Conventional pollutants, mg/L:			
Total phenol	0.017	0.017	0
Toxic pollutants, μg/L:			
Arsenic	83	42	49
Copper	120	<80	>33
Cyanide	260	40	85
Zinc	400	120	70
Bis(2-ethylhexyl) phthalate	14	26	0[a]
Anthracene/phenanthrene	0.2	0.1	50
Benzo(a)pyrene	0.2	<0.02	>90
Benzo(k)fluoranthene	0.1	<0.02	>80
Fluoranthene	0.2	<0.02	>90
Pyrene	0.3	<0.01	>97
Methylene chloride[b]	4.8	1.8	62

[a]Actual data indicate negative removal.

[b]Presence may be due to sample contamination.

Note: Blanks indicate information was not specified.

TREATMENT TECHNOLOGY: Granular Activated Carbon Adsorption

Data source: Government report
Point source category:
Subcategory:
Plant: Reichhold Chemicals, Inc.
References: B4,pp. 66-85

Data source status:
Engineering estimate ___
Bench scale ___
Pilot scale ___
Full scale ___

Use in system:
Pretreatment of influent: Clarification

DESIGN OR OPERATING PARAMETERS

Unit configuration: See below
Wastewater flow: 20 mL/min
Contact time: 25.3 min/m of bed depth
Hydraulic loading:
Organic loading:
Bed depth: See below
Total carbon inventory: See below
Carbon exhaustion rate: See below
Backwash rate:
Air scour rate:

Regeneration technique:
Carbon makeup rate:
Carbon type/ characteristics: Calgon Filtrasorb 300 GAC

REMOVAL DATA

Sampling period: 24-hour composites

Data source status	Unit configuration: number of 25.4-mm (1-in.) diameter columns in series	Bed depth, m (ft)	Total carbon inventory, g	TOC Concentration, mg/L Influent	TOC Concentration, mg/L Effluent	Percent removal
Bench scale	1	0.305(1)	66	2,150[a]	1,950[a]	9[a]
Pilot scale	4	2.75(9)	597	2,150	989	54
Pilot scale	5	3.68(2)	797	2,150	831	61
Bench scale	2	0.92(3)	197	2,150	1,580	27
Pilot scale	3	1.83(6)	397	2,150	1,120	48

[a]Average concentrations listed.

Note: Blanks indicate information was not specified.

6.2 POWDERED CARBON ADDITION [1]

6.2.1 Function

Powdered activated carbon is used in wastewater facilities to adsorb soluble organic materials and to aid in the clarification process.

6.2.2 Description

Powdered carbon is fed to a treatment system using chemical feed equipment similar to that used for other chemicals that are purchased in dry form. The spent carbon is removed with the sludge and then discarded or regenerated. Regeneration can be accomplished in a furnace or wet air oxidation system.

Powdered carbon can be fed to primary clarifiers directly, or to a separate sludge recirculation-type clarifier that enhances the contact between the carbon and the wastewater. Powdered carbon can also be fed to tertiary clarifiers to remove additional amounts of soluble organics. Powdered carbon, when added to a sludge recirculation-type clarifier, has been shown to be capable of achieving secondary removal efficiencies.

Powdered carbon can be fed in the dry state using volumetric or gravimetric feeders or it can be fed in slurry form.

6.2.3 Common Modifications

A new technology has been developed over the past several years that consists of the addition of powdered activated carbon to the aeration basins of biological systems. This application is capable of the following: high BOD_5 and COD reduction, although effluent concentration may still exceed limitations, despite hydraulic and organic overloading; aiding solids settling in the clarifiers; a high degree of nitrification due to extended sludge age; a substantial reduction in phosphorus; adsorbing coloring materials such as dyes and toxic compounds; and adsorbing detergents and reducing foam.

6.2.4 Technology Status

Powdered carbon addition is used mostly in municipal applications at the present time. Two new municipal plants using powdered carbon addition to activated sludge are under construction, and several more are planned.

6.2.5 Applications

Has been used in clarifiers and has potential use in aeration basins to adsorb soluble organic materials, thus removing BOD_5 and COD, as well as some toxic materials.

6.2.6 Limitations

Will increase the amount of sludge generated; regeneration will be necessary at higher dosages in order to maintain reasonable costs; most powdered carbon systems will require post-filtration to capture any residual carbon particles; some sort of flocculating agent, such as an organic polyelectrolyte, is usually required to maintain efficient solids captured in the clarifier.

6.2.7 Chemicals Required

Powdered activated carbon and polyelectrolytes.

6.2.8 Residuals Generated

One pound of dry sludge is generated per pound of carbon added; if regeneration is practiced, carbon sludge is reactivated and reused with only a small portion removed to prevent buildup of inerts.

6.2.9 Reliability

Powdered activated carbon systems are reasonably reliable from both a unit and process standpoint; in fact, powdered carbon systems can be used to improve process reliability of existing systems.

6.2.10 Environmental Impact

Land use requirements vary with application; air pollution may result from regeneration; spent carbon may be a land disposal problem unless regenerated.

6.2.11 Design Criteria

The amount of powdered carbon fed to a system greatly depends on the characteristics of the wastewater and the desired effluent quality; however, powdered carbon will generally be fed at a rate between 50 and 300 mg/L.

6.2.12 Flow Diagram

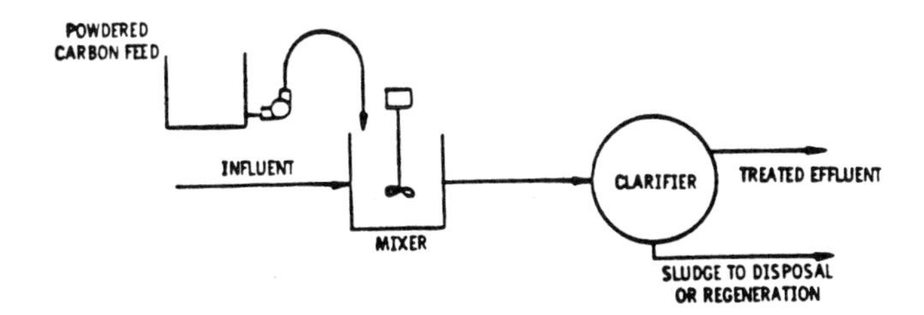

6.2.13 Performance

Subsequent data sheet provide performance data from studies on the following industries and/or wastesteams:

Petroleum refining

Pharmaceuticals and fine organic chemicals production

Pulp, paper, and paperboard production

Textile milling
- Carpet finishing
- Knit fabric finishing
- Stock and yarn finishing
- Wool finishing
- Wool scouring
- Woven fabric finishing

6.2.14 References

1. Innovative and Alternative Technology Assessment Manual. EPA-430/9-78-009 (draft), U.S. Environmental Protection Agency, Cincinnati, Ohio, 1978. 252 pp.

CONTROL TECHNOLOGY SUMMARY FOR POWDERED CARBON ADDITION (WITH ACTIVATED SLUDGE)

Pollutant	Number of data points	Effluent concentration				Removal efficiency, %			
		Minimum	Maximum	Median	Mean	Minimum	Maximum	Median	Mean
Conventional pollutants, mg/L:									
BOD_5	24	4	54	13	17	<90	>99	96	96
COD	26	33	563	98	160	60	98	91	87
TOC	25	9	387	38	67	64	97	90	86
TSS	4	17	83	54	52	0[a]	96	0[a]	24
Oil and grease	4	11	57	13	23	8	96	54	53
Total phenol	4	<0.010	0.058	0.013	<0.023	99	>99	>99	>99
TKN	1				28				96
Toxic pollutants, μg/L:									
Antimony	2	41	150		96	0[a]	5		2.5
Cadmium	1				10				0[a]
Chromium	4	24	90	53	55	73	97	88	86
Chromium (+6)	3	<20	20	<20	<20	0[a]	>64	>60	41
Copper	3	7	29	14	17	0[a]	96	61	52
Cyanide	3	<20	45	20	<28	50	69	>67	>62
Lead	2	<18	38		28	0[a]	>78		39
Mercury	1				0.6				0[a]
Nickel	3	<10	22	<10	<14	0[a]	>58	>0	19
Selenium	2	<20	40		<30	0[a]	13		6

(continued)

CONTROL TECHNOLOGY SUMMARY FOR POWDERED CARBON ADDITION (WITH ACTIVATED SLUDGE) (continued)

Pollutant	Number of data points	Effluent concentration				Removal efficiency, %			
		Minimum	Maximum	Median	Mean	Minimum	Maximum	Median	Mean
Toxic pollutants (cont'd), µg/L:									
Zinc	4	78	140	95	100	0[a]	98	38	44
Bis(2-chloroethyl) ether	1				44				53
Bis(2-ethylhexyl) phthalate	1				<10[b]				>97
2-Chlorophenol	1				190,000				81
Phenol	2	<10[b]	190,000		95,000	81	>85		>83
Benzene	1				21,000				95
Ethylbenzene	1				18,000				84
Toluene	1				67,000				79
Naphthalene	1				<10[b]				>96
1,2-Dichloroethane	1				190,000				81
1,2-Dichloropropane	1				70,000				93
Acrolein	1				700,000				30
Isophorone	1				30,000				97

[a] Actual data indicate negative removal.

[b] Below detection limit, assumed to be <10 µg/L.

TREATMENT TECHNOLOGY: Powdered Activated Carbon Adsorption (With Activated Sludge)

Data source: Effluent Guidelines
Point source category: Petroleum refining
Subcategory:
Plant: B
References: A3, pp. VI-43 to 45

Data source status:
Engineering estimate ___
Bench scale ___
Pilot scale x
Full scale ___

Use in system: Secondary
Pretreatment of influent: Dissolved air flotation

DESIGN OR OPERATING PARAMETERS

Carbon dosage:
Carbon type/characteristics:
Flocculent dosage:
Clarifier configuration:
Depth:
Hydraulic detention time:
Hydraulic loading:
Weir loading:

Sludge underflow:
Percent solids in sludge:
Carbon regeneration technique:
Carbon makeup rate:

REMOVAL DATA

Sampling period:

Pollutant/parameter	Concentration Influent	Effluent	Percent removal
Conventional pollutants, mg/L:			
COD	420	100	76
TOC	100	30	70
TSS	36	56	0[a]
Oil and grease	25	9	64
Total phenol	24	<0.01	>99
Toxic pollutants, μg/L:			
Chromium	90	24	73
Chromium (+6)	55	<20	>64
Cyanide	60	<20	>67
Selenium	<20	40	0[a]

[a]Actual data indicate negative removal.

Note: Blanks indicate information was not specified.

TREATMENT TECHNOLOGY: Powdered Activated Carbon Adsorption (With Activated Sludge)

Data source: Conference paper
Point source category: Petroleum refining
Subcategory:
Plant: First of four refinery and/or petrochemical plants
References: D2, pp. 225-230

Data source status:
- Engineering estimate ___
- Bench scale ___
- Pilot scale ___
- Full scale x

Use in system: Primary
Pretreatment of influent:

DESIGN OR OPERATING PARAMETERS

Carbon dosage:
Carbon type/characteristics: Hydrodarco C (high density, lignite based)
Flocculent dosage: 20 mg/L cationic polymer for secondary solids capture
Depth:
Hydraulic detention time:
Hydraulic loading: 17.2 $m^3/m^2/d$ (432 gpd/ft^2)
Weir loading:
MLSS: 3,600 mg/L

Sludge underflow:
Percent solids in sludge:
Carbon regeneration technique:
Carbon makeup rate:
Wastewater flow: 3,790 m^3/day (2.2 MGD)

REMOVAL DATA

Sampling period:

Pollutant/parameter	Concentration, mg/L Influent	Effluent	Percent removal
Conventional pollutants:			
BOD_5	300	<30	>90
COD	1,180	350	70
TOC	420	100	76

Note: Blanks indicate information was not specified.

TREATMENT TECHNOLOGY: Powdered Activated Carbon Adsorption (With Activated Sludge)

Data source: Effluent Guidelines
Point source category: Petroleum refining
Subcategory:
Plant: K
References: A3, pp. Vl-43 to 45

Data source status:
- Engineering estimate ___
- Bench scale ___
- Pilot scale x
- Full scale ___

Use in system: Secondary
Pretreatment of influent: Dissolved air flotation

DESIGN OR OPERATING PARAMETERS

Carbon dosage:
Carbon type/characteristics:
Flocculent dosage:
Clarifier configuration:
Depth:
Hydraulic detention time:
Hydraulic loading:
Weir loading:

Sludge underflow:
Percent solids in sludge:
Carbon regeneration technique:
Carbon makeup rate:

REMOVAL DATA

Sampling period:

Pollutant/parameter	Concentration Influent	Effluent	Percent removal
Conventional pollutants, mg/L:			
COD	900	53	94
TOC	250	20	92
TSS	430	17	96
Oil and grease	270	11	96
Total phenol	1.4	0.012	99
Toxic pollutants, μg/L:			
Chromium	1,800	60	97
Chromium (+6)	50	<20	>60
Copper	380	14	96
Lead	82	<18	>78
Mercury	<0.5	0.6	0[a]
Nickel	24	<10	>58
Zinc	5,900	110	98

[a]Actual data indicate negative removal.

Note: Blanks indicate information was not specified.

TREATMENT TECHNOLOGY: Powdered Activated Carbon Adsorption (With Activated Sludge)

Data source: Effluent Guidelines
Point source category: Petroleum refining
Subcategory:
Plant: M
References: A3, pp. V1-43 to 45

Data source status:
Engineering estimate ___
Bench scale ___
Pilot scale _x_
Full scale ___

Use in system: Secondary
Pretreatment of influent: Dissolved air flotation

DESIGN OR OPERATING PARAMETERS

Carbon dosage:
Carbon type/characteristics:
Flocculent dosage:
Clarifier configuration:
Depth:
Hydraulic detention time:
Hydraulic loading:
Weir loading:

Sludge underflow:
Percent solids in sludge:
Carbon regeneration technique:
Carbon makeup rate:

REMOVAL DATA

Sampling period:

Pollutant/parameter	Concentration Influent	Effluent	Percent removal
Conventional pollutants, mg/L:			
COD	300	106	65
TOC	77	23	70
TSS	29	52	0[a]
Oil and grease	23	16	43
Total phenol	6.0	0.013	>99
Toxic pollutants, μg/L:			
Cadmium	<1	10	0[a]
Chromium	450	46	90
Copper	18	7	61
Cyanide	140	45	69
Lead	<18	38	0[a]
Nickel	10	<10	>0
Selenium	23	<20	>13
Zinc	280	140	50

[a]Actual data indicate negative removal.

Note: Blanks indicate information was not specified.

TREATMENT TECHNOLOGY: Powdered Activated Carbon Adsorption (With Activated Sludge)

Data source: Effluent Guidelines
Point source category: Petroleum refining
Subcategory:
Plant: P
References: A3, pp. V1-43 to 45

Data source status:
- Engineering estimate ___
- Bench scale ___
- Pilot scale x
- Full scale ___

Use in system: Secondary
Pretreatment of influent: API design gravity oil separator

DESIGN OR OPERATING PARAMETERS

Carbon dosage:
Carbon type/characteristics:
Flocculent dosage:
Clarifier configuration:
Depth:
Hydraulic detention time:
Hydraulic loading:
Weir loading:
Sludge underflow:
Percent solids in sludge:
Carbon regeneration technique:
Carbon makeup rate:

REMOVAL DATA

Sampling period:

Pollutant/parameter	Concentration Influent	Concentration Effluent	Percent removal
Conventional pollutants, mg/L:			
COD	400	160	60
TOC	120	43	64
TSS	62	83	0[a]
Oil and grease	62	57	8
Total phenol	55	0.058	>99
Toxic pollutants, μg/L:			
Antimony	43	41	5
Chromium	660	90	86
Chromium (+6)	<20	20	0[a]
Copper	10	29	0[a]
Cyanide	40	20	50
Nickel	10	22	0[a]
Zinc	100	78	26

[a]Actual data indicate negative removal.

Note: Blanks indicate information was not specified.

TREATMENT TECHNOLOGY: Powdered Activated Carbon Adsorption (With Activated Sludge)

Data source: Journal article
Point source category: Pharmaceuticals
Subcategory: Pharmaceuticals and fine organic chemicals
Plant: Texas plant
References: C2, pp. 854-855

Data source status:
Engineering estimate ___
Bench scale ___
Pilot scale ___
Full scale x

Use in system: Primary
Pretreatment of influent:

DESIGN OR OPERATING PARAMETERS

Carbon dosage:
Carbon type/characteristics:
Flocculent dosage:
Clarifier configuration:
Depth:
Hydraulic detention time:
Hydraulic loading:
Weir loading:
Wastewater flow: 946 m^3/d (0.25 MGD)

Sludge underflow:
Percent solids in sludge:
Carbon regeneration technique: Wet air oxidation
Carbon makeup rate: 90% of carbon recovered

REMOVAL DATA

Sampling period:

Pollutant/parameter	Concentration, mg/L Influent	Concentration, mg/L Effluent	Percent removal
Conventional pollutants:			
BOD_5	7,470	11	>99
COD	14,800	280	98
TKN	690	28[a]	96

[a]Calculated from influent concentration and percent removal.

Note: Blanks indicate information was not specified.

TREATMENT TECHNOLOGY: Powdered Activated Carbon Adsorption (With Activated Sludge)

Data source: Effluent Guidelines
Point source category: Pulp, paper, and paperboard
Subcategory:
Plant:
References: A26, pp. VII 24, 25

Data source status: See below
Engineering estimate ___
Bench scale ___
Pilot scale ___
Full scale ___

Use in system: Secondary
Pretreatment of influent:

DESIGN OR OPERATING PARAMETERS

Carbon dosage: See below
Carbon type/characteristics:
Flocculent dosage:
Clarifier configuration:
Depth:
Hydraulic detention time: See below
Hydraulic loading:
Weir loading:

Sludge underflow:
Percent solids in sludge:
Carbon regeneration technique: See below
Carbon makeup rate:

REMOVAL DATA

Scale	Carbon dosage, mg/L	Carbon regeneration technique	Hydraulic detention time, hr	BOD_5 Concentration, mg/L Influent	Effluent	Percent removal
Bench	160	Thermally regenerated and acid washed	6.1	300	23	92
Full	182		14.6	500	15.2	95

Note: Blanks indicate information was not specified.

TREATMENT TECHNOLOGY: Powdered Activated Carbon Adsorption (With Activated Sludge)

Data source: Government report
Point source category:
Subcategory:
Plant:
References: B20, pp. 24, 27, 30, 33, 41

Data source status:[a]
Engineering estimate ___
Bench scale ___
Pilot scale ___
Full scale ___

Use in system:
Pretreatment of influent:

[a]Each pollutant looked at separately in unspecified test.

DESIGN OR OPERATING PARAMETERS

Carbon dosage: 5,000 mg/L
Carbon type/characteristics:
Flocculent dosage:
Clarifier configuration:
Depth:
Hydraulic detention time:
Hydraulic loading:
Weir loading:

Sludge underflow:
Percent solids in sludge:
Carbon regeneration technique:
Carbon makeup rate:

REMOVAL DATA

Sampling period:

Pollutant/parameter	Concentration		Percent
	Influent	Effluent[a]	removal
Toxic pollutants, μg/L:			
Bis(2-chloroethyl) ether	94	44	53
2-Chlorophenol	1,000,000	190,000	81
Phenol	1,000,000	190,000	81
Benzene	416,000	21,000	95
Ethylbenzene	115,000	18,000	84
Toluene	317,000	67,000	79
1,2-Dichloroethane	1,000,000	190,000	81
1,2-Dichloropropane	1,000,000	70,000	93
Acrolein	1,000,000	700,000	30
Isophorone	1,000,000	30,000	97

[a]Calculated from influent and percent removal.

Note: Blanks indicate information was not specified.

TREATMENT TECHNOLOGY: **Powdered Activated Carbon Adsorption (With Activated Sludge)**

Data source: Effluent Guidelines
Point source category: Textile mills
Subcategory: Carpet finishing
Plant:
References: A6, p. VII-97

Data source status:
- Engineering estimate ___
- Bench scale ___
- Pilot scale ___
- Full scale x

Use in system: Primary
Pretreatment of influent: Screening, equalization

DESIGN OR OPERATING PARAMETERS

System configuration: Mix tank and filter press for solids removal
Wastewater flow: 757 m^3/day (0.2 MGD)
Carbon dosage:
Carbon type/characteristics:
Flocculent dosage:
Clarifier configuration:
Depth:
Hydraulic detention time:
Hydraulic loading:
Weir loading:

Sludge underflow:
Percent solids in sludge:
Carbon regeneration technique:
Carbon makeup rate:

REMOVAL DATA

Sampling period:

Pollutant/parameter	Concentration, µg/L Influent	Effluent	Percent removal
Toxic pollutants:			
Antimony	<12	150	0[a]
Zinc	20	80	0[a]
Bis(2-ethylhexyl) phthalate	400	BDL[b]	>97
Phenol	67	BDL	>85
Naphthalene	240	BDL	>96

[a]Data indicate negative removal.

[b]Below detectable limits; assumed to be <10 µg/L.

Note: Blanks indicate information was not specified.

TREATMENT TECHNOLOGY: Powdered Activated Carbon Adsorption (With Activated Sludge)

Data source: Effluent Guidelines
Point source category: Textile mills
Subcategory: Carpet finishing
Plant: F
References: A6, p. VII-102

Data source status:
- Engineering estimate ___
- Bench scale x
- Pilot scale ___
- Full scale ___

Use in system: Secondary
Pretreatment of influent:

DESIGN OR OPERATING PARAMETERS

Carbon dosage: 2,000-5,000 mg/L in aeration basin
Carbon type/characteristics: ICI-KB
Flocculent dosage:
Clarifier configuration:
Depth:
Hydraulic detention time:
Hydraulic loading:
Weir loading:

Sludge underflow:
Percent solids in sludge:
Carbon regeneration technique:
Carbon makeup rate: 277-694 mg/L/d

REMOVAL DATA

Sampling period: Two weeks

Pollutant/parameter	Concentration, mg/L Influent	Concentration, mg/L Effluent	Percent removal	Carbon dosage, mg/L
Conventional pollutants:				
BOD_5	471	6	99	2,000
	471	4	99	5,000
COD	1,450	67	95	2,000
	1,450	40	97	5,000
TOC	390	35	91	2,000
	390	18	95	5,000

Note: Blanks indicate information was not specified.

TREATMENT TECHNOLOGY: Powdered Activated Carbon Adsorption (With Activated Sludge)

Data source: Effluent Guidelines
Point source category: Textile mills
Subcategory: Knit fabric finishing
Plant: E
References: A6, p. VII-101

Data source status:
- Engineering estimate ___
- Bench scale x
- Pilot scale ___
- Full scale ___

Use in system: Secondary
Pretreatment of influent:

DESIGN OR OPERATING PARAMETERS

Carbon dosage: 2,000-5,000 mg/L in aeration basin
Carbon type/characteristics: Westvaco "SC"
Flocculent dosage:
Clarifier configuration:
Depth:
Hydraulic detention time:
Hydraulic loading:
Weir loading:
Sludge underflow:
Percent solids in sludge:
Carbon regeneration technique:
Carbon makeup rate: 216-540 mg/L/d

REMOVAL DATA

Sampling period: Two weeks

Pollutant/parameter	Concentration, mg/L Influent	Effluent	Percent removal	Carbon dosage, mg/L
Conventional pollutants:				
BOD_5	505	21	96	2,000
	505	21	96	5,000
COD	1,740	103	94	2,000
	1,740	69	96	5,000
TOC	446	52	88	2,000
	446	40	91	5,000

Note: Blanks indicate information was not specified.

TREATMENT TECHNOLOGY: Powdered Activated Carbon Adsorption (With Activated Sludge)

Data source: Effluent Guidelines
Point source category: Textile mills
Subcategory: Knit fabric finishing
Plant: Q
References: A6, p. VII-100

Data source status:
- Engineering estimate ___
- Bench scale x
- Pilot scale ___
- Full scale ___

Use in system: Secondary
Pretreatment of influent:

DESIGN OR OPERATING PARAMETERS

Carbon dosage: 1,000-5,000 mg/L in aeration basin
Carbon type/characteristics: Westvaco "SC"
Flocculent dosage:
Clarifier configuration:
Depth:
Hydraulic detention time:
Hydraulic loading:
Weir loading:
Sludge underflow:
Percent solids in sludge:
Carbon regeneration technique:
Carbon makeup rate: 35-173 mg/L/d

REMOVAL DATA

Sampling period: Two weeks

Pollutant/parameter	Concentration, mg/L Influent	Concentration, mg/L Effluent	Percent removal	Carbon dosage, mg/L
Conventional pollutants:				
BOD_5	318	14	96	1,000
	318	11	97	5,000
COD	963	175	82	1,000
	963	119	88	5,000
TOC	383	56	85	1,000
	383	44	89	5,000

Note: Blanks indicate information was not specified.

TREATMENT TECHNOLOGY: Powdered Activated Carbon Adsorption (With Activated Sludge)

Data source: Effluent Guidelines
Point source category: Textile mills
Subcategory: Stock and yarn finishing
Plant: S
References: A6, p. VII-103

Data source status:
- Engineering estimate ___
- Bench scale x
- Pilot scale ___
- Full scale ___

Use in system: Secondary
Pretreatment of influent:

DESIGN OR OPERATING PARAMETERS

Carbon dosage: 2,000-5,000 mg/L in aeration basin
Carbon type/characteristics: Westvaco "SC"
Flocculent dosage:
Clarifier configuration:
Depth:
Hydraulic detention time:
Hydraulic loading:
Weir loading:
Sludge underflow:
Percent solids in sludge:
Carbon regeneration technique:
Carbon makeup rate: 122-304 mg/L/d

REMOVAL DATA

Sampling period: Two weeks

Pollutant/parameter	Concentration, mg/L Influent	Concentration, mg/L Effluent	Percent removal	Carbon dosage, mg/L
Conventional pollutants:				
BOD_5	95	8.5	91	2,000
	95	6	94	5,000
COD	956	74	92	2,000
	956	35	96	5,000
TOC	390	35	91	2,000
	390	18	95	5,000

Note: Blanks indicate information was not specified.

TREATMENT TECHNOLOGY: Powdered Activated Carbon Adsorption (With Activated Sludge)

Data source: Effluent Guidelines
Point source category: Textile mills
Subcategory: Wool finishing
Plant: B
References: A6, p. VII-99

Data source status:
Engineering estimate ___
Bench scale x
Pilot scale ___
Full scale ___

Use in system: Secondary
Pretreatment of influent:

DESIGN OR OPERATING PARAMETERS

Carbon dosage: 2,000-8,000 mg/L in aeration basin
Carbon type/characteristics: Westvaco "SA"
Flocculent dosage:
Clarifier configuration:
Depth:
Hydraulic detention time:
Hydraulic loading:
Weir loading:
Sludge underflow:
Percent solids in sludge:
Carbon regeneration technique:
Carbon makeup rate: 97-388 mg/L/d

REMOVAL DATA

Sampling period: Two weeks

Pollutant/parameter	Concentration, mg/L Influent	Effluent	Percent removal	Carbon dosage, mg/L
Conventional pollutants:				
BOD_5	407	29	93	2,000
	407	18	96	8,000
COD	1,920	107	94	2,000
	1,920	73	96	8,000
TOC	461	44	90	2,000
	461	38	92	8,000

Note: Blanks indicate information was not specified.

TREATMENT TECHNOLOGY: Powdered Activated Carbon Adsorption (With Activated Sludge)

Data source: Effluent Guidelines
Point source category: Textile mills
Subcategory: Woven fabric finishing
Plant: D
References: A6, p. VII-99

Data source status:
- Engineering estimate ___
- Bench scale x
- Pilot scale ___
- Full scale ___

Use in system: Secondary
Pretreatment of influent:

DESIGN OR OPERATING PARAMETERS

Carbon dosage: 3,000-6,000 mg/L in aeration basin
Carbon type/characteristics: Westvaco "SA"
Flocculent dosage:
Clarifier configuration:
Depth:
Hydraulic detention time:
Hydraulic loading:
Weir loading:
Sludge underflow:
Percent solids in sludge:
Carbon regeneration technique:
Carbon makeup rate: 105-210 mg/L/d

REMOVAL DATA

Sampling period: Two weeks

Pollutant/parameter	Concentration, mg/L Influent	Effluent	Percent removal	Carbon dosage, mg/L
Conventional pollutants:				
BOD_5	1,170	24	98	3,000
	1,170	24	98	6,000
COD	2,115	390	82	3,000
	2,115	447	79	6,000
TOC	624	113	82	3,000
	624	105	83	6,000

Note: Blanks indicate information was not specified.

TREATMENT TECHNOLOGY: Powdered Activated Carbon Adsorption (With Activated Sludge)

Data source: Effluent Guidelines
Point source category: Textile mills
Subcategory: Wool finishing
Plant: O
References: A6, p. VII-102

Data source status:
- Engineering estimate ___
- Bench scale x
- Pilot scale ___
- Full scale ___

Use in system: Secondary
Pretreatment of influent:

DESIGN OR OPERATING PARAMETERS

Carbon dosage: 1,000-5,000 mg/L in aeration basin
Carbon type/characteristics: Westvaco "SC"
Flocculent dosage:
Clarifier configuration:
Depth:
Hydraulic detention time:
Hydraulic loading:
Weir loading:
Sludge underflow:
Percent solids in sludge:
Carbon regeneration technique:
Carbon makeup rate: 25-125 mg/L/d

REMOVAL DATA

Sampling period: Two weeks

Pollutant/parameter	Concentration, mg/L Influent	Effluent	Percent removal	Carbon dosage, mg/L
Conventional poolutants:				
BOD_5	247	8	97	1,000
	247	6.5	97	5,000
COD	1,100	63	94	1,000
	1,100	33	97	5,000
TOC	344	23	93	1,000
	344	11	97	5,000

Note: Blanks indicate information was not specified.

TREATMENT TECHNOLOGY: Powdered Activated Carbon Adsorption (With Activated Sludge)

Data source: Effluent Guidelines
Point source category: Textile mills
Subcategory: Woven fabric finishing
Plant: P
References: A6, p. VII-100

Data source status:
- Engineering estimate ___
- Bench scale x
- Pilot scale ___
- Full scale ___

Use in system: Secondary
Pretreatment of influent:

DESIGN OR OPERATING PARAMETERS

Carbon dosage: 1,000-5,000 mg/L in aeration basin
Carbon type/characteristics: Westvaco "SC"
Flocculent dosage:
Clarifier configuration:
Depth:
Hydraulic detention time:
Hydraulic loading:
Weir loading:
Sludge underflow:
Percent solids in sludge:
Carbon regeneration technique:
Carbon makeup rate: 122-608 mg/L/d

REMOVAL DATA

Sampling period: Two weeks

Pollutant/parameter	Concentration, mg/L		Percent	Carbon
	Influent	Effluent	removal	dosage, mg/L
Conventional pollutants:				
BOD_5	400	8	98	1,000
	400	8.5	98	5,000
COD	572	96	83	1,000
	572	82	86	5,000
TOC	243	42	83	1,000
	243	34	86	5,000

Note: Blanks indicate information was not specified.

TREATMENT TECHNOLOGY: Powdered Activated Carbon Adsorption (With Activated Sludge)

Data source: Effluent Guidelines
Point source category: Textile mills
Subcategory: Woven fabric finishing
Plant: Y
References: A6, p. VII-103

Data source status:
- Engineering estimate ___
- Bench scale x
- Pilot scale ___
- Full scale ___

Use in system: Secondary
Pretreatment of influent:

DESIGN OR OPERATING PARAMETERS

Carbon dosage: 2,000-5,000 mg/L in aeration basin
Carbon type/characteristics: ICI-Hydrodarco
Flocculent dosage:
Clarifier configuration:
Depth:
Hydraulic detention time:
Hydraulic loading:
Weir loading:
Sludge underflow:
Percent solids in sludge:
Carbon regeneration technique:
Carbon makeup rate: 210-526 mg/L/d

REMOVAL DATA

Sampling period: Two weeks

Pollutant/parameter	Concentration, mg/L Influent	Concentration, mg/L Effluent	Percent removal	Carbon dosage, mg/L
Conventional pollutants:				
BOD_5	114	5	96	2,000
	114	4	96	5,000
COD	301	60	80	2,000
	301	37	88	5,000
TOC	91	12	87	2,000
	91	9	90	5,000

Note: Blanks indicate information was not specified.

TREATMENT TECHNOLOGY: Powdered Activated Carbon Adsorption (With Activated Sludge)

Data source: Effluent Guidelines
Point source category: Textile mills
Subcategory: Wool scouring
Plant: A
References: A6, p. VII-101

Data source status:
Engineering estimate ___
Bench scale x
Pilot scale ___
Full scale ___

Use in system: Secondary
Pretreatment of influent:

DESIGN OR OPERATING PARAMETERS

Carbon dosage: 2,000 - 10,000 mg/L in aeration basin
Carbon type/characteristics: Westvaco "SC"
Flocculent dosage:
Clarifier configuration:
Depth:
Hydraulic detention time:
Hydraulic loading:
Weir loading:

Sludge underflow:
Percent solids in sludge:
Carbon regeneration technique:
Carbon makeup rate: 139-694 mg/L/d

REMOVAL DATA

Sampling period: Two weeks

Pollutant/parameter	Concentration, mg/L Influent	Concentration, mg/L Effluent	Percent removal	Carbon dosage, mg/L
Conventional pollutants:				
BOD_5	2,580	54	98	2,000
	2,580	51	98	10,000
COD	5,540	563	90	2,000
	5,540	457	92	10,000
TOC	1,780	387	78	2,000
	1.780	336	81	10,000

Note: Blanks indicate information was not specified.

6.3 CHEMICAL OXIDATION [1]

6.3.1 Function

The chemical oxidation process involves the chemical rather than the biological oxidation of dissolved organics in wastewater.

6.3.2 Description

The processes discussed here are based on chemical oxidation as differentiated from thermal, electrolytic, and biological oxidation. Ozonation, a commonly used chemical method of oxidation for waste treatment, and another oxidation process, chlorination, are discussed elsewhere in this volume. The oxidation reactions discussed here should be distinguished from the higher temperature, and typically pressurized, wet oxidation processes, such as the Zimpro process, which are also discussed in a separate section of this volume.

Oxidation-reduction or "redox" reactions are those in which the oxidation state of at least one reactant is raised while that of another is lowered. In reaction (1) in alkaline solution:

$$2MnO_4^- + CN^- + 2OH^- \rightleftarrows 2MnO_4^{2-} + CNO^- + H_2O \quad (1)$$

the oxidation state of the cyanide ion is raised from -1 to +1 (the cyanide is oxidized as it combines with an atom of oxygen to form cyanate); the oxidation state of the permanganate decreases from -1 to -2 (permanganate is reduced to manganate). This change in oxidation state implies that an electron was transferred from the cyanide ion to the permanganate. The increase in the positive valence (or decrease in the negative valence) with oxidation takes place simultaneously with reduction in chemically equivalent ratios.

There are many oxidizing agents; however, only a few are convenient to use. Those more commonly used in waste treatment are shown in the following table.

Some oxidations proceed readily to CO_2. In other cases, the oxidation is not carried as far perhaps because of the dosage of the oxidant, the pH of the reaction medium, the oxidation potential of the oxidant, or the formation of stable intermediates. The primary function performed by oxidation in the treatment of hazardous wastes is essentially detoxification. For instance, oxidants are used to convert cyanide to the less toxic cyanate or completely to carbon dioxide and nitrogen. The oxidant itself is reduced. For example, in the potassium permanganate treatment of phenolics, the permanganate is reduced to manganese dioxide. A secondary function is to assure complete precipitation, as in the oxidation of Fe^{++} to Fe^{+++} and similar reactions.

WASTE TREATMENT APPLICATIONS OF OXIDATION IDENTIFIED

Oxidant	Waste
Ozone[a]	--
Air (atmospheric oxygen)	Sulfites ($SO_3^=$)
	Sulfides ($S^=$)
	Ferrous iron (Fe^{++}) (very slow)
Chlorine gas	Sulfide
	Mercaptans
Chlorine gas and caustic[b]	Cyanide (CN^-)
Chlorine dioxide	Cyanide
	Diquat, Paraquat — pesticides
Sodium hypochlorite	Cyanide
	Lead
Calcium hypochlorite	Cyanide
Potassium permanganate	Cyanide (organic odors)
	Lead
	Phenol
	Diquat, Paraquat — pesticides
Oxidants that are present in trace quantities only	Organic sulfur compounds
	Rotenone
	Formaldehyde
Permanganate	Manganese
Hydrogen peroxide	Phenol
	Cyanide
	Sulfur compounds
	Lead
Nitrous acid	Benzidene

[a]Discussed in another section of this volume.

[b]Alkaline chlorination.

The first step of the chemical oxidation process is the adjustment of the pH of the solution to be treated. In the use of chlorine gas to treat cyanides, for instance, this adjustment is required because acid pH has the effect of producing hydrogen cyanide and/or cyanogen chloride, both of which are poisonous gases. The pH adjustment is done with an appropriate Alkali (e.g., sodium hydroxide). This is followed by the addition of the oxidizing agent. Mixing is provided to contact the oxidizing agent and the waste. Because some heat is often liberated, more concentrated solutions will require cooling. The agent can be in the form of a gas (chlorine gas), a solution (hydrogen peroxide) or perhaps a solid if there is adequate mixing. Reaction times vary but are in the order of seconds and minutes for most of the commercial-scale installations. Additional time is allowed to ensure complete mixing and oxidation. At this point, additional oxidation may be desired and, as with cyanide destruction, often requires the readjustment of the pH followed by the addition of more oxidant. Once reacted, this final oxidized solution is then generally subjected to some form of treatment to settle or precipitate any insoluble oxidized material, metals, and other residues. A treatment for the removal of what remains of the oxidizing agent (both reacted and unreacted) may be required. A product of potassium permanganate oxidation is manganese dioxide (MnO_2), which is insoluble and can be settled or filtered for removal.

The characteristics of a number of common oxidizing agents are described in the following paragraphs.

- Potassium Permanganate

Potassium permanganate ($KMnO_4$) has been used for destruction of organic residues in wastewater and in potable water. Its usual reduced form, manganese dioxide (MnO_2), can be removed by filtration. $KMnO_4$ reacts with aldehydes, mercaptans, phenols, and unsaturated acids. It is considered a relatively powerful oxidizing agent.

- Hydrogen Peroxide

Hydrogen peroxide (H_2O_2) has been used for the separation of metal ions by selective oxidation. In this way it helps remove iron from combined streams by oxidizing the ferrous ion to ferric, which is then precipitated by the addition of the appropriate base. In dilute solution (<30%), the decomposition of hydrogen peroxide is accelerated by the presence of metal ion contaminants. At higher concentrations of hydrogen peroxide, these contaminants can catalyze its violent decomposition. Hydrogen peroxides should be added slowly to the solution with good mixing. This caution relates to other oxidants as well. If the follow-on treatment involves distillation or crystallization, the absence of all unspent peroxides must be confirmed since these techniques tend to concentrate the unused reagent. Hydrogen peroxide has also been used as an "anti-chlor" to remove residual chlorine following chlorination treatment.

- Chromic Acid

Chromium trioxide (CrO_3) commercially called chromic acid, is used as an oxidizing agent in the preparation of organic compounds. It is often regenerated afterward by electrolytic oxidation. In the oxidation of organic compounds, chromic acid in a solution of sulfuric acid is reduced and forms chromium sulfate [$Cr_2(SO_4)_3$].

6.3.3 Technology Status

Technology for large-scale application of chemical oxidation is well developed. Application to industrial wastes is well developed for cyanides and for other hazardous species in dilute waste streams (phenols, organic sulfur compounds, etc.).

6.3.4 Applications

The following are selected examples of the application of chemical oxidation to hazardous waste management problems.

- Oxidation of Cyanide Effluents

Numerous plating and metal finishing plants use chemical oxidation methods to treat their cyanide wastes. Cyanides and heavy metals are often present together in plating industry wastes. Their concentration and their value influence the selection of the treatment process. If the cyanide and heavy metal are not economically recoverable by a method such as ion exchange, the cyanide radical is converted either to the less toxic cyanate or to CO_2 and N_2 by oxidation, while the heavy metal is precipitated and removed as a sludge.

Chemical oxidation is applicable to both concentrated and dilute waste streams, but the competing processes are more numerous for the concentrated streams. These methods include thermal and catalytic decomposition of the cyanide and decomposition using acidification.

In treating cyanide waste by oxidation, hypochlorite or caustic plus chlorine (alkaline chlorination) may be used to oxidize the cyanide to cyanate or to oxidize it completely to nitrogen and carbon dioxide. It is a fast reaction that is adaptable to either batch or continuous operation. Smaller volumes would be treated in a batch system for simplicity and safety. The destruction of cyanide is believed to proceed according to the following equations:

$$NaCN + Cl_2 \rightarrow CNCl + NaCl \quad (2)$$

$$CNCl + 2NaOH \rightarrow NaCNO + NaCl + H_2O \quad (3)$$

$$2NaCNO + 4NaOH + 3Cl_2 \rightarrow 6NaCl + 2CO_2 + N_2 + 2H_2O \quad (4)$$

The rate of the second reaction is dependent upon pH and proceeds rapidly at a pH of 11 or higher. About 8 parts chlorine and 7.3 parts sodium hydroxide are required per part of cyanide. Neutralization is required after treatment because the waste is generally alkaline. Calcium, magnesium, and sodium hypochlorite are frequently used in place of gaseous chlorine even though the chlorine is more rapid and costs about half as much as the hypochlorites. This is because they are easier and safer to use and do not require the addition of supplementary alkali. Calcium hypochlorite will give more sludge than the sodium hypochlorite if certain anions such as sulfate are present.

There are problems associated with alkaline chlorination of cyanide if soluble iron or certain other transition metal ions are present. The iron forms very stable ferrocyanide complexes which prevent the cyanide from being oxidized. Potassium permanganate and hydrogen peroxide are also used to oxidize cyanide wastes. Potassium permanganate ($KMnO_4$) is not used widely for the destruction of cyanide. One advantage of the use of permanganate is that there is no need to monitor pH. Once the pH adjustment has been made there is continuous formation of the hydroxide ion.

$$KMnO_4 + 3CN^- + H_2O \rightarrow 3CNO^- + 2MnO_2 + 2OH^- \qquad (5)$$

to constantly keep the reaction medium on the alkaline side. This is fortunate because otherwise there is the danger that if the pH drops to between 6 and 9, hydrogen cyanide and/or cyanogen, both of which are poisonous gases, may be formed. With other oxidative methods the reaction medium is kept alkaline by the addition of alkali. The use of permanganate oxidizes the waste cyanide only to the cyanate. Simple acid hydrolysis can be used to further treat the cyanate, converting it to CO_2 and N_2.

- Oxidation of Phenol

Oxidation reactions involving phenol are often complex, since the reaction products depend upon the substituents. The reactions are believed to involve as a first step in the removal of the hydroxyl hydrogen to yield a phenoxy radical. The eventual reaction products can include quinone, which is considered more toxic than phenol. In one commercial reaction, for instance, the oxidation of phenol with chromic acid is designed to yield quinone.

Chemical oxidation of phenols has found application to date only on dilute waste streams. Potassium permanganate, one of the oxidants used, is reduced to manganese dioxide (MnO_2), which is a filterable solid. In one application, the product MnO_2 has been found to act also as a coagulant aid to settle other material from the waste stream. Because of the high potential of formation of chlorophenols, chlorine gas is not frequently used.

When phenol is present only in trace quantities, the economics appear favorable for chemical oxidation. It has been used in the treatment of potable water. Removal of 1 ppm phenol in this application can be accomplished by the addition of 6 to 7 ppm potassium permanganate.

• Oxidation of Other Organics

Chemical oxidizing agents have been used for the control of organic residues in wastewaters and in potable water treatment. Among the organics for which oxidative treatment has been reported are aldehydes, mercaptans, phenols, benzidine, and unsaturated acids. For these applications sodium hypochlorite, calcium hypochlorite, potassium permanganate and hydrogen peroxide have been reported as oxidants. In one application nitrous acid was used.

Benzidine, an organic used in the manufacture of dyes, is considered a carcinogen. Its concentration is generally reduced to ppb in wastewaters prior to discharge for this reason. Nitrous acid oxidation is used to achieve this effluent quality. While biodegradation, carbon adsorption, radiation, and oxidation by ozone and by other chemicals such as hydrogen peroxide has been suggested, only the oxidation (commonly called diazotization) using nitrous acid has been used on a full scale basis. The reaction of benzidine with an excess amount of nitrous acid in a strong acid reaction medium yields the quinone form, 4,4'-dihydroxybiphenyl and/or similar products. The reaction products cannot revert to benzidine. The quinone product is also toxic but considered less so than the reactant, benzidine. Since the effluent stream is very dilute, no secondary treatment is required.

• Oxidation of Sulfur Compounds

Much of the work on oxidative treatment of sulfur compounds is centered on the problem of odor removal. Scrubbers using oxidizing solutions of potassium permanganate, for example, have been used to remove organic sulfur compounds from air. Thiophene, one of these compounds, in which the molecule is unsaturated, is susceptible to complete degradation.

Chlorine and calcium hypochlorite have been used to prevent accumulation of soluble sulfides in sewer lines. If an excess of chlorine is added to a wastewater containing sulfide, the sulfide will be oxidized to sulfate.

$$HS^- + 4Cl_2 + 4H_2O \rightarrow SO_4^= + 9H^+ + 8Cl^- \quad (6)$$

On a pure waste stream containing only small concentrations of sulfide, the chlorine requirement would be nearly 9 parts (by weight) for each part of sulfide. In streams where there are other oxidizable constituents, this requirement may actually be in the order of 15 to 20 parts.

Hydrogen peroxide has also been used for this application of sulfide oxidation. In a wastewater which contained about 6 mg/L total sulfide, the addition of 30 mg/L hydrogen peroxide (H_2O_2)

reduced the concentration of sulfide to less than 1 mg/L. The average retention time was about two hours.

Although later developed into a catalyzed, two-stage, higher temperature system, the initial concept of the Sulfox® system for control of sulfur emissions was to convert hydrogen sulfide to elemental sulfur by oxidation with atmospheric oxygen. Knowing that there was a strong tendency for sulfide reactions to go to the thiosulfate and sulfite stages, attempts were made to find the kind of solutions that could regulate the extent of the oxidation. Caustic solutions were not favorable. Ammoniacal solutions gave improved selectivity. The availability of byproduct ammonia at refineries that had sulfur emission problems made the use of ammoniacal solutions appear promising. Later improvements to the system involved the use of a cobalt catalyst.

- Oxidation of Pesticides

Because of the resistance of pesticides to biodegradation, chemical oxidative methods have been investigated to remove pesticide residues from water. Work has been completed to study the use of chemical oxidation for the removal of residual diquat and paraquat from water.

With potassium permanganate oxidation, manganese dioxide was precipitated as expected. The application of $KMnO_4$ at a molar concentration 25 times that of the two pesticides causes fairly complete oxidation to oxalate, ammonia, and water. The reaction is said to go through several intermediate reactions and the reaction rates are pH dependent, being faster above pH 8. In an alkaline medium

$$3(C_{12}H_{12}N_2)^{2+} + 40MnO_4^- + 20H^- \rightleftarrows 40MnO_2 + 18C_2O_4^= + 6NH_3 + 10H_2O \qquad (7)$$
(Diquat)

$$(C_{12}H_{14}N_2)^{2+} + 14MnO_4^- \rightleftarrows 14MnO_2 + 6C_2O_4^= + 2NH_3 + 4H_2O \qquad (8)$$
(Paraquat)

When using chlorine dioxide as the oxidizing agent on these substances in concentrations of 15 and 30 mg/L, the reactions were complete in less than one minute. These rates were observed at pH values above 8. At pH 9.04, for example, 15 mg/L of Diquat treated with 6.75 mg/L of chlorine dioxide had a residual Diquat of 0.00 and a residual chlorine dioxide of 2.61.

- Oxidation of Lead

Although for a particular application other methods were considered more practicable, the use of chemical oxidative techniques for the removal of trace quantities of soluble lead from an effluent was investigated on a laboratory scale. In this particular application, the insoluble lead was already removable by other techniques to acceptable levels. However, in order to meet effluent regulations, more of the soluble lead had to be removed. Potassium permanganate, hydrogen peroxide, and sodium hypochlorite were tested and found to convert portions of the soluble lead as described below:

Oxidizing agent	Initial soluble lead concentration, ppm	Final soluble lead concentration, ppm
Potassium permanganate	14	4 to 7
Hydrogen peroxide	14	9
Sodium hypochlorite	14	9 to 10

6.3.5 Limitations

Oxidation has limited application to slurries, tars, and sludges. Because other components of the sludge, as well as the material to be oxidized, may be attacked indiscriminately by oxidizing agents, careful control of the treatment via multistaging of the reaction, careful control of pH, etc., are required.

6.3.6 Typical Equipment

Only very simple equipment is required for chemical oxidation. This includes storage vessels for the oxidizing agents and perhaps for the wastes, metering equipment for both streams, and contact vessels with agitators to provide suitable contact of oxidant and waste. Some instrumentation is required to determine the concentration and pH of the water and the degree of completion of the oxidation reaction. The oxidation process may be monitored by an oxidation-reduction potential (ORP) electrode. This electrode is generally a piece of noble metal (often platinum) that is exposed to the reaction medium, and which produces an EMF output that is empirically related to the reaction condition by revealing the ratio of the oxidized to the reduced constituents.

6.3.7 Residuals Generated/Environmental Impact

One disadvantage of chemical oxidation for waste treatment is that it introduces new metal ions into the effluent. If the level of these new contaminants is high enough to exceed effluent regulations, additional treatment steps will be required. Often these are steps such as filtration or sedimentation. Potassium permanganate used to treat wastes will be reduced to MnO_2 in the process. This can be reduced by filtration to levels less than 0.05 mg/L in the final effluent. On the other hand, oxidation with hydrogen peroxide adds no harmful species to the final effluent (except perhaps excess peroxide) since its product is water.

Whether the products of incomplete oxidation are an environmental hazard depends upon the specific situation. Cyanate, the product of potassium permanganate oxidation of cyanide, is not completely oxidized. Treatment with another oxidant, or acid hydrolysis after permanganate oxidation, can oxidize the cyanide completely to CO_2 and N_2. Cyanate, however, is at least a thousand times less toxic than free cyanide. The conversion of benzidine to the products of diazotization is another case in which the treated waste is less hazardous than the first, but still is considered a problem.

Often the extent to which excess chlorine must be added for waste oxidation is such that the residual chlorine in the effluent becomes a problem. Careful in-process control or recycling of the oxidizing solution may be necessary to reduce this level to meet regulation limits. Also, hydrogen peroxide has been used as a reducing agent in some applications as an "anti-chlor" to destroy the chlorine remaining in the stream after purification.

With the exception of escape of chlorine, which is a potential hazard wherever chlorine is used, the only other air emission problem is the possible production of HCN from the destruction of cyanide wastes when the reaction medium is allowed to become acidic.

From most chemical oxidations, there will be a residue for disposal unless the concentration of the waste constituent is so low that the oxidant waste products (if any) and the oxidized (and de-toxified) waste can be carried away with the effluent. Most of the residue develops from the use of caustic or lime slurry with chlorine gas in alkaline chlorination. Smaller amounts of residue result from oxidations using hypochlorites. The only waste that appears particularly troublesome is the sludge, which can develop in the oxidation treatment of cyanides when iron and certain other transition metal ions are present. In this form (ferrocyanide, for example), the cyanide cannot be easily reached for further oxidation.

6.3.8. Reliability

The process has proven to be highly reliable for demonstrated applications.

6.3.9 Flow Diagram

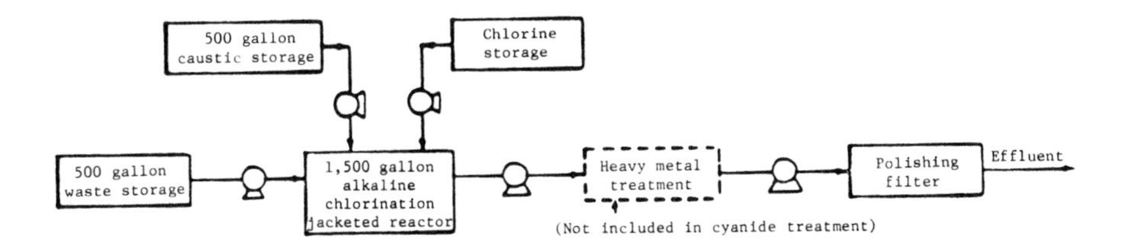

Treatment Batch
Waste Concentrated Cyanide Waste
7,000 ppm Copper Cyanide
1,000 ppm Sodium Cyanide
Waste Processing Capacity: 1,000 gal/d
Operating Period: 240 d/yr
8 hr/d

Utilities Summary
1,500 gal/d Cooling Water
Raw Materials:
95 lb/d NaOH
227 lb/d Chlorine

Example Process Flowsheet - Oxidation

6.3.10 Performance

Performance data presented on the following data sheets includes information on the listed industries and/or wastestreams:

Industries

Inorganic chemicals production
- Hydrogen cyanide
- Sodium bisulfite

Ore mining and dressing
- Ferroalloy mining/milling
- Lead/zinc milling

Organic and inorganic wastes

6.3.11 References

1. Physical, Chemical, and Biological Treatment Techniques for Industrial Wastes, PB 275 287, U.S. Environmental Protection Agency, Washington, D.C., November 1976. pp. 35-1 through 35-19.

CONTROL TECHNOLOGY SUMMARY FOR CHEMICAL OXIDATION (CHLORINATION)

Pollutant	Number of data points	Effluent concentration				Removal efficiency, %			
		Minimum	Maximum	Median	Mean	Minimum	Maximum	Median	Mean
Conventional pollutants, mg/L:									
COD	7	441	978	565	632	7	39	28	26
TSS	2	33.3	159		96	0[a]	97		48
Toxic pollutants, µg/L:									
Copper	1				320				14
Cyanide	17	<2	130	30	38	58	>99	84	84
Lead	1				2,500				0[a]
Other pollutants, mg/L:									
NH_3-N	1				120				36

[a]Actual data indicate negative removal.

TREATMENT TECHNOLOGY: Chemical Oxidation (Chlorination)

Data source: Effluent Guidelines
Point source category: Inorganic chemicals
Subcategory: Hydrogen cyanide
Plant: 765
References: A29, pp. 427-428

Data source status:
- Engineering estimate ___
- Bench scale ___
- Pilot scale ___
- Full scale x

Use in system: Primary
Pretreatment of influent: pH adjustment

DESIGN OR OPERATING PARAMETERS

Unit configuration: Two ponds in parallel where sodium hypochlorite is added, then caustic and chlorine are added in another treatment pond
Wastewater flow:
Chemical dosage(s):
Contact time:
pH:

REMOVAL DATA

Sampling period: 72-hr composite

Pollutant/parameter	Concentration[a] Influent	Effluent	Percent removal
Conventional pollutants, mg/L:			
TSS	979	33.3	97
Toxic pollutants, µg/L:			
Cyanide	6,800	<2	>99
Other pollutants, mg/L:			
NH_3-N	194	124	36

[a]Concentration is calculated from the wastewater flow in m^3/kkg of HCN and the pollutant load in kg/kkg. Pollutant load was calculated by approtioning the mass emitted between the two waste streams on the basis of measured flows. This is a very approximate process.

Note: Blanks indicate information was not specified.

TREATMENT TECHNOLOGY: Chemical Oxidation (Chlorination)

Data source: Effluent Guidelines
Point source category: Inorganic chemical
Subcategory: Sodium bisulfite
Plant: 282
References: A29, pp. 555-556

Data source status:
Engineering estimate ___
Bench scale ___
Pilot scale ___
Full scale x

Use in system: Primary
Pretreatment of influent:

DESIGN OR OPERATING PARAMETERS

Unit configuration: Single reactor tank
Wastewater flow:
Chemical dosage(s): NaOCl (unspecified dosage)
Contact time:
pH:

REMOVAL DATA

Sampling period:

Pollutant/parameter	Concentration[a]		Percent
	Influent	Effluent	removal
Conventional pollutants, mg/L:			
COD	1,510	978	35
TSS	88.8	159	0[b]
Toxic pollutants, μg/L:			
Copper	370	320	14
Lead	2,500	2,500	0[b]

[a]Concentration is calculated from pollutant flow in m^3/kkg and pollutant load in kg/kkg.

[b]Actual data indicate negative removal.

Note: Blanks indicate information was not specified.

TREATMENT TECHNOLOGY: Chemical Oxidation (Chlorination)

Data source: Effluent Guidelines
Point source category: Ore mining and dressing
Subcategory: Ferroalloy mine/mill
Plant: 6102
References: A2, p. VI-26

Data source status:
Engineering estimate ___
Bench scale ___
Pilot scale x
Full scale ___

Use in system: Tertiary
Pretreatment of influent:

DESIGN OR OPERATING PARAMETERS

Unit configuration:
Wastewater flow:
Chemical dosage: 10-20 mg/L NaOCl
Contact time: 30-90 min
pH: 8.8-11.0

REMOVAL DATA

Sampling period:

Pollutant/parameter	Concentration, µg/L Influent	Concentration, µg/L Effluent	Percent removal	NaOCl dosage, mg/L	Contact time, min	pH
Toxic pollutants:						
Cyanide	190	80	58	20	30	8.8
	190	50	74	20	60	8.8
	190	70	63	20	90	8.8
	190	40	79	10	30	10.6
	190	30	84	10	60	10.6
	190	40	79	10	90	10.6
	190	30	84	20	30	10.6
	190	20	89	20	60	10.6
	190	20	89	20	90	10.6
	190	30	84	10	30	11.0
	190	30	84	10	60	11.0
	190	30	84	10	90	11.0
	190	10	95	20	30	11.0
	190	20	89	20	60	11.0
	190	20	89	20	90	11.0

Note: Blanks indicate information was not specified.

TREATMENT TECHNOLOGY: Chemical Oxidation (Chlorination)

Data source: Effluent Guidelines
Point source category: Ore mining and dressing
Subcategory: Lead/zinc mill
Plant: 3144
References: A2, p. VI-28

Data source status:
Engineering estimate ___
Bench scale ___
Pilot scale ___
Full scale x

Use in system: Tertiary
Pretreatment of influent:

DESIGN OR OPERATING PARAMETERS

Unit configuration: Three FRP reactor tanks in series plus chlorination and lime slaker
Wastewater flow:
Chemical dosage: 1,200-1,500 lb/d Cl_2 Lime to pH of 11-12

REMOVAL DATA

Sampling period:

Pollutant/parameter	Concentration, µg/L Influent	Effluent	Percent removal
Toxic pollutants:			
Cyanide	68,300	130	>99

Note: Blanks indicate information was not specified.

TREATMENT TECHNOLOGY: Chemical Oxidation (Chlorination)

Data source: Government report
Point source category:[a]
Subcategory:
Plant: Reichhold Chemical, Inc.
References: B4, p. 55

Data source status:
Engineering estimate ___
Bench scale x
Pilot scale ___
Full scale ___

Use in system: Tertiary
Pretreatment of influent:

[a]Organic and inorganic wastes.

DESIGN OR OPERATING PARAMETERS (Also see removal data)

Contact time: 15 min
Chemical dosage (initial): 5.25% aqueous solution of NaOCl

REMOVAL DATA

Sampling period:

Pollutant/parameter	Concentration, mg/L Influent	Concentration, mg/L Effluent	Percent removal	NaOCl dosage, weight %
Conventional pollutants:				
COD	777	717	7	0.5
COD	777	706	9	1.0
COD[a]	753	565	25	2
COD[a]	753	505	28	3
COD[b]	822	510	38	4
COD	724	441	39	5

[a]Average of 9 samples.
[b]Average of 3 samples.

Note: Blanks indicate information was not specified.

6.4 AIR STRIPPING [1,2]

6.4.1 Function

Air stripping of wastewater removes volatile organics and ammonia nitrogen from the wastewater and discharges it to the air.

6.4.2 Description

Ammonia is quite soluble in water, but this solubility is temperature dependent. The relationship between temperature and the solubility of ammonia for dilute ammonia solution is expressed by Henry's Law:

$$y = Mx$$

where y = mole fraction NH_3 in the vapor
x = mole fraction NH_3 in the liquid
M = Henry's constant

Henry's constant is a function of temperature. By raising the temperature of the wastewater the vapor pressure of the ammonia is increased, and ammonia removal efficiency increased.

Another factor in ammonia removal efficiency is the pH of the wastewater. A portion of the ammonia dissolved in the water reacts with the water to give the following equilibrium:

$$NH_3 + H_2O \rightleftharpoons NH_4^+ + OH^- \quad (1)$$

By increasing the pH (concentration of OH^-), the equilibrium is shifted to the left, reducing the concentration of NH_4^+ and increasing the concentration of free dissolved ammonia.

In air stripping of ammonia from dilute wastewater, the air temperature limits the effectiveness of heating the wastewater. Ammonia removal efficiency is enhanced instead by increasing the pH, usually by the addition of lime. The ammonia-containing wastewater and the lime slurry are fed to a rapid mix tank. Following the rapid mix tank are flocculators and a settling basin, where calcium phosphate precipitates and recirculated calcium carbonate settle out. The clarified, lime-treated, wastewater is pumped to the top of packed towers. In each tower, fans draw air up through the tower countercurrent to the falling wastewater. The "packing" in the tower is actually a series of bundles of pipe with the pipe sections spaced 2 to 3 inches on center. The pipe sections are horizontal, and the direction of each row alternates. After the wastewater has been air stripped of ammonia, it flows into the recarbonation basin where compressed carbon dioxide rich gas from the lime reclacining furnace is bubbled through it to precipitate calcium carbonate. Some of the calcium carbonate sludge is returned to the rapid mix tank to enhance flocculation while the remainder of the calcium carbonate sludge and the phosphate sludge from the settling basins are sent to centrifuges. The sludges can be fractionally centrifuged to yield two dewatered sludges, one rich in calcium carbonate and one containing phosphate.

6.4.3 Technology Status

The future application of air stripping of volatiles from wastewater will be limited to those volatiles that will not cause an air emission problem. Air stripping of ammonia from treated wastewater dilute solutions of ammonia (with no other volatiles) is a good application. It is unlikely that many applications other than this one will be found for air stripping of wastewater.

6.4.4 Applications

Several studies have been reported in which ammonia was removed from petroleum refinery wastewater by stripping with air. The concentrations of ammonia-nitrogen in the untreated wastewater averaged slightly more than 100 mg/L. When 300 ft^3 of air were applied per gallon of wastewater, the ammonia removal was found to be 85% at a pH of 10.5, and 34% at a pH of 9.4. In another study, in which the wastewater was passed through a closely packed aeration tower with 480 ft^3 of air supplied per gallon, ammonia-nitrogen removal by air stripping was found to be very effective (more than 95% removal) at any pH above 9.0. When the pH fell below 9.0, the ammonia-nitrogen removal decreased sharply. The removal fell to 91% at a pH of 8.9, and to 58% at a pH of 8.8.

At the low concentration of ammonia cited in these studies ($\sim$100 ppm), air stripping would indeed be a practical means for NH_3 removal. For the high concentrations of ammonia typically present in refinery "sour water" (2,000 to 10,000 ppm), air stripping could result in serious air emission problems.

6.4.5 Limitations

Air stripping has one major industrial application: the stripping of ammonia from wastewater. The application of air stripping to the removal of other gases or volatile components from dilute aqueous streams would depend on the environmental impact of the air emissions that resulted. If sufficiently low concentrations are involved, the gaseous compounds can be emitted directly to the air. Otherwise, air pollution control devices may be needed - making the economics less favorable.

6.4.6 Residuals Generated/Environmental Impact

When the concentration of ammonia in the wastewater is about 23 ppm and the air-to-water ratio is 500 ft^3/gal, the concentration of ammonia in the saturated air leaving the tower is about 6 mg/m³. This is well below the odor threshold concentration of 35 mg/m³. There are no U.S. standards for ammonia emissions, but Czechoslovakia and the U.S.S.R. have established limitations of 100 and 200 mg/m³, respectively.

Calculations for the ammonia washout in a rainfall rate of 3 mm/hr (0.12 in./hr) have been made. The concentrations of ammonia in the rainfall would approach natural background levels within 16,000 feet of the tower. Of course, the ammonia discharge during dry periods diffuses into the atmosphere quickly so that the background concentration and resulting washout rate of ammonia at greater distances from the tower are not affected during a subsequent storm. The ultimate fate of the ammonia that is washed out by rainfall within the 16,000-foot downwind distance depends on the nature of the surface upon which it falls. Most soils will retain the ammonia. That portion which lands on paved areas or directly on a stream surface will appear in the runoff from that area. Even though a portion of the ammonia washed out by precipitation will find its way into surface runoff, the net discharge of ammonia to the aquatic environment in the vicinity of the plant would be very substantially reduced.

The treated wastew ter should be low enough in residual ammonia (<5 ppm) to allow safe discharge to a receiving body of water.

About 25 tons per day of dewatered calcium phosphate, magnesium carbonate, and calcium carbonate sludge must be disposed of by landfill for a 15 M gal/d plant. This sludge disposal will require a significant amount of land, but should not pose any environmental hazard. In lime applications for pH control and phosphate removal, there will be sludge disposal requirements.

6.4.7 Reliability

Reliability has been a problem for installations where cold weather operation is required; freezing and scaling of $CaCO_3$ have occurred.

6.4.8 Chemicals Required

Lime or caustic soda is needed to raise the pH of the wastewater to the range of 10.8 to 11.5. For wastewater with high calcium content, an inhibiting polymer may be added to ease the scaling problem. Effluent from the stripping may need pH readjustment to neutral condition with an acid (H_2SO_4 at 1.75 parts for one part of lime added) or recarbonation followed by clarification.

6.4.9 Design Criteria

Wastewater loading: 1 to 2 gpm/ft²
Stripping air flow rate: 300 to 500 ft³/gal

Packing depth: 20 to 25 ft
pH of wastewater: 10.8 to 11.5
Air pressure drop: 0.015 in. to 0.019 in. of water/ft
Packing material: Plastic or wood
Packing spacing: Approximately 2 in. horizontal and vertical
Must provide: Uniform water distribution, and scale removal and cleanup
Land requirement: Small

6.4.10 Flow Diagram

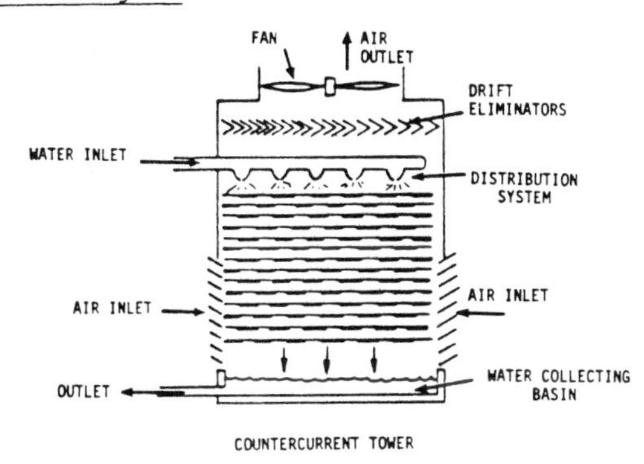

COUNTERCURRENT TOWER

6.4.11 Performance

Subsequent data sheets provide performance data from studies on the following industries and/or wastestreams:

Industries	Wastestreams
Inorganic chemicals production	
Hydrogen cyanide	

6.4.12 References

1. Physical, Chemical, and Biological Treatment Techniques for Industrial Wastes, PB 275 287, U.S. Environmental Protection Agency, Washington, D.C., November 1976. pp. 41-1 through 41-13.

2. Innovative and Alternative Technology Assessment Manual. EPA-430/9-78-009 (draft) U.S. Environmental Protection Agency, Cincinnati, Ohio, 1978. 252 pp.

TREATMENT TECHNOLOGY: Air Stripping

Data source: Effluent Guidelines
Point source category: Inorganic chemicals
Subcategory: Hydrogen cyanide
Plant: 782
References: A29, pp. 430-431

Data source status:
Engineering estimate ___
Bench scale ___
Pilot scale ___
Full scale x

Use in system: Primary
Pretreatment of influent:

DESIGN OR OPERATING PARAMETERS

Unit configuration: Ammonia stripper
Flow--wastewater: 1,140 m^3/day
Flow--air:
Temperature--wastewater:
Temperature--air:
Pressure drop:
Power requirement:
Packing material:
Packing depth:
Packing spacing:

REMOVAL DATA

Sampling period: Three 24-hr composite samples

Pollutant/parameter	Concentration Influent	Concentration Effluent	Percent removal
Conventional pollutants, mg/L:			
TSS	76	162	0[a]
Toxic pollutants, µg/L:			
Cyanide	170,000	51,000	60
Other pollutants, mg/L:			
NH_3-N	410	41	90

[a]Actual data indicate negative removal.

Note: Blanks indicate information was not specified.

6.5 NITRIFICATION [1,2]

6.5.1 Function

Nitrification is used for the biological oxidation of ammonia to nitrates and nitrites.

6.5.2 Description

Nitrification is achieved by the autotrophic microorganisms, *nitrosomonas*, which converts ammonia to nitrite, and *nitrobactor*, which oxidizes the nitrite to nitrate. The organisms require inorganic carbon for synthesis and oxygen consumption approximates the stoichiometric oxygen requirement for the oxidation of ammonia to nitrate (4.57 g O_2/g N). Sufficient alkalinity must be available to offset production of the nitrous and nitric acids. Optimal pH conditions are in the range of 7.2 to 8.5.

The growth rate of the nitrifiers is significantly lower than that of the heterotrophic bacteria used in the breakdown of carbonaceous organic matter. Thus in the presence of high carbonaceous organic material concentrations (BOD_5) the nitrifiers are unable to compete successfully with the heterotrophs and cannot accumulate as significant populations. Single stage BOD_5 removal and nitrification is practicable only when the organic loading is kept sufficiently low (generally in the range of 0.05 to 0.2 lb BOD_5/lb MLVSS/day). In effect, this implies maintaining a sludge retention time (SRT) high enough for development and maintenance of the nitrifying bacteria. This is typically greater than 5 days and is optimum at about 15 days.

Separate stage nitrification relies on pretreatment of the wastewater to remove the carbonaceous demand; this is typically carried out by conventional or high-rate activated sludge systems although fixed growth systems and physical/chemical systems are also appropriate. In the two-stage process, interstage clarification is provided to allow segragation of the two types of bacteria. In general, if the pretreatment effluent has a BOD_5/TKN ratio of less than 3.0, sufficient carbonaceous removal has occurred such that the subsequent nitrification step may be classified as a separate stage.

6.5.3 Common Modifications

Any low-rate modification of the activated sludge process such as extended aeration and the oxidation ditch can be used in a single stage application. In addition, the use of the powdered activated carbon has the potential to enhance ammonia removal, although its application is in a state of infancy.

Separate stage attached growth processes are used, analagous to trickling filters, packed or fluidized beds, or rotating biological contractor systems.

6.5.4 Technology Status

Overall, the single stage process is fully demonstrated. There are nearly 650 shallow oxidation ditch installations in the United States and Canada. In addition, pre-enginnered extended aeration plants are also widely used.

Separate stage nitrification has been well demonstrated throughout the United States and England in numerous pilot plant studies and several full-scale designs. Separate stage suspended growth systems outnumber separate stage attached growth systems in these applications by about 4 to 1.

6.5.5 Applications

The single stage process is applicable during warm weather if levels of 1 to 3 mg/L of ammonia nitrogen in effluent are permitted. Separate stage nitrification is often an advanced treatment added on to a secondary plant.

6.5.6 Limitations

Biological nitrification is very sensitive to temperature, resulting in poor reduction in colder months; heavy metals such as Cd, Cr, Cu, Ni, Pb and Zn, phenolic compounds, cyanide and halogenated compounds can inhibit nitrification reactions.

6.5.7 Reliability

Process reliability is good.

6.5.8 Residuals Generated/Environmental Impact

Process produces no primary sludge; in a single stage system secondary sludge is lesser in quantity and better stabilized than the high-rate and conventional activated sludge process, which minimizes the magnitude of the disposal problem considerably. A separate nitrification sludge is generated as a result of separate stage nitrification.

From the solid waste point of view, the impact is very minimal compared to high-rate and conventional activated sludge processes; however, odor and air pollution problems are very similar to other activated sludge processes.

6.5.9 Design Criteria

Criteria	Units	Single stage	Separate stage
Suspended growth:			
F/M	g BOD_5/g MLVSS/d	0.05-0.15	<0.15
Sludge retention			
time	days	20-30	10-20
MLVSS	mg/L	2,000-3,000	1,500-2,500
pH	pH units	7.2-8.5	7.2-8.5
Attached growth:			
Media area	ft^2/lb NH_3-N		3,000-10,000

6.5.10 Flow Diagram

Single Stage System

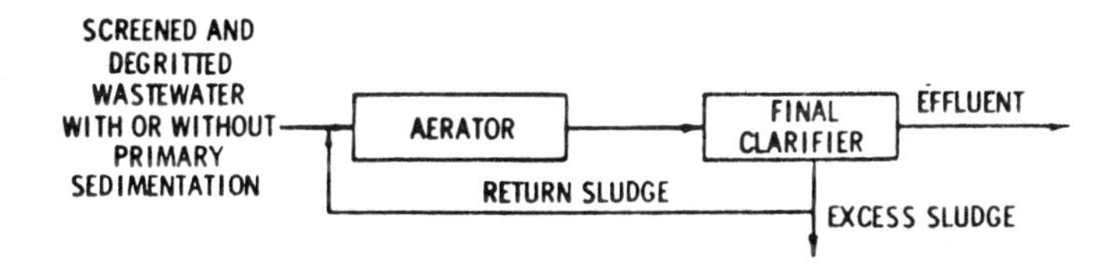

Separate Stage System

6.5.11 Performance

Conversions of ammonia (and nitrite) to nitrate of up to 98 percent are achievable. Properly designed systems have effluent ammonia in the 1 to 3 mg/L range. BOD_5 reductions are generally 70 to 80 percent (influent of BOD_5 assumed to be approximately 50 mg/L).

6.5.12 References

1. **Innovative and Alternative Technology Assessment Manual. EPA-430/9-78-009 (draft), U.S. Environmental Protection Agency, Cincinnati, Ohio, 1978. 252 pp.**

2. **Metcalf, & Eddy. Wastewater Engineering: Collection, Treatment, Disposal. McGraw-Hill Book Co., New York, New York, 1972. pp. 662-667.**

6.6 DENITRIFICATION [1]

6.6.1 Function

Denitrification is used for the reduction of nitrates and nitrites to nitrogen gas.

6.6.2 Description

Denitrification involves the reduction of nitrates and nitrites to nitrogen gas through the action of facultative heterotrophic bacteria. In suspended growth, separate stage denitrification processes, nitrified wastewater containing primarily nitrates is passed through a mixed anoxic vessel containing denitrifying bacteria. Because the nitrified feedwater contains very little carbonaceous material, a supplemental source of carbon is required to maintain the denitrifying biomass. This supplemental energy is provided by feeding methanol to the biological reactor along with the nitrified wastewater. Mixing in the anoxic denitrification reaction vessel may be accomplished using low-speed paddles analogous to standard flocculation equipment. The denitrified effluent is aerated for a short period (5 to 10 min) prior to clarification to strip out gaseous nitrogen formed in the previous step that might otherwise inhibit sludge settling. Clarification follows the stripping step with the collected sludge being either returned to the head end of the denitrification system or wasted.

6.6.3 Common Modifications

Common modifications include the use of alternate energy sources such as sugars, acetic acid, ethanol or other compounds. Nitrogen-deficient materials such as brewery wastewater may also be used. An intermediate aeration step for stabilization (about 50 min) between the denitrification reactor and the clarification step may be used to guard against carryover of carbonaceous materials. The denitrification reactor may be covered but not air tight to assure anoxic conditions by minizing surface reaeration.

Coarse media dentrification filters are attached growth biological processes in which nitrified wastewater is passed through submerged beds containing natural (gravel or stone) or synthetic (plastic) media. The systems may be pressure or gravity. Minimum media diameter is about 15 mm. The coarse media filter effluent is usually moderately high in suspended solids (20 to 40 mg/L) and may require a final polishing step. A wide variety of media types may be used as long as high void volume and low specific volume are maintained. Backwashing is infrequent and is usually done to control effluent suspended solids.

Five attached growth filters may be operated in either a pressurized downflow or a fluidized upflow mode. Various media are used, such as garnet sand, silica sand, or anthracite coal with varying size distributions, typically less than 15 mm in diameter.

In downflow systems, final clarification is not required. Backwashing is needed to maintain an acceptable pressure drop. An air scour may be incorporated into the backwashing cycle.

6.6.4 Technology Status

Denitrification technology is well developed at full scale but is not in widespread use.

6.6.5 Applications

Used almost exclusively to denitrify municipal wastewaters that have undergone carbon oxidation and nitrification; may also be used to reduce nitrate in industrial wastewaters.

6.6.6 Limitations

Specifically acts on nitrate and nitrite; will not affect other forms of nitrogen.

6.6.7 Chemicals Required

An energy source is needed and usually supplied in the form of methanol; methanol feed concentration may be estimated using the following values per mg/L of the material at the inlet to the process:

2.47 mg/L CH_3OH per mg/L of NO_3-N
1.53 mg/L CH_3OH per mg/L of NO_2-N
0.87 mg/L CH_3OH per mg/L of D.O.

6.6.8 Reliability

High levels of reliability are achievable under controlled pH, temperature, loading, and chemical feed.

6.6.9 Residuals Generated/Environmental Impact

If supplemental energy feed rates are controlled, very little excess sludge is generated; sludge production 0.6 to 0.8 lb/lb NH_3-N reduced; reduces the nitrogen loading on receiving streams.

6.6.10 Design Criteria

Criteria	Units	Value/range
Flow scheme	-	Plug flow (preferable)
Optimum pH	-	6.5 - 7.5
MLVSS	mg/L	1,000 - 3,000
Mixer power requirement	hp/1,000 ft^3	0.25 - 0.5
Clarifier depth	ft	12 - 15
Clarifier surface loading rate	gpd/ft^2	400 - 600
Solids loading	$lb/d/ft^2$	20 - 30
Return sludge rate	percent	50 - 100
Sludge generation	lb/lb CH_3OH	0.2
	lb/lb NH_3-N reduced	0.7
Hydraulic detention time	hr	0.2 - 2
Mean cell residence time	d	1 - 5

6.6.11 Flow Diagram

Fine Media Dentrification Filter

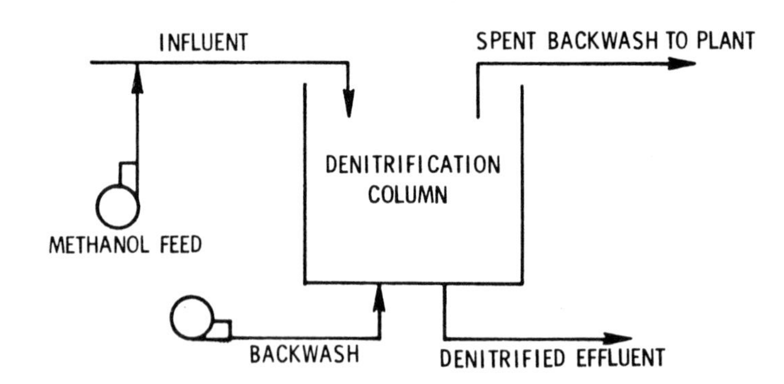

Coarse Media Dentrification Filter

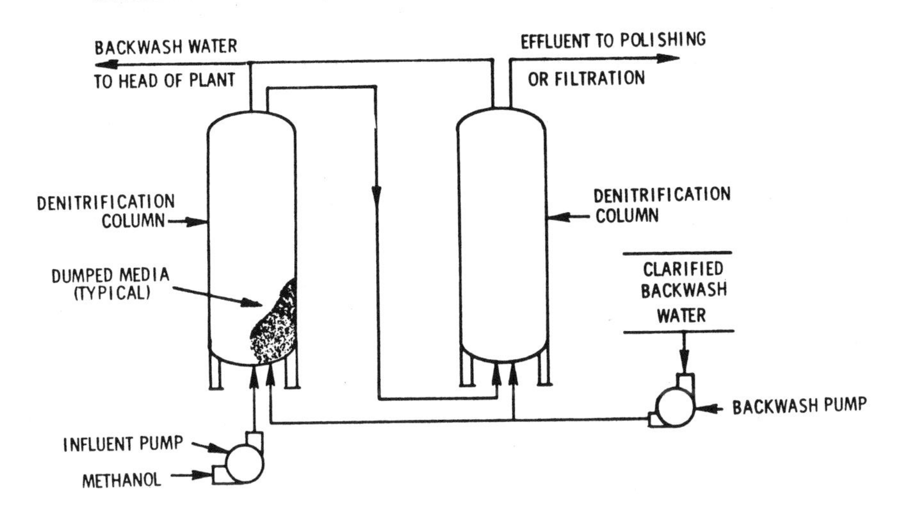

6.6.12 Performance

Capable of reducing 80 to 90 percent of the nitrate and nitrite entering the system to gaseous nitrogen. Overall nitrogen removals of 70 to 95 percent are achievable.

6.6.13 References

1. Innovative and Alternative Technology Assessment Manual. EPA-430/9-009 (draft), U.S. Environmental Protection Agency, Cincinnati, Ohio, 1978. 252 pp.

6.7 ION EXCHANGE [1]

6.7.1 Function

Ion exchange involves the removal of ionic species, principally inorganic, from an aqueous or partially aqueous phase.

6.7.2 Description

In simplest terms, ion exchange may be thought of as the reversible interchange of ions between an insoluble, solid salt (the "ion exchanger") and a solution of electrolyte in contact with that solid.

In the customary mode of usage, the ion exchanger is contacted with the solution containing the ion to be removed until the active sites in the exchanger are partially or completely used up ("exhausted") by that ion. The exchanger is then contacted with a sufficiently concentrated solution of the ion originally associated with it to convert ("regenerate") it back to its original form.

The ion exchange process works well with cations (including, of course, the hydrogen ion) and anions, both inorganic and organic. However, the organic species frequently interact with the exchangers (particularly the organic resins) via both absorption and ion exchange reactions, often necessitating the use of extremely high regenerant concentrations and/or the use of organic solvents to remove the organics. Consequently, most of the applications of ion exchange of interest have involved inorganic species.

There are a variety of different cation and anion exchangers that form salts of more or less different stabilities with a particular ion. Thus, knowledgeable choice of a particular ion exchange material will often allow selective separation of one ion in solution from another, and afford selective removal of an undesirable ion from a number of innocuous ones. As a general rule, ions with a higher charge will form more stable salts with the exchanger that those with a lower charge, and hence polyvalent species can frequently be selectively removed from a solution of monovalent ones.

In carrying out ion exchange reactions in a column or bed operation, (as opposed to a stirred batch operation which is occasionally used in chemical processing), there are four operations carried out in a complete cycle: service (exhaustion), backwash, regeneration, and rinse. The service and regeneration steps have been described above. The backwash step is one in which the bed is washed (generally with water) in reverse direction to the service cycle in order to expand and resettle the resin bed. This step eliminates channeling which might have occurred during service and removes fines or other material that may be clogging the bed. The rinse step removes the excess regeneration solution prior to the next service step.

There are three principal operating modes in use today: cocurrent fixed-bed, countercurrent fixed-bed and continuous countercurrent. A comparison summary is presented in the following table.

COMPARISON OF ION EXCHANGE OPERATING MODES

Criteria	Cocurrent Fixed Bed	Countercurrent Fixed Bed	Countercurrent Continuous
Capacity for high feed flow and concentration	Least	Middle	Highest
Effluent quality	Flucturates with bed exhaustion	High, minor fluctuations	High
Regenerant and rinse requirements	Highest	Somewhat less than cocurrent	Least, yields most concentration regenerant waste
Equipment complexity	Simplest, can use manual operation	More complex, automatic controls for regeneration	Most complex, completely automated
Equipment for continuous operation	Multiple beds, single regeneration equipment	Multiple beds, single regeneration equipment	Provides continuous service
Relative costs (per unit volume)			
Investment	Least	Middle	Highest
Operating	Highest chemicals and labor, highest resin inventory	Less chemicals, water and labor than cocurrent	Least chemical and labor, lowest resin inventory

Most ion exchange installations in use today are of the fixed-bed type, with countercurrent operation coming more into favor, especially for removal (polishing) of traces of hazardous species from the stream prior to reuse or discharge.

In order to minimize regeneration chemical requirements (i.e., to make most efficient use of regenerant), many fixed-bed installations use a technique termed "staged," or "proportional," regeneration. The first part of the regeneration solution to exit from the ion exchange bed is the most enriched in the component being removed; the concentration of that component decreases in succeeding portions of the exiting regeneration solution. In staged regeneration, the solution is divided (generally in separate tanks) into two or more portions. The first portion through the bed is "discarded" (i.e., sent for subsequent treatment), while the second and succeeding postions (less rich in the species being removed) are retained. On the next regeneration cycle, the second portion from the preceding cycle is passed through the bed first (and then "discarded"), followed by the succeeding portions, the last of which is a portion of fresh regenerant. In this way, regenerant utilization can be maximized.

6.7.3 Technology Status

The earliest applications of ion exchange were "water softening" - the substitution of sodium for calcium and magnesium in water, and the reverse substitution in sugar solutions to promote better crystallization. These applications were initiated in the late 1800's and early 1900's, using natural and synthetic zeolites (aluminosilicate minerals). Synthetic ion exchange resins were discovered in the late 1930's and were developed rapidly, particulary after World War II. Applications broadened rapidly into diverse areas such as hydrometallurgy (separations of uranium elements and the rare earth series, for example), and waste treatment (recovery and removal of chromium species). Deionization applications, especially for high quality process water (nuclear power and conventional steam generators) is probably still the most widespread application.

6.7.4 Applications

• Deionization. Industrial deionization, which in its broadest meaning includes processes yielding products ranging from potable water to boiler water for steam production, is by far the most frequent application of ion exchange, apart from domestic softening. (This latter area involves only exchange of sodium for calcium and magnesium under ambient conditions and affords little information for waste treatment application.) Deionization applications generally operate on a relatively clean feed, at worst brackish water, which has been pretreated where necessary to remove most foulants. The product must often meet stringent quality standards, particularly for newer boiler-water applications. Information on reliability of equipment operation can be obtained from the manufacturers of ion exchange equipment. Since this application is generally a steady-state operation, such information can be used to set upper limits on the reliability of equipment, particularly for newer modes of operation such as continuous countercurrent.

• Electroplating Wastewaters and Resins. Ion exchange is used extensively in the electroplating industry, especially in large installations, to remove ionic impurities from rinse water enabling re-use of the water and for further treatment of the impurities prior to disposal or recycle. Some new installations are being designed to meet the "zero discharge" requirements anticipated in the near future. In certain cases, the electroplating bath itself may require a cleanup treatment, but this is not usually done directly via ion exchange.

Ion exchange is used most frequently in combination with other techniques such as reverse osmosis or precipitation to yield an optimal solution for the particular application; in general, ion exchange is employed as the final or "polish" step, particularly if the stream to be treated contains higher concentrations of the species to be removed than can be easily handled by this process. Small-scale portable (skid-mounted) units incorporating carbon adsorption filters with series and parallel beds of appropriate ion exchange resins (cation, anion and chelating) have been marketed for cleaning up individual rinse tanks on-site. These units are regenerated separately off-site.

The rinse solutions from electroplating operations are for the most part fairly dilute mixtures of components that might well be found in the effluent from a hazardous waste treatment facility - chromium (VI and III), cyanide, nickel, etc. Thus information on ion exchange applications in this area may well be directly applicable to waste treatment processes involving reclamation of hazardous components and rectification of water prior to discharge. The equipment used is virtually all simple batch-type, and operation is often intermittent. Information on equipment and material reliability under conditions approximating batch waste disposal should be available from the users.

• Mixed Waste Streams. In the general metals finishing business, it is quite common to have a single solution waste handling system that can only be described as "mixed wastes." Obviously a variety of waste treatment schemes would be needed in order to be able to treat mixtures with constituents including suspended metal particulates, oil and grease, chromium (III and VI), iron phosphate, cyanide, zinc, etc. A common thread among most treatment schemes is the frequent use of some sort of ion exchange step for final treatment before re-use or discharge. The major amounts of materials in mixed wastes are removed or destroyed by precipitation, filtration, or a membrane separation and ion exchange is used as the "polishing" step.

• Other Metal Finishing Streams. In addition to treating dilute aqueous streams, ion exchange is being used to remove low concentrations of undesirable impurities from relatively highly concentrated aqueous streams. The object of treatment in most cases is to recycle or reclaim the active materials while ridding the bath of unwanted impurities. Frequently ion exchange is the sole separation step, with other post-treatment steps being carried out on the spent regenerant solution.

Minor concentrations of cations such as iron, aluminum and chromium (III) are removed from chromic acid plating bath liquors via cation exchange, after dilution of the chromic acid content of the liquor from 250 g/L down to 100 g/L. The dilution is necessary in order to obtain efficient exchange and to minimize oxidative damage to the sulfonated styrene-divinyl benzene resins used.

• Applications in Hydrometallurgical Processing. Ion exchange has been used for recovery of valuable metals such as copper, molybdenum, cobalt and nickel, especially from dilute leach liquors from tailings or dump piles. Liquid ion exchange has been more widely used in general in the areas; however, the advent of new, more-selective resins coupled with the increased cost of solvent losses (which are at present unavoidable in liquid ion exchange) result in increased interest in the solid exchangers.

Uranium processing and extraction are active fields for both solid and liquid ion exchange. Solid ion exchange is being used for recovery of carbonate leaches from *in situ* uranium mining in Texas.

Information in this field may have direct application to treatment of certain waste streams and should be useful for comparison of solid and liquid ion exchange.

• Removal and Isolation of Radioactive Wastes. A great deal of work has been reported on removal of traces of radioactive species from solutions of various kinds. Of particular interest to waste treatment is a summary of the performance of ion exchange systems in operational nuclear power plants, which indicated that the severe conditions of radiation and heat resulted in attrition rates higher than those expected in nonnuclear service. Even under those conditions, operating capacity varied from 50 to 75% of theoretical.

Experience over long service lives in nuclear operations may provide some useful information on the long term behavior of ion exchange materials. Equipment reliability is normally extremely good in nuclear service, having been deliberately designed that way because of the extreme necessity to avoid trace ion leakage and equipment downtime.

6.7.5 Limitations

The upper concentration limit for the exchangeable ions for efficient operation is generally 2,500 mg/L, expressed as calcium carbonate (or 0.05 equivalents/L). This upper limit is due primarily to the time requirements of the operation cycle. A high concentration of exchangeable ion results in rapid exhaustion during the service cycle, with the result that regeneration requirements, for both equipment and of the percentage of resin inventory undergoing regeneration at any time, become inordinately high.

There is also an upper concentration limit (around 10,000-20,000 mg/L), which is governed by the properties of the ion exchangers themselves, in that the selectivity (preference for one ion over another) begins to decrease as the total concentration of dissolved salts (ionic strength) increases.

Synthetic resins can be damaged by oxidizing agents and heat. In addition, the stream to be treated should contain no suspended matter or other materials that will foul the resin and that cannot be removed by the backwash operation. Some organic compounds, particularly aromatics, will be irreversibly absorbed by the resins, and this will result in a decreased capacity, as for example in the case of electroplating bath additives.

6.7.6 Typical Equipment

Fixed-bed ion exchange operations are straightforward systems, requiring a cylindrical ion exchange bed, tanks for solution storage, and pumps. The choice of materials is governed by the chemical environment. Continuous ion exchange systems are much more complex, requiring solids handling equipment and more intricate control systems. Apparently only one company (Chemical Separations Corp.) has been truly successful in the design and fabrication of continuous ion exchange systems, and it should be consulted if the use of such a system is contemplated.

6.7.7 Residuals Generated/Environmental Impact

Ion exchange is a solution(aqueous) phase process. The dilute, purified product stream can be suitable for discharge to sewers. The concentrated regeneration stream requires further treatments for recovery and/or safe disposal of its components. Emissions to air will be essentially zero. Emissions to water will be significant only if the regenerant solution is discharged inadvertently to ground or surface water. In normal operation, emissions will be within environmental discharge limits. Emissions to land will be insignificant, except for spills from process accidents, or improper disposal of solids exchangers loaded with hazardous substances that would be leachable under the landfill conditions.

The above points address only the ion exchange process itself, and not disposal of spent or degraded ion exchange materials. These materials should be disposed of (after proper cleaning to remove the hazardous substances) with other solid industrial wastes of similar composition.

There are no special land use factors associated with ion exchange processes. Fixed-bed operations are run with the beds next to each other, with intermediate pumping. Continuous systems do require some overhead height for the loop, but have greatly decreased floor space requirements.

The only safety problems that might arise involve handling and processing the spent regenerant liquor with its potentially high concentrations of hazardous substances.

6.7.8 Reliability

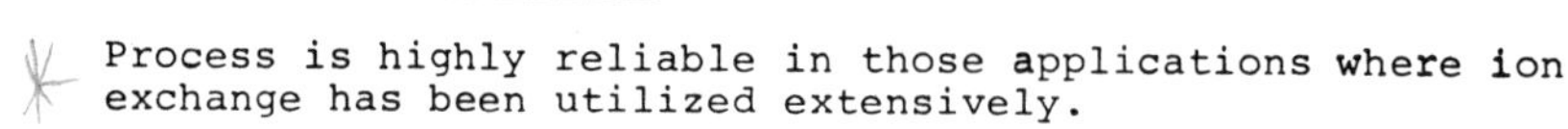

Process is highly reliable in those applications where ion exchange has been utilized extensively.

6.7.9 Design Criteria

Ion Exchange Operation

Clinoptilolite size = 20 x 50 mesh
Bed height = 4 to 6 ft
Wastewater suspended solids = 35 mg/L max.
Wastewater loading rate = 7.5 to 20 bed volume/h
Pressure drop = 8.4 in. of water/ft
Cycle time = 100 to 150 bed volumes for one 6 ft bed; 200 to 250 bed volumes for two 6 ft beds in series

Regeneration

Solution = 2% NaCl
Solution flow rate = 4 to 10 bed volume/h or 4 to 8 gal/min/ft^2
Total solution volume = 2.5 to 5% of treated wastewater or 10 bed volumes
Cycle time = 1 to 3 hours
Backwash = 8 gal/min/ft^2

6.7.10 Flow Diagram

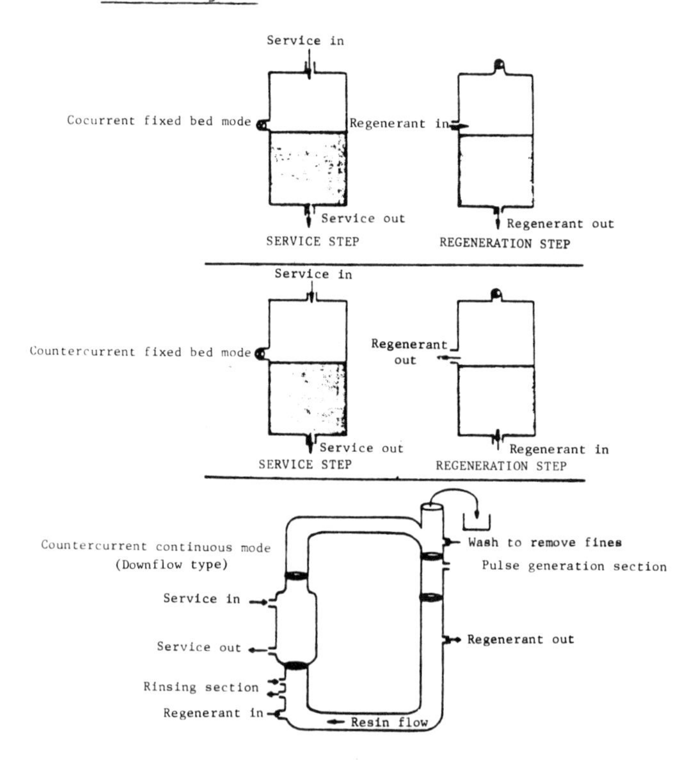

6.7.11 Performance

Subsequent data sheets provide performance data from studies on the following industries and/or wastestreams:

- Electroplating
- Ore mining and dressing
  Ferroalloy mine/mill
  Uranium mine

6.7.12 References

1. Physical, Chemical, and Biological Treatment Techniques for Industrial Wastes, PB 275 287, U.S. Environmental Protection Agency, Washington, D.C., November 1976. pp. 30-1 through 30-26.

CONTROL TECHNOLOGY SUMMARY FOR ION EXCHANGE

Pollutant	Number of data points	Effluent concentration				Removal efficiency, %			
		Minimum	Maximum	Median	Mean	Minimum	Maximum	Median	Mean
Toxic pollutants, µg/L:									
Cadmium	2	10[a]	<10[a]		<10[b]	>99	>99		>99
Chromium	2	10	60		35	99	>99		>99
Chromium (+6)	1				10				>99
Copper	2	90	90		90	98	98		98
Cyanide	2	40	200		120	75	99		87
Nickel	2	<10[a]	<10[a]		<10[b]	99	>99		>99
Silver	2	<10[a]	<10[a]		<10[b]	>99	>99		>99
Zinc	1				400				97
Other pollutants, µg/L:									
Molybdenum	1				1,300				94
Radium (total)	1				7.2				99
Radium (dissolved)	1				<1				>99

Note: Blanks indicate data not applicable.

[a]Not detected, assumed to be <10 µg/L.

[b]Below detection limit, assumed to be <10 µg/L.

[c]Concentration reported in picoCi/L

TREATMENT TECHNOLOGY: Ion Exchange

Data source: Effluent Guidelines
Point source category: Electroplating
Subcategory:
Plant:
References: A14, p. 181

Data source status:
Engineering estimate ___
Bench scale ___
Pilot scale ___
Full scale x

Use in system: Tertiary
Pretreatment of influent:

DESIGN OR OPERATING PARAMETERS

Wastewater flow:
Solids loading rate:
Bed height:
Pressure drop:
Resin type:
Run length:
Regenerant used:
Cycle time:
Backwash rate:
Resin pulse volume:
Unit configuration:

REMOVAL DATA

Sampling period:

Pollutant/parameter	Concentration, μg/L Influent	Concentration, μg/L Effluent	Percent removal
Toxic pollutants:			
Cadmium	1,000	ND[a]	>99
Chromium	7,600	60	99
Copper	4,400	90	98
Cyanide	800	200	75
Nickel	6,200	ND[a]	>99
Silver	1,500	ND[a]	>99

[a]Not detected, assumed to be <10 μg/L.

Note: Blanks indicate information was not specified.

TREATMENT TECHNOLOGY: Ion Exchange

Data source: Effluent Guidelines
Point source category: Electroplating
Subcategory:
Plant:
References: A14, p. 144

Data source status:
Engineering estimate ___
Bench scale ___
Pilot scale ___
Full scale x

Use in system: Tertiary
Pretreatment of influent:

DESIGN OR OPERATING PARAMETERS

Wastewater flow:
Solids loading rate:
Bed height:
Pressure drop:
Resin type:
Run length:
Regenerant used:
Cycle time:
Backwash rate:
Resin pulse volume:
Unit configuration:

REMOVAL DATA

Sampling period:

Pollutant/parameter	Concentration, μg/L Influent	Effluent	Percent removal
Toxic pollutants:			
Cadmium	5,700	BDL[a]	>99
Chromium	3,100	10	>99
Chromium (+6)	7,100	10	>99
Copper	4,500	90	98
Cyanide	9,800	40	99
Nickel	6,200	BDL	>99
Silver	1,500	BDL	>99
Zinc	15,000	400	97

[a]Below detectable limits; assumed to be <10 μ/L.

Note: Blanks indicate information was not specified.

TREATMENT TECHNOLOGY: Ion Exchange

Data source: Effluent Guidelines
Point source category: Ore mining and dressing
Subcategory: Ferroalloy mine/mill
Plant: 6102
References: A2, p. VI-59

Data source status:
Engineering estimate ___
Bench scale ___
Pilot scale x
Full scale ___

Use in system: Tertiary
Pretreatment of influent:

DESIGN OR OPERATING PARAMETERS

Wastewater flow: 0.121-0.125 m^3/min
Solids loading rate:
Bed height:
Pressure drop:
Resin type:
Run length: 41 min
Regenerant used:
Cycle time:
Backwash rate:
Resin pulse volume: 1.73 L
Unit configuration: Pulsed bed, counter flow ion exchange unit

REMOVAL DATA

Sampling period: Average of six two-day samples

Pollutant/parameter	Concentration, µg/L Influent	Concentration, µg/L Effluent	Percent removal
Other pollutants:			
Molybdenum	22,000	1,290	94

Note: Blanks indicate information was not specified.

TREATMENT TECHNOLOGY: Ion Exchange

Data source: Effluent Guidelines
Point source category: Ore mining and dressing
Subcategory: Uranium mine
Plant: 9452
References: A2, p. VI-48

Data source status:
Engineering estimate ___
Bench scale ___
Pilot scale ___
Full scale x

Use in system: Tertiary
Pretreatment of influent: Flocculation, barium chloride co-precipitation, twc settling ponds in series

DESIGN OR OPERATING PARAMETERS

Wastewater flow:
Solids loading rate:
Bed height:
Pressure drop:
Resin type:
Run length:
Regenerant used:
Cycle time:
Backwash rate:
Resin pulse volume:
Unit configuration: Two upflow ion exchange columns operating in parallel each consisting of fiber-reinforced plastic
Resin volume: 11.3 m^3 (400 ft^3)

REMOVAL DATA

Sampling period:

Pollutant/parameter	Concentration, picoCi/L Influent	Concentration, picoCi/L Effluent	Percent removal
Other pollutants:			
Radium (total)	955	7.2	99
Radium (dissolved)	93.4	<1	>99

Note: Blanks indicate information was not specified.

6.8 POLYMERIC (RESIN) ADSORPTION [1]

6.8.1 Function

Adsorption on synthetic resins is considered here primarily as a process for the removal of organic chemicals from liquid waste streams; a separate section on ion-exchange resins, which are used for inorganic ion removal and/or recovery, appears elsewhere in this volume.

6.8.2 Description

Waste treatment by resin adsorption involves two basic steps: (1) contacting the liquid waste stream with the resins and allowing the resins to adsorb the solutes from the solution; and, (2) subsequently regenerating the resins by removing the adsorbed chemicals, often effected by simply washing with the proper solvent.

The chemical nature of the various commercially available resins can be quite different; perhaps the most important variable in this respect is the degree of their hydrophilicity. The adsorption of a nonpolar molecule on to a hydrophobic resin (e.g., a styrene-divinyl benzene based resin) results primarily from the effect of Van der Waal's forces. In other cases, other types of interactions such as dipole-dipole interaction and hydrogen bonding are also important. In a few cases, an ion-exchange merchanism may be involved; this is thought to be true, for example, in the adsorption of alkylbenzene sulfonates from aqueous solution on to weakly basic resins; e.g., a phenol-formaldehyde-amine based resin.

Resin adsorbents are used in much the same way as granular carbon. Commonly, a typical system for treating low volume waste streams will consist of two fixed beds of resin. One bed will be on-stream for adsorption while the second is being regenerated. In cases where the adsorption time is very much longer than regeneration time (as might be when solute concentrations are very low), one resin bed plus a hold-up storage tank could suffice.

The adsorption bed is usually fed downflow at flow rates in the range of 0.25 to 2 gpm per cubic foot of resin; this is equivalent to 2 to 16 bed volumes/hr, and thus contact times are in the range of 3 to 30 minutes. Linear flow rates are in the range of 1 to 10 gpm/ft^2. Adsorption is stopped when the bed is fully loaded and/or the concentration in the effluent rises above a certain level.

Regeneration of the resin bed is performed *in situ* with basic, acidic, and salt solutions or regenerable nonaqueous solvents being most commonly used. Basic solutions may be used for the removal of weakly acidic solutes and acidic solutions for the removal of weakly basic solutes; hot water or steam could be used for volatile solutes; methanol and acetone are often used for the removal of nonionic organic solutes. A prerinse and/or a post-rinse with water will be required in some cases. As a rule, about three bed volumes of regenerant will be required for resin regeneration; as little as one-and-a-half bed volumes may suffice in certain applications.

Solvent regeneration will be required unless (1) the solute-laden solvent can be used as a feed stream in some industrial process at the plant, or (2) the cost of the solvent is low enough so that it may be disposed of after a single use. Solvent recovery, usually by distillation, is thus most common when organic solvents are used. Distillation will allow solute recovery for reuse if such is desired.

Resin lifetimes may vary considerably depending on the nature of the feed and regenerant streams. Regeneration with caustic is estimated to cause a loss of 0.1 to 1% of the resin per cycle; replacement of resins at such installations may be necessary every two to five years. Regeneration with hot water, steam, or organic solvent should not affect the resins, and, in this case, lifetimes will be limited by slow fouling or oxidation resulting in a loss of capacity; actual experience indicates that lifetimes of more than five years are obtainable.

6.8.3 Technology Status

Relatively little information is available on the few systems that are currently in operation. Thus there are areas of uncertainty concerning practicability, start-up problems, realistic operating costs, etc.

6.8.4 Applications

Little publicly available information exists on current or proposed industrial applications of resin adsorption systems; several current applications of resin adsorptions, for which some information is available are discussed below.

- Color Removal

A dual resin adsorption system is being used to remove color associated with metal complexes and other organics from a 300,000 gpd waste stream from a dyestuff production plant; color is reduced from an average of 75,000 to 500 APHA units on the Pt-Co scale, and COD is reduced from an average of 5,280,000 to 2,600 ppm. The system also removes copper and chromium present in the influent waste stream both as salts and as organic chelates. While there have been some problems with this system, the effluent does meet the NPDES requirements.

Two large systems are also currently operating to remove colored pollutants (derived from lignin) from paper mill bleach plant effluents in Sweden and in Japan. The Swedish plant, which produces 300 tons of pulp/day, installed its system about three years ago. The resin adsorption system removes 92 to 96% of the color (initially 30 to 40,000 Pt-Co units), 80 to 90% of the COD and 40 to 60% of the BOD_5 from the effluent of the caustic extraction stage in the bleach plant. The system consists of three resin columns, each containing about 20 cubic meters of resin. The system in Japan is for a 420 metric ton/day pulp plant and consists of four resin columns, each with about 30 cubic meters of resin. In both cases, the resins are regenerated with a caustic wash followed by a reactivation with an acid stream (e.g., H_2SO_4).

Some resin adsorption units in operation are used to remove color in water supply systems; others are used to decolorize sugar, glycerol, wines, milk whey, pharmaceuticals, and similar products. One plant in Louisiana, which removes color from an organic product stream, is said to have been in operation for eight years now without replacement of the initial resin charge.

- Phenol Removal

One plant in Indiana currently uses a resin system to recover phenol from a waste stream. This unit had been operating for about nine months as of March, 1976, and is said to be performing satisfactorily. A dual resin system is currently being installed at a coal liquefaction plant in West Virginia to remove phenol and high molecular-weight polycyclic hydrocarbons from a 10-gpm waste stream; methanol will be used as the regenerant for the primary resin adsorbent.

- Miscellaneous Applications

One resin adsorption system, in operation for five years, is removing fat from the wastewaters of a meat production plant. Other applications include the recovery of antibiotics from a fermentation broth, the removal of organics from brine, and the removal of drugs from urine for subsequent analysis. Adsorbent resins are also currently being used on a commercial scale for screening out organic foulants prior to deionization in the production of extremely high purity water.

6.8.5 Limitations

Feed stream into a resin adsorption system must be a single liquid phase; in most cases, this will be an aqueous solution, but there is no basic reason that an organic solution could not be treated so long as the resin is not chemically or physically harmed by the solution; other limitations include the following:

- Suspended solids should be no higher than 50 ppm and may have to be kept below 10 ppm in some cases to prevent clogging of the resin bed.

- pH may vary widely; some resins have been able to operate as low as pH 1-2 and as high as pH 11-12, in many cases, adsorption will be pH dependent, and will thus require pH control.

- Temperature may also vary significantly; resins have been used in applications where the influent temperature was as high as 80°C; adsorption will, however, be favored by lower temperatures; conversely, regeneration will be aided by higher temperatures.

- High levels of total dissolved solids (particularly inorganic salts) do not interfere with the action of resin adsorbents on organic solutes; there are clear indications that some organic chemicals are more easily removed from solutions with high concentrations of dissolved salts than from salt-free solutions; in some cases of high salt content, the adsorbent may have to be prerinsed before regeneration.

- Concentration of organic solute(s) in the feed stream should probably be at least a factor of ten less than the maximum amount that can be adsorbed in a resin bed divided by three bed volumes; this will allow a reasonably long cycle time; higher influent concentrations may be treated when special provisions are made.

6.8.6 Typical Equipment

Equipment for resin adsorption systems is relatively simple. The system will generally consist of two or more steel tanks (stainless or rubber-lined) with associated piping, pumps, and (perhaps) influent hold-up tank. Regeneration takes place in the same tanks, and thus the extra equipment needs for regeneration will consist only of such items as solvent storage tanks, associated solvent piping and pumps, and solvent (and perhaps solute) recovery equipment; e.g., a still. Up to three stills may be required in some systems.

Materials needed include a regenerant solution (e.g., aqueous caustic solution or organic solvent), and resin. In one full-scale installation for the removal of organic dye wastes from water, two different resins are employed. In this case, the waste stream is first contacted with a normal polymeric adsorbent and then with an anion exchange resin.

Features of a few currently available resin adsorbents are given in the following table. Surface areas of resin adsorbents are generally in the range of 100 to 700 m^2/g; this is below the typical range for activated carbons (800 to 1,200 m^2/g) and, in general, indicates lower adsorptive capacities, although the chemical nature and pore structure of the resin may be more important factors. This has been demonstrated in one application relating to color removal.

Tests should be run on several resins when evaluating a new application. Important properties are the degree of hydrophilicity and polarity, particle shape (granular versus spherical), size, porosity, and surface area.

It is frequently possible to "tailor" a resin for specific applications because much greater control over the chemical and surface nature can be achieved in resin production than in activated carbon manufacture. The cost of developing a totally new resin would be prohibitive for most applications, but minor modifications of currently available resins are often feasible.

Name[a]	Base	Specific gravity (wet)	Void volume, %	Particle size mesh	Bulk density, lb/ft^3	Surface area, m^2/g	Average pore size, Å
XAD-1		1.02	37	20 - 50	-	100	200
XAD-2	Styrene-divinylbenzene	1.02	42	20 - 50	40 - 44	300	90
XAD-4		1.02	51	20 - 50	39	780	50
XAD-7		1.05	55	20 - 50	41	450	90
XAD-8	Acrylic ester	1.09	52	20 - 50	43	140	235
Dow XFS 4256[b]	Styrene-divinylbenzene	-	40	+10	27	400	110
Dow XFS 4022		-	35	20 - 50	-	100	200
Dow XFS 4257		-	40	20 - 50	-	400	110
Duolite S-30		1.11	35	16 - 50	~30	128	-
Duolite S-37		1.12	35 - 40	16 - 50	40	-	-
Duolote ES-561	Phenol-formaldehyde[c]	1.12	35 - 40	18 - 50	40 - 45	-	-
Duolite A-7D		-	-	-	-	24	-
Duolite A-7		1.12	35 - 40	16 - 50	~40	-	-

[a]XAD resins manufactured by Rohm and Haas Company; Dow XFS resins manufactured by Dow Chemical U.S.A.; Duolite resins manufactured by Diamond Shamrock Chemical Company.

[b]Resin designed for use in vapor phase adsorption applications.

[c]Functional groups such as phenolic hydroxyl groups, secondary and tertiary amines are present on the basic phenol-formaldehyde structure; physical form of these resins is granular as opposed to a bead form for the other brands.

6.8.7 Residuals Generated/Environmental Impact

The only major environmental impacts resulting from the use of resin adsorption systems are related to the disposal of the used regenerant solution or extracted solutes when they are not recycled. For example, when highly colored wastewaters are treated, the used regenerant solution (containing 2 to 4% caustic plus the eluted wastes) is not recycled and must be disposed of, usually by evaporation and incineration. A second example is the removal of pesticides from water, with regeneration being affected by an organic solvent. In this case, the solvent is recovered, probably by distillation, resulting in a concentrated waste (still bottoms) to be disposed of, probably by incineration. In both of these examples where incineration is used for the eventual destruction of the wastes, the environmental impacts would be on air quality (from incinerator emissions), energy use (for the incinerator fuel), and land use (from the disposal of unburned residues).

Only minor environmental impacts might be associated with the rinse waters discharged. In most cases, these effluents can be adequately treated by conventional means or safely discharged to surface waters.

Resin adsorption systems are relatively compact and thus require little space. The systems do not have any known health or safety problems associated with their operation.

6.8.8 Reliability

Reliability is still uncertain for this technology.

6.8.9 Design Criteria

Criteria have not yet been developed; design is application specific.

6.8.10 Flow Diagram

DIAGRAM OF A RESIN ADSORPTION SYSTEM FOR THE REMOVAL AND RECOVERY OF PHENOL FROM WATER

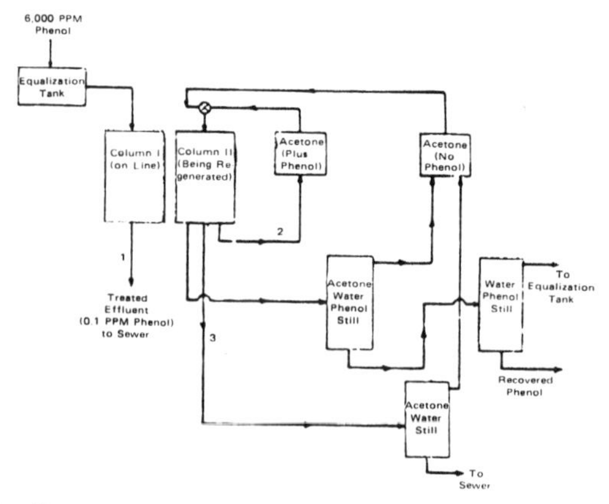

6.8.11 Performance

Subsequent data sheets provide performance data from studies on the listed industries and/or wastestreams.

6.8.12 References

1. Physical, Chemical, and Biological Treatment Techniques for Industrial Wastes, PB 275 287, U.S. Environmental Protection Agency, Washington, D.C. November 1976. pp.2-1 to 2-26.

6.9 REVERSE OSMOSIS

Function. Reverse osmosis is used for the removal of dissolved organic and inorganic materials and control of such wastewater parameters as soluble metals, TDS, and TOC.

Description. Reverse Osmosis (RO) separates dissolved materials in solution by filtration through a semipermeable membrane at a pressure greater than the osmotic pressure caused by the dissolved materials in the wastewater. With existing membranes and equipment, operating pressures vary from atmospheric to 1,500 psi. Products from the process are (1) the permeate or product stream with dissolved material removed, and (2) concentrate stream containing all removed material. Removal levels obtainable are dependent on membrane type, operating pressure, and the specific pollutant of concern. Removal of multicharged cations and anions is normally very high, while most low molecular weight dissolved organics are not removed or are only partially removed.

Technology Status. RO has been commercially available since the mid-1960's. Originally developed for desalination of seawater, it is seeing broader acceptance as a wastewater treatment tool, especially when a wastestream has pollutants with recoverable value.

Applications. Recovery of silver, concentration of dilute wastestreams, metals recovery, radioactive waste treatment, and water reuse and recycle.

Limitations. Concentration polarization (decreased water production with time per square meter of membrane); pretreatment is necessary for removal of solids (colloidal and suspended). Dechlorination required when using polyamide membranes. Membrane fouling results from precipitation of insoluble salts. Concentrated steam is necessary for the treatment and disposal of these precipitates.

Typical Equipment. Membrane modules; feed, product, concentrate tanks; high pressure pump; prefilter plus pump; stainless steel piping; heat exchanger; flow and pressure instrumentation.

Design Criteria. Membrane type: cellulose acetate (also di- and triacetate), polyamide, polysulfone; flux (product) rate at 600 psi, 5,000 ppm NaCl solution, and 25°C: 6 to 10 gpd/ft^2 membrane or 25 to 100 gpd/ft^3 module; rejection at 600 psi, 5,000 ppm NaCl solution, and 25°C: 70% to 99% depending on membrane specification; operating pressure: 250 to 1,500 psi; membrane configuration: plate, tubular, spiral, or hollow fiber; water recovery: 50% to 85% depending on minimum solubility.

Side Streams. Concentrate (15% to 30% of initial feed volume); rinse, clean (10% to 20% of final product volume or additional distilled/deionized water); rinse, chemical - dependent on application.

Chemicals Required. Sodium tripolyphosphate to increase water recovery; chlorine as biocide when using cellulose-based membranes.

Reliability. Dependent on wastestream being treated. Fouling and membrane deterioration have been common in past. Recent applications have shown reliability to be improving with vendors willing to issue guarantees on membrane life.

Toxics Management. Removes substantially all soluble heavy metals and many, but not all, high molecular weight organics.

Environmental Impact. The concentrate stream must be disposed of or treated further.

Flow Diagram.

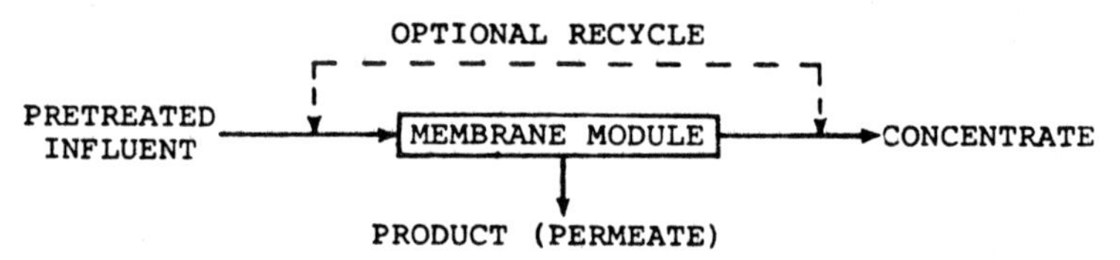

Performance. Performance data presented on the following data sheets include information from studies on the following industries and/or wastestreams:

Industries	Wastestreams
Brass finishing	Cooling tower blowdown
Synthetic rubber	Synthetic laboratory
Pulp and paper	Sanitary
Textiles	Acid mine drainage

CONTROL TECHNOLOGY SUMMARY FOR REVERSE OSMOSIS

Pollutant	Number of data points	Effluent concentration				Removal efficiency, %			
		Minimum	Maximum	Median	Mean	Minimum	Maximum	Median	Mean
Conventional pollutants, mg/L:									
BOD_5	11	1	429	2.7	43	64	92	87	83
COD	18	6	736	25.5	73	0[a]	>99	91.5	87
TOC	18	5	50	8	<11	>5	96	90	84
TSS	2	<4	<5		<4.5	>85	>90		>88
Oil and grease	5	<4	17	7	<7	>20	>72	>50	>40
Total phenol	6	<0.001	0.020	0.014	0.013	0[a]	81	2.5	27
TKN	1				<3.26				>0
Toxic pollutants, μg/L:									
Antimony	11	2	200	90	77	0[a]	60	30	26
Arsenic	10	<1	49	1	7.7	0[a]	>99	>92	>79
Beryllium	2	<0.5	5		<2.8	0[a]	>85		42
Cadmium	11	<0.5	48	14	13	0[a]	50	0	13
Chromium	13	<1	1,500	520	460	0[a]	>99	67	44
Chromium (+3)	1				15				>99
Chromium (+6)	1				10				0[a]
Copper	17	9	28,000	40	2,300	0[a]	>99	82	74
Cyanide	10	1	22,000	22	2,200	0[a]	97	>42	43
Lead	11	<3	520	250	210	0[a]	>99	>25	31
Mercury	3	<0.2	0.53	0.3	<0.34	0[a]	>60	4	>21
Nickel	13	<1	210	<10	70	0[a]	>98	47	46
Selenium	4	<1	13	4	<5.5	>67	85	77	>76
Silver	13	<0.2	78	9	25	0[a]	76	17	31
Thallium	3	1	4	2	2.3	0[a]	89	50	46
Zinc	30	<2	8,600	57	640	0[a]	>99	97	79
Bis(2-ethylhexyl) phthalate	5	2	31	3	8.4	0[a]	96	67	51
Di-n-butyl phthalate	3	0.8	1	1	0.93	20	83	75	59
Dimethyl phthalate	2	45	170		110	18	41		30
Phenol	4	0.2	10	0.7	2.9	0[a]	80	25	32
Benzene	3	0.4	3.0	1	1.5	0[a]	80	50	43
Toluene	6	0.7	29	20	18	0[a]	12	0[a]	6
Acenaphthene	3	0.8	3	0.8	1.5	57	99	73	76
Anthracene	1				0.7				77
Pyrene	1				18				0
Chloroform	4	6	31	13	13	0[a]	79	0[a]	20
Methyl chloride	1				45				0[a]
Methylene chloride	4	4	5	5	4.8	0[a]	64	10	21
Trichloroethylene	1				0.4				60

[a]Actual data indicate negative removal.

[b]Note detected, assumed to be less than influent concentration.

TREATMENT TECHNOLOGY: Reverse Osmosis

Data source: Government report
Point source category: Adhesives and sealants
Subcategory:
Plant: Grace Chicago
References: B10, p. 75

Data source status:
- Engineering estimate ___
- Bench scale ___
- Pilot scale x
- Full scale ___

Use in system: Tertiary
Pretreatment of influent: Primary settling, ultrafiltration, 5 μ and 1 μ spiral wound cartridge filters in series

DESIGN OR OPERATING PARAMETERS

Product flow rate:
Flux rate:
Membrane configuration: Hollow fiber
Membrane type: Du Pont B-9 polyamide
Retentate (concentrate) flow rate:
Recycle flow rate:
Operating temperature: 27-30°C (81-86°F)
Rated production capacity:
Operating pressure: 2,700 kPa (400 psig)
Feed flow rate: 27.3 m^3/d (5 gpm)

REMOVAL DATA

Sampling period: Equal volume grab samples collected throughout an 8-hr day

Pollutant/parameter	Concentration,[a] mg/L Influent	Effluent	Percent removal
Conventional pollutants:			
BOD_5	1,280	429	66
COD	7,040	736	90

[a]Average of two samples.

Note: Blanks indicate information was not specified.

TREATMENT TECHNOLOGY: Reverse Osmosis

Data source: Government report
Point source category: Electroplating
Subcategory: Copper acid plating bath
Plant: Precious Metal Platers, Inc.,
Hopkins, Minnesota
References: B13, pp. 25-26

Data source status:
- Engineering estimate ___
- Bench scale ___
- Pilot scale x
- Full scale ___

Use in system: Primary
Pretreatment of influent: Acid bath was diluted to one-tenth of full strength

DESIGN OR OPERATING PARAMETERS

Product flow rate:
Flux rate (average): 0.023 $m^3/hr/m^2$
Membrane configuration: Eight, 0.6 m (2 ft) tubular membranes
Membrane type: NS-101 polyethylenimine, isophthalal chloride support layer
Retentate (concentrate) flow rate:
Recycle flow rate:
Operating temperature:
Rated production capacity:
pH: 1.18

REMOVAL DATA

Sampling period: Average values, samples taken over 1,200-hr period

Pollutant/parameter	Concentration Influent	Concentration Effluent	Percent removal
Conventional pollutants, mg/L:			
TOC	23	7.4	68
Toxic pollutants, μg/L:			
Copper	4,900	50	99

Note: Blanks indicate information was not specified.

TREATMENT TECHNOLOGY: Reverse Osmosis

Data source: Government report
Point source category: Electroplating
Subcategory: Copper plating
Plant: New England Plating Co.
(Worchester, Mass.)
References: B11, p. 65

Data source status:
Engineering estimate ___
Bench scale ___
Pilot scale x
Full scale ___

Use in system: Tertiary
Pretreatment of influent:

DESIGN OR OPERATING PARAMETERS

Product flow rate: 0.008 m^3/min (∿2 gpm)
Flux rate:
Membrane configuration:
Membrane type:
Retentate (concentrate) flow rate:
Recycle flow rate:
Operating temperature: 25°C (77°F)
Rated production capacity:
Feed pressure (average): 1,240 kPa (180 psi)
Percent conversion (average): 84
Total feed concentration: 1.5 μg/L

REMOVAL DATA

Sampling period: Average of 17 samples taken over a 1,130-hr period for copper, average of 9 samples taken in the latter part of the 1,130-hr period for cyanide

Pollutant/parameter	Concentration, μg/L Influent	Effluent	Percent removal
Toxic pollutants:			
Copper	170,000	28,000	84
Cyanide	240,000	22,000	91

Note: Blanks indicate information was not specified.

TREATMENT TECHNOLOGY: Reverse Osmosis

Data source: Government report
Point source category: Electroplating
Subcategory: Zinc cyanide plating bath
Plant: Superior Plating, Inc., (Minneapolis, Minnesota)
References: B13, pp. 31-33

Data source status:
- Engineering estimate ___
- Bench scale ___
- Pilot scale x
- Full scale ___

Use in system: Primary
Pretreatment of influent: Bath diluted to one-tenth of original strength

DESIGN OR OPERATING PARAMETERS

Product flow rate:
Flux rate (average): 0.16 $m^3/hr/m^2$
Membrane configuration: Ten, 0.61 m (2 st) tubular membranes
Membrane type: NS-100 polyethylenimine toluene diisocyanate
Retentate (concentrate) flow rate:
Recycle flow rate:
Operating temperature: 25°C
Rated production capacity:
pH: 12.8

REMOVAL DATA

Sampling period: Average values, samples taken over 1,044-hr period

Pollutant/parameter	Concentration		Percent
	Influent	Effluent	removal
Conventional pollutants, mg/L:			
TOC	1,250	50	96
Toxic pollutants, µg/L:			
Cyanide	2,800	80	97
Zinc	1,700	30	98

Note: Blanks indicate information was not specified.

TREATMENT TECHNOLOGY: Reverse Osmosis

Data source: Effluent Guidelines
Point source category: Steam electric power generating
Subcategory:
Plant: 0630
References: A31, pp. 242-245

Data source status:
Engineering estimate ___
Bench scale ___
Pilot scale ___
Full scale x

Use in system:
Pretreatment of influent:

DESIGN OR OPERATING PARAMETERS

Product flow rate:
Flux rate:
Membrane configuration:
Membrane type:
Retentate (concentrate) flow rate:
Recycle flow rate:
Operating temperature:
Rated production capacity:

REMOVAL DATA

Sampling period:

Pollutant/parameter	Concentration, µg/L Influent	Effluent	Percent removal
Toxic pollutants:			
Copper	30	10	67
Cyanide	30	20	33
Mercury	0.55	0.53	4
Selenium	58	13	78
Silver	<5	9	0[a]
Zinc	14	13	7
Phenol	20	10	50
Benzene	2.0	3.0	0[a]
Toluene	28	25	11

[a]Actual data indicate negative removal.

Note: Blanks indicate information was not specified.

TREATMENT TECHNOLOGY: Reverse Osmosis

Data source: Effluent Guidelines
Point source category: Steam electric power generating
Subcategory:
Plant: 1226
References: A31, pp. 15-17

Data source status:
Engineering estimate ___
Bench scale ___
Pilot scale x
Full scale ___

Use in system: Primary
Pretreatment of influent:

DESIGN OR OPERATING PARAMETERS

Product flow rate:
Flux rate:
Membrane configuration:
Membrane type:
Retentate (concentrate) flow rate:
Recycle flow rate:
Operating temperature:
Rated production capacity:
pH: 6.8

REMOVAL DATA

Sampling period:

Pollutant/parameter	Concentration, µg/L Influent	Concentration, µg/L Effluent	Percent removal
Toxic pollutants:			
Antimony	7	10	0[a]
Arsenic	4	1	75
Cadmium	1.8	2.5	0[a]
Chromium	5	<2	>60
Copper	47	10	79
Cyanide	5	1	80
Lead	3	<3	>0
Mercury	0.2	0.3	0[a]
Nickel	6.0	3.0	50
Silver	0.7	0.6	14
Zinc	27	<2	>92

[a]Actual data indicate negative removal.

Note: Blanks indicate information was not specified.

TREATMENT TECHNOLOGY: Reverse Osmosis

Data source: Effluent Guidelines
Point source category: Steam electric power generatingSubcategor
Subcategory:
Plant: 1226
References: A31, pp. 15-17

Data source status:

Engineering estimate	___
Bench scale	___
Pilot scale	x
Full scale	___

Use in system: Tertiary
Pretreatment of influent: Ash pond

DESIGN OR OPERATING PARAMETERS

Product flow rate:
Flux rate:
Membrane configuration:
Membrane type:
Retentate (concentrate) flow rate:
Recycle flow rate:
Operating temperature:
Rated production capacity:
pH: 9.1

REMOVAL DATA

Sampling period:

Pollutant/parameter	Concentration, μg/L		Percent
	Influent	Effluent	removal
Toxic pollutants:			
Antimony	7	BDL[a]	>0
Arsenic	9	<1	>89
Cadmium	2.0	1.3	35
Chromium	6	<2	>67
Copper	14	10	29
Cyanide	<1	8	0[b]
Lead	4	<3	>25
Nickel	5.5	5.0	9
Selenium	8	2	75
Silver	0.5	<0.2	>60
Zinc	7	<2	>57

[a]Below detection limit; assumed to be less than the corresponding influent concentration.

[b]Actual data indicate negative removal.

Note: Blanks indicate information was not specified.

TREATMENT TECHNOLOGY: Reverse Osmosis

Data source: Effluent Guidelines
Point source category: Steam electric power generating
Subcategory:
Plant: 5409
References: A31, pp. 16-19

Data source status:
Engineering estimate ___
Bench scale ___
Pilot scale x
Full scale ___

Use in system: Primary
Pretreatment of influent:

DESIGN OR OPERATING PARAMETERS

Product flow rate:
Flux rate:
Membrane configuration:
Membrane type:
Retentate (concentrate) flow rate:
Recycle flow rate:
Operating temperature:
Rated production capacity:

REMOVAL DATA

Sampling period:

Pollutant/parameter	Concentration		Percent removal
	Influent	Effluent	
Conventional pollutants, mg/L:			
TOC	21	<20	>5
Toxic pollutants, μg/L:			
Beryllium	3.4	<0.5	>85
Cadmium	0.8	<0.5	>37
Chromium	37	<2	>94
Copper	620	51	92
Cyanide	5	24	0[a]
Lead	70	<3	>96
Mercury	0.5	<0.2	>60
Nickel	4.0	3.6	10
Silver	14	1.1	92
Thallium	8	4	50
Zinc	61	<2	>97

[a]Actual data indicate negative removal.

Note: Blanks indicate information was not specified.

TREATMENT TECHNOLOGY: Reverse Osmosis

Data source: Effluent Guidelines
Point source category: Steam electric power generating
Subcategory:
Plant: 5409
References: A31, pp. 16-19

Data source status:
Engineering estimate ___
Bench scale ___
Pilot scale x
Full scale ___

Use in system: Tertiary
Pretreatment of influent: Ash pond

DESIGN OR OPERATING PARAMETERS

Product flow rate:
Flux rate:
Membrane configuration:
Membrane type:
Retentate (concentrate) flow rate:
Recycle flow rate:
Operating temperature:
Rated production capacity:

REMOVAL DATA

Sampling period:

Pollutant/parameter	Concentration, µg/L Influent	Concentration, µg/L Effluent	Percent removal
Toxic pollutants:			
Antimony	5	2.5	50
Arsenic	74	<1	>99
Copper	26	9	65
Nickel	25	1.5	40
Selenium	42	6.1	85
Silver	1	1	0
Thallium	9	1	89
Zinc	11	2	82

Note: Blanks indicate information was not specified.

TREATMENT TECHNOLOGY: Reverse Osmosis

Data source: Effluent Guidelines
Point source category: Steam electric power generating
Subcategory:
Plant: 5604
References: A31, pp. 14-17 (Appendix)

Data source status:
- Engineering estimate ___
- Bench scale ___
- Pilot scale x
- Full scale ___

Use in system: Primary
Pretreatment of influent:

DESIGN OR OPERATING PARAMETERS

Product flow rate:
Flux rate:
Membrane configuration:
Membrane type:
Retentate (concentrate) flow rate:
Recycle flow rate:
Operating temperature:
Rated production capacity:

REMOVAL DATA

Sampling period:

Pollutant/parameter	Concentration, μg/L Influent	Effluent	Percent removal
Toxic pollutants:			
Antimony	5	2	60
Arsenic	7	49	0[a]
Cadmium	<0.5	2	0[a]
Copper	180	32	82
Lead	<3	20	0[a]
Nickel	6	<1	>83
Silver	3	4	0[a]
Zinc	780	3	99

[a]Actual data indicate negative removal.

Note: Blanks indicate information was not specified.

TREATMENT TECHNOLOGY: Reverse Osmosis

Data source: Effluent Guidelines
Point source category: Steam electric power generating
Subcategory:
Plant: 5604
References: A31, pp. 16-20 (Appendix)

Data source status:
- Engineering estimate ___
- Bench scale ___
- Pilot scale x
- Full scale ___

Use in system: Tertiary
Pretreatment of influent: Ash pond

DESIGN OR OPERATING PARAMETERS

Product flow rate:
Flux rate:
Membrane configuration:
Membrane type:
Retentate (concentrate) flow rate:
Recycle flow rate:
Operating temperature:
Rated production capacity:

REMOVAL DATA

Sampling period:

Pollutant/parameter	Concentration, μg/L Influent	Effluent	Percent removal
Toxic pollutants:			
Antimony	6	3	50
Beryllium	2.5	5	0[a]
Cadmium	1	<1	>0
Chromium	4	<1	>75
Copper	80	9	89
Cyanide	22	4	82
Nickel	9.5	<1	>89
Selenium	3	<1	>67
Silver	5.5	2	64
Thallium	<1	2	0[a]
Zinc	300	53	82

[a]Actual data indicate negative removal.

Note: Blanks indicate information was not specified.

TREATMENT TECHNOLOGY: Reverse Osmosis

Data source: EPA report
Point source category: Synthetic rubber
Subcategory: Emulsion crumb unless otherwise specified
Plant:
References:

Data source status:
Engineering estimate ___
Bench scale x
Pilot scale ___
Full scale ___

Use in system: Tertiary
Pretreatment of influent: None

DESIGN OR OPERATING PARAMETERS

Product flow rate:
Flux rate:
Membrane configuration: Hollow fiber
Membrane type: Polyamide
Retentate (concentrate) flow rate:
Recycle flow rate:
Operating temperature:
Rated production capacity:

REMOVAL DATA

Sampling period:

	BOD_5			COD			TOC		
	Concentration, mg/L		Percent	Concentration, mg/L		Percent	Concentration, mg/L		Percent
Plant	Influent	Effluent	removal	Influent	Effluent	removal	Influent	Effluent	removal
1[a]	11	4	64	511	6	99	66	8	88
2[b]	30	4	87	444	36	92	122	10	92
3[c]	12	1	92	830	20	98	246	8	97

	TSS			Oil and grease		
	Concentration, mg/L		Percent	Concentration, mg/L		Percent
Plant	Influent	Effluent	removal	Influent	Effluent	removal
1[a]	27	<4	>85	8	<4	>50
2[b]				11	7	36
3[c]	48	<5	>90	5	<4	>20

[a]Influent is from secondary effluent that has been filtered.

[b]Influent is from ultrafiltrate of secondary effluent; subcategory: solution crumb.

[c]Influent is from ultrafiltrate of final effluent from emulsion process.

Note: Blanks indicate information was not specified.

TREATMENT TECHNOLOGY: Reverse Osmosis

Data source: Government report
Point source category: Textile mills
Subcategory:[a]
Plant:
References: B17, pp. 4-7

Data source status:
- Engineering estimate ___
- Bench scale ___
- Pilot scale X
- Full scale ___

Use in system:
Pretreatment of influent:

[a]Wastewater from scour bath wash.

DESIGN OR OPERATING PARAMETERS

Product flow rate:
Flux rate:
Membrane configuration:
Membrane type: Cellulose acetate
Retentate (concentrate) flow rate:
Recycle flow rate:
Operating temperature:
Rated production capacity:

REMOVAL DATA

Sampling period:

Pollutant/parameter	Concentration		Percent
	Influent	Effluent	removal
Conventional pollutants, mg/L:			
Total phenol	0.006	0.016	0[a]
Toxic pollutants, μg/L:			
Antimony	100	132	0[a]
Arsenic	19	1	95
Cadmium	15	14	7
Chromium	640	620	3
Copper	90	32	64
Cyanide	<4	30	0[a]
Lead	380	340	11
Nickel	130	100	24
Silver	42	42	0
Zinc	520	8,600	0[a]
Bis(2-ethylhexyl) phthalate	9	3	67
Toluene	0.8	29	0[a]
Chloroform	5	22	0[a]
Methylene chloride	0.3	5	0[a]
Trichloroethylene	1	0.4	60

[a]Actual data indicate negative removal.

Note: Blanks indicate information was not specified.

TREATMENT TECHNOLOGY: Reverse Osmosis

Data source: Government report
Point source category: Textile mills
Subcategory:[a]
Plant:
References: B17, pp. 4-7

Data source status:
- Engineering estimate ___
- Bench scale ___
- Pilot scale x
- Full scale ___

Use in system: Tertiary
Pretreatment of influent:

[a]Dye waste wastewater.

DESIGN OR OPERATING PARAMETERS

Product flow rate:
Flux rate:
Membrane configuration:
Membrane type: Cellulose acetate
Retentate (concentrate) flow rate:
Recycle flow rate:
Operating temperature:
Rated production capacity:

REMOVAL DATA

Sampling period:

Pollutant/parameter	Concentration Influent	Concentration Effluent	Percent removal
Conventional pollutants, mg/L:			
Total phenol	0.019	0.018	5
Toxic pollutants, μg/L:			
Antimony	190	120	40
Arsenic	35	<1	>97
Cadmium	22	48	0[a]
Chromium	540	520	4
Copper	480	50	90
Lead	520	380	27
Nickel	220	62	72
Silver	82	20	76
Zinc	7,200	140	98
Bis(2-ethylhexyl) phthalate	4	3	25
Phenol	0.2	0.7	0[a]
Benzene	2	1	50
Toluene	10	24	0[a]
Chloroform	19	4	79
Methylene chloride	5	4	20

[a]Actual data indicate negative removal.

Note: Blanks indicate information was not specified.

TREATMENT TECHNOLOGY: Reverse Osmosis

Data source: Government report
Point source category: Textile mills
Subcategory:[a]
Plant:
References: B17, pp. 4-7

Data source status:
Engineering estimate ___
Bench scale ___
Pilot scale _x_
Full scale ___

Use in system: Tertiary
Pretreatment of influent:

[a]Dye waste wastewater.

DESIGN OR OPERATING PARAMETERS

Product flow rate:
Flux rate:
Membrane configuration: Dual-layer hydrous ZR (IV) oxide-polyacrylate dynamic membrane
Membrane type:
Retentate (concentrate) flow rate:
Recycle flow rate:
Operating temperature:
Rated production capacity:

REMOVAL DATA

Sampling period:

Pollutant/parameter	Concentration Influent	Effluent	Percent removal
Conventional pollutants, mg/L:			
Total phenol	0.064	0.012	81
Toxic pollutants, µg/L:			
Antimony	280	200	30
Arsenic	220	2	99
Cadmium	40	20	50
Chromium	1,000	900	10
Copper	3,100	11,000	0[a]
Cyanide	8	<4	>50
Lead	700	520	26
Nickel	480	190	61
Silver	120	70	40
Zinc	5,400	6,600	0[a]
Bis(2-ethylhexyl) phthalate	51	2	96
Di-n-butyl phthalate	6	1	83
Dimethyl phthalate	290	170	41
Phenol	1	0.2	80
Acenaphthene	7	3	57
Anthracene	3	0.7	77
Methylene chloride	14	5	64

[a]Actual data indicate negative removal.

Note: Blanks indicate information was not specified.

TREATMENT TECHNOLOGY: Reverse Osmosis

Data source: Government report
Point source category: Textile mills
Subcategory:[a]
Plant:
References: B17, pp. 4-7

Data source status:
Engineering estimate ___
Bench scale ___
Pilot scale x
Full scale ___

Use in system: Tertiary
Pretreatment of influent:

[a]Wastewater from scour bath wash.

DESIGN OR OPERATING PARAMETERS

Product flow rate:
Flux rate:
Membrane configuration:
Membrane type: Dual-layer hydrous Zr (IV) oxide-polyacrylate dynamic membrane
Retentate (concentrate) flow rate:
Recycle flow rate:
Operating temperature:
Rated production capacity:

REMOVAL DATA

Sampling period:

Pollutant/parameter	Concentration		Percent removal
	Influent	Effluent	
Conventional pollutants, mg/L:			
Total phenol	0.004	<0.001	>75
Toxic pollutants, μg/L:			
Antimony	170	150	14
Arsenic	35	5	86
Cadmium	16	20	0[a]
Chromium	760	800	0[a]
Lead	400	410	0[a]
Nickel	200	210	0[a]
Silver	62	78	0[a]
Zinc	460	250	46
Toluene	0.8	0.7	12
Methylene chloride	4	5	0[a]

[a]Actual data indicate negative removal.

Note: Blanks indicate information was not specified.

TREATMENT TECHNOLOGY: Reverse Osmosis

Data source: Government report
Point source category: Textile mills
Subcategory:[a]
Plant:
References: B17, pp. 4-7

Data source status:
- Engineering estimate ___
- Bench scale ___
- Pilot scale x
- Full scale ___

Use in system: Tertiary
Pretreatment of influent:

[a]Wastewater from scour bath wash.

DESIGN OR OPERATING PARAMETERS

Product flow rate:
Flux rate:
Membrane configuration:
Membrane type: Polyamide
Retentate (concentrate) flow rate:
Recycle flow rate:
Operating temperature:
Rated production capacity:

REMOVAL DATA

Sampling period:

Pollutant/parameter	Concentration Influent	Concentration Effluent	Percent removal
Conventional pollutants, mg/L:			
Total phenol	0.006	0.012	0[a]
Toxic pollutants, µg/L:			
Antimony	100	90	10
Arsenic	19	<1	>95
Cadmium	15	15	0
Chromium	640	720	0[a]
Copper	90	26	71
Cyanide	<4	72	0[a]
Lead	380	250	34
Nickel	130	70	47
Silver	42	26	38
Zinc	520	360	31
Bis(2-ethylhexyl) phthalate	9	3	67
Di-n-butyl phthalate	4	1	75
Toluene	0.8	15	0[a]
Acenaphthene	7	0.8	99
Pyrene	18	18	0
Chloroform	5	6	0[a]

[a]Actual data indicate negative removal.

Note: Blanks indicate information was not specified.

TREATMENT TECHNOLOGY: Reverse Osmosis

Data source: Government report
Point source category: Textile mills
Subcategory:[a]
Plant:
References: B17, pp. 4-7

Data source status:
Engineering estimate ___
Bench scale ___
Pilot scale x
Full scale ___

Use in system: Tertiary
Pretreatment of influent:

[a]Dye waste wastewater.

DESIGN OR OPERATING PARAMETERS

Product flow rate:
Flux rate:
Membrane configuration:
Membrane type: Polyamide
Retentate (concentrate) flow rate:
Recycle flow rate:
Operating temperature:
Rated production capacity:

REMOVAL DATA

Sampling period:

Pollutant/parameter	Concentration		Percent
	Influent	Effluent	removal
Conventional pollutants, mg/L:			
Total phenol	0.019	0.02	0[a]
Toxic pollutants, μg/L:			
Antimony	190	130	31
Arsenic	35	15	57
Cadmium	22	20	9
Chromium	540	760	0[a]
Copper	480	46	90
Lead	520	400	22
Nickel	220	200	9
Silver	82	68	17
Zinc	7,200	360	95
Bis(2-ethylhexyl) phthalate	4	31	0[a]
Di-n-butyl phthalate	1	0.8	20
Dimethyl phthalate	55	45	18
Phenol	0.2	0.7	0[a]
Benzene	2	0.4	80
Toluene	10	11	0[a]
Acenaphthene	3	0.8	73
Chloroform	19	31	0[a]
Methyl chloride	5	45	0[a]

[a]Actual data indicate negative removal.

Note: Blanks indicate information was not specified.

TREATMENT TECHNOLOGY: Reverse Osmosis

Data source: Conference paper
Point source category: Textiles
Subcategory:
Plant: Dye waste
References:

Data source status:
- Engineering estimate ___
- Bench scale ___
- Pilot scale x
- Full scale ___

Use in system: Tertiary
Pretreatment of influent:

DESIGN OR OPERATING PARAMETERS

Membrane configuration:
Flow rate:
Water recovery:
Membrane type:
Flux:

Influent pressure:

REMOVAL DATA

Sampling period:

	Concentration, mg/L			
	Influent			Percent
Pollutant/parameter	Feed	Brine	Effluent	removal
Conventional pollutants:				
TOC	140	670	12.5	92

Note: Blanks indicate information was not specified.

TREATMENT TECHNOLOGY: Reverse Osmosis

Data source: Government report
Point source category: Textile mills
Subcategory: Dyeing and finishing
Plant: LaFrance Industries
References: B12, pp. 119, 126, 141

Data source status:
- Engineering estimate ___
- Bench scale ___
- Pilot scale x
- Full scale ___

Use in system: Secondary
Pretreatment of influent: Filtration (250-μ screen)

DESIGN OR OPERATING PARAMETERS

Product flow rate:
Flux rate:
Membrane configuration: Eight externally coated 19-tube bundles in series
Membrane type: Selas Flotronics Zr (IV) - PAA
Retentate (concentrate) flow rate:
Recycle flow rate:
Operating temperature: 20-90°C
Rated production capacity:
Membrane inlet pressure: 2,400 - 7,200 kPa

REMOVAL DATA

Sampling period: Composite of several daily samples taken in 1-week period

Pollutant/parameter	Concentration Influent	Concentration Effluent	Percent removal
Conventional pollutants, mg/L:			
BOD_5	20[a]	2[a]	90[a]
COD	248[b]	14[b]	94[b]
	160	15	91
TOC	83[c]	6[c]	93[c]
	30	5	83
Toxic pollutants, μg/L:			
Zinc	1,400	30	98
	940	20	98

[a]Only one sample.
[b]Average of five samples.
[c]Average of six samples.

Note: Blanks indicate information was not specified.

TREATMENT TECHNOLOGY: Reverse Osmosis

Data source: Government report
Point source category: Textile mills
Subcategory: Dyeing and finishing
Plant: LaFrance Industries
References: B12, pp. 115, 125, 140

Data source status:
- Engineering estimate ___
- Bench scale ___
- Pilot scale x
- Full scale ___

Use in system: Secondary
Pretreatment of influent: Filtration (25-μ and 1-μ cartridge filters and diatomaceous earth filter when needed)

DESIGN OR OPERATING PARAMETERS

Product flow rate:
Flux rate:
Membrane configuration: Hollow polyamide filter
Membrane type: DuPont #400600
Retentate (concentrate) flow rate:
Recycle flow rate:
Operating temperature: 11-32°C
Rated production capacity:
Membrane inlet pressure: 2,400 kPa

REMOVAL DATA

Sampling period: Composite of several daily samples taken in 1-week period

Pollutant/parameter	Concentration Influent	Concentration Effluent	Percent removal
Conventional pollutants, mg/L:			
BOD_5	15	2	87
COD	110	10	91
	253[a]	32[a]	87[a]
TOC	47[b]	6[b]	87[b]
Toxic pollutants, μg/L:			
Zinc	3,600	500	86
	4,100	180	96

[a]Average of 14 samples.

[b]Average of 12 samples.

Note: Blanks indicate information was not specified.

TREATMENT TECHNOLOGY: Reverse Osmosis

Data source: Government report
Point source category: Textile mills
Subcategory: Dyeing and finishing
Plant: LaFrance Industries
References: B12, pp. 117, 126, 141

Data source status:
- Engineering estimate ___
- Bench scale ___
- Pilot scale x
- Full scale ___

Use in system: Secondary
Pretreatment of influent: Filtration (25-μ and 1-μ cartridge filter when necessary)

DESIGN OR OPERATING PARAMETERS

Product flow rate:
Flux rate:
Membrane configuration: Spiral-wound
Membrane type: Gulf, cellulose acetate
Retentate (concentrate) flow rate:
Recycle flow rate:
Operating temperature: 15-26°C
Rated production capacity:
Membrane inlet pressure: 2,800 kPa

REMOVAL DATA

Sampling period:

Pollutant/parameter	Concentration Influent	Concentration Effluent	Percent removal
Conventional pollutants, mg/L:			
BOD_5	10	1	90
	104[a]	18[a]	83[a]
COD	160	25	84
	590[b]	26[b]	96[b]
TOC	35	5	86
	109[c]	7[c]	94[c]
Toxic pollutants, μg/L:			
Chromium	300	100	67
Copper	120	40	67
	1,000[b]	71[b]	93[b]
Zinc	960	40	96
	1,200[b]	22[b]	98[b]

[a]Average of four samples.

[b]Average of 13 samples.

[c]Average of 12 samples.

Note: Blanks indicate information was not specified.

TREATMENT TECHNOLOGY: Reverse Osmosis

Data source: Government report
Point source category: Textile mills
Subcategory: Dyeing and finishing
Plant: LaFrance Industries
References: B12, pp. 113, 125, 140

Data source status:
- Engineering estimate ___
- Bench scale ___
- Pilot scale x
- Full scale ___

Use in system: Secondary
Pretreatment of influent: Filtration (25-μ cartridge filter)

DESIGN OR OPERATING PARAMETERS

Product flow rate:
Flux rate:
Membrane configuration: Tubular (18 in series)
Membrane type: Westinghouse #4-291
Retentate (concentrate) flow rate:
Recycle flow rate:
Operating temperature: <32°C
Rated production capacity:
Membrane inlet pressure: 2,100 - 3,100 kPa
Tube diameter: 13 mm

REMOVAL DATA

Sampling period: Composite of several daily samples taken in 1-week period

Pollutant/parameter	Concentration		Percent removal
	Influent	Effluent	
Conventional pollutants, mg/L:			
BOD_5	15	1.3	91
COD	320[a]	19[a]	94[a]
	891[b]	36[b]	96[b]
	150	200	0[c]
TOC	100[a]	7[a]	93[a]
	138[b]	9[b]	95[b]
Toxic pollutants, μg/L:			
Zinc	14,000[a]	230[a]	98[a]
	24,000[b]	430[b]	98[b]
	6,000	820	86

[a]Average of three samples.

[b]Average of eight samples.

[c]Actual data indicate negative removal.

Note: Blanks indicate information was not specified.

TREATMENT TECHNOLOGY: Reverse Osmosis

Data source: Government report
Point source category: Textile mills
Subcategory: Dyeing and finishing
Plant: LaFrance Industries
References: B12, pp. 111, 124, 139-140

Data source status:
- Engineering estimate ___
- Bench scale ___
- Pilot scale x
- Full scale ___

Use in system: Secondary
Pretreatment of influent: Filtration

DESIGN OR OPERATING PARAMETERS

Product flow rate:
Flux rate:
Membrane configuration:
Membrane type:
Retentate (concentrate) flow rate:
Recycle flow rate:
Operating temperature:
Rated production capacity:

REMOVAL DATA

Sampling period: Composite of several daily samples taken in 1-week period

Pollutant/parameter	Concentration Influent	Concentration Effluent	Percent removal
Conventional pollutants, mg/L:			
BOD_5	35	2.7	92
	16[a]	4[a]	75[a]
COD	599[b]	37[b]	94[b]
	230	30	87
	272[c]	42[c]	85[c]
	164[d]	13[d]	92[d]
TOC	153[e]	10[e]	93[e]
	50[c]	8[c]	84[c]
	24[d]	6[d]	75[d]
Toxic pollutants, µg/L:			
Zinc	9,700[f]	37[f]	>99[f]
	5,200	60	99
	2,500	20	99
	1,500[d]	38[d]	98[d]

[a]Average of two samples.
[b]Average of 13 samples.
[c]Average of six samples.
[d]Average of five samples.
[e]Average of 11 samples.
[f]Average of nine samples.

Note: Blanks indicate information was not specified.

TREATMENT TECHNOLOGY: Reverse Osmosis

Data source: Effluent Guidelines
Point source category: Timber products (pentachlorophenol wastewater)
Subcategory:
Plant:
References: Al, p. E-4

Data source status:
- Engineering estimate ___
- Bench scale ___
- Pilot scale x
- Full scale ___

Use in system: Secondary
Pretreatment of influent: Ultrafiltration

DESIGN OR OPERATING PARAMETERS

Product flow rate:
Flux rate:
Membrane configuration:
Membrane type:
Retentate (concentrate) flow rate:
Recycle flow rate:
Operating temperature:
Rated production capacity:

REMOVAL DATA

Sampling period:

Pollutant/parameter	Concentration, mg/L Influent	Concentration, mg/L Effluent	Percent removal
Conventional pollutants:			
Oil and grease	55	17	69

Note: Blanks indicate information was not specified.

TREATMENT TECHNOLOGY: Reverse Osmosis

Data source: Conference paper
Point source category:
Subcategory:
Plant: Brass finishing
References:

Data source status:
Engineering estimate ___
Bench scale ___
Pilot scale x
Full scale ___

Use in system: Tertiary
Pretreatment of influent:

DESIGN OR OPERATING PARAMETERS

Membrane configuration: Tubular
Flow rate:
Water recovery: 95%
Membrane type:
Flux:

Influent pressure:

REMOVAL DATA

Sampling period:

Pollutant/parameter	Concentration Influent	Concentration Effluent	Percent removal
Conventional pollutants, mg/L:			
COD	1,050	ND[a]	>99
Oil and grease	35.6	ND[a]	>72
Nitrate as N	3	0.2	90
TKN	3.2	ND[b]	>0
Toxic pollutants, µg/L:			
Chromium	10,000	25	>99
Chromium (+6)	ND	10	0[c]
Chromium (+3)	10,000	15	>99
Copper	120,000	90	>99
Lead	1,400	<10	>99
Nickel	600	<10	>98
Zinc	110,000	90	>99

[a]Not detected; assumed to be <10 µg/L.

[b]Not detected; assumed to be less than influent concentration.

[c]Actual data indicate negative removal.

Note: Blanks indicate information was not specified.

TREATMENT TECHNOLOGY: Reverse Osmosis

Data source: Journal article
Point source category:
Subcategory:
Plant: Municipal sewage (pretreated)
References:

Data source status:
- Engineering estimate ___
- Bench scale x
- Pilot scale ___
- Full scale ___

Use in system: Tertiary
Pretreatment of influent:

DESIGN OR OPERATING PARAMETERS

Membrane configuration:
Influent pressure: 4,140 kPa (600 psi)
Flow rate:
Water recovery: 95%
Membrane type:
Flux:
Temperature: 25°C

REMOVAL DATA

Sampling period:

Pollutant/parameter	Concentration, mg/L Influent	Effluent	Percent removal
Conventional pollutants:			
TOC	67.0	11.1	84

Note: Blanks indicate information was not specified.

TREATMENT TECHNOLOGY: Reverse Osmosis

Data source: Symposium article
Point source category:
Subcategory:
Plant: Sanitary waste
References:

Data source status:
Engineering estimate ___
Bench scale ___
Pilot scale x
Full scale ___

Use in system: Tertiary
Pretreatment of influent:

DESIGN OR OPERATING PARAMETERS

Membrane configuration: Hollow fiber
Influent pressure:
Flow rate: 0.076 m^3/min feed (20 gpm), 0.066 m^3/min product (175 gpm)
Water recovery: 89%
Membrane type: Cellulose triacetate
Flux:

REMOVAL DATA

Sampling period:

Pollutant/parameter	Concentration Influent (average)	Concentration Effluent (average)	Percent removal
Toxic pollutants, µg/L:			
Zinc	200	100	50

Note: Blanks indicate information was not specified.

TREATMENT TECHNOLOGY: Reverse Osmosis

Data source: Technical literature
Point source category:
Subcategory:
Plant: Cooling tower water-chromate removal
References:

Data source status:
- Engineering estimate ___
- Bench scale x
- Pilot scale ___
- Full scale ___

Use in system: Tertiary
Pretreatment of influent:

DESIGN OR OPERATING PARAMETERS

Membrane configuration:
Flow rate:
Water recovery:
Membrane type:
Flux:

Influent pressure:

REMOVAL DATA

Sampling period:

Pollutant/parameter	Concentration		Percent removal
	Influent	Effluent	
Toxic pollutants, µg/L:			
Chromium	35,500	1,500	96
Zinc	10,000	300	97.0

Note: Blanks indicate information was not specified.

6.10 ELECTRODIALYSIS [1]

6.10.1 Function

The general function of electrodialysis is the separation of an aqueous stream under the action of an electric field into two streams: an enriched stream (more concentrated in electrolyte than the original), and a depleted stream. Success of the process depends on special synthetic membranes, usually based on ion exchange resins, which are permeable only to a single charge type of ion. Cation exchange membranes permit passage only of positive ions under the influence of the electric field; anion exchange membranes permit passage only of negatively charged ions.

6.10.2 Description

In the electrodialysis process, feed water passes through compartments formed by the spaces between alternating cation-permeable and anion-permeable membranes held in a stack. At each end of the stack is an electrode that has the same area as the membranes. A dc potential applied across the stack causes the positive and negative ions to migrate in opposite directions. Because of the semipermeability of the membranes, a given ion will either migrate to the adjacent compartment or be confined to its original compartment, depending on whether or not the first membrane it encounters is permeable to it. As a result, salts are concentrated or diluted in alternate compartments.

To achieve high throughput, electrodialysis cells in practice are made very thin and assembled in stacks of cells in series. Each stack often consists of more than 100 cells. Feed material is first filtered to remove suspended particulate matter that could clog the system or foul the membrane and, if required, is given a pretreatment to remove oxidizing materials and ferrous or manganous ions, which would damage the membranes. Very high organic levels may also lead to membrane fouling. The catholyte stream is commonly acidified to offset the increase in pH that would normally occur within the cell, and an antiscaling additive may be required as well. An operating plant usually contains many recirculation, feedback, and control loops and pumps to optimize the concentrations and pH's at different points and thus maximize the overall efficiency. Although a certain amount of water transfer (electrosmosis) does occur, the process can be categorized with ion exchange, solvent extraction, or adsorbent processes as one in which solutes are removed from the solvent, rather than with distillation, freezing, or reverse osmossis in which the solvent is transported.

All ionized species are not removed in proportion to their concentration because of different mobilities and equilibrium concentrations within the membrane. Therefore, a solution partially deionized or concentrated by electrodialysis may contain significantly different proportions of ionized species than the original feed.

Many colloids and polyanions have a net negative charge. For this reason they may collect upon or foul anion exchange membranes because of their positively charged functional groups. This problem may be avoided to some extent using an electrodialysis cell that consists of alternating cation and "neutral" membranes. Such systems utilizing a porous "neutral" membrane to avoid convective flow or mixing, frequently perform very well from a separation standpoint although they are not common commercially because of their higher electrical power requirements.

Generally, electrodialysis works best on acidic streams containing a single principal metal ion (such as acid nickel baths). At alkaline pH's membrane life may diminish, but the system has been reported useable up to pH 14 under special circumstances. Mixed metals may not be concentrated in the same ratio as that in the feed, leading to problems in recycle. In addition, although a sodium and copper cyanide stream may perform as expected under electrodialysis, the presence of zinc (a common occurrence, especially in brass plating) can foul the anion membrane by the $(ZnCl)^-$ ion and partially convert that membrane to the cation form, with significant loss in system performance. If strongly alkaline, the feed streams are generally neutralized or rendered slightly acidic to prevent degradation of the anion membrane, which usually contains quaternary ammonium groups. Iron and manganese in the feed water also degrade most common membranes and must be removed if their total concenration in the feed water is greater than about 0.3 mg/L.

Calcium sulfate scale can also accumulate if the calcium concentration in the concentrated stream is allowed to exceed about 400 mg/L. Addition of a sequestering agent to the feed permits operation to a higher calcium concentration, but generally not above 900 mg/L. For this reason, the brine rarely constitutes less than 10 to 15% of the feed water volume (a concentration factor of 6 to 10).

Because the process depends on electrolytic conductance through the various liquid streams, it is rarely practical to produce product water of less than about 250 ppm total dissolved solids. For the same reason, it is often desirable to operate an electrodialysis system at a slightly elevated temperature. As a rule of thumb, a temperature increase of 17°C reduces the power consumption by 1%.

Membrane life, although dependent upon service conditions, is frequently five years. Other components are generally long lived, because the system, although somewhat corrosive perhaps, operates at a modest or ambient temperatures and pressures, and abrasives and particulates normally will have been removed from the feed water.

6.10.3 Technology Status

Electrodialysis is a mature technology with well-known performance characteristics and prices; it can be easily evaluated as a potential component of any multiprocess treatment being considered. However, its success may be determined to a large extent by whether it can be made sufficiently reliable and attention free to be offered as a "black box" treatment package.

6.10.4 Applications

Industrial applications are widespread but varied and include the use of the process to remove the mineral constituents or contaminants from process streams that contain large amounts of organic products, e.g., de-ashing of sugars, washing of photographic emultions, and demineralization of whey. It frequently is used in the production of potable water from brackish waters, for the desalting of food products such as whey, and in the chemical industry for a variety of solution enrichment or depletion purposes.

Pilot operations have been carried out on the desalting of sewage plant effluent, sulfite-liquor recovery, acid mine drainage treatment, the desalting of cooling tower waters, and numerous other industrial applications. Treatment of plating wastes and rinses has been studied and piloted with encouraging but generally modest results. Recent work at General Motors suggests use of the process to salvage chromium wastes from chromic plating rinses.

At least two facilities have installed electrodialysis units to treat the hydrogen fluoride and ammonium fluoride effluents from glass and quartz etching facilities. Starting with a feed stream that contains 400 to 500 ppm fluorides, it is possible to produce a dischargeable dilute stream and a low-volume concentrate stream that may be recycled or economically treated.

An interesting example exists of the use of electrodialysis in series with reverse osmossis for the treatment of a concentrated salt (NaCl) stream. Such a system is presently in the pilot-plant stage. Although cost data are not yet available, this application shows how a system utilizing more than one type of process may be arranged. Here electrodialysis is chosen for the salt-rich end of the system where it can operate at high current efficiency.

6.10.5 Limitations

Electrodialysis is not available as standard "turnkey" equipment for pollution control, and its design and operation may require more skill and care than that of other systems with which it may compete. It will probably continue as a viable process in those applications for which it is especially suitable, but it does not appear to have general utility as a waste treatment tool.

6.10.6 Residuals Generated/Environmental Impact

An electrodialysis plant produces two product streams, one concentrated and one dilute in the original contaminants; these must be either recycled, sold, or disposed of in some other manner. Electrodialysis may cause some local air pollution, because both H_2 and a Cl_2O_2 mix may be generated at the electrode surfaces. These represent a hazard if permitted to collect in an enclosed space; therefore, they generally are vented to the outside and allowed to escape into the atmosphere.

6.10.7 Reliability

For this technology, reliability is highly dependent on operator skill and the specific application.

6.10.8 Flow Diagram

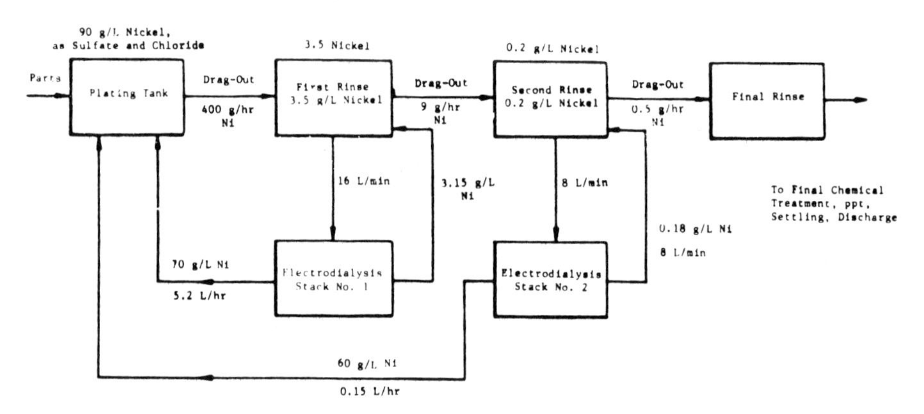

6.10.9 References

1. Physical, Chemical, and Biological Treatment Techniques for Industrial Wastes, PB 275 287, U.S. Environmental Protection Agency, Washington, D.C., November 1976. pp. 18-1 through 18-14.

6.11 DISTILLATION [1]

6.11.1 Function

Distillation is a unit operational process that is most often employed in industry to segregate, separate, or purify liquid organic product streams, some of which contain aqueous fractions. Sometimes the operation is used to recover one product; sometimes it is used to produce many desirable fractions from a process stream. Distillation is usually nondestructive and can produce products of any desired composition.

6.11.2 Description

Distillation is the boiling of a liquid solution and condensation of the vapor for the purpose of separating the components. In the distillation process there are two phases, the liquid phase and the vapor phase. The components that are to be separated by distillation are present in both phases but in different concentrations. If there are only two components in the liquid, one concentrates in the condensed vapor (condensate) and the other in the residual liquid. If there are more than two components, the less volatile components concentrate in the residual liquid and the more volatile in the vapor or vapor condensate. The ease with which a component is vaporized is called its volatility, and the relative volatilities (ratio of equilibrium ratios) of the components determine their vapor-liquid equilibrium relationships.

There are five general types of distillation, and a general description of each type is provided below.

- Batch Distillation. The simplest form of distillation is a single equilibrium stage operation. It is carried out in a "still" in which the reboiler equivalent consists of a steam jacket or a heating coil. The liquid is "boiled"; the vapor is driven off, condensed, and collected in an accumulator (a condensed vapor collector) until the desired concentration of the "product" has been reached. As the remaining liquid becomes leaner in the volatile component and richer in the less volatile component, its volume diminishes. If the residual liquid is the product, then "bottoms" concentration will be the controlling parameter. The batch still, as previously described, consists of a vessel that provides one equilibrium stage. By adding a condenser and recycling some of the condensed vapor, a second vapor/liquid equilibrium stage is added, and the separation is improved.

- Continuous Fractional Distillation. In continuous fractional distillation, a steady stream feed enters the column, which contains plates or packing (packing is normally used only in small-scale equipment) that provide additional vapor/liquid contact (equilibrium) stages. Overhead vapors and bottoms are continuously withdrawn. Vapor from the top plate is condensed and collected in a vessel known as an accumulator. Some of the liquid in the accumulator is continuously returned to the top plate of the column as reflux while the remainder of the liquid is continuously withdrawn as the overhead product stream. At the bottom of the column the liquid collects in the reboiler, where it is heated by steam coils or a steam jacket. The function of the reboiler is to receive the liquid overflow from the lowest plate and return a protion of this as a vapor stream, while the remainder is withdrawn continuously as a liquid bottom product.

- Azeotropic Distillation. An azeotrope is a liquid mixture that maintains a constant boiling point and produces a vapor of the same composition of the mixture when boiled. Because the composition of the vapor produced from an azeotrope is the same as that of the liquid, an azeotrope may be boiled away at a constant pressure, without change in concentration in either liquid or vapor. Since the temperature cannot vary under these conditions, azeotropes are also called constant boiling mixtures.

An azeotrope cannot be separated by constant pressure distillation into its components. Furthermore, a mixture on one side of the azeotrope composition cannot be transformed by distillation to a mixture on the other side of the azeotrope. If the total pressure is changed, the azeotropic composition is usually shifted. Sometimes this principle can be applied to obtain separations under pressure or vacuum that cannot be obtained under atmospheric pressure conditions. Most often, however, a third component - an additive, sometimes called an entrainer - is added to the binary (two-component) mixture to form a new boiling-point azeotrope with one of the original constituents. The volatility of the new azeotrope is such that it may be easily separated from the other original constituents.

- Extractive Distillation. Extractive distillation is a multi-component rectification method of distillation. A solvent is added to a binary mixture that is difficult or impossible to separate by ordinary means. This solvent alters the relative volatility of the original constituents, thus permitting separation. The added solvent is of low volatility and is not appreciably vaporized in the fractionator.

- Molecular Distillation. Molecular distillation is a form of a very low pressure distillation conducted at absolute pressures of the order of 0.003 mm of mercury suitable for heat-sensitive substances. Ordinarily, the net rate of evaporation is very low, at a save temperature, owing to the fact the evaporated molecules are reflected back to the liquid after collisions occurring in the vapor. By reducing the absolute pressure to values used in the molecular distillation, the mean free path of the molecules becomes very large (in the order of 1 cm). If the condensing surface is then placed at a distance not exceeding a few centimeters from the vaporing liquid surface, very few molecules will return to the liquid and the net rate of evaporation is substantially improved.

6.11.3 Technology Status

The process is well developed for processing applications. Wastewater applications are less numerous and less demonstrated.

6.11.4 Applications

Treatment of waste by distillation is not widespread, perhaps because of the cost of the energy requirements. The only hazardous waste materials that can be feasibly and practicably treated are liquid organics, including organic solvents and halogenated organics, which do not contain appreciable quantities of materials that would cause operational or equipment problems.

There are a number of manufacturers of chemicals and chemical products who have always recovered solvent streams by distillation for internal reuse. There are independent operators and companies that specialize in solvent or chemical reclamation by distillation. Historically, distillable solvents have been recovered primarily as an economic consideration, but with imposition of more stringent government regulations for the disposal of hazardous wastes and increases in the cost of petrochemicals, by-product credits will become even more important. Thus, the recovery of organic solvents should become more prevalent. If by-product credits offset the higher cost of distillation, vs the cost of other recovery methods, distillation will become a more competitive means of waste solvent recovery.

The solvent reclaiming industry pertains to those private contractors engaged in the reprocessing of organic solvents. In many cases, these operations also include other means of reclamation such as steam-stripping evaporation, filtration, etc.

Typical industrial wastes which can be handled by distillation are listed below:

- Plating wastes containing an organic component - usually the solvents are evaporated and the organic vapors distilled.

- Organic effluents from printed circuit boards are adsorbed on activated carbon. Regeneration of the activated carbon gives a liquid which is distillable for recovery of the organic component.

- Phenol recovery from aqueous solutions is a major waste treatment problem. The recovery process uses a polymeric adsorber, which is regenerated using a vaporized organic solvent. A complex distillation system is used to recover both the regeneration solvent and the phenol.

- Methylene chloride that contains contaminants is a disposal problem, but it can be salvaged for industrial application by distilling.
- Methylene chloride can be recovered from polyurethane waste.
- The separation of ethylbenzene from styrene and recovery of both.
- Waste solvents for reuse in cleaning industrial equipment; this is usually a mixture of acetone (ketones) (alcohols) and some aromatics.
- Recovery of acetone from a waste stream that was created by the regeneration of a carbon adsorption bed used to remove acetone vapor from the offgas in plastic filter products.
- The production of (penicillin) antibiotics results in the generation of large quantities of wastes containing butyl acetate. The waste is distilled, and a portion of the butyl acetate can be recycled. The still bottoms, however, are hazardous wastes, which contain 50% butyl acetate and 50% dissolved organics (fats and protein). These are disposed of by incineration.
- Waste motor oil from local service stations and from industrial locations can be re-refined to produce regenerated lube oil or fuel oil with the aid of distillation.

6.11.5 Limitations

Equipment and auxiliaries are usually comparatively large; they can have heights up to 200 ft and cover large land areas.

The equipment is expensive, and capital recovery changes usually constitute the major portion of solvent recovery cost.

Recovery is energy-intensive and is a close second to capital recovery charges; energy requirements are nominally 250 to 1,200 Btu/lb of feed.

Application to feed is limited in that it will handle only liquid solutions that are relatively "clean."

Equipment is often complex and requires operation by highly skilled personnel.

6.11.6 Residuals Generated/Environmental Impact

Waste treatment by distillation creates no air or liquid effluent problems that cannot be easily averted. Still bottoms may present a waste disposal problem, because they sometimes contain considerable quantities of tars and sludges that are usually incinerated. Vacuum distillation using steam or water eductors, yields volatile impurities in the condensed steam or water used to produce the vacuum. Disposal of this water is always a problem. Where disposal or treatment of this waste is a major problem, mechanical vacuum pumps might be considered as an alternative to the eductor.

6.11.7 Reliability

Process is highly reliable for proven applications and when properly operated and maintained.

6.11.8 Flow Diagram

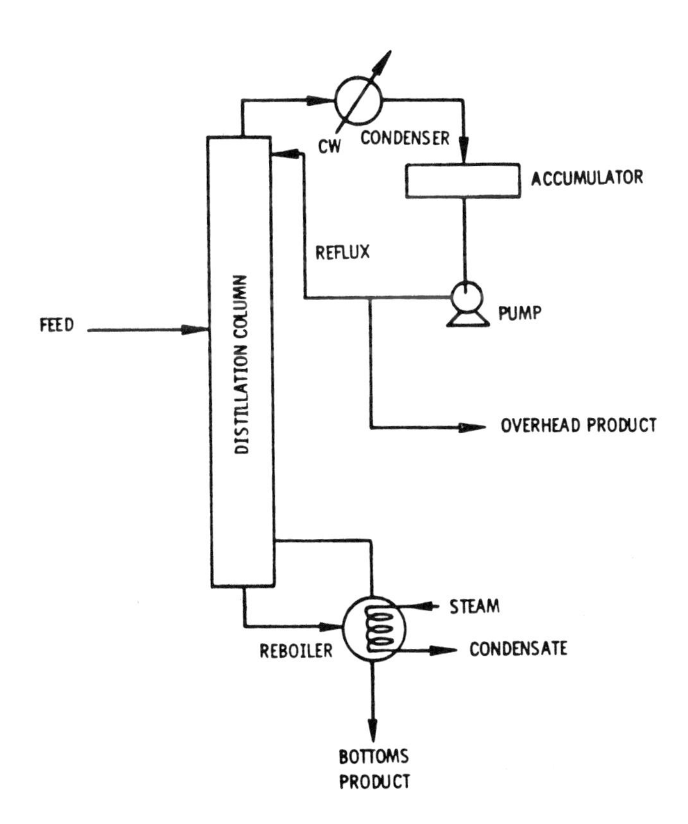

6.11.9 References

1. Physical, Chemical, and Biological Treatment Techniques for Industrial Wastes, PB 275 287, U.S. Environmental Protection Agency, Washington, D.C., November 1976. pp. 17-1 through 17-35.

6.12 Chlorination (Disinfection) [1]

6.12.1 Function

Chlorination is the most commonly used disinfection process; it is especially used for the removal of pathogens and other disease causing organisms.

6.12.2 Description

The chlorination process involves the addition of elemental chlorine or hypochlorites to the wastewater. Chlorine combines with water to form hypochlorous and hydrochloric acids:

$$Cl_2 + H_2O \rightleftarrows HOCl + H^+ + Cl^-$$

Hydrolyses of the hypochlorous acid to the hypochlorite ion

$$HOCl \rightleftarrows H^+ + OCL$$

is pH dependent. The equilibrium at pH 8 is approximately 80% OCl^-. Since $HOCl^-$ is more effective, a pH of 7.5 or less is best.

Chlorine and hychlorous acid react within a wide variety of substances, including ammonia. This, in effect, causes a chlorine demand which must be satisifed before the disinfecting ability of the ions can be utilized. In potable water systems this demand is satisfied quickly and the $HOCl^-$ and OCl^- ions become available for disinfection. In wastewater which contains ammonia, hypochlorous acid reacts to form mono, di, and trichloramines:

$$NH_3 + HOCl \longrightarrow NH_2Cl + H_2O$$
$$NH_2Cl + HOCl \longrightarrow NHCl_2 + H_2O$$
$$NHCl_2 + HOCl \longrightarrow NCl_3 + H_2O$$

Continued chlorination would take the NCl_3 to nitrogen gas. Since monochloramine and dichloramine have significant disinfecting power, it is often best to rely on these ions for disinfection rather than attempting to satisfy the total demand and make available the HOCl and OCl^- ions.

The amount of chlorine added is determined by cylinder weight loss. Chlorine demand is determined by the difference between the chlorine added and the measured residual concentration after a certain period has passed from the time of addition; this detention (or exposure) is usually 15-30 minutes.

6.12.3 Common Modifications

Chlorine or hypochlorite salts can be used. The two most common hypochlorite salts are calcium and sodium hypochlorite. In areas where saline water is available, electrolytic processes are available for on-site generation of hypochlorite. Dechlorination may be required subsequent to disinfection if only very low chlorine residuals are allowed; this generally involves the addition of sulfur dioxide, aeration, or even activated carbon adsorption.

6.12.4 Technology Status

Chlorination of water supplies on an emergency basis has been practiced since about 1850. Presently, chlorination of both water supplies and wastewaters is an extremely wide-spread practice.

6.12.5 Applications

Used to prevent the spread of waterborne diseases and to control algae growth and odors.

6.12.6 Limitations

May cause the formation of chlorinated hydrocarbons, some of which are known to be carcinogenic compounds. The effectiveness of chlorination is greatly dependent on pH and temperature of the wastewater. Chlorine gas is a hazardous material, and requires sophisticated handling procedures. Chlorine will react with certain chemicals in the wastewater, leaving only the residual amounts of chlorine for disinfection. Chlorine will oxidize ammonia, hydrogen sulfide, as well as metals present in their reduced states.

6.12.7 Chemicals Required

Chlorine, sodium hypochlorite, or calcium hypochlorite.

6.12.8 Design Criteria

Generally a contact period of 15 to 30 minutes at peak flow is required. Detention tanks should be designed to prevent short circuiting; this usually involves the use of baffling. Baffles can either be the over-and-under or the end-around varieties. Residuals of at least 0.5 mg/L are generally required. The following table, found on the next page, presents typical dosages for disinfection:

Effluent from	Dosage range, mg/L
Untreated wastewater (prechlorination)	6 to 25
Primary sedimentation	5 to 20
Chemical-precipitation plant	3 to 10
Trickling-filter plant	3 to 10
Activated-sludge plant	2 to 8
Multimedia filter following activated-sludge plant	1 to 5

6.12.9 Reliability

Process is extremely reliable.

6.12.10 Environmental Impact

Can cause the formation of chlorinated hydrocarbons; chlorine gas may be released to the atmosphere; relatively small land requirements.

6.12.11 Flow Diagram

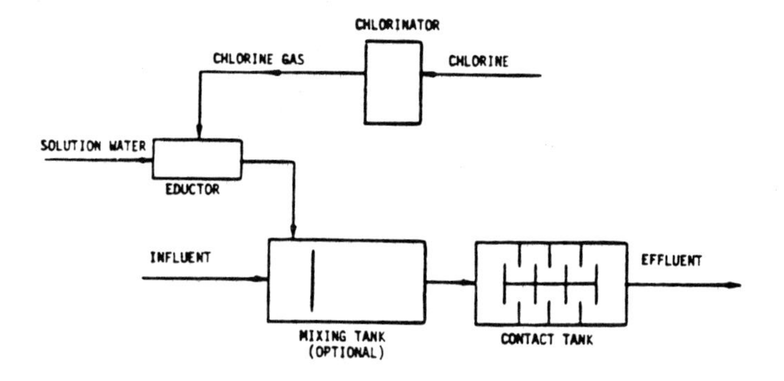

6.12.12 Performance

It should be noted that disinfection is designed to kill harmful organisms, and generally does not result in a sterile water (free of all microorganisms). The following table presents coliform remaining after 30 minutes of chlorine contact time.

	Coliform MPN/100 mL	
Chlorine residual, mg/L	Primary effluent	Secondary effluent
1	60,000	23,000
2	1,700	550
3	300	85
4	150	30

In normal low dose disinfection treatment, the COD, BOD_5, and TOC of the treated wastewater are not measurably changed.

6.12.13 References

1. Innovative and Alternative Technology Assessment Manual. EPA-430/9-78-009 (draft) U.S. Environmental Protection Agency, Cincinnati, Ohio, 1978. 252 pp.

6.13 DECHLORINATION [1]

6.13.1 Function

Dechlorination is used to remove free and combined chlorine.

6.13.2 Description

Since about 1970, much attention has been focused on the toxic effects of chlorinated effluents. Both free chlorine and chloramine residuals are toxic to fish and other aquatic organisms. Dechlorination involves the addition of sulfur dioxide to wastewater, whereby the following reactions occur:

$$SO_2 + HOCl + H_2O = SO_4^{+2} + Cl^- + 3H^+ \text{ (For free chlorine)} \quad (1)$$

$$SO_2 + NH_2Cl + 2H_2O = SO_4^{+2} + Cl^- + 2H^+ + NH_4^+ \text{ (For combined chlorine)} \quad (2)$$

As noted, small amounts of sulfuric and hydrochloric acids are formed; however, they are generally neutralized by the buffering capacity of the wastewater. Dechlorination can also be used in conjunction with superchlorination. Because superchlorination involves the addition of excess chlorine, dechlorination is required to eliminate this residual. Sulfur dioxide, the most common chemical used for dechlorination, is fed as a gas, using the same equipment as chlorine systems. Because the reaction of sulfur dioxide with free or combined chlorine is practically instantaneous, the design of contact systems is less critical than that of chlorine contact systems. Detention of less than 5 minutes is quite adequate, and in-line feed arrangements may also be acceptable under certain conditions.

6.13.3 Common Modifications

Metabisulfite, bisulfite, or sulfite salts can be used, as can automatic or manually fed systems. If chlorine is used at the site, sulfur dioxide is preferred, because identical equipment can be used for the addition of both chemicals. Alternative dechlorination systems include activated carbon, H_2O_2, and ponds (sunlight and aeration).

6.13.4 Technology Status

The technology of dechlorination with sulfur dioxide is established but is not in widespread use. A few plants in California and at least one in New York are known to be practicing effluent dechlorination with SO_2 on either a continuous or intermittant basis.

6.13.5 Applications

Dechlorination can be used whenever a chlorine residual is undesirable. This usually occurs when the receiving water contains aquatic life sensitive to free chlorine. Dechlorination is generally required when superchlorination is practiced or stringent effluent chlorine residuals are dictated.

6.13.6 Limitations

The process will not destroy chlorinated hydrocarbons already formed in the wastewater. It has been reported that about 1 percent of the chlorine ends up in a variety of stable organic compounds.

6.13.7 Chemicals Required

Sulfur dioxide (SO_2) and sulfite salts are the most common chemicals used; sodium metabisulfite ($Na_2S_2O_5$) can also be used, but is much less common; infact, any reducing agent can be considered, depending on cost and availability.

6.13.8 Reliability

Sulfur dioxide addition for dechlorination purposes is reasonably reliable from a mechanical standpoint; the greatest problems are experienced with analytical control which may lower the process reliability.

6.13.9 Environmental Impact

Requires very little use of land, and no residuals are generated; is used to eliminate the environmental impact of chlorine residuals; overdosing can result in low pH and low DO effluents, however.

6.13.10 Design Criteria

Contact time:	1 to 5 min
Sulfur dioxide feed rate:	1.1 lb/lb residual chlorine
Sodium sulfite feed rate:	0.57 lb/lb chlorine
Sodium bisulfite feed rate:	0.68 lb/lb chlorine
Sodium thiosulfate feed rate:	1.43 lb/lb chlorine

6.13.11 Flow Diagram

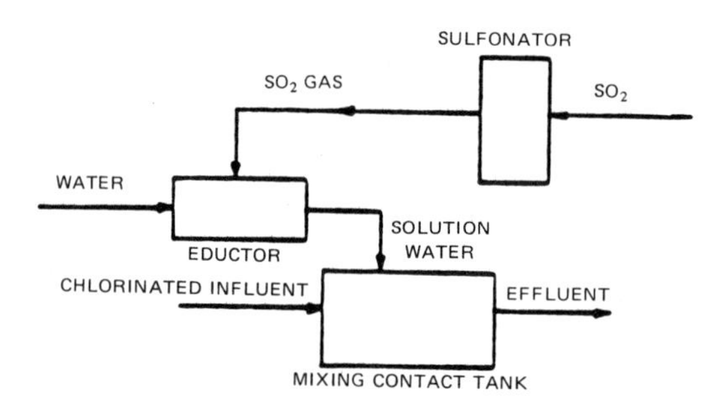

6.13.12 Performance

Available chlorine residuals can be reduced to essentially zero by sulfur dioxide dechlorination.

6.13.13 References

1. Innovative and Alternative Technology Assessment Manual. EPA-430/9-78-009 (draft) U.S. Environmental Protection Agency, Cincinnati, Ohio, 1978. 252 pp.

6.14 OZONATION [1]

6.14.1 Function

Ozonation is a process for oxidizing organics using ozone (O_3).

6.14.2 Description

Ozone is a powerful oxidizing agent, as illustrated by the following comparison of redox potentials of chlorine, ozone, and permanganate.

$$O_3 + 2H^+ + 2e^- \longrightarrow O_2 + H_2O \quad E_o = 2.07v \tag{1}$$

$$MnO_4^- + 4H^+ + 3e^- \longrightarrow MnO_2 + 2H_2O \quad E_o = 1.70v \tag{2}$$

$$1/2\ Cl_2 + e^- \longrightarrow Cl^- \quad E_o = 1.36v \tag{3}$$

Ozone is sufficiently strong to break many carbon-carbon bonds and even to cleave aromatic ring systems (e.g., conversion of phenol to three molecules of oxalic acid). Complete oxidation of an organic species to CO_2, H_2O, etc., is not improbable if the ozone dosage is sufficiently high.

Ozone dosage is commonly expressed in two ways: ppm of ozone, and pounds of ozone per pound of stream contaminant treated. The ozone dosage in ppm ozone is obtained by multiplying the flow rate of ozonized gas by the concentration of ozone in the gas and dividing by the flow rate of the waste stream. In disinfection applications, ozone doses of <4 ppm are typical for secondary treated streams. In industrial waste treatment applications, it is more usual to supply ozone at 10, 20, or 40 ppm. In the second measure of ozone dosage, the weight ratio of ozone to contaminant treated is obtained from the ppm ozone applied, the residence time of the waste stream in the ozone contact chamber, and the concentrations of contaminant in the influent and effluent streams. The ratio can vary from less than one (0.33 parts ozone per part of cyanide under optimum conditions) to very large values (approximately 80 parts ozone per part of phenol for very low concentrations of phenol). In most applications, the amount of ozone applied is 1.5 to 3 pounds of ozone per pound of contaminant removed.

The two measures of ozone dosage are clearly not entirely independent. However, it should be noted that 40 min of treatment at 10 ppm ozone will not, *a priori*, produce the same result as 10 min of treatment of 40 ppm ozone. The optimum combination of instantaneous ozone dose (ppm) and contact time must be determined for each case.

The extent of oxidation obtained will increase as either the weight ratio or the instantaneous dose is increased, up to certain limits defined by the fundamental chemistry of the ozonation reaction(s). However, there are practical and economic constraints on the amount of ozone that can actually be applied. Ozone is generally produced at a concentration of about 1% by weight in air (2% maximum) or 2 to 3% by weight in oxygen (6% maximum). This corresponds to 650 ft^3 of air, or to 325 ft^3 of oxygen, per pound of ozone delivered. To produce an instantaneous dose of 40 ppm O_3 in a waste stream, one would have to supply 208 ft^3 of ozonized air per 1,000 gallons (133 ft^3) of waste. This would require very efficient mixing indeed to achieve effective mass transfer. With a Venturi mixer, for example, the maximum ozone dose obtainable from ozonized air is 15 ppm. These calculations indicate why there is intense interest in design and development of more efficient ozone delivery systems.

Ozone is more soluble and more stable in acidic than in basic solutions. However, the rate of most ozonation reactions is relatively insensitive to pH, and it is rarely worthwhile to adjust pH prior to ozonation. The cost of the neutralization process will frequently offset any gains in ozonation efficiency. One exception to this generalization is cyanide ozonation. The cyanate formed initially hydrolyzes more rapidly in alkaline media. If complete conversion of cyanide to CO_2 is required, acidic streams should be adjusted to a pH of about 9 before ozonation. (Ammonia ozonation is also more effective in alkaline solution, but ozonation is unlikely to be the treatment method of choice for this species.)

6.14.3 Technology Status

Technology for large-scale ozone application is well developed. Applications to industrial wastes are not numerous, but feasibility has been demonstrated for cyanides and for phenols. Laboratory and pilot studies have demonstrated potential for ozone treatment of other oxidizable hazardous species including chlorinated hydrocarbons polynuclear aromatics, and pesticides. Several research studies and small scale applications have been reported on ultraviolet light catalyzed ozonation of organics.

6.14.4 Applications

Ozone treatment has been used in Europe and elsewhere in large-scale installations for years, for disinfection of water supplies. Over 500 such installations are in use worldwide. Within the past few years, there have been a number of pilot- and full-scale applications of ozone to treatment of municipal sewage plant effluents in the United States. The following are some selected examples of application of ozone to hazardous waste problems:

Liquid Effluents: Cyanide

- At an installation in Kansas, 350 lb/day of ozone are used to treat effluent containing cyanides, sulfides, sulfites, and other hazardous components; this ozonation follows biological waste treatment.
- At the Michelin tire factories in Clermont-Ferrand, France, 3.5 lb ozone/lb cyanide are used to reduce cyanide levels in effluent from 25 mg/L to 0 mg/L; a flow of 90 gpm is being treated.
- At a large chemical plant in France, UV/ozonation (Houston Research) is being used to treat several hundred gpm of waste containing 6 mg/L of $Fe(CN)_6^{-3}$; the effluent stream meets the standard of <0.1 mg/L total cyanide; in the process, oxidation of organics also occurs; the influent stream has TOC of 800 ppm, and the effluent has a BOD of 30 ppm.
- In the northeastern U.S., a small (<10,000 gpd) electroplating facility uses ozonation as the primary treatment process in reducing CN^- levels (60 ppm) in plating wastes to below detectable levels.

Liquid Effluents: Dyestuffs

- In Japan, a combination of ozonation and activated carbon adsorption has been used to remove color, BOD and COD from 3,300 m^3/d of waste dyeing water.

Liquid Effluents: Phenols

- An ozonation process for oxidation of phenols in coke oven wastes is in the start-up phase at Allen Wood Steel in Conshohocken, Pennsylvania; an ozone dosage of 10 ppm reduces phenol from 10 ppb to <0.5 ppb.
- An ozonation process for oxidation of phenols in wood products waste is in the start-up phase at Blandon F.P., Grand Rapids, Michigan; an ozone dosage of 10 ppm reduces phenol from 300 ppb to <10 ppb.
- At the Cities Service Refinery in Bronte, Ontario, Canada, ozonation is used for final removal of phenols from biologically treated effluents; the ozone is applied at 20 ppm to 40 ppm and reduces phenol from 380 ppb to 12 ppb.

6.14.5 Limitations

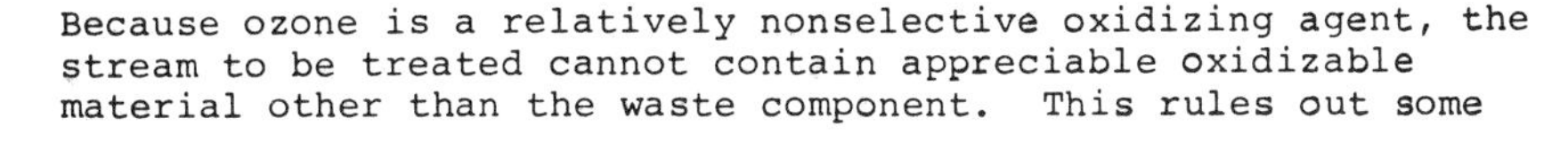

Because ozone is a relatively nonselective oxidizing agent, the stream to be treated cannot contain appreciable oxidizable material other than the waste component. This rules out some nonaqueous solvents, which are easily oxidizable organics. Sludges, slurries, and tars are also unlikely candidates for ozonation, unless they could be greatly diluted and dispersed in water. Even if these wastes were primarily composed of nonoxidizable material, the problems of mixing ozonized air with the semisolid waste would be formidable. It might be possible to ozonate suitable bulk solids or powders in a fluidized bed reactor, but we know of no attempts to do so. Of course, the waste would still be subject to the restriction of low levels of oxidizable material. (It should be noted that the ozonized air produced by modern generators is at low pressure (approximately 8 psi) and would not suffice to fluidize the waste.)

6.14.6 Typical Equipment

Ozone is produced by the facile reaction of oxygen molecules with oxygen atoms that are produced from oxygen by the action of ultraviolet light or an electric discharge. The photochemical production of ozone is important in stratospheric chemistry, but commercial ozone generators are all of the electric-discharge type.

In an electric-discharge ozone generator, an oxygen-containing gas is passed between two electrodes, coated with a dielectric material such as borosilicate glass. A high voltage (5 to 20 kilovolts) ac (50 to 10,000 Hz) potential is maintained across the electrodes. Generator output is varied according to signals from control instrumentation, by modulating voltage or frequency. The dielectric material provides a uniform-glow discharge across the electrode gap, preventing an arc discharge. The geometry of the electrode system is variable; electrodes may be tubular or flat and may be mounted either horizontally or vertically. Tubular generators are used for most high capacity systems, although one manufacturer uses a Lowther Plate type for all sizes of generator. Materials that come in contact with ozone must be corrosion-resistant; stainless steel, unplasticized PVC, aluminum, Teflon and chromium-plated brass or bronze are all suitable.

Ozone production is inherently inefficient; about 10% of the ac energy supplied is used in formation of ozone. In order to maximize efficiency, the oxygen-containing gas must be free of dust and organic matter and must be dry (dew point -50°C) because water accelerates the decomposition of ozone. Ozone is also thermally unstable; hence, provision must be made for air or water cooling of the high voltage electrodes. This requires about 1/3 gpm of cooling water at 21°C per lb O_3/d.

Most efficient ozone production is obtained when oxygen is used as the feed gas to the ozonizer, and such feed may be required for some hazardous waste treatment. With air as feed gas, output of ozone is about two times lower in quantity and concentration; maximum yields from air are about 25 g/m³ or 2% by weight. Choice of oxygen, air, or some intermediate oxygen concentration for the feed gas will depend on economic factors. Oxygen is a viable choice only for fairly large-scale systems (>0.5 mgd) or those where inexpensive oxygen is already available (steel mills, for example, and some biological treatment plants). The availability of pressure-swing oxygen enrichment systems may make oxygen feed more practical in the future.

Venturi mixers and porous diffusers are the two ozone/water mixing systems in most widespread use. With the Venturi mixer, ozonized gas and waste flow cocurrently, and ozonized gas flow is limited to 30 to 60% of the liquid volume flow. In a porous diffuser system, a countercurrent flow is usual, and gas flow may be up to twenty times the liquid flow.

In some systems the contact column is a packed bed. This increases surface area and increases the rate of mass transfer of ozone into solution. One equipment manufacturer, TII Ecology, has been using ultrasonics in conjunction with ozonation; this also increases surface area. Depending on the extent of treatment required, it may be necessary to incorporate two or more contact stages, which may be of different types. Where oxygen is used as a feed gas to the ozonizer, it is usual to recycle the effluent from the contact chamber.

Modern ozone systems are completely automated. An ozone monitor provides continuous on-line monitoring of the ozone concentration in the gaseous effluent from the contactor. If the concentration of ozone exceeds a preset level, usually 0.05 ppm, the voltage or frequency of the ozone generator is reduced. Depending on the characteristics of the waste, the system may also include on-line monitoring for hazardous species concentration in the liquid effluent. When appropriate instruments exist, the output signals may feed back to the ozonator to increase ozone dosage as necessary. The system also includes automatic shutoff provisions in the event of loss of ozonator coolant. Finally, an ambient air ozone monitor is used to sound an alarm and shut off power to the ozonator in the event of gross leaks of ozonized air.

6.14.7 Reliability

Reliability of this process is dependent on the application.

6.14.8 Residuals Generated/Environmental Impact

One advantage of ozonation is that the process leaves no inherent harmful residue. In aqueous "ozone demand free" solution, ozone decomposes to oxygen with a half-life of 20 to 30 minutes. For aqueous streams, the residual oxygen produced by ozone decomposition may be considered a beneficial residue. Ozone lifetime in a gaseous stream is somewhat longer, but in practice, stack effluents from gas ozonation processes are easily controlled to <0.04 ppm of ozone.

Whether products of incomplete oxidation constitute an environmental hazard must be assessed for each waste stream. In a number of cases, it has been found that these products are less toxic and more biodegradable than the original waste components.

One of the advantages of ozonation systems over competitive processes is that they are relatively compact. This is partly due to the fairly short detention time required in the ozone contact chamber. This feature can be particularly attractive when a treatment process is to be installed in a pre-existing facility.

Ozone is recognized to be a toxic substance. The OSHA Threshold Limit Value (which represents an airborne concentration to which it is believed that nearly all workers can be exposed day after day without adverse effect) is 0.1 ppm of ozone. The odor of ozone is distinctive and serves as an effective warning signal at levels well below the toxic level; the threshold odor level is 0.01 to 0.02 ppm. Furthermore, all ozonation systems are equipped with monitors to detect ozone in gaseous effluents; the monitors reduce power to the ozone generator if effluent levels exceed 0.05 ppm of ozone. Since ozone is generated at the same rate as it is applied to the waste and at low pressure (<15 psi), the risk of exposure to high ozone levels is extremely small.

6.14.9 Flow Diagram

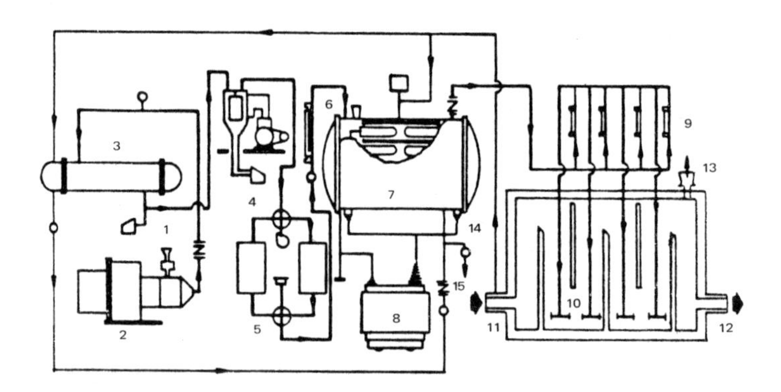

1 AIR INLET
2 ROTARY AIR COMPRESSOR
3 AIR COOLER
4 REFRIGERATOR
5 AIR DRIER
6 AIR FLOW MEASUREMENT
7 OZONIZER
8 H.T. TRANSFORMER
9 OZONIZED-AIR MEASUREMENT
10 POROUS DIFFUSERS
11 INLET OZONIZED-AIR-WATER EMULSIFICATION TANK
12 OUTLET OZONIZED-AIR-WATER EMULSIFICATION
13 AIR RETURN TO ATMOSPHERE
14 COOLING WATER SUPPLY
15 COOLING WATER DISCHARGE

6.14.10 Performance

Subsequent data sheets provide performance data on the following industries and/or wastestreams:

Adhesives and sealants production

Electroplating

Ore mining and dressing
- Gold mining/milling

Organic chemicals production
- Ethylene dichloride
- Ethylene glycol
- Toluene diisocyanate

Textile milling
- Knit fabric finishing
- Wool scouring
- Woven fabric finishing

6.14.11 References

1. Physical, Chemical, and Biological Treatment Techniques for Industrial Wastes, PB 275 287, U.S. Environmental Protection Agency, Washington, D.C., November 1976. pp. 36-1 through 36-28.

CONTROL TECHNOLOGY SUMMARY FOR OZONATION

Pollutant	Number of data points	Effluent concentration				Removal efficiency, %			
		Minimum	Maximum	Median	Mean	Minimum	Maximum	Median	Mean
Conventional pollutants, mg/L:									
BOD_5	4	4.9	5,190	330	1,460	0[a]	10	0[a]	2.5
COD	4	17	12,100	213	3,140	0[a]	92	51	48
TOC	33	15	2,840	543	682	0[a]	>75	10	12
TSS	4	3	140	14	43	0[a]	33	15	32
Oil and grease	1				4.0				97
Total phenol	3	0.013	0.13	0.021	0.055	0[a]	>99	24	41
Total phosphorus	1				1.1				0
Toxic pollutants, μg/L:									
Antimony	2	25	1,200		610				0[a]
Arsenic	2	4	43		24	0	48		24
Cadmium	1				250				0[a]
Chromium	1				6.3				0[a]
Copper	2	89	590		340				0[a]
Cyanide	50	<2	12,000	<320	750	0[a]	>99	99	97
Lead	1				<22				>29
Nickel	2	66	5,000		2,500				0[a]
Silver	2	16	1,300		660				0[a]
Zinc	3	90	460	240	260	0[a]	0[a]	96	32
Bis(2-ethylhexyl) phthalate	2	90	110		100				0[a]
Butyl benzyl phthalate	1				<0.03				>97
Di-n-butyl phthalate	2	1.2	2.7		2.0	0[a]	77		38
Toluene	1				0.9				31
Anthracene/phenanthrene	2	<0.01	0.4		<0.2	0[a]	>97		49
Benzo(a)pyrene	1				<0.02				>90
Benzo(k)fluoranthene	1				<0.02				>80
Fluoranthene	1				0.1				50
Pyrene	1				0.1				67
1,2-*Trans*-dichloroethylene	1				2.1				0[a]
Methylene chloride	2	15	61		38				0[a]
Trichloroethylene	1				0.9				0[a]

Note: Blanks indicate data not applicable.

[a]Actual data indicate negative removal.

TREATMENT TECHNOLOGY: Ozonation[a]

Data source: Government report
Point source category: Adhesives and sealants
Subcategory:
Plant: San Leandro
References: B10, p. 81

Data source status:
- Engineering estimate ___
- Bench scale x
- Pilot scale ___
- Full scale ___

Use in system: Tertiary
Pretreatment of influent: Settling, ultrafiltration

[a]Catalysts also used.

DESIGN OR OPERATING PARAMETERS

Unit configuration:
Wastewater flow:
Air/oxygen consumption:
Ozone generation rate:
Ozone concentration (in air/oxygen):
Ozone utilization:
Contact time:
Power consumption:

REMOVAL DATA

Sampling period: Equal volume grab samples collected throughout an 8-hr day; average of 2 days sampling

Pollutant/parameter	Concentration		Percent removal
	Influent	Effluent	
Conventional pollutants, mg/L:			
BOD_5	5,780	5,190	10
COD	76,700	12,100	84
TSS	64	140	0[a]
Oil and grease	140	4.0	97
Total phenol	47	0.13	>99
Toxic pollutants, µg/L:			
Cyanide	560	1,500	0[a]
Zinc	2,200	90	96

[a]Actual data indicate negative removal.

Note: Blanks indicate information was not specified.

TREATMENT TECHNOLOGY: Ozonation

Data source: Government report
Point source category: Electroplating
Subcategory:
Plant: Sealectro Corp.
References: B8, pp. 22-26

Data source status:
Engineering estimate ___
Bench scale ___
Pilot scale ___
Full scale x

Use in system: Primary
Pretreatment of influent: None

DESIGN OR OPERATING PARAMETERS

pH: See below
Ozone concentration: See below
Weight ratio required for complete oxidation:
Gas feed rate:
Ozone, wt. % of feed:
Turbine speed:
Unit configuration:
Mole ratio (O_3/CN): See below

REMOVAL DATA

Sampling period:

pH	Ozone concentration, mg/L	Mole ratio (O_3/CN^-)	Cyanide Concentration, µg/L Influent	Cyanide Concentration, µg/L Effluent	Cyanide Percent removal
10.9		0.35-0.5	37,500	350	99
9.4		0.35-0.5	75,000	11,500	85
9.5		1.3-1.6	12,900	80	99
11.0		1.3-1.6	63,000	520	99
7.7		2.3-2.7	13,000	20	>99
11.9		2.3-2.7	32,000	380	99
9.6		2.3-2.7	34,200	600	98
10.8		3.64	29,000	80	>99
10.9		2.81	37,500	350	99
11.9		2.42	32,000	380	99
9.6		2.26	34,200	600	98
9.5		2.01	38,400	620	98
11.0		1.33	63,000	520	99
9.4		0.35	75,000	11,500	85
8.0		11.37	2,100	80	96
12.6		6.62	5,300	40	99
10.1		5.39	6,500	80	99
7.5		1.48	12,900	80	99
7.7		2.69	13,000	20	>99
9.5		1.05	15,200	80	99
7.9	143	2.0	38,400	220	99
9.4	143	2.3	34,200	340	99
8.9	143	2.3	33,000	600	98
	143	2.4	32,800	380	99
	143	2.4	32,500	600	98
10.1	143	2.4	32,100	550	98
11.8	143	2.4	32,000	540	98
8.8	143	2.5	31,500	520	98
9.1	143	2.6	30,200	600	98
	143	2.6	29,500	620	98
8.3	64.4	2.7	13,000	20	>99
11.3	143	2.7	28,500	620	98
9.8	64.4	5.4	6,500	80	99
10.8	64.4	6.3	5,600	120	98
9.0	64.4	6.6	5,300	300	94
11.9	64.4	6.6	5,300	40	99
10.0	173	1.33	63,000	520	99
8.9	130	1.84	38,000	230	99
9.8	195	2.81	37,500	350	99
9.8	195	2.91	36,300	210	99
8.7	195	3.64	29,000	80	>99
7.0-8.0	29.7	1.05	15,200	80	99
9.5	35.2	1.48	12,900	80	99

Note: Blanks indicate information was not specified.

TREATMENT TECHNOLOGY: Ozonation

Data source: Effluent Guidelines
Point source category: Ore mining and dressing
Subcategory: Gold mine/mill
Plant: 4105
References: A2, pp. VI-29, 58

Data source status:
- Engineering estimate ___
- Bench scale ___
- Pilot scale x
- Full scale ___

Use in system: Tertiary
Pretreatment of influent: Carbon adsorption unless otherwise specified

DESIGN OR OPERATING PARAMETERS

pH:
Ozonation time:
Weight ratio required for complete oxidation:
Flow rate: See below
Ozone feed rate: See below
Turbine speed:
Unit configuration:
Mole ratio (O_3/CN):

REMOVAL DATA

Sampling period:

Flow rate, L/min	Ozone feed rate, g/hr	Cyanide Concentration, µg/L Influent	Effluent	Percent removal
3,200		900	<20	>97
9.5	3	355[a]	20	94
9.5	3	163[b]	18	89
4.9	6	195	95	51

[a]Copper ion was used as catalyst.

[b]Copper wire was used as catalyst.

Note: Blanks indicate information was not specified.

TREATMENT TECHNOLOGY: Ozonation

Data source: Government report
Point source category:[a] Organic chemicals
Subcategory:
Plant:
References: B2, p. 160

Data source status:
Engineering estimate ___
Bench scale ___
Pilot scale x
Full scale ___

Use in system: Primary
Pretreatment of influent:

[a]Wastewater from a toluene diisocyanate process used in the manufacture of polyurethane.

DESIGN OR OPERATING PARAMETERS (also see removal data)

pH:
Ozonation time:
Weight ratio required for complete oxidation:
Gas feed rate:
Ozone, wt. % of feed:
Turbine speed:
Unit configuration: Tubular reactor with static mixers[a]
Mole ratio:
Liquid flow: 1.5 L/min

[a]Tubular reactors were eventually abandoned because of the inefficiency of mixing the gas and liquid.

TOC REMOVAL DATA

Sampling period:

pH	Gas flow, L/min	Residence time, min	Mole ratio[a]	Influent TOC Concentration, mg/L	Effluent TOC Concentration, mg/L	Percent removal
1	10	1.5	0.176	1,070	970	9
1	10	3.0	0.176	1,070	938	12
1	24	1.0	0.424	1,070	965	10
1	24	2.0	0.424	1,070	933	13
1	26	0.7	0.451	1,070	965	10
1	26	1.4	0.459	1,070	965	10
8	10	1.5	0.22	1,070	1,120	0[b]
8	10	3.0	0.200	1,070	1,050	2
8	20	0.8	0.396	1,070	946	10
8	20	1.6	0.396	1,070	1,030	4

[a]Mole ratio (Ozone to TDA) calculated on the basis of the TOC being pure TDA.
[b]Actual data indicate negative removal.

Note: Blanks indicate information was not specified.

TREATMENT TECHNOLOGY: Ozonation

Data source: Government report
Point source category:[a] Organic chemicals
Subcategory:
Plant:
References: B2, p. 159

Data source status:
- Engineering estimate ___
- Bench scale ___
- Pilot scale x
- Full scale ___

Use in system: Primary
Pretreatment of influent:

[a]Wastewater from a toluene diisocyanate process used in the manufacture of polyurethane.

DESIGN OR OPERATING PARAMETERS

pH:
Ozonation time:
Weight ratio required for complete oxidation:
Gas feed rate:
Ozone, wt. % of feed:
Turbine speed:
Unit configuration: Tubular reactor,[a] dispersion of the gas and liquid was achieved with a nozzle.
Mole ratio:
Liquid flow: 1.75 L/min

[a]The tubular reactor was eventually abandoned because of the inefficiency of mixing the gas and liquid.

TOC REMOVAL DATA

Sampling period:

pH	Gas flow, L/min	Residence time, min	Mole ratio[a]	Influent TOC Concentration, mg/L	Effluent TOC Concentration, mg/L	Percent removal
11	3.54	1.8	0.059	560	586	0[b]
11	3.54	3.7	0.059	560	561	0[b]
11	6.04	1.3	0.120	560	528	6
11	6.04	2.6	0.102	560	549	2
11	8.0	1.0	0.127	560	520	6
11	8.0	2.0	0.127	560	512	9
8	4.0	1.7	0.064	560	491	12
8	4.0	3.4	0.064	560	544	3
8	8.0	1.0	0.127	560	491	12
8	8.0	2.0	0.127	560	481	14
1	4.0	1.7	0.068	560	538	4
1	4.0	3.4	0.068	560	530	5
1	8.0	1.0	0.135	560	527	6
1	8.0	2.0	0.135	560	541	2
6	8.0	1.0	0.135	560	663	0[b]
6	8.0	2.0	0.135	560	538	5

[a]Mole ratio (Ozone to TDA) is calculated on the basis of the TOC being pure TDA.
[b]Actual data indicate negative removal.

Note: Blanks indicate information was not specified.

TREATMENT TECHNOLOGY: Ozonation

Data source: Government report
Point source category: Organic chemicals
Subcategory:
Plant:
References: B2, pp. 160, 163, 166, 169

Data source status:
- Engineering estimate ___
- Bench scale ___
- Pilot scale x
- Full scale ___

Use in system: See below
Pretreatment of influent: See below

DESIGN OR OPERATING PARAMETERS

pH: >10 unless otherwise specified
Ozonation time: See below
Weight ratio required for complete oxidation: See below
Gas feed rate: 11.5 L/min
Ozone, wt. % of feed: 1.0 - 1.2 wt.%
Turbine speed: 700 rpm
Unit configuration: Stirred tank reactor
Mole ratio (O_3/CN):

REMOVAL DATA

Sampling period:

Use in system	Pretreatment of influent	Ozonation time, min	Weight ratio required for complete oxidation, mg O_3/mg TOC	BOD_5 Concentration, mg/L Influent	BOD_5 Concentration, mg/L Effluent	BOD_5 Percent removal	TOC Concentration, mg/L Influent	TOC Concentration, mg/L Effluent	TOC Percent removal
Secondary[a]	Steam stripping	180					400	<100	>75
Primary[b]		360	7.0				3,360	2,840	16
Secondary[a]	Air stripping		5.6				409	286	30
Primary[c]		330	7.3	93.1	614	0[d]	830	626	25
Primary[c]		180	7.3				100	50	50

[a]Wastewater from an ethylene dichloride process.
[b]Wastewater from a toluene diisocyanata process used in the manufacture of polyurethane.
[c]Polyol wastewater was taken from an ethylene glycol process plant.
[d]Actual data indicate negative removal.

Note: Blanks indicate information was not specified.

TREATMENT TECHNOLOGY: Ozonation

Data source: Government report
Point source category: Textile mills
Subcategory: Woven fabric finishing
Plant: V
References: B3, pp. 70-75

Data source status:
Engineering estimate ___
Bench scale ___
Pilot scale x
Full scale ___

Use in system: Tertiary
Pretreatment of influent: Screening, activated sludge, multimedia filtration

DESIGN OR OPERATING PARAMETERS

Unit configuration: Contactor - 2.0 m (77 in.); 1.58 m^3 (416 gal) contactor
Generator - PCI Ozone Corporation Model C2P-3C

Wastewater flow:
Air/oxygen consumption:
Ozone generation rate: 6 g/hr (capacity with pure oxygen feed)
Ozone concentration (in air/oxygen):
Ozone utilization:
Contact time:
Power consumption:

REMOVAL DATA

Sampling period: 24-hr composite, volatile organics were grab-sampled

Pollutant/parameter	Concentration		Percent
	Influent	Effluent	removal
Conventional pollutants, mg/L:			
COD	72	76	0[a]
TSS	4	12	0[a]
Total phenol	0.013	0.021	0[a]
Total phosphorus	1.1	1.1	0
Toxic pollutants, µg/L			
Antimony	<10	25	0[a]
Arsenic	4	4	0
Chromium	<4	6.3	0[a]
Copper	75	89	0[a]
Cyanide	3	<2	>33
Lead	31	<22	>29
Nickel	<36	66	0[a]
Silver	<5	16	0[a]
Zinc	190	240	0[a]
Bis(2-ethylhexyl) phthalate	16	90	0[a]
Butyl benzyl phthalate	0.9	<0.03	>97
Di-n-butyl phthalate	12	2.7	77
Toluene	1.3	0.9	31
Anthracene/phenanthrene	0.3	<0.01	>97
1,2-*Trans*-dichloroethylene	<2.0	2.1	0[a]
Methylene chloride[b]	13	15	0[a]
Trichloroethylene	0.4	0.9	0[a]

[a]Actual data indicate negative removal.

[b]Presence may be due to sample contamination.

Note: Blanks indicate information was not specified.

TREATMENT TECHNOLOGY: Ozonation

Data source: Effluent Guidelines and Government report
Point source category: Textile mills
Subcategory: Wool scouring
Plant: A, W (different references)
References: A6, p. VII-55; B3, pp. 50-54

Data source status:
Engineering estimate ___
Bench scale ___
Pilot scale x
Full scale ___

Use in system: Tertiary
Pretreatment of influent: Grit removal, sedimentation, activated sludge, multimedia filtration

DESIGN OR OPERATING PARAMETERS

Unit configuration: Contactor - 2.0 m (77 in.); 1.58 m^3 (416 gal) column
Generator - PCI Ozone Corporation Model C2P-3C
Wastewater flow:
Air/oxygen consumption:
Ozone generation rate: 6 g/hr (capacity with pure oxygen feed)
Ozone concentration (in air/oxygen):
Ozone utilization:
Contact time:
Power consumption:

REMOVAL DATA

Sampling period: 24-hr composite, volatile organics were grab-sampled

Pollutant/parameter	Concentration Influent	Concentration Effluent	Percent removal
Conventional pollutants, mg/L:			
Total phenol	0.017	0.013	24
Toxic pollutants, µg/L:			
Antimony	<200	1,200	0[a]
Arsenic	83	43	48
Cadmium	<40	250	0[a]
Copper	120	590	0[a]
Cyanide	260	<4	>98
Nickel	<700	5,000	0[a]
Silver	<100	1,300	0[a]
Zinc	400	460	0[a]
Bis(2-ethylhexyl) phthalate	14	110	0[a]
Toluene	<0.1	1.2	0[a]
Anthracene/phenanthrene	0.2	0.4	0[a]
Benzo(a)pyrene	0.2	<0.02	>90
Benzo(k)fluoranthene	0.1	<0.02	>80
Fluoranthene	0.2	0.1	50
Pyrene	0.3	0.1	67
Methylene chloride[b]	4.8	61	0[a]

[a]Actual data indicate negative removal.
[b]Presence may be due to sample contamination.

Note: Blanks indicate information was not specified.

TREATMENT TECHNOLOGY: Ozonation

Data source: Effluent Guidelines
Point source category: Textile mills
Subcategory: Knit and woven fabric finishing
Plant: Q and D
References: A6, pp. VII 52-54

Data source status:
- Engineering estimate ___
- Bench scale ___
- Pilot scale x
- Full scale ___

Use in system: Tertiary
Pretreatment of influent:

DESIGN OR OPERATING PARAMETERS

Unit configuration: Contactor - 2.0 m (77 in.); 1.58 m^3 (416 gal) column
Generator - PCI Ozone Corporation Model C2P-3C (continuous operation)

Wastewater flow:
Air/oxygen consumption:
Ozone generation rate: 6 g/hr (capacity with pure oxygen feed)
Ozone concentration (in air/oxygen)
Ozone utilization: See below
Contact time:
Power consumption:

REMOVAL DATA

Sampling period:

	BOD_5			COD		
	Concentration, mg/L		Percent	Concentration, mg/L		Percent
Utilization, mg/L	Influent	Effluent	removal	Influent	Effluent	removal
Ozone,[a] 1,300-1,500	4.2	4.9	0[b]	206	17	92
Ozone,[c] 427	13	47	0[b]	422	349	17

	TOC			TSS		
	Concentration, mg/L		Percent	Concentration, mg/L		Percent
	Influent	Effluent	removal	Influent	Effluent	removal
Ozone,[a] 1,300-1,500	22	15	32	4.5	3	33
Ozone,[c] 427	101	106	0[b]	23	16	30

[a]Pretreatment of influent: screening, equalization, activated sludge, multimedia filtration.

[b]Actual data indicate negative removal.

[c]Pretreatment of influent: screening, neutralization, activated sludge, multimedia filtration, granular activated carbon adsorbtion.

Note: Blanks indicate information was not specified.

6.15 CHEMICAL REDUCTION [1]

6.15.1 Function

Chemical reduction is used to reduce metals to less toxic oxidation states.

6.15.2 Description

Reduction-oxidation, or "Redox" reactions are those in which the oxidation state of at least one reactant is raised while that of another is lowered. In the reaction

$$2H_2CrO_4 + 3SO_2 + 3H_2O \rightarrow Cr_2(SO_4)_3 + 5H_2O \qquad (1)$$

the oxidation state of Cr changes from 6^+ to 3^+ (Cr is reduced); the oxidation state of S increased from 2^+ to 3^+ (S is oxidized). This change of oxidation state implies that an electron was transferred from S to Cr(VI). The decrease in the positive valence (or increase in the negative valence) with reduction takes place simultaneously with oxidation in chemically equivalent ratios. Reduction is used to treat wastes in such a way that the reducing agent lowers the oxidation state of a substance in order to reduce its toxicity, reduce its solubility, or transform it into a form that can be more easily handled.

The base metals are good reducing agents, as evidenced by the use of iron, aluminum, zinc, and sodium compounds for reduction treatments. In addition, sulfur compounds also appear among the more common reducing agents.

Liquids are the primary waste form treatable by chemical reduction. The most powerful reductants are relatively nonselective; therefore, any easily reducible material in the waste stream will be treated. For example, in reducing heavy metals to remove them from a waste oil, quantities of esters large enough to cause odor problems may also be formed by the reduction.

Gases such as chlorine dioxide and chlorine have been treated by reducing solutions for the small-scale disposal of gas in laboratories. For reduction of fluorine, instead of a solution, a scrubber filled with solid bicarbonate, soda lime or granulated carbon is recommended. Reduction has limited application to slurries, tars, and sludges, because of the difficulties of achieving intimate contact between the reducing agent and the hazardous constituent; consequently the reduction process would be very inefficient.

In general, hazardous materials occurring as powders or other solids usually have to be solubilized prior to chemical reduction.

The first step of the chemical reduction process is usually the adjustment of the pH of the solution to be treated. With sulfur dioxide treatment of chromium (VI), for instance, the reaction requires a pH in the range of 2 to 3. The pH adjustment is done with the appropriate acid (e.g., sulfuric). This is followed by addition of the reducing agent. Mixing is provided to improve contact between the reducing agent and the waste. The agent can be in the form of a gas (sulfur dioxide) or solution (sodium borohydride) or perhaps finely divided powder if there is adequate mixing. Reaction times vary for different wastes, reducing agents, temperatures, pH, and concentration. For commercial-scale operations for treating chromium wastes, reaction times are in the order of minutes. Additional time is usually allowed to ensure complete mixing and reduction. Once reacted, the reduced solution is generally subjected to some form of treatment to settle or precipitate the reduced material. A treatment for the removal of what remains of the reducing agent may be included. This can be unused reducing agent or the reducing agent in its oxidized state. Unused alkali metal hydrides are decomposed by the addition of a small quantity of acid. The pH of the reaction medium is typically increased so that the reduced material will precipitate out of solution. Filters or clarifiers are often used to improve separation.

While some stream components may be added or removed, the output stream from a chemical reduction treatment is not very different from the input stream. Reducing agents, such as sodium borohyride and zinc, introduce to the reaction mixture ions that are not easily separable from the product streams. The effluent solution is typically acidic and must be neutralized prior to discharge with materials such as hydrated lime, caustic soda, or soda ash.

6.15.3 Technology Status

Technology for large-scale application of chemical reduction is well developed.

6.15.4 Applications

The following paragraphs describe some selected examples of the application of chemical reduction to hazardous waste management problems.

- Reduction of Chromium (VI) to Chromium (III) in Effluents

Numerous plating and metal finishing plants treat their chromium (VI) wastes using chemical reduction methods. Cyanides and chromium are often present together in plating industry wastes. The concentrations of these substances and their potential recovery value influence the selection of the treatment process. If the cyanide and chromium are not economically recoverable by a method such as ion exchange, the cyanide radical is first destroyed or converted to the less toxic cyanate by oxidation, and the chromium (VI) is converted, by subsequent reduction, to chromium (III), which precipitates and is removed as a sludge.

Hexavalent chromium can be reduced to chromium (III) by a variety of reducing agents including sulfur dioxide, sulfite salts, and ferrous sulfate. In industry, sulfur dioxide is the most widely used reducing agent for this purpose. Because soluble chromium (III) compounds are themselves toxic, chromium reduction processes are usually followed by a precipitation operation in which the chromium (III) is precipitated as $Cr(OH)_3$ with either lime or sodium carbonate. In the tanning and plating industries, sludges containing from 10 to 80% solids obtained from prior concentration of chromates are often redissolved by acidification and then subjected to reduction followed by precipitation to obtain the chromium in an insoluble, concentrated form.

- Reduction Using Sulfur Dioxide

In the chromium waste treatment using sulfur dioxide, the reaction equations are as follows:

$$SO_2 + H_2O \rightarrow H_2SO_3 \quad (2)$$

$$2H_2CrO_4 + 3H_2SO_3 \rightarrow Cr_2(SO_4)_3 + 5H_2O \quad (3)$$

Using hydrated lime, the neutralization is:

$$Cr_2(SO_4)_3 + 3Ca(OH)_2 \rightarrow 2Cr(OH)_3 + 3CaSO_4 \quad (4)$$

Hexavalent chromium can be reduced to the range of 0.7 to 1 mg/L in the effluent by using such a treatment including reduction, chemical precipitation and sedimentation.

- Reduction with Sodium Metabisulfite (and Bisulfite)

About three pounds of sodium metabisulfite ($Na_2S_2O_5$) are required to reduce one pound of hexavalent chromium using the following reaction:

$$4H_2CrO_4 + 3Na_2S_2O_5 + 3H_2O + 6H_2SO_4 \rightarrow 2Cr_2(SO_4)_3 + 6NaHSO_4 + 10H_2O \quad (5)$$

- Reduction with Ferrous Sulfate

Because of the sludge volume produced, ferrous sulfate is rarely used in larger-scale treatment facilities according to the following reaction:

$$2H_2CrO_4 + 6FeSO_4 + 7H_2O + 6H_2SO_4 \rightarrow Cr_2(SO_4)_3 + 3Fe_2(SO_4)_3 + 15H_2O \quad (6)$$

- Removal of Mercury from Effluents

Reduction/precipitation processes are being used increasingly to treat wastewater containing mercury when the flowrate is relatively small and intermittent. Because of its value and because it is not amenable to disposal, the elemental mercury produced by reduction processes is usually recovered for recycle. Depending upon the process, a cyclone, filter or perhaps a furnace and mercury condenser may be used.

In a recently commercialized reduction/precipitation process, a caustic solution of sodium borohydride ($NaBH_4$) is mixed with mercury-containing wastewater. The ionic mercury is reduced to metallic mercury, which precipitates out of solution, and the following reaction occurs:

$$4Hg^{2+} + BH_4^- + 8OH^- = 4Hg + B(OH)_4^- + 4H_2O \quad (7)$$

In theory, 1.0 pound of sodium borohydride can reduce 21 pounds of mercury; in actual operations, this is closer to 10 pounds of mercury. If the mercury solution is in the form of an organic complex, the driving force of the reduction reaction may not be sufficient to break the complex. In that case, the wastewater must be chlorinated prior to the reduction step in order to break down the metal-organic bond.

- Removal of Lead

Removal of dissolved lead compounds, including organo-lead salts, in wastewater from the manufacture of tetraalkyl lead compounds is now being done on a commercial scale. The reduction process, using an alkali metal hydride as reductant, lowers the lead content in the waste stream by altering the chemical form of the lead so that it can be precipitated. The reaction is believed to go partially to elemental lead and partially to an alkyl-lead compound that is not stable over long periods of time, some of which is eventually converted spontaneously to elemental lead. As the element, the lead precipitates and can be removed by techniques such as settling or by filtration.

The concentration range in the effluents to the reduction process are 2 to 300 ppm. The lead is mostly in the form of soluble organo-lead compounds, which will not precipitate with pH adjustment alone, together with some other lead in the form of soluble inorganic lead compounds.

After treatment with an alkali metal hydride (sodium borohydride is preferred in this reaction), insoluble lead products are formed. They include hexaalkyl-dilead compounds (that may with time decompose to elemental lead), which are formed from the soluble alkyl-lead compounds, and elemental lead from the soluble inorganic lead components.

Low concentrations of the borohydride are preferred because one of the characteristics of the material is that it hydrolyzes with evolution of hydrogen and with an accompanying loss in its reductive properties. This is particularly true at higher temperatures, pH below 8 or 9, and in the presence of certain catalysts. For this reaction, a pH of 8 to 11 is preferred.

6.15.5 Limitations

Introduction of foreign ions into the waste is a real or potential disadvantage with many of the reducing agents.

6.15.6 Typical Equipment

Very simple equipment is required for chemical reduction including storage vessels for the reducing agents and perhaps for the wastes, metering equipment for both streams, and contact vessels with agitators to provide suitable contact of reducing agent and waste. Some instrumentation is required to determine the concentration and pH of the waste and the degree of completion of the reduction reaction. The reduction process may be monitored by an oxidation-reduction potential electrode. This electrode is generally a piece of noble metal (often platinum) that is exposed to the reaction medium and produces an EMF output that is empirically relatable to the reaction condition by revealing the ratio of the oxidized and reduced constituents. Section III.6.15.9 shows a process flow diagram for a typical chemical system.

Numerous companies have commercial units for the treatment of chromium (VI) in industrial effluents. All of these units offer the user a pre-engineered system for a specific waste or range of waste streams.

6.15.7 Reliability

The chemical reduction process is well developed and reliable for chrome and mercury applications.

6.15.8 Environmental Impact

One disadvantage of chemical reduction for waste treatment is that it may introduce new ions into the effluent. If the level of these new contaminants is high enough to exceed effluent regulations, additional treatment operations will be required. Often these treatments are precipitation, filtration, or sedimentation.

- Air emissions are not expected to be significant from these processes.

After chromium (VI) reduction, the treated solution will be acidic and will also contain the reduced chromium and any other metals present in the original waste stream. Because this solution is corrosive, it may require neutralization prior to discharge or further treatment. Precipitation will occur because of the chemical nature of the materials used and, therefore, settling basins or clarifiers will be required to reduce the solids carry-over.

Small amounts of sulfate resulting from the use of sulfur dioxide on dilute wastes pose no problem, but the zinc ion can be of concern. Reduction with sodium borohydride results in the formation of greater-than-stoichiometric amounts of soluble borate in the effluent solution; borate at sufficiently high levels could also be of environmental concern. When the waste constituents are present only in very small concentrations, these materials in the effluents are of little concern; however, if the processes are extended to more concentrated waste streams, additional treatment steps may be needed.

Most chemical reductions will produce a residue for disposal, unless the concentration of the waste constituent is so low that the reducing agent and the reduced waste can be carried away with the effluent. Residues for eventual disposal on land can be a problem with this treatment process. The sludges formed in follow-up treatment may cause disposal problems because the metal hydroxides they contain may be susceptible to acid leaching. Because the common alkalies used are sodium hydroxide and hydrated lime, a large portion of the sludge will be excess lime and calcium sulfate.

Lesser amounts of waste residues will be produced from the use of sodium borohydride because the metal can often be precipitated in the form of the element or another form that can be processed for recovery.

6.15.9 Flow Diagram

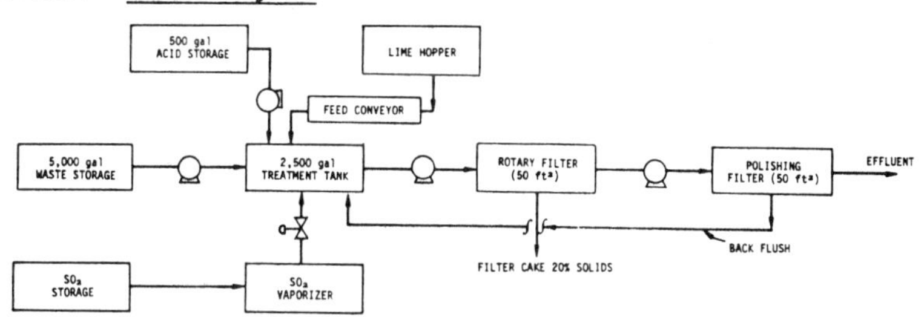

Treatment: Batch Waste: Concentrated Chrome Waste 100,000 ppm CrO_3; 85% as Cr^{+3} in 20% H_2SO_4

Waste Processing Capacity: 2,000 gal/shift

Operating Period: 240 days/yr, 8 hr/day

Raw Materials: 240 lb/day SO_2; 2,065 lb/day lime

6.15.10 Performance

Subsequent data sheets provide performance data on the following industries and/or wastestreams:

Inorganic chemicals production
  Chrome pigments

6.15.11 References

1. Physical, Chemical, and Biological Treatment Techniques for Industrial Wastes, PB 275 287, U.S. Environmental Protection Agency, Washington, D.C. November 1976. pp. 38-1 through 38-13.

TREATMENT TECHNOLOGY: Chemical Reduction (SO_2, Acid, Caustic)

Data source: Effluent Guidelines
Point source category: Inorganic chemicals
Subcategory: Chrome pigment
Plant: 002
References: A29, pp. 396-397

Data source status:
Engineering estimate ___
Bench scale ___
Pilot scale ___
Full scale x

Use in system: Primary
Pretreatment of influent:

DESIGN OR OPERATING PARAMETERS

Unit configuration:
Wastewater flow:
Chemical dosage(s):
pH in clarifier:
Clarifier detention time:
Hydraulic loading:
Weir loading:
Sludge underflow:
Percent solids in sludge:
Scum overflow:

REMOVAL DATA

Pollutant/parameter	Concentration,[a] µg/L		Percent
	Influent	Effluent	removal
Toxic pollutants:			
Chromium	310,000	130,000	58
Lead	160,000	120,000	25
Zinc	54,000	1,500	97

[a]Concentration is calculated from pollutant flow in m^3/kkg and pollutant loading in kg/kkg.

Note: Blanks indicate information was not specified.

7. Sludge Treatment

7.1 GRAVITY THICKENING [1]

7.1.1 Function

Thickening of sludge consists of the removal of supernatant, thereby reducing the volume of sludge that requires disposal or further treatment. Gravity thickening takes advantage of the difference in specific gravity between the solids and water.

7.1.2 Description

A gravity thickener normally consists of two truss-type steel scraper arms mounted on a hollow pipe shaft keyed to a motorized hoist mechanism. A truss-type bridge is fastened to the tank walls or to steel or concrete columns. The bridge spans the tank and supports the entire mechanism. The thickener resembles a conventional circular clarifier with the exception of having a greater bottom slope. Sludge enters at the middle of the thickener, and the solids settle into a sludge blanket at the bottom. The concentrated sludge is very gently agitated by the moving rake, which dislodges gas bubbles and prevents bridging of the sludge solids. It also keeps the sludge moving toward the center well from which it is removed. Supernatant liquor passes over an effluent weir around the circumference of the thickener. In the operation of gravity thickeners, it is desirable to keep a sufficiently high flow of fresh liquid entering the concentrator to prevent the development of septic conditions and resulting odors.

Gravity thickening is characterized by zone settling. The four basic settling zones in a thickener are:

- The clarification zone at the top containing the relatively clear supernatant.
- The hindered settling zone where the suspension moves downward at a constant rate and a layer of settled solids begins building from the bottom of the zone.
- The transition zone characterized by a decreasing solids settling rate.
- The compression zone where consolidation of sludge results solely from liquid being forced upward around the solids.

7.1.3 Common Modifications

Tanks can be square or round, with the round variety being much more prevalent. Tanks can be manufactured of concrete or steel. Chemicals can be added to aid in the sludge dewatering.

7.1.4 Technology Status

Gravity thickening has been in wide use for many years.

7.1.5 Applications

Used to thicken primary, secondary, and digested sludges.

7.1.6 Limitations

Does not perform satisfactorily on most waste activated, mixed primary-waste activated, and alum or iron sludges; is highly dependent on the dewaterability of the sludges being treated.

7.1.7 Chemicals Required

Lime (CaO) and/or polymers may be added to aid in the dewatering and settling of the sludge; chlorine can be added to prevent septicity.

7.1.8 Residuals Generated

Supernatant volume is directly related to the increase in solids concentration in the thickener; supernatant will contain varying amounts of solids, ranging from tens to hundreds of milligrams per liter.

7.1.9 Design Criteria

See Section 7.1.13; detentions of one to three days are usually used; sludge blankets of at least three feet are common; side water depths of at least ten feet are general practice.

7.1.10 Environmental Impact

Requires relatively little use of land; supernatant will need disposal, which can be accomplished by recycling it to the head end of the plant for further treatment; odor problems frequently result from septic conditions.

7.1.11 Reliability

Gravity thickeners are mechanically reliable, but are greatly affected by the quality of sludge received; therefore, they may be upset due to a radical change in the raw wastewater or digested sludge quality.

7.1.12 Flow Diagram

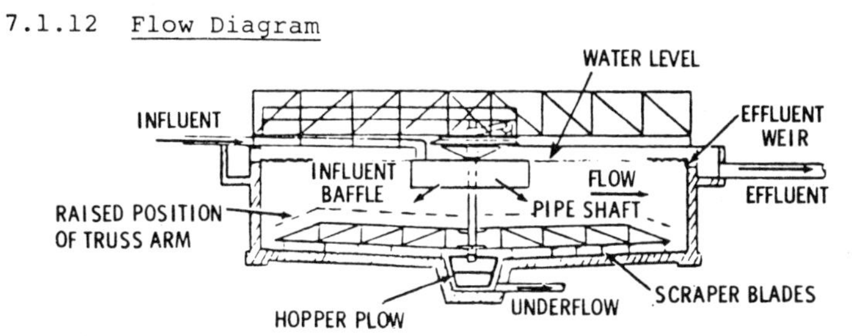

7.1.13 Performance

(No chemical conditioning)

Type of sludge	Solids surface loading, lb/d/ft²	Thickened sludge solids concentration, %
Primary	20 to 30	8 to 10
Waste activated	5 to 6	2.5 to 3
Trickling filter	8 to 10	7 to 9
Limed tertiary	60	12 to 15
Primary and activated	6 to 10	4 to 7
Primary and trickling filter	10 to 12	7 to 9
Limed primary	20 to 25	7 to 12

7.1.14 References

1. Innovative and Alternative Technology Assessment Manual. EPA-430/9-78-009 (draft), U.S. Environmental Protection Agency, Cincinnati, Ohio, 1978. 252 pp.

7.2 FLOTATION THICKENING [1]

7.2.1 Function

Flotation (Dissolved Air Flotation) thickening utilizes air to float sludge to the surface of the thickener, thereby reducing the water content and volume of the sludge.

7.2.2 Description

In a Dissolved Air Flotation (DAF) system, a recycled subnatant flow is pressurized from 30 to 70 $lb/in.^2$ (gage) and then saturated with air in a pressure tank. The pressurized effluent is then mixed with the influent sludge and subsequently released into the flotation tank. The excess dissolved air then separates from solution, which is now under atmospheric pressure, and the minute (average diameter 80 mm) rising gas bubbles attach themselves to particles that form the floating sludge blanket. The thickened blanket is skimmed off and pumped to the downstream sludge handling facilities while the subnatant is returned to the plant. Polyelectrolytes are frequently used as flotation aids to enhance performance and create a thicker sludge blanket. A description of the DAF process in general is presented in Section 4.4.

7.2.3 Technology Status

DAF is the most common form of flotation thickening in use in the United States, has been used for many years to thicken waste activated sludges, and to a lesser degree to thicken combined sludges. DAF has widespread industrial wastewater applications.

7.2.4 Applications

The use of air flotation is limited primarily to thickening of sludges prior to dewatering or digestion. Used in this way, the efficiency of the subsequent dewatering units can be increased, and the volume of supernatant from the subsequent digestion units can be decreased. Existing air flotation thickening units can be upgraded by the optimization of process variables, and by the utilization of polyelectrolytes. Air flotation thickening is best applied to waste activated sludge. With this process, it is possible to thicken the sludge to 6 percent solids, while the maximum concentration attainable by gravity thickening without chemical addition is 2 to 3 percent solids. The DAF process can also be applied to mixtures of primary and waste activated sludge. DAF also maintains the sludge in aerobic condition and potentially has a better solids capture than gravity thickening. There is some evidence that activated sludges from pure oxygen systems are more amenable to flotation thickening than sludges from conventional systems.

7.2.5 Limitations

DAF has high operating costs (primarily for power for aeration and chemicals) and is therefore generally limited to waste activated sludges. The variability of sludge characteristics requires that some pilot work be done prior to design of a DAF system.

7.2.6 Chemicals Required

Flotation aids (generally polyelectrolytes) are usually used to enhance performance.

7.2.7 Residuals Generated

Supernatant (effluent) quality is approximately 150 mg/L SS, returned to mainstream of STP.

7.2.8 Performance

Data from various air flotation units indicate that solids recovery ranges from 83 to 99 percent at solids loadings rates of 7 to 48 $lb/ft^2/d$.

Operating data from 14 sewage treatment plants showed the following: influent suspended solids, 3,000 to 20,000 mg/L (median 7,300); supernatant suspended solids, 31 to 460 mg/L (median 144); suspended solids removal, 94 to 99+ percent (median 98.7); float solids, 2.8 to 12.4 percent (median 5.0); loading, 1.3 to 7.7 $lb/h/ft^2$ (median 3.1); flow 0.4 to 1.8 gpm/ft^2 (median 1.0).

7.2.9 Environmental Impact

Requires less land than gravity thickeners; subnatant stream is returned to the head of the treatment plant, although it should be compatible with other wastewater; air released to the atmosphere may strip volatile organic material from the sludge; volume of sludge requiring ultimate disposal may be reduced, although its composition will be altered if chemical flotation aids are used; air compressors will require shielding to control the noise generated.

7.2.10 Reliability

DAF systems are reliable from a mechanical standpoint; variations in sludge characteristics can affect process (treatment) reliability, and may require operator attention.

7.2.11 Flow Diagram

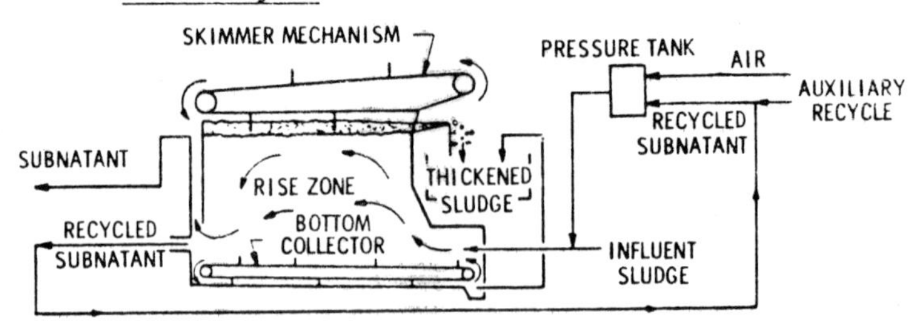

7.2.12 Design Criteria

Pressure, 30 to 70 lb/in²g; effluent recycle ratio, 30 to 150 percent of influent flow; air-to-solids ratio, 0.02 lb air/lb solids; solids loading, 5 to 55 lb/ft²/d (depending on sludge type and whether flotation aids are used); polyelectrolyte addition (when used), 5 to 10 lb/ton of dry solids; solids capture, 70 to 98+ percent; total solids in thickened solids, 3 to 12 percent; hydraulic loading, 0.4 to 2.0 gpm/ft².

Sludge type	Feed solids concentration, %	Typical loading rate without polymer, lb/ft²/d	Typical loading rate with polymer, lb/ft²/d	Float solids concentration, %
Primary + WAS	2.0	20	60	5.5
Primary + (WAS + $FeCl_3$)	1.5	15	45	3.5
(Primary + $FeCl_3$) + WAS	1.8	15	45	4.0
WAS	1.0	10	30	3.0
WAS + $FeCl_3$	1.0	10	30	2.5
Digested primary + WAS	4.0	20	60	10.0
Digested primary + (WAS + $FeCl_3$)	4.0	15	45	8.0
Tertiary, alum	1.0	8	24	2.0

7.2.13 References

1. Innovative and Alternative Technology Assessment Manual. EPA-430/9-78-009 (draft) U.S. Environmental Protection Agency, Cincinnati, Ohio, 1978. 252 pp.

7.3 CENTRIFUGAL THICKENING [1]

7.3.1 Function

Centrifugal thickening is the thickening of sludges using disc, basket, or solid bowl centrifuges.

7.3.2 Description

Centrifuges may be used to thicken sludges by the use of centrifugal force to increase the sedimentation rate of sludge solids. The three most common types of units are the continuous solid bowl type, the disc type, and the basket type. Refer to Section 7.12 for unit descriptions.

7.3.3 Technology Status

Centrifuges have had limited use in thickening excess activated sludges (EAS). Field trials have been conducted at two facilities. Disc-type units have been selected for three treatment plants.

7.3.4 Applications

Centrifuges may be used for thickening excess activated sludge where space limitations or sludge characteristics make other methods unsuitable. Further, if a particular sludge can be effectively thickened by gravity or by flotation thickening without chemicals, centrifuge thickening is not economically feasible.

7.3.5 Limitations

Centrifugal thickening processes can have significant maintenance and power costs; adequate chemical conditioning may be required in order to achieve 90 percent solids capture and 4 percent solids concentration with activated sludge in a bowl-type unit; disc-type units require prescreening to prevent pluggage of discharge nozzles, especially if flow is interrupted or reduced; rotating parts of disc units must be manually cleaned every two weeks.

7.3.6 Design Criteria

See Section 7.12; maximum available capacity per unit is 500 to 600 gpm for disc units and 400 gpm for solid-bowl units.

7.3.7 Environmental Impact

For some sludges, odor controls may be required; noise control is always required.

7.8 Reliability

Pluggage of discharge orifices is a problem on disc-type units if feed to the centrifuge is stopped, interrupted, or reduced below a minimum value.

7.3.9 Flow Diagram

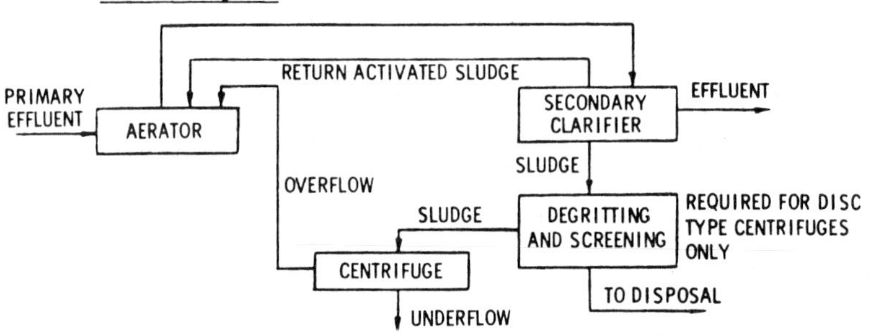

7.3.10 Performance

Typical performance data are presented below for the disc, basket, and solid bowl centrifuges when they are employed in the thickening of EAS. Note that chemical addition is not always required. In general, underflow solids concentration from disc units is lower than from solid bowl units (3 to 5 percent versus 5 to 7 percent).

Type of sludge	Centrifuge type	Capacity, gpm	Feed solids, %	Underflow solids, %	Solids recovery, %	Polymer requirement, lb/ton
EAS	Disc	150	0.75 to 1.0	5 to 5.5	90+	None
EAS	Disc	400	-	4.0	80	None
EAS (after roughing filter)	Disc	50 to 80	0.7	5 to 7	93 to 87	None
EAS (after roughing filter)	Disc	60 to 270	0.7	6.1	97 to 80	None
EAS	Basket	30 to 70	0.7	9 to 10	90 to 70	None
EAS	Solid bowl	10 to 12	1.5	9 to 13	90	-
EAS	Solid bowl	75 to 100	0.44 to 0.78	5 to 7	90 to 80	None
EAS	Solid bowl	110 to 160	0.5 to 0.7	5 to 8	65	None
					85	<5
					90	5 to 10
					95	10 to 15

7.3.11 References

1. Innovative and Alternative Technology Assessment Manual. EPA-430/9-78-009 (draft) U.S. Environmental Protection Agency, Cincinnati, Ohio, 1978. 252 pp.

7.4 AEROBIC DIGESTION [1]

7.4.1 Function

Aerobic digestion is a method of sludge stabilization in an open tank that can be regarded as a modification of the activated sludge process.

7.4.2 Description

Microbiologicical activity beyond cell synthesis is stimulated by aeration, oxidizing both the biodegradable organic matter and some cellular material into CO_2, H_2O, and NO_3. The oxidation of cellular matter is called endogenous respiration and is normally the predominant reaction occurring in aerobic digestion. Stabilization is not complete until there has been an extended period of primarily endogenous respiration (typically 15 to 20 days). Major objectives of aerobic digestion include odor reduction, reduction of biodegradable solids, and improved sludge dewaterability. Aerobic bacteria stabilize the sludge more rapidly than anaerobic bacteria, although a less complete breakdown of cells is usually achieved. Oxygen can be supplied by surface aerators, turbines, or by diffusers. Other equipment may include sludge recirculation pumps and piping, mixers, and scum collection baffles. Aerobic digestors are designed similar to rectangular aeration tanks and use conventional aeration systems, or employ circular tanks and use an eductor tube for deep tank aeration.

7.4.3 Common Modifications

Both one- and two-tank systems are used. Small plants often use a one-tank batch system with a complete mix cycle followed by settling and decanting (to help thicken the sludge). Larger plants may consider a separate sedimentation tank to allow continuous flow and facilitate decanting and thickening. Air may be replaced with oxygen.

7.4.4 Technology Status

Aerobic digestion is primarily used in small plants and rural plants, especially where extended aeration or contact stabilization is practiced.

7.4.5 Applications

Suitable for waste primary sludge, waste biological sludges (activated sludge or trickling filter sludge), or a combination of any of these. Advantages of aerobic digestion over anaerobic digestion include simplicity of operation, lower capital cost, lower BOD concentrations in supernatant liquid, recovery of more of the fertilizer value of sludge, fewer effects from interfering substances (such as heavy metals), and no danger of methane explosions. The process also reduces grease content and the level of pathogenic organisms, reduces the volume of the sludge, and sometimes produces a more easily dewatered sludge (although it may have poor characteristics for vacuum filters). Volatile solids reduction is generally not as good as anaerobic digestion.

7.4.6 Limitations

High operating costs (primarily to supply oxygen) make the process less competitive at large plants; required stabilization time is highly temperature sensitive, and aerobic stabilization may require excessive periods in cold areas or will require sludge heating, further increasing its cost; no useful byproducts, such as methane, are produced; process efficiency also varies according to sludge age and sludge characteristics, and pilot work should be conducted prior to design; improvement in dewaterability frequently does not occur.

7.4.7 Residuals Generated

Supernatant typical quality is SS, 100 to 12,000 mg/L; BOD_5, 50 to 1,700 mg/L; soluble BOD_5, 4 to 200 mg/L; COD, 200 to 8,000 mg/L; Kjeldahl nitrogen, 10 to 400 mg/L; total phosphorus, 20 to 250 mg/L; soluble phosphorus, 2 to 60 mg/L, pH, 5.5 to 7.7; digested sludge.

7.4.8 Design Criteria

Solids retention time (SRT) required for 40% VSS reduction is 18 to 20 days at 20°C for mixed sludges from AS to TF plant, 10 to 16 days for waste activated sludge only, 16 to 18 days average for activated sludge from plants without primary settling; volume allowance, 3 to 4 ft^3/capita; VSS loading, 0.02 to 0.4 $lb/ft^3/d$; air requirements, 20 to 60 $ft^3/min/1,000\ ft^3$; minimum DO, 1 to 2 mg/L; energy for mechanical mixing, 0.75 to 1.25 $hp/1,000\ ft^3$; oxygen requirements, 2 lb/lb of cell tissue destroyed (includes nitrification demand) and 1.6 to 1.9 lb/lb of BOD removed in primary sludge.

7.4.9 Environmental Impact

Supernatant stream is returned to head of plant with high organic loadings; sludge stabilization reduces the adverse impact of land disposal of sludge; process has high power requirements; odor controls may be required.

7.4.10 Reliability

Less sensitive to environmental factors than anaerobic digestion; requires less laboratory control and daily maintenance; relatively resistant to variations in loading, pH, and metals interference; lower temperatures require much longer detention times to achieve a fixed level of VSS reduction; however, performance loss does not necessarily cause an odorous product; maintenance of the DO at 1 to 2 mg/L with adequate detention results in a sludge that is often easier to dewater (except on vacuum filters).

7.4.11 Flow Diagram

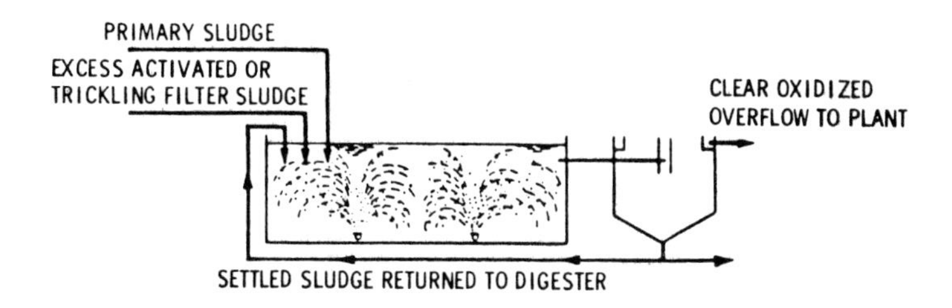

7.4.12 Performance

Material	Influent, %	Effluent, %	Reduction, %
Total solids	2 to 7	3 to 12	--
Volatile solids	50 to 80 (of above)		30 to 70 (typical 35 to 45)
Pathogens			Up to 85

7.4.13 References

1. Innovative and Alternative Technology Assessment Manual. EPA-430/9-78-009 (draft), U.S. Environmental Protection Agency, Cincinnati, Ohio, 1978. 252 pp.

7.5 ANAEROBIC (TWO-STAGE) DIGESTION [1]

7.5.1 Function

Anaerobic digestion is a process for breakdown of sludge into methane, carbon dioxide, unusable intermediate organics, and a relatively small amount of cellular protoplasm.

7.5.2 Description

A two-vessel system is used for sludge stabilization. The first tank, used for digestion, is equipped with one or more of the following: heater, sludge recirculation pumps, methane gas recirculation, mixers, and scum breaking mechanisms. The second tank is used to store and concentrate the digested sludge and to form a supernatant.

The anaerobic digestion process consists of two distinct simultaneous stages of conversion of organic material by acid-forming bacteria and gasification of the organic acids by methane-forming bacteria. The methane-producing bacteria are very sensitive to conditions of their environment and require careful control of temperature, pH, excess concentrations of soluble salts, metal cations, oxidizing compounds, and volatile acids. They also show an extreme substrate specificity. The digester requires periodic cleanout (from 1 to 2 years) due to buildup of sand and gravel on the digester bottom.

7.5.3 Technology Status

Anaerobic digestion is in widespread use (60 to 70 percent) for primary and secondary sludge in plants having a capacity of 1 Mgal/d or more.

7.5.4 Applications

This process is suitable for primary sludge or combinations of primary sludge and limited amounts of secondary sludges. Digested sludge is reduced in volume and pathogenic organism content; it is less odorous and easily de-watered, and it is suitable for ultimate disposal. Advantages over single-stage digestion include increased gas production, a clearer supernatant liquor, necessity for heating a smaller primary tank thus economizing in heat, and more complete digestion. The process also lends itself to modification changes, such as to high-rate digestion.

7.5.5 Limitations

Process is relatively expensive, about twice the capital cost of single-stage digestion. It is the most sensitive operation in the treatment plant and is subject to upsets by interfering substances, e.g., excessive quantities of heavy metals, sulfides, and chlorinated hydrocarbons. The addition of activated and advanced waste treatment sludges can cause high operating costs and poor plant efficiencies. The additional solids do not readily settle after digestion. The digester requires periodic cleanout due to buildup of sand and gravel on digester bottom.

7.5.6 Chemicals Required

The pH must be maintained using lime, ammonia, soda ash, bicarbonate of soda, or lye; addition of powder activated carbon may improve stability of over stressed digesters; heavy metals are precipitated with ferrous or ferric sulfate; odors are controlled with hydrogen peroxide; heat must be provided.

7.5.7 Residuals Generated

Supernatant contains 200 to 15,000 mg/L suspended solids; 500 to 10,000 mg/L BOD_5; 1,000 to 30,000 mg/L COD; 300 to 1,000 mg/L TKN; 50 to 1,000 mg/L total phosphorus; scum; sludge; and gas.

7.5.8 Environmental Impact

Return of supernatant to head of plant may cause plant upsets; adverse environmental impact of sludge disposal on land is reduced as a result of the process.

Digester gas can be used for on-site generation of electricity and/or for any in-plant purpose requiring fuel; can also be used off-site in a natural gas supply system; off-site use usually requires treatment to remove impurities such as hydrogen sulfide and moisture; removal of CO_2 further increases the heat value of the gas; utilization is more successful when a gas holder is provided.

7.5.9 Reliability

Successful operation subject to a variety of physical, chemical, and biological phenomena, e.g., pH, alkalinity, temperature and concentrations of toxic substances of digester contents. Sludge digester biomass is relatively intolerant to changing environmental conditions. Under one set of conditions, particular concentrations of a substance can cause upsets, while under another set of conditions higher concentrations of the same substance are harmless. Process requires careful monitoring of pH, gas production, and volatile acids.

7.5.10 Design Criteria

Solids Retention Times (SRT) required at various temperatures are shown below:

	Mesophilic Range				
Temperature, °F	50	67	75	85	95
SRT, days	55	40	30	25	20

Volume criteria (ft^3/capita): primary sludge, 1.3/3; primary and trickling filter sludges, 2.6/5; primary and waste activated sludges, 2.6/6.

Tank size: diameter, 20 to 115 ft; depth, 25 to 45 ft; bottom slope, 1 vertical/4 horizontal.

Solids loading, 0.04 to 0.40 lb VSS/ft^3/d; volumetric loading, 0.038 to 0.1 ft^3/cap/d; wet sludge loading, 0.12 to 0.19 lb/cap/d; pH 6.7 to 7.6.

7.5.11 Flow Diagram

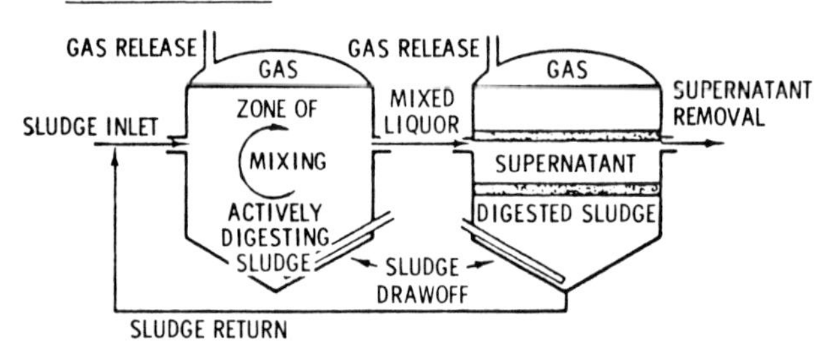

7.5.12 Performance

	Influent	Effluent	Reduction
Total solids	2 to 7%	2.5 to 12%	33 to 58%
Volatile solids			35 to 50%
Pathogen			85 to <100%
Odor reduction			--

Sidestream - gas production

Quantity - 8 to 12 ft^3/lb volatile solids added, or 12 to 18 ft^3/lb volatile solids destroyed, or 11 to 12 ft^3/lb total solids digested.

Quality - 65 to 70% methane; trace N_2, H_2, H_2S, and NH_3; 25 to 30% CO_2; 550 to 600 Btu/ft^3.

7.5.13 References

1. Innovative and Alternative Technology Assessment Manual. EPA-430/9-78-009 (draft) U.S. Environmental Protection Agency, Cincinnati, Ohio 1978. 252 pp.

7.6 CHEMICAL CONDITIONING [1]

7.6.1 Function

Chemical conditioning is a process for coagulating sludge solids and releasing absorbed water.

7.6.2 Description

The use of chemicals to condition sludge for dewatering is economical because of the increased yields and greater flexibility obtained.

Chemicals are most easily applied and metered in liquid form. Dissolving tanks are needed if the chemicals are received as dry powder. These tanks should be large enough for at least one-day's supply of chemicals and should be furnished in duplicate. They must be fabricated or lined with corrosion-resistant material. Polyvinyl chloride, polyethylene, and rubber are suitable materials for tank and pipe linings for handling acid solutions. Metering pumps, which must be corrosion resistant, are generally of the positive-displacement type with variable-speed or variable-stroke drives to control the flowrate. Another metering system consists of a constant-head tank supplied by a centrifugal pump. A rotameter and throttling valve are used to meter the flow.

The chemical dosage required for any sludge is determined in the laboratory. Filter-leaf test kits are used to determine chemical doses, filter yields, and the suitability of various filtering media. These kits have several advantages over the Büchner funnel procedure. In general, it has been observed that the type of sludge has the greatest impact on the quantity of chemical required. Difficult-to-dewater sludges require larger doses of chemicals and generally do not yield as dry a cake. Sludge types, listed in the approximate order of increasing chemical requirements for conditioning, are as follows:

- Untreated (raw) primary sludge
- Untreated mixed primary and trickling-filter sludge
- Untreated mixed primary and waste activated sludge
- Anaerobically digested primary sludge
- Anaerobically digested mixed primary and waste activated sludge
- Aerobically digested sludge (normally dewatered on drying beds without the use of chemicals for conditioning).

Intimate admixing of sludge and coagulant is essential for proper conditioning. The mixing must not break the floc after it has formed, and the detention time is kept to a minimum so that sludge reaches the filter as soon after conditioning as possible. Mixing tanks are generally of the vertical type for small plants and of the horizontal type for large plants. They are ordinarily built of welded steel and lined with rubber or other acid-proof coating. A typical layout for a mixing or conditioning tank has a horizontal agitator driven by a variable-speed motor to provide a shaft speed of 4 to 10 r/min. Overflow from the tank is adjustable to vary the detention period. Vertical cylindrical tanks with propeller mixers are also used.

7.6.3 Common Modifications

Elutriation is a unit operation in which a solid or a solid-liquid mixture is intimately mixed with a liquid for the purpose of transferring certain components to the liquid. A typical example is the washing of digested wastewater sludge before chemical conditioning to remove certain soluble organic and inorganic components that would consume large amounts of chemicals. The cost of washing the sludge is, in general, more than compensated for by the savings that result from a lower demand for conditioning chemicals.

The usual leaching operation consists of two steps: (1) a thorough mixing of the solid or solid-liquid mixture with the leaching liquid, and (2) separation of the leaching liquid. Each combination of mixing and washing is called a stage. A stage is said to be ideal if the concentration of the component being leached is the same in the separating liquid as it is in the liquid that remains with the solids. Mixing and separating can be carried out either in the same tank or in separate tanks. In sanitary engineering, separate tanks are usually used for each stage.

Since alkalinity is usually present in high concentrations in digested sludge, it is commonly used to measure leaching efficiency. A decrease in the quantity of chemicals required to condition sludge has been correlated with the decrease in alkalinity that results from elutriation.

7.6.4 Technology Status

The technology of chemical conditioning is well-developed.

7.6.5 Applications

Conditioning is used in advance of sludge dewatering.

7.6.6 Limitations

Although elutriation was used commonly in the past, it has fallen into disfavor because of the concern that the finely divided solids washed out of the sludge may not be fully captured in the main wastewater treatment facilities. In fact, the U.S. Environmental Protection Agency has stated that sludge elutriation is not considered desirable and its use will not be approved without adequate safeguards.

7.6.7 Chemicals Required

Chemicals used in chemical conditioning include ferric chloride, lime, alum, and organic polymers.

7.6.8 Design Criteria

The dosage of chemicals for various types of sludges for vacuum filtration is shown below (conditioners are shown in percentage of dry sludge).

Type of sludge	Fresh solids		Digested		Elutriated, digested	
	$FeCl_3$	CaO	$FeCl_3$	CaO	$FeCl_3$	CaO
Primary	1-2	6-8	1.5-3.5	6-10	2-4	
Primary and trickling filter	2-3	6-8	1.5-3.5	6-10	2-4	
Primary and activated	1.5-2.5	7-9	1.5-4	6-12	2-4	
Activated (alone)	4-6					

7.6.9 References

1. Metcalf and Eddy, Wastewater Engineering - Treatment, Disposal, Reuse, McGraw-Hill, Inc., 1979. pp. 634-636.

7.7 THERMAL CONDITIONING (HEAT TREATMENT) [1]

7.7.1 Function

Heat treatment is essentially a conditioning process that prepares sludge for dewatering on vacuum filters or filter presses without the use of chemicals.

7.7.2 Description

The heat treatment process involves heating sludge to 144°C to 210°C for short periods of time under pressure of 150 to 400 lb/in² gage. In addition, the sludge is sterilized and generally stabilized and rendered inoffensive. Heat treatment results in coagulation of solids, a breakdown in the cell structure of sludge, and a reduction of the water affinity of sludge solids.

Several proprietary variations exist for heat treatment. In these systems, sludge is passed through a heat exchanger into a reactor vessel, where steam is injected directly into the sludge to bring the temperature and pressure into the necessary ranges. In one variation, air is also injected into the reactor vessel with the sludge. The detention time in the reactor is approximately 30 minutes. After heat treatment, the sludge passes back through the heat exchanger to recover heat, and then is discharged to a thickener-decant tank. The thickened sludge may be dewatered by filtration or centrifugation to a solids content of 30 to 50 percent. The sludge may be ground prior to heat treatment.

7.7.3 Technology Status

The process of heat treating sludge, first introduced in 1935, has become common during the last decade. About 100 units are currently in operation in the United States.

7.7.4 Applications

Heat treatment is practiced as a sludge conditioning method to reduce the costs of sludge dewatering and ultimate disposal. The benefits of heat treatment include (1) improved dewatering characteristics of treated sludge without chemical conditioning; (2) generally innocuous and sterilized sludge suitable for ultimate disposal by a variety of methods including land application in some cases; (3) few nuisance problems; (4) a product suitable for many types of sludge that cannot be stabilized biologically; (5) reduction in subsequent incineration energy requirements; and (6) reduction in size of subsequent vacuum filters and incinerators.

7.7.5 Limitations

The thermal conditioning process has very high capital and operating costs, and may not be economical at small treatment plants. Specialized supervision and maintenance are required due to the high temperatures and pressures involved. Expensive material costs are necessary to prevent corrosion and withstand the operating conditions. Heavy metal concentrations in sludges are not reduced by heat treatment, and further treatment of sludges with high metals concentrations may be required if the sludge is to be applied to crop land. The sludge supernatant and filtrate recycle liquor are strongly colored and contain a very high concentration of soluble organic compounds and ammonia nitrogen, and in some cases must be pretreated prior to return to the head of the treatment plant.

7.7.6 Chemicals Required

Chemicals are not normally required for dewatering; corrosion control aids may be required for the boiler and/or the process; heat must be provided.

7.7.7 Residuals Generated

Sidestream (recycle liquor) contains 50 percent of the sludge flow (by volume); stream quality: BOD, 5,000 to 15,000 mg/L; COD, 10,000 to 30,000 mg/L; NH_3-N, 500 to 800 mg/L; phosphorus, 140 to 250 mg/L; total suspended solids, 9,000 to 12,000 mg/L; volatile suspended solids, 8,000 to 10,000 mg/L; pH, 4 to 6.

This stream is generally amenable to biological treatment but can contribute up to 30 to 50 percent of the organic loading to a treatment plant. If the plant has not been designed for this additional load, pretreatment prior to return may be necessary. Some noncondensable gases may be generated that will require combustion or disposal. Boiler breakdown and/or water treatment residuals (for boiler feedwater) may result.

7.7.8 Environmental Impact

Recycle liquor sent to head of plant can cause plant upsets due to very high organic loadings. The process can result in offensive odor production if proper odor control is not practiced. A colored effluent may also result, requiring additional processing where discharge standards prohibit this condition.

The composition of the recycle liquor can vary among the various processes. Some liquors may contain a high proportion of nonbiodegradable matter. This matter is largely humic acids, which can give rise to unpleasant odors and taste if present in water that has been chlorinated prior to use for domestic supply. If industrial wastes of various types are included in the wastewater to be treated, the actual chemical composition of the liquor resulting from heat treatment of the sludge should be determined by a detailed chemical activated carbon adsorption for nonbiodegradable organics.

7.7.9 Reliability

Limited operating data are available; mechanical and process reliability appear adequate after some initial operational problems; careful operator attention is required.

7.7.10 Design Criteria

Temperature, 140 to 210°C; pressure, 150 to 400 lb/in² gage; detention time, 30 to 90 min; steam consumption, 600lb/1,000 gal of sludge.

7.7.11 Flow Diagram

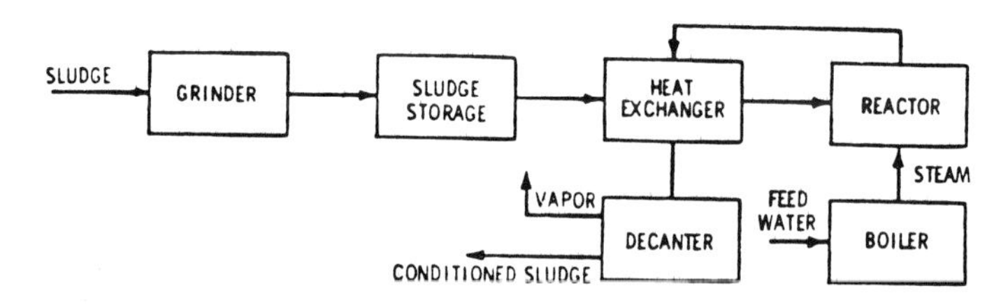

7.7.12 Performance

Heat treatment is a conditioning process intended to enhance the performance of subsequent operations. Within the process itself, pathogens are destroyed and 30 to 40 percent of the volatile suspended solids are solubilized. Dewatering efficiency can be increased to a solids capture of over 95 percent and a solids content of up to 50 percent.

7.7.13 References

1. Innovative and Alternative Technology Assessment Manual, EPA-430/9-78-009 (draft), U.S. Environmental Protection Agency, Cincinnati, Ohio, 1978. 252 pp.

7.8 DISINFECTION (HEAT) [1]

7.8.1 Function

Heating to pasteurization temperatures is a well known method of destroying pathogenic organisms that has been applied sucessfully to disinfecting sludge.

7.8.2 Description

Pasteruization implies heating to a specific temperature for a time period sufficient to destroy undesirable organisms in sludge and to make sludge suitable for land disposal on cropland. Usually heat is applied at 70 to 75°C for 20 to 60 minutes. Treatment can be applied to raw liquid sludge (thickened or unthickened), or stabilized or digested sludge.

Pasteurization is usually a batch process, consisting of a reactor to hold sludge, a heat source, and heat exchange equipment, pumping and piping, and instrumentation for automated operation. Pasteurization has little effect on sludge composition or structure because the sludge is only heated to a relatively moderate temperature.

7.8.3 Technology Status

Heating to pasteurization temperature is not widely used; the process is more common in Europe than in the United States. In West Germany and Switzerland, there are regulations (actually seldom followed) that require pasteurization when sludge is spread on pastures during summer growth periods. The process may find increased application with the renewed interest of land disposal of sludges.

7.8.4 Application

Disinfection can be applied to a wide variety of sludges in various forms. Pasteurization may be redundant where sludges are treated by other processes which destroy pathogenic matter. The largest potential application is to otherwise untreated sludges that are disposed of on land. Studies show that liquid sludge need only be cooled to 60°C for application to land with no adverse effects from temperature. Small treatment plants can pasteurize liquid digested sludge in a tank truck with steam injection.

7.8.5 Limitations

Pasteurization has little or no effect on metals or other toxic materials. Pasteurized but undigested sludges still have considerable risk of foul smelling fermentation after land applications. Limited data are available on interferences and other process controls required for optimizing the process. Heating unthickened sludge requires excessive amounts of heat. Because of the low temperatures involved, heat recovery is not cost effective unless the sludge flow is at least 50,000 gal/d. At this level, one-stage heat recuperation may be cost effective. Two-stage recuperation is not cost effective until a flow of over 100,000 gal/d of sludge is reached.

7.8.6 Chemicals Required

Typical boiler feedwater pretreatment chemicals are used to prevent scale and/or corrosion; heat must be provided.

7.8.7 Residuals Generated

Boiler blowdown and air pollution from the boiler are generated.

7.8.8 Environmental Impact

Reduces the adverse impact of sludge disposal to cropland. If steam injection is used to heat the sludge, chemicals used for feedwater pretreatment must be acceptable for land spreading of sludge.

Digested sludge heat can reduce the need for supplemental energy. Methane from anaerobic digestion can provide the required fuel for pasteurization.

7.8.9 Reliability

Mechanical and process reliability are high; pasteurization can be fully automated and requires minimum operator attention; there is little operating experience in the United States.

7.8.10 Design Criteria

Temperature, 70 to 75°C; time, 20 to 60 minutes; heat required, 4-6 x 10^6 Btu/ton of sludge solids. Two units or more are usually designed in parallel so that one unit can be filling while the other is holding sludge for the required length of time. Units can share a common boiler.

7.8.11 Flow Diagram

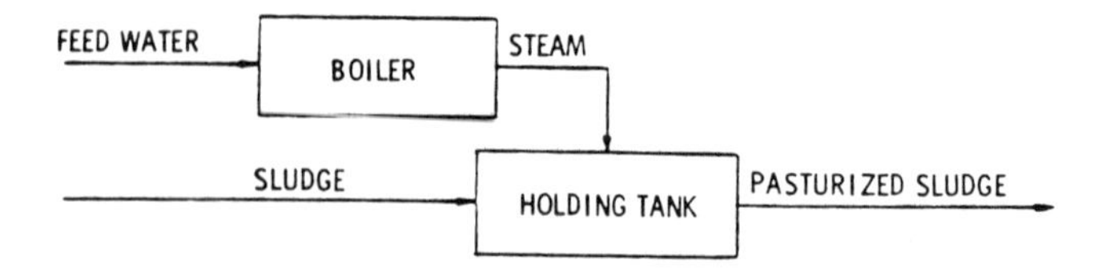

7.8.12 Performance

Seventy-five degrees Centigrade for 60 minutes will reduce coliform indicators below 1,000 counts per 100 mL. Seventy degrees Centigrade for 30 to 60 minutes is effective for destroying pathogens in digested sludge. Seventy degrees Centigrade for 20 minutes is effective for destroying pathogens in raw sludge. Heat treatment also appears to destroy viruses. The table below indicates the time required for 100 percent elimination of various typical pathogenic organisms found in sludge at various temperatures:

	Time, min				
Organism	50°C	55°C	60°C	65°C	70°C
Time required for 100% reduction (minutes)					
Cysts of entamoeba histolytica	5				
Eggs of ascaris lumbricoides	60	7			
Brucella abortis		60		3	
Corynebacterium diptheriae		45			4
Salmonella typhosa			30		4
Escherichia coli			60		5
Micrococcus pyrogene var. aureus					20
Mycobacterium tuberculosis var.					20
Viruses					25

7.8.13 References

1. Innovative and Alternative Technology Assessment Manual, EPA-430/9-78-009 (draft), U.S. Environmental Protection Agency, Cincinnati, Ohio, 1978. 252 pp.

7.9 VACUUM FILTRATION [1]

7.9.1 Function

Vacuum filters are used to dewater sludges so as to produce a cake having the physical handling characteristics and moisture contents required for subsequent processing.

7.9.2 Description

A rotary vacuum filter consists of a cylindrical drum rotating partially submerged in a vat or pan of conditioned sludge. The drum is divided radially into a number of sections, which are connected through internal piping to ports in a valve body (plate) at the hub. This plate rotates in contact with a fixed valve plate with similar ports, which are connected to a vacuum supply, a compressed air supply, and an atmospheric vent. As the drum rotates, each section is thus connected to the appropriate service. Various operating zones are encountered during a complete revolution of the drum. In the pickup or form section, vacuum is applied to draw liquid through the filter covering (media) and form a cake of partially dewatered sludge. As the drum rotates, the cake emerges from the liquid sludge pool, while suction is maintained to promote further dewatering. A lower level of vacuum often exists in the cake drying zone. If the cake tends to adhere to the media, a scraper blade may be provided to assist removal.

The three principal types of rotary vacuum filters are the drum type, coil type, and the belt type. The filters differ primarily in the type of covering used and the cake discharge mechanism employed. Cloth media are used on drum and belt types; stainless steel springs are used on the coil type. Infrequently, a metal media is used on belt types. The drum filter also differs from the other two in that the cloth covering does not leave the drum but is washed in place, when necessary. The design of the drum filter provides considerable latitude in the amount of cycle time devoted to cake formation, washing, and dewatering; the design also minimizes inactive time.

The top feed drum filter is a variation of the conventional drum filter. In this case, sludge is fed to the vacuum filter through a hopper located above the filter. The potential advantages of the top feed drum filter are that gravity aids in cake formation; capital costs may be lower since the feed hopper is smaller and no sludge agitator and related drive equipment are required; and "blinding" of the media may be reduced.

The coil-type vacuum filter uses two layers of stainless steel coils arranged in corduroy fashion around the drum. After a dewatering cycle, the two layers of springs leave the drum and are separated from each other so that the cake is lifted off the lower layer of springs and discharged from the upper layer. Cake release is essentially free of problems. The coils are then washed and reapplied to the drum. The coil filter has been and is widely used for all types of sludge. However, sludges with particles that are both extremely fine and resistant to flocculation dewater poorly on coil filters.

Media on the belt-type filter leaves the drum surface at the end of the drying zone and passes over a small diameter discharge roll to facilitate cake discharge. Washing of the media next occurs before it returns to the drum and to the vat for another cycle. This type filter normally has a small diameter curved bar between the point where the belt leaves the drum and the discharge roll that aids in maintaining belt dimensional stability. In practice, it is frequently used to insure adequate cake discharge.

Many types of filter media are available for belt and drum filters. There is some question whether increases in yield due to operating vacuums greater than 15 inches of mercury are justifiable. The cost of a greater filter area must be balanced against the higher power costs for higher vacuums. An increase from 15 to 20 inches of vacuum is reported to have provided about 10 percent greater yield in three full-scale installations.

7.9.3 Common Modifications

Chemical conditioning is often employed to agglomerate a large number of small particles. It is almost universally applied with mixed sludges.

7.9.4 Technology Status

Vacuum filtration is the most common method of mechanical sludge dewatering utilized in the United States.

7.9.5 Applications

Generally used in larger facilities where space is limited, or when incineration is necessary for maximum volume reduction.

7.9.6 Limitations

Relatively high operating skill required; operation is sensitive to type of sludge and conditioning procedures. As raw sludge ages (3 to 4 hours) after thickening, vacuum filter performance decreases. Poor release of the filter cake from the belt is occasionally encountered. Chemical conditioning costs can sometimes be extremely large if a sludge is hard to dewater.

7.9.7 Chemicals Required

$FeCl_3$ and/or lime, or polymer dosing is a function of type of sludge and vacuum filter characteristics.

7.9.8 Environmental Impact

Vacuum filtration involves relatively high chemical and energy requirements.

7.9.9 Reliability

Large doses of lime may require frequent washings of drum filter media; remedial measures are frequently required to obtain operable cake releases from belt filters; high operating skill is required to maintain high level of reliability.

7.9.10 Design Criteria

Typical loads are shown below. The loading is a function of feed solids concentrations, subsequent processing requirements, and chemical preconditioning.

Sludge type	Typical loading, lb dry solids/hr-ft^2
Raw primary	7 to 15
Digested primary	4 to 7
Mixed digested	3.5 to 5

7.9.11 Flow Diagram

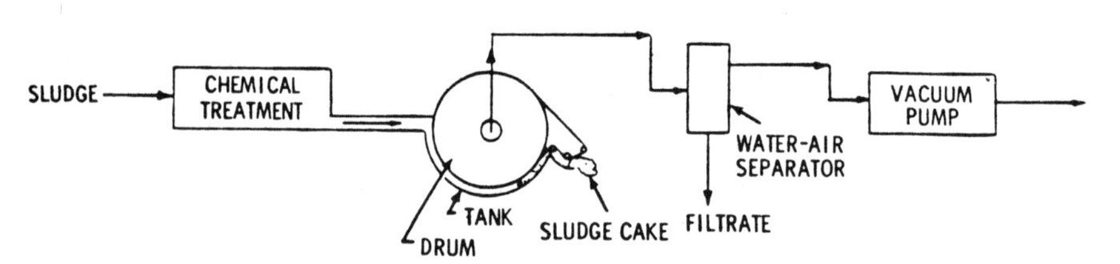

7.9.12 Performance

Solids capture ranges from 85 to 99.5 percent; cake moisture is usually 60 to 90 percent, depending on feed type, solids concentration, chemical conditioning, machine operation and management; dewatered cake is suitable for landfill, heat drying, incineration or land spreading.

7.9.13 References

1. Innovative and Alternative Technology Assessment Manual. EPA-430/9-78-009 (draft), U.S. Environmental Protection Agency, Cincinnati, Ohio, 1978. 252 pp.

7.10 FILTER PRESS DEWATERING [1]

7.10.1 Function

Filter press dewatering is the removal of water from sludge using conventional filter presses.

7.10.2 Description

The recessed plate press is the conventional filter press used for dewatering sewage sludges. This press consists of vertical recessed plates up to 5 ft in diameter (or 5 ft on a side, if square) that are held rigidly in a frame and pressed together between a fixed and moving end. A filter cloth is mounted on the face of each individual plate. The sludge is fed into the press at pressures up to 225 psi gage and passes through feed holes in the trays along the length of the press. The water passes through the cloth; the solids are retained and form a cake on the surface of the cloth. Sludge feeding is stopped when the cavities or chambers between the plates are completely filled. Drainage ports are provided at the bottom of each press chamber. The filtrate is collected in these ports, taken to the end of the press, and discharged to a common drain. At the commencement of a processing cycle, the drainage from a large press can be in the order of 2,000 to 3,000 gph. This rate falls rapidly to about 500 gph as the cake begins to form, when the cake completely fills the chamber, the rate is virtually zero. The dewatering step is complete when the filtrate is near zero. At this point, the pump feeding sludge to the press is stopped, and any back pressure in the piping is released through the bypass valve. The electrical closing gear is then operated to open the press. The individual plates are then moved in turn over the gap between the plates and the moving end; this allows the filter cakes to fall out. The plate-moving step can be either manual or automatic. When all of the plates have been moved and the cakes released, the complete pack of plates is pushed back by the moving end and closed by the electrical closing gear. The valve to the press is then opened, the sludge feed pump started, and the next dewatering cycle commences. Thus, a cycle includes the time required for filling, pressing, cake removal, media washing, and press closing.

A monofilament filter media is now used which, unlike multifilament filter cloth, resists blinding in service. Many systems utilize an efficient precoat system that deposits a protective layer of porous material (fly ash, cement kiln dust, buffing dust) on the filter media to prevent blinding and to facilitate cake release.

While pressure filters with a total effective filtration area of 2,5000 ft² were once considered large, today's units with an effective filtration area of 4,500 ft² are not uncommon.

Until recently, pressure filters, with few exceptions, have operated at a maximum pressure differential of 100 lb/in². Extensive studies during the early 1960's showed that pressure differentials of up to 225 psi produced filter cake solids concentration well in excess of 50 percent. Some commercially available systems now operate near these pressures. As a result of these greater pressures, filter presses offer several advantages, such as higher cake solids concentrations, improved filtrate clarity, improved solids capture, and reduced chemical consumption.

7.10.3 Common Modifications

Modifications to filter press dewatering include various weaves and materials for the filter media, precoating materials, and methods, mechanical plate shifting, and washing devices.

7.10.4 Technology Status

Experience in United States with pressure filtration of wastewater sludges is limited. Plate presses have been used in European wastewater plants for many years. Industry has made use of the process for many years.

7.10.5 Applications

Filter press dewatering is used for sludges prior to incineration and for hard-to-handle sludges; the process is used where a large filtration area is required in a minimum floor area.

7.10.6 Limitations

Batch discharge requires equalization of pressed cake production prior to incineration; life of filter cloth is limited; presses must normally be installed well above floor level so that cakes can drop onto conveyors or trailers; cake must be delumped prior to incineration.

7.10.7 Reliability

Pressure filter plate warpage has been a major problem; plate gasket deterioration (sometimes caused by plate warpage) has also been a problem requiring maintenance.

7.10.8 Design Criteria

Chamber volume	0.75 to 2.8 ft^3/chamber
Filter areas	14.5 to 45 ft^2/chamber
Number of chambers	Up to 100
Sludge cake thickness	1 to 1 1/2 in
Sludge feed rate	Approximately 2 lb/cycle - ft^2 (dry solids basis)

7.10.9 Flow Diagram

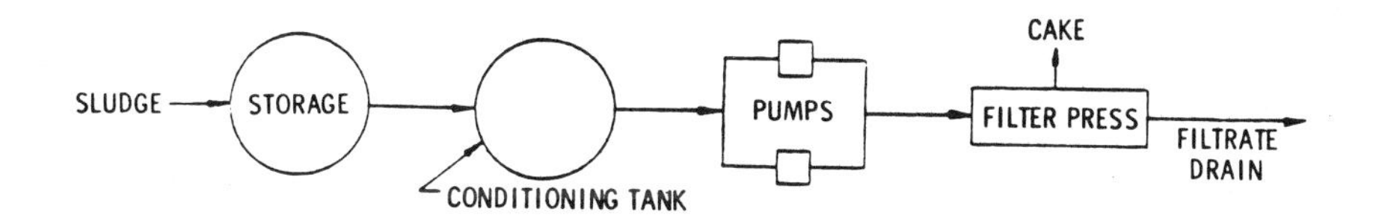

7.10.10 Performance

With input sludges of varying types having a TSS of 1 to 10 percent, typical filter press production data show cake solids concentrations of 50 percent with 100 to 250 percent (on dry solids basis) fly ash conditioning and cycle times of 1.5 to 2.0 h. Cake solids concentrations of 45 percent have been achieved with chemical conditioning (5 to 7.5 percent $FeCl_3$ and 10 to 15 percent lime) and cycle times of 1.0 to 2.0 h. In general, cakes of 25 to 50 percent solids concentrations are achieved.

7.10.11 References

1. Innovative and Alternative Technology Assessment Manual, EPA-430/9-78-009 (draft), U.S. Environmental Protection Agency, Cincinnati, Ohio, 1978. 252 pp.

7.11 BELT FILTER DEWATERING [1]

7.11.1 Function

Belt filter dewater is the removal of water from sludge using filtration in the form of rolling belts.

7.11.2 Description

A belt filter consists of an endless filter belt that runs over a drive and guide roller at each end like a conveyor belt. The upper side of the filter belt is supported by several rollers. Above the filter belt is a press belt that runs in the same direction and at the same speed; its drive roller is coupled with the drive roller of the filter belt. The press belt can be pressed on the filter belt by means of a pressure roller system whose rollers can be individually adjusted horizontally and vertically. The sludge to be dewatered is fed on the upper face of the filter belt and is continuously dewatered between the filter and press belts. After having passed the pressure zone, further dewatering in a reasonable time cannot be achieved by only applying static pressures; however, a superimposition of shear forces can effect this further dewatering. The supporting rollers of the filter belt and the pressure rollers of the pressure belt are adjusted in such a way that the belts and the sludge between them describe an S-shaped curve. Thus, there is a parallel displacement of the belts relative to each other due to the differences in the radii. After further dewatering in the shear zone, the sludge is removed by a scraper.

Some units consist of two stages; the initial draining zone is on the top level, followed by an additional lower section wherein pressing and shearing occur. A significant feature of the belt filter press is that it employs a coarse-mesh, relatively open weave, metal-medium fabric. This is feasible because of the rapid and complete cake formation obtainable when proper flocculation is achieved. Belt filters do not need vacuum systems and do not have the sludge pickup problem occasionally experienced with rotary vacuum filters. The belt filter press system includes auxiliaries such as polymer solution preparation equipment and automatic process controls.

7.11.3 Common Modifications

Some belt filters include the added feature of vacuum boxes in the free drainage zone. To obtain higher cake solids, a vacuum of about 6 in Hg is applied. A "second generation" of belt filters has extended shearing or pressure stages that produce substantial increases in cake solids but are more costly.

7.11.4 Technology Status

As of 1971, 67 units were installed in Europe. At that time, several units were also being installed in the United States. In 1975, a belt filter press was installed in a 0.9 Mgal/d (average) plant in Medford Township, NJ.

7.11.5 Applications

Hard-to-dewater sludges can be handled more readily; low cake moisture permits incineration of primary/secondary sludge combinations without auxiliary fuel; large filtration area can be installed in a minimum floor area.

7.11.6 Limitations

To avoid penetration of the filter belt by sludge, it is usually necessary to coagulate the sludge (generally with synthetic, high polymeric flocculants).

7.11.7 Environmental Impact

Belt filter dewatering involves relatively high chemical and energy requirements.

7.11.8 Reliability

Almost one year of trouble-free operation had been achieved on the Medford, NJ plant as of October, 1977. The two-meter-wide filter belt showed only slight discoloration and remained cleaned and free from blinding or other signs of wear.

7.11.9 Design Criteria

The loadings shown below are based on active belt area:

Sludge type	Sludge loading, $gal/ft^2/h$	Dry solids loading, $lb/ft^2/h$
Raw primary	27-34	13.5-17
Digested primary	20-24	20.5-24
Digested mixed/secondary	13-17	6.7-8.4

7.11.10 Flow Diagram

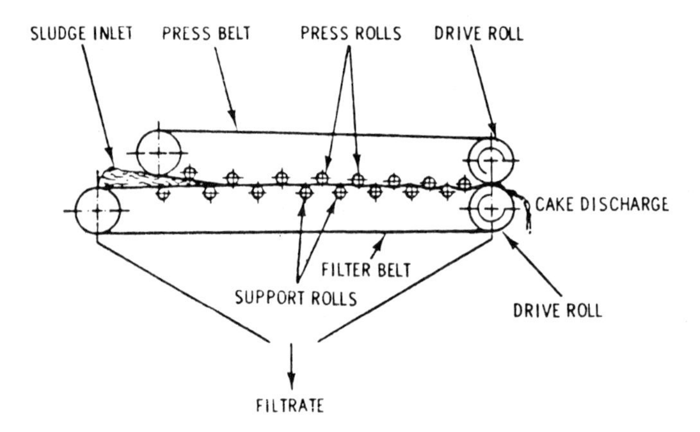

7.11.11 Performance

The table below shows performance achieved in pilot studies.

Feed solids, %	Secondary/ primary ratio	Polymer dosage[a]	Pressure, psi gage[b]	Cake solids, %	Solids recovery, %	Capacity[c]
9.5	100% primary	1.6	100	41	97-99	2,706
8.5	1/5	2.4	100	38	97-99	2,706
7.5	1/2	2.7	25-100	33-38	95-97	1,485
6.8	1/1	2.9	25	31	95	898
6.5	2/1	3.1	25	31	95	858
6.1	3/1	4.1	25	28	90-95	605
5.5	100% secondary	5.5	25	25	95	546

[a] lb/ton dry solids.
[b] psi, gauge.
[c] lb dry solids/hr-m.

7.11.12 References

1. Innovative and Alternative Technology Assessment Manual. EPA-930/9-78-009 (draft), U.S. Environmental Protection Agency, Cincinnati, Ohio, 1978. 252 pp.

7.12 CENTRIFUGAL DEWATERING [1]

7.12.1 Function

Centrifuges are used to dewater sludges using centrifugal force to increase the sedimentation rate of sludge solids. The solid bowl, the disc, and the basket are the three most common types of units.

7.12.2 Description

The solid-bowl continuous centrifuge assembly consists of a bowl and conveyor joined through a planetary gear system, designed to rotate the bowl and the conveyor at slightly different speeds. The solid cylindrical bowl, or shell, is supported between two sets of bearings and includes a conical section at one end. This section forms the dewatering beach over which the helical conveyor screw pushes the sludge solids to outlet ports and then to a sludge cake discharge hopper. The opposite end of the bowl is fitted with an adjustable outlet weir plate to regulate the level of the sludge pool in the bowl. The centrate flows through outlet ports either by gravity or by a centrate pump attached to the shaft at one end of the bowl. Sludge slurry enters the unit through a stationary feed pipe extending into the hollow shaft of the rotating bowl and passes to a baffled, abrasion-protected chamber for acceleration before discharge through the feed ports in the rotating conveyor hub into the sludge pool. Due to the centrifugal forces, the sludge pool takes the form of a concentric annular ring on the inside of the bowl. Solids settle through this ring to the wall of the bowl where they are picked up by the conveyor scroll. Separate motor sheaves or a variable speed drive can be used to adjust the bowl speed for optimum performance.

Bowls and conveyors can be constructed from a large variety of metals and alloys to suit special application. For dewatering of wastewater sludges, mild steel or stainless steel has been used normally. Because of the abrasive nature of many sludges, hardfacing materials are applied to the leading edges and tips of the conveyor blades, the discharge ports, and other wearing surfaces. Such wearing surfaces may be replaced by welding when required.

In the continuous concurrent solid-bowl centrifuge, incoming sludge is carried by the feed pipe to the end of the bowl opposite the discharge. Centrate is skimmed off and cake proceeds up the beach for removal. As a result, settled solids are not disturbed by incoming feed.

In the disc-type centrifuge, the incoming stream is distributed between a multitude of narrow channels formed by stacked conical discs. Suspended particles have only a short distance to settle, so that small and low density particles are readily collected and discharged continuously through fairly small orifices in the bowl wall. The clarification capability and throughput range are high, but sludge concentration is limited by the necessity of discharging through orifices 0.050 in to 0.100 in in diameter. Therefore, it is generally considered a thickener rather than a dewatering device.

In the basket-type centrifuge, flow enters the machine at the bottom and is directed toward the outer wall of the basket. Cake continually builds up within the basket until the centrate, which overflows a weir at the top of the unit, begins to increase in solids. At that point, feed to the unit is shut off, the machine decelerates, and a skimmer enters the bowl to remove the liquid layer remaining in the unit. A knife is then moved into the bowl to cut out the cake, which falls out of the open bottom of the machine. The unit is a batch device with alternate charging of feed sludge and discharging of dewatered cake.

7.12.3 Technology Status

Solid-bowl and disc-type centrifuges are in widespread use; basket-type centrifuges are fully demonstrated for small plants but not widely used.

7.12.4 Applications

Solid-bowl and disc-type centrifuges are generally used for dewatering sludge in larger facilities where space is limited or where sludge incineration is required. Basket-type units are used primarily for partial dewatering at small plants. Disc-type centrifuges are more useful for thickening and clarification than dewatering.

7.12.5 Limitations

Centrifugation requires a sturdy foundation because of the vibration and noise that result from centrifuge operation. Adequate electric power must also be provided because large motors are required. The major difficulty encountered in the operation of centrifuges has been the disposal of the centrate, which is relatively high in suspended nonsettling solids. With disc-type units, the feed must be degritted and screened to prevent pluggage of discharge orifices.

7.12.6 Environmental Impact

Centrate is relatively high in suspended nonsettling solids which, if returned to treatment units, could reduce effluent quality from primary settling system; noise may require some control measures.

7.12.7 Reliability

Pluggage of discharge orifices is a problem on disc-type units if feed to the centrifuge is stopped, interrupted, or reduced below a minimum value; wear is a serious problem with solid-bowl centrifuges.

7.12.8 Design Criteria

Each installation is site specific and dependent upon a manufacturers' product line. Maximum capacities of about 100 tons/h of dry solids are available in solid-bowl units with diameters up to 54 in and power requirements up to 175 hp. Disc-type units are available with capacities up to 400 gpm of concentrate.

7.12.9 Flow Diagram

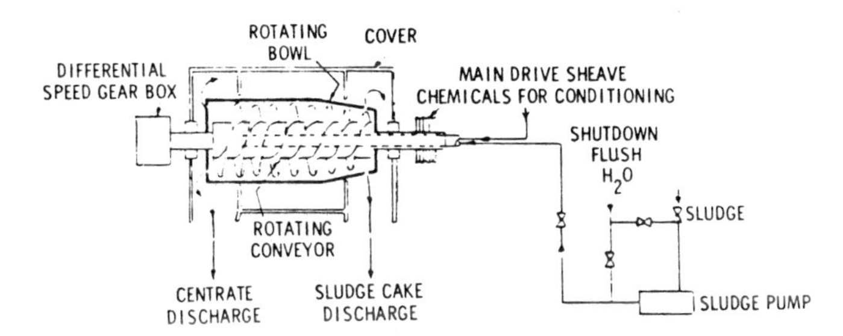

7.12.10 Performance

Solids recovery in solid-bowl centrifuges is 50 to 75 percent without chemical addition, and 80 to 95 percent with chemical addition. Solids concentration is 15 to 40 percent depending on type of sludge. For basket-type centrifuges, solids capture is 90 to 97 percent without chemical addition, and cake solids concentration is 9 to 14 percent. Disc-type centrifuges can dewater a 1-percent sludge to 6-percent solids concentration.

7.12.11 References

1. Innovative and Alternative Technology Assessment Manual. EPA-430/9-78-009 (draft), U.S. Environmental Protection Agency, Cincinnati, Ohio, 1978. 252 pp.

7.13 THERMAL DRYING [1]

7.13.1 Function

Thermal drying is the process of reducing the moisture in sludge by evaporation to 8 to 10 percent using hot air, without combusting the solid materials. For economic reasons, the moisture content of the sludge must be reduced as much as possible through mechanical means prior to heat drying. The five available heat treating techniques are flash, rotary, toroidal, multiple hearth and atomizing spray.

7.13.2 Description

Flash drying is the instantaneous vaporization of moisture from solids by introducing the sludge into a hot gas stream. The system is based on several distinct cycles that can be adjusted for different drying arrangements. The wet sludge cake is first blended with some previously dried sludge in a mixer to improve pneumatic conveyance. Blended sludge and hot gases from the furnace at about 1200°F to 1400°F (650 to 760°C) are mixed and fed into a cage mill in which the mixture is agitated and the water vapor flashed. Residence time in the cage mill is only a matter of seconds. Dry sludge with eight-to-ten percent moisture is separated from the spent drying gases in a cyclone, with part of it recycled with incoming wet sludge cake and another part screened and sent to storage.

A rotary dryer consists of a cylinder that is slightly inclined from the horizontal and revolves at about five-to-eight rpm. The inside of the dryer is equipped usually with flights or baffles throughout its length to break up the sludge. Wet cake is mixed with previously heat dried sludge in a pug mill. The system may include cyclones for sludge and gas separation, dust collection scrubbers, and a gas incineration step.

The toroidal dryer uses the jet mill principle, which has no moving parts, dries, and classifies sludge solids simultaneously. Dewatered sludge is pumped into a mixer where it is blended with previously dried sludge. Blended material is fed into a doughnut-shaped dryer, where it comes into contact with heated air at a temperature of 800°F to 1100°F. Particles are dried, broken up into fine pieces, and carried out of the dryer by the air stream. The dried, powdered sludge is supplemented with nitrogen and phosphorus and formed into briquettes, which are crushed and screened to produce final products.

The multiple hearth furnace is adapted for heat drying of sludge by incorporating fuel burners at the top and bottom hearths, plus down draft of the gases. The dewatered sludge cake is mixed in a pug mill with previously dried sludges before entering the furnace. At the point of exit from the furnace, the solids temperature is about 100°F, and the gas temperature is about 325°F.

Atomizing drying involves spraying liquid sludge in a vertical tower through which hot gases pass downward. Dust carried with hot gases is removed by a wet scrubber or dry dust collector. A high-speed centrifugal bowl can also be used to atomize the liquid sludge into fine particles and to spray them into the top of the drying chamber where moisture is transferred to the hot gases.

7.13.3 Technology Status

Heat drying of sludge was developed more than 50 years ago; however, it is not widely used.

7.13.4 Applications

Thermal drying is an effective way for ultimate sludge disposal and resource conservation when the end products are applied on land for agricultural and horticultural uses. Although an expensive process, it can become a viable alternative if the product can be successfully marketed.

7.13.5 Limitations

Cost and high operator skill are limitations of thermal drying.

7.13.6 Chemicals Required

Nitrogen and phosphorus may be added to increase nutrient values of the dried sludge; heat must be provided.

7.13.7 Residuals Generated

All solids captured in the wet scrubbers and dry solids collectors are recycled and incorporated in the end products.

7.13.8 Environmental Impact

Potential exists for explosion and air pollution if the system is not properly operated and maintained.

7.13.9 Design Criteria

Approximately 1,400 Btu are needed to vaporize one pound of water, based on a thermal efficiency of 72 percent. Less fuel would be required with additional heat recovery. Chemical scrubbers are used, or chemicals are added prior to heat drying. Excessive drying tends to produce a sludge that is dusty or contains many fine particles; this is less acceptable for marketing and should be avoided. Wet scrubbers and/or solids collectors are needed. Standby heat-drying equipment is needed for continuous operation.

7.13.10 Flow Diagram

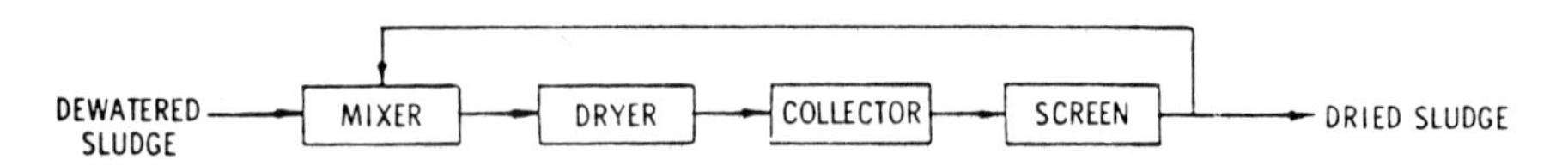

7.13.11 Performance

Heat drying destroys most of the bacteria in the sludge; however, undigested heat dried sludge is susceptible to putrefaction if allowed to get wet in thick layers on the ground. Heat drying does not cause any significant decrease of the heavy metals concentration in the sludge. In general, heat-dried sludge contains nutrients that are only about one-fifth of those contained in chemical fertilizers. Heat-dried sludge is therefore useful only as a fertilizer supplement and a soil conditioner.

7.13.12 References

1. Innovative and Alternative Technology Assessment Manual. EPA-430/9-78-009 (draft), U.S. Environmental Protection Agency, Cincinnati, Ohio, 1978. 252 pp.

7.14 DRYING BEDS [1]

7.14.1 Function

Drying beds are used to dewater sludge both by drainage through the sludge mass and by evaporation from the surface exposed to the air. Collected filtrate is usually returned to the treatment plant.

7.14.2 Description

Drying beds usually consist of 4 to 9 inches of sand, which is placed over 8 to 18 inches of graded gravel or stone. The sand typically has an effective size of 0.3 to 1.2 mm and a uniformity coefficient of less than 5.0. Gravel is normally graded from 1/8 to 1.0 inch. Drying beds have underdrains that are spaced from 8 to 20 feet apart. Underdrain piping is often vitrified clay laid with open joints and having a minimum diameter of 4 inches and a minimum slope of about 1%.

Sludge is placed on the beds in an 8- to 12-inch layer. The drying area is partitioned into individual beds, approximately 20 feet wide by 20 to 100 feet long, of a convenient size so that one or two beds will be filled by a normal withdrawal of sludge from the digesters. The interior partitions commonly consist of two or three creosoted planks, one on top of the other, to a height of 15 to 18 inches, stretching between slots in precast concrete posts. The outer boundaries may be of similar construction or earthen embankments for open beds, but concrete foundation walls are required if the beds are to be covered.

Piping to the sludge beds is generally made of cast iron and designed for a minimum velocity of 2.5 feet/second. It is arranged to drain into the beds and provisions are made to flush the lines and prevent freezing in cold climates. Distribution boxes are provided to divert sludge flow to the selected bed. Splash plates are used at the sludge inlets to distribute the sludge over the bed and prevent erosion of the sand.

Sludge can be removed from the drying bed after it has drained and dried sufficiently to be spadable. Sludge removal is accomplished by manual shoveling into wheelbarrows or trucks, or by a scraper or front-end loader. Provisions should be made for driving a truck onto or along the bed to facilitate loading. Mechanical devices can remove sludges of 20% to 30% solids while cakes of 30% to 40% are generally required for hand removal.

Paved drying beds with limited drainage systems permit the use of mechanical equipment for cleaning. Field experience indicates that the use of paved drying beds results in shorter drying times as well as more economical operation when compared with conventional sandbeds because, as indicated above, the use of mechanical equipment for cleaning permits the removal of sludge with a higher moisture content than does hand cleaning. Paved beds have worked successfully with anaerobically digested sludges but are less desirable than sandbeds for aerobically digested activated sludge.

7.14.3 Common Modifications

Sandbeds can be enclosed by glass. Glass enclosures (1) protect the drying sludge from rain, (2) control odors and insects, (3) reduce the drying periods during cold weather, and (4) can improve the appearance of a waste treatment plant.

Wedge-wire drying beds have been used successfully in England. This approach prevents the rising of water by capillary action through the media, and the construction lends itself well to mechanical cleaning. The first U.S. installations have been made at Rollinsford, New Hampshire, and in Florida. In small plants, it is possible to place the entire dewatering bed in a tiltable unit from which sludge may be removed merely by tilting the entire unit mechanically.

7.14.4 Technology Status

Over 6,000 plants use open or covered sandbeds.

7.14.5 Applications

Sandbeds are generally used to dewater sludges in small plants; they require little operator attention or skill.

7.14.6 Limitations

Air drying is normally restricted to well digested or stabilized sludge, because raw sludge is odorous, attracts insects, and does not dry satisfactorily when applied at reasonable depths. Oil and grease clog sandbed pores and thereby seriously retard drainage. The design and use of drying beds are affected by weather conditions, sludge characteristics, land values, and proximity of residences. Operation is severely restricted during periods of prolonged freezing and rain.

7.14.7 Environmental Impact

Land requirements are large; odors can be a problem with poorly digested sludges and inadequate buffer zone areas.

7.14.8 Design Criteria

Open bed area for various sludge types is shown below.

Sludge type	Open bed area, ft²/capita
Primary digested sludge	1.0 - 1.5
Primary and activated sludge	1.75 - 2.5
Alum or iron precipitated sludge	2.0 - 2.5

Experience has shown that enclosed beds require 60% to 75% of the open bed area. Solids loading rates vary from 10 to 28 lb/ft²/yr for open beds and 12 to 40 lb/ft²/yr for closed beds. Sludge beds should be located at least 200 feet from dwellings to avoid odor complaints due to poorly digested sludges.

7.14.9 Flow Diagram

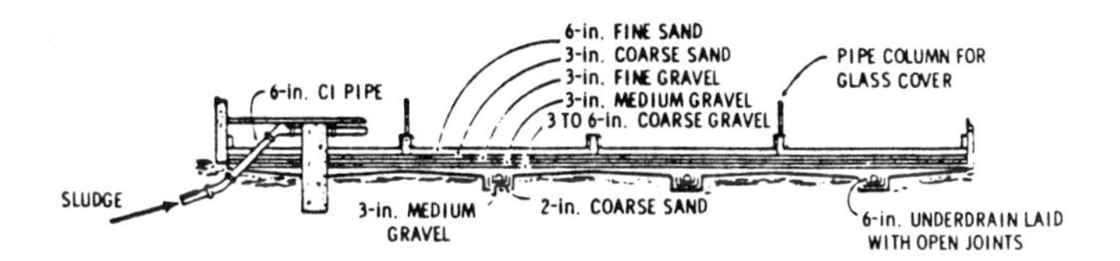

7.14.10 Performance

A cake of 40% to 45% solids may be achieved in two to six weeks in good weather and with a well digested waste activated, primary or mixed sludge. With chemical conditioning, dewatering time may be reduced by 50% or more. Solids contents of 85% to 90% have been achieved on sand beds, but normally the times required are impractical.

7.14.11 References

1. Innovative and Alternative Technology Assessment Manual, EPA-430/9-78-009 (draft), U.S. Environmental Protection Agency, Cincinnati, Ohio, 1978. 252 pp.

7.15 LAGOONS [1]

7.15.1 Function

Digested sludge has often been applied to sludge lagoons adjacent to or in the proximity of treatment facilities. These sludge lagoons are primarily designed to accomplish long-term drying of the digested sludge through the physical processes of percolation and evaporation, primarily the latter.

7.15.2 Description

This method of sludge processing has been extremely popular in the U.S. due to its relatively low cost (when inexpensive land is plentiful) and minimal operation and maintenance requirements, especially at smaller wastewater treatment facilities. The process is relatively simple, requiring periodic decanting of supernatant back to the head of the plant and occasional mechanical excavation of dewatered or dried sludge for transportation to its ultimate disposal location. Lagoons can be a very useful process step. Lagoon supernatant is far better (low SS) than supernatant from a secondary digester or even a thickener. Ultimate disposal of the product solids often is as a soil conditioner or landfill.

Sludge lagoons may also be used as contingency units at treatment plants to store and/or process sludges when normal processing units are either overloaded or out of service.

The drying time to 30 % solids is generally quite lengthy and may require years. Climatic conditions and pre-lagoon sludge processing greatly influence lagoon performance. In warmer, drier climates well-digested sludges are economically and satisfactorily treated by sludge-drying lagoons because of their inherent simplicity of operation and flexibility. Complete freezing causes sludge to agglomerate; hence, when it thaws, the supernatant decants or drains away easily. Well digested sludges minimize potential odor problems that are inherent in this type of system. Multiple-cells are required for efficient operation.

7.15.3 Common Modifications

Methods and patterns of loading, supernatant recycling techniques, and mechanical cleaning techniques vary with location, climate, and type of sludge to be processed.

7.15.4 Technology Status

Lagoon technology is widely used for industrial and municipal sludge processing throughout the world.

7.15.5 Applications

The use of lagoons is a simple sludge drying method for digested sludge in smaller plants because large inexpensive land areas are required.

7.15.6 Limitations

There is a high potential for odors and nuisance insect breeding if feed sludges are not well-digested. Odor and nuisance control chemicals are not entirely satisfactory; also, definitive data on performance and design parameters are lacking despite the popularity of this approach.

7.15.7 Chemicals Required

Lime or other odor control chemicals may be required if digestion is incomplete.

7.15.8 Residuals Generated

Generally, the residuals resulting from a well-operated lagoon will be in the range of 30% solids and are suitable for use as a soil conditioner or landfill.

7.15.9 Environmental Impact

Odor and vector portential are high unless unit is properly designed and operated; land-use requirement is high; groundwater pollution potential is high unless proper site characterization is incorporated into design.

7.15.10 Reliability

Where properly designed, process reliability is a function of upstream processing (digestion).

7.15.11 Design Criteria

Item	Criteria
Dikes:	Slopes of 1:2 exterior and 1:3 interior are needed to permit maintenance and mowing and to prevent erosion; width must be sufficient to allow vehicle transport during cleaning.
Depth:	1.5 to 4.0 ft of sludge depth (depending upon climate).
Bottom:	Separation from groundwater is dependent upon application depths and soil characteristics, but should not be less than 4 ft to prevent groundwater contamination.
Cells:	A minimum of two cells is required.
Loading rates:	2.2 to 2.4 lb solids/yr/ft^3 of capacity; 1.7 to 3.3 lb solids/ft^2 of surface/30 days of bed use; 1 to 4 ft^2/capita (depending on climate).
Decant:	Single- or multiple-level decant for periodic returning supernatant to head of plant.
Sludge removal:	Approximately 1.5 to 3 yr intervals.

7.15.12 Flow Diagram

7.15.13 References

1. Innovative and Alternative Technology Assessment Manual, EPA-430/9-78-009 (draft), U.S. Environmental Protection Agency, Cincinnati, Ohio, 1978. 252 pp.

8. Disposal

8.1 EVAPORATION LAGOONS [1]

8.1.1 Function

The evaporation lagoon is an open holding facility that depends solely on climatic conditions such as evaporation, precipitation, temperature, humidity, and wind velocity to effect dissipation (evaporation) of on-site sludge.

8.1.2 Description

Individual lagoons may be considered as an alternate means of sludge disposal on individual pieces of property. The basic impetus to consider this system is to allow building and other land uses on properties that have soil conditions not conducive to the workability and acceptability of the conventional on-site drain-field or leachbed disposal systems.

If the annual evaporation rate exceeds the annual precipitation, evaporation lagoons may at least be considered as a method of disposal. The deciding factor then becomes the required land area and holding volume. For on-site installations such as small industrial applications, there may also be a certain amount of infiltration or percolation in the initial period of operation. However, after a time, solids deposition may be expected to eventually clog the surface to the point where infiltration is eliminated. The potential impact of wastewater infiltration to the groundwater, and particularly on-site water supplies, should be evaluated in any event and, if necessary, lagoon lining may be utilized to alleviate the problem.

8.1.3 Technology Status

The technology of evaporation is well developed in terms of our scientific understanding and application of climatological and meteorological data.

8.1.4 Applications

The on-site utilization of evaporation lagoons for the disposal of sludge from smaller industrial or commercial facilities may be applicable where access to a municipal sanitary sewer is not available, where subsurface methods are not feasible, and where effluent polishing for surface discharge is not practical.

8.1.5 Limitations

Local health ordinances may limit the use of evaporation lagoons; lagoons represent a potential health hazard when not properly disinfected and controlled; facilities require land area and depend on meteorologic and climatological conditions; may require provision to add makeup water to maintain a minimum depth during dry, hot seasons; public access restrictions are likely.

8.1.6 Residuals Generated

Periodic pump out of accumulated sludge is required.

8.1.7 Environmental Impact

Potential odors; potential health hazard; land area requirements may be large; may adversely affect surrounding property values.

8.1.8 Reliability

Good reliability; however, should be closely controlled to prevent health hazard.

8.1.9 Design Criteria

Hydraulic loading is the primary sizing criteria for an individual total retention lagoon. In order to size the system properly, the following information is needed: (1) anticipated flow of sludge, (2) evaporation rates (10-yr minimum of monthly data), and (3) precipitation rates (10-yr minimum of monthly data).

8.1.10 Flow Diagram

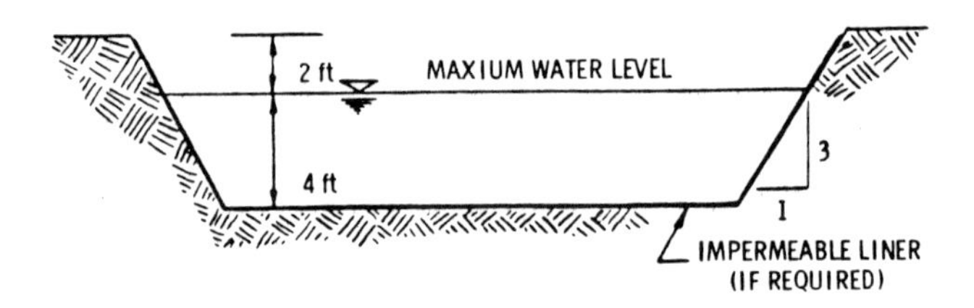

8.1.11 Performance

The performance of evaporation lagoons is necessarily site-specific; therefore, the following data are presented on the basis of net annual evaporation rate that may exist in a certain area:

Net annual evaporation[a], in.	Lagoon Performance, gal water evaporated/ft²/yr
5	3.1
10	6.2
15	9.4
20	12.5
40	24.9
60	37.4

[a]Net annual evaporation = true annual evaporation - annual precipitation.

8.1.12 References

1. Innovative and Alternative Technology Assessment Manual, EPA-430/9-78-009 (draft), U.S. Environmental Protection Agency, Cincinnati, Ohio, 1978. 252 pp.

8.2 INCINERATION [1]

8.2.1 Function

Sludge incineration is a two-step process involving drying and combustion after preliminary dewatering. A typical sludge contains 75% water and 75% volatiles in dry solids. Self-sustained combustion without supplementary fuel is often possible with dewatered raw sludges having a solids concentration greater than 30%.

8.2.2 Description

Two types of incinerator furnaces are descriped: the fluidized bed furnace, and the multiple hearth furnace.

Fluidized Bed Furnace. The fluidized bed furnace (FBF) is a vertically oriented, cylindrically shaped, refractory-lined steel shell that contains a sand bed and fluidizing air distributor. The FBF is normally available in diameters of 9 to 25 feet and heights of 20 to 60 feet. There is one industrial unit operating with a diameter of 53 feet. The sand bed is approximately 2.5 feet thick and rests on a refractory-lined air-distribution grid containing tuyeres through which air is injected at a pressure of 3 to 5 psi to fluidize the bed. Bed expansion is approximately 80% to 100%. Bed temperature is controlled between 1,400°F and 1,500°F by auxiliary burners and/or a water spray or heat removal system above the bed. Ash is carried out the top of the furnace and removed by air pollution control devices, usually wet venturi scrubbers. Sand is lost by attrition at an approximate rate of 5% of the bed volume every 300 hours of operation. Furnace feed can be introduced either above or directly into the bed depending on the type of feed. Generally, sewage sludge is fed directly into the bed.

Excess air requirements for the FBF vary from 20% to 40%. It requires less supplementary fuel than a multiple hearth furnace. An oxygen analyzer in the stack controls the air flow into the reactor, and the auxiliary fuel feed rate is controlled by a bed-temperature controller.

Multiple Hearth Furnace. The multiple hearth furnace (MHF) is a vertically oriented, cylindrically shaped, refractory-lined steel shell having a diameter of 4 to 25 feet and containing 4 to 13 horizontal hearths positioned one above the other. The hearths are constructed of high heat duty fire brick and special fire brick shapes. Sludge is raked radially across the hearths by rabble arms that are supported by a central rotating shaft that runs the height of the furnace. The cast iron shaft is motor driven with provision for speed adjustment from 1/2 to 1-1/2 r/min. Sludge is fed to the top hearth and proceeds downward through the furnace from hearth to hearth. Inflow hearths have a central port through which sludge passes to the next lower hearth. Outflow hearths have ports on their periphery that also tend to regulate gas velocities. The central shaft contains internal concentric flow passages through which air is routed to cool the shaft and rabble arms. The flow of combustion air is countercurrent to that of the sludge. Gas or oil burners are provided on some hearths for start-up and/or supplemental use as required.

The rabble arms provide mixing action as well as movement to the sludge so that a maximum sludge surface is exposed to the hot furnace gases. Because of the irregular surface left by the rabbling action, the surface area of sludge exposed to the hot gases is as much as 130% of the hearth area. While there is significant solids-gas contact time on the hearths, the overall contact time is actually still greater, due to the fall of the sludge from hearth to hearth through the countercurrent flow of hot gases.

The various phases of the incineration process occur in three zones of the MHF. The drying zone consists of the upper hearths, the combustion zone consists of the central hearths, and the lower hearths comprise the cooling zone. Temperatures in each zone are shown below.

	Temperature, °F	
Zone	Sludge	Air
Drying	∿100	∿800
Burning	∿1,500	∿1,500
Cooling	∿400	∿350

8.2.3 Common Modifications

Fluidized Bed Furnace. An air preheater is used in conjunction with a fluidized bed to reduce fuel costs. Also, cooling tubes may be submerged in the bed for energy recovery.

Multiple Hearth Furnace. An afterburner fired with oil or gas is provided where required by local air pollution regulations to eliminate unburned hydrocarbons and other combustibles.

8.2.4 Technology Status

Fluidized Bed Furnace. The first fluidized bed wastewater sludge incinerator was installed in 1962. Many units are now operating in the U.S. with capacities of 200 to 1,000 lb/h of dry solids.

Multiple Hearth Furnace. The MHF is the most widely used wastewater sludge incinerator in the U.S. today. As of 1970, 120 units have been installed.

8.2.5 Applications

Fluidized Bed Furnace. The fluidized bed furnace is used for reduction of sludge volume, thereby reducing land requirements for disposal; unit has energy recovery potential and is suitable for plants where hauling distances to disposal sites are long, or where regulations concerning these alternative methods are prohibitive.

Multiple Hearth Furnace. Same as for fluidized bed furnace.

8.2.6 Limitations

Fluidized Bed Furnace. Because a minimum amount of air is always required for bed fluidization, fan energy savings during load turndown (i.e., sludge feed reduction) are minor. FBF is generally not cost effective for small plants.

Multiple Hearth Furnace. Capacities of MHF's vary from 200 to 8,000 lb/h of dry sludge. Maximum operating temperatures are limited to 1,700°F. There may be operational problems with high-energy feeds. MHF requires 24 to 30 hours for furnace warm-up or cool-down to avoid refractory problems. Failure of rabble arms and hearths have been encountered; nuisance shutdowns have occurred due to ultraviolet flame scanner malfunctions. Thickening and dewatering pretreatment is required.

8.2.7 Environmental Impact

Fluidized Bed Furnace. Particulate collection efficiencies of 96% to 97% are required to meet current standards. There are very few data on the amount of toxic metals that are volatilized and discharged. Limited test data indicate that 4% to 35% of the mercury entering an incinerator with emission controls will volatilize and be emitted to the atmosphere (excluding particulate forms). Gaseous emissions of CO, HCl, SO_2 and NO_2 may be appreciable; additional air pollution control measures may be necessary. Pesticides and PCB's are found in the sludge, but tests indicate that they can be destroyed during incineration and should not be problematical.

Multiple Hearth Furnace. Same as for fluidized bed furnace.

8.2.8 Design Criteria

Fluidized Bed Furnace. Design criteria for FBF are shown below. Concerning actual operations, some extensive maintenance problems have occurred with air preheaters. Scaling of the venturi scrubbers has also been a problem. Screw feeds and screw pump feeds are both subject to jamming because of either overdrying of the sludge feed at the incinerator or because of silt carried into the feed system with the sludge. Another frequent problem has been the burnout of spray nozzles or thermocouples in the bed.

Parameter	Design criteria
Bed loading rate	50 - 60 lb wet solids/ft^2/hr
Superficial bed velocity	0.4 - 0.6 ft/s
Sand effective size	0.2 - 0.3 mm (uniformity coefficient = 1.8)
Operating temperature	1,400 - 1,500°F (normal); 2,200°F (maximum)
Bed expansion	80 - 100%
Sand loss	5% of bed volume per 300 hr of operation

Multiple Hearth Furnace. Design criteria for MHF are shown below.

Parameter	Design criteria
Maximum operating temperature	1,700°F
Hearth loading rate	6 - 10 lb wet solids/ft^2/hr with a dry solids concentration of 20 - 40%
Combustion air flow	12 - 13 lb/lb dry solids
Shaft cooling air flow	1/3 - 1/2 of combustion air flow
Excess air	75 - 100%

8.2.9 Flow Diagrams

Fluidized Bed Furnace.

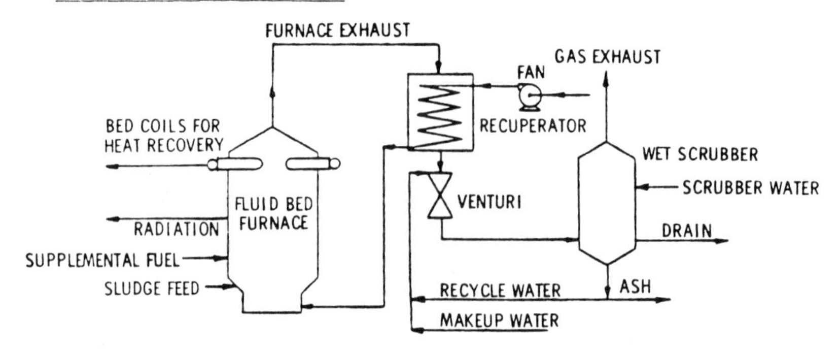

Multiple Hearth Furnace.

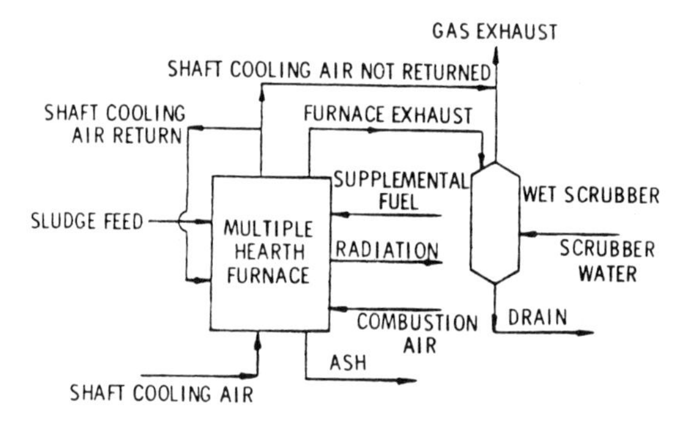

8.2.10 Performance

Fluidized Bed Furnace. The mass of dry solids is reduced to 25% to 35% of the amount entering the unit.

Multiple Hearth Furnace. Dry solids are reduced to 20% to 25% of the mass entering the unit. The recoverable heat ranges from 18% of the total heat input (sludge and supplementary fuel) at 20% solids concentration to 45% of the total heat input at 40% solids concentration.

8.2.11 References

1. Innovative and Alternative Technology Assessment Manual, EPA-430/9-78-009 (draft), U.S. Environmental Protection Agency, Cincinnati, Ohio, 1978. 252 pp.

8.3 STARVED AIR COMBUSTION [1]

8.3.1 Function

Starved air combustion is used for the volumetric and organic reduction of sludge solids.

8.3.2 Description

The process utilizes equipment and process flows similar to incineration except that less than the theoretical amount of air for complete combustion is supplied. Autogenous starved air combustion (SAC) can be achieved with a sludge solids concentration greater than 25%. For lower concentrations, an auxiliary fuel may be required, depending on the percent volatiles in the solids. High temperatures decompose or vaporize the solid components of this sludge. The gas phase reactions are pyrolytic or oxidative, depending on the concentration of oxygen remaining in the stream. Under proper control, the gas leaving the vessel is a low-Btu fuel gas that can be burned in an afterburner to produce power and/or thermal energy. Some processes utilize pure oxygen instead of air and thus produce a higher-Btu fuel gas. The solid residue is a char with more or less residual carbon, depending on how much combustion air had to be supplied to reach the proper operating temperatures. Because the process is neither purely pyrolytic nor purely oxidative, it is called starved-air combustion or thermal gasification, rather than pyrolysis. Other processes still in the development stage use indirect heating, rather than the partial combustion. These are true pyrolysis processes. SAC reduces the sludge volumes and sterilizes the end product. Unlike incineration, it offers the potential advantages of producing useful by-products and of reducing the volume of sludge without large amounts of supplementary fuels. The gas that is produced has a heating value up to 130 Btu/standard dry cubic foot using air for combustion and is suitable for use in local applications, such as combustion in an afterburner or boiler or for fuel in another furnace. SAC has a higher thermal efficiency than incineration due to the lower quantity of air required for the process. In addition, capital economies can be realized due to the smaller gas handling requirements.

Furnaces may be operated in one of three modes resulting in substantially different heat generation and residue characteristics. The low temperature char (LTC) mode only pyrolyzes the volatile material thereby producing a charcoal-like residue with a high ash content. The high temperature char (HTC) mode produces a charcoal-like material converted to fixed carbon and ash. The char burned (CB) mode reacts away all carbon and produces ash as a residue. Heat recovered is maximum for the CB mode, less for the HTC mode, and substantially less for the LTC mode of operation.

SAC operation has shown the following advantages in addition to tnose discussed above: (1) it is easier to control than a standard incinerator; (2) it is a more stable operation with little response to changes in feed; (3) it has more feed capacity compared to an equal area for incineration; (4) all equipment used is currently being manufactured; (5) less air pollutants are generated and air pollution control is easier to manage; and (6) the process uses lower sludge solids content for autogenous operation.

8.3.3 Technology Status

Autogenous SAC of sludge has been demonstrated at a full-scale multiple hearth furnaces (MHF) project at the Central Contra Costa Sanitary District in California. One SAC unit for disposal of sludge from a 40 Mgal/d industrial wastewater treatment plant is reported to have gone on stream in 1978 and other units were contemplated.

8.3.4 Applications

Starved air combustion is used for the reduction of sludge volume and production of fuel gas for a nearby combustor or furnace; most existing MHF's can easily be retrofitted to operate in the SAC mode.

8.3.5 Limitations

There are significant disadvantages to starved air combusion including: (1) the need for an afterburner may limit use in existing installations due to space problems; (2) relatively large amount of instrumentation is required; (3) one must be very careful of bypass stack exhaust since furnace exhaust is high in hydrocarbons and may be combustible in air (this may result in bypassing only after afterburning with appropriate emergency controls in some areas); (4) furnace exhaust gases are corrosive; (5) combustibles in ash may create ultimate disposal problems; (6) sludge volume reduction is lower than with incineration; and (7) the process requires recovery of the energy in the product gas to fully realize the improved efficiency.

8.3.6 Environmental Impact

Air pollution can be expected to be less of a problem due to the lower air flows and the potential for particulate carryover. Data to date indicate conventional equipment can achieve acceptable controls. Depending upon the mode of operation, heavy metals in the sludge can be retained in the residue.

8.3.7 Reliability

Mechanical function of MHF units under the SAC mode is expected to be similar to the conventional operating modes. Increased operating stability is expected to result in higher process reliability.

8.3.8 Design Criteria

In MHF systems, hearth loadings are 9 to 15 lb wet (22 percent) solids/ft²/h; for autogenous combustion, sludge solids content is 25% to 39% depending upon volatility. The off-gas heating value is dependent upon operating mode.

8.3.9 Flow Diagram

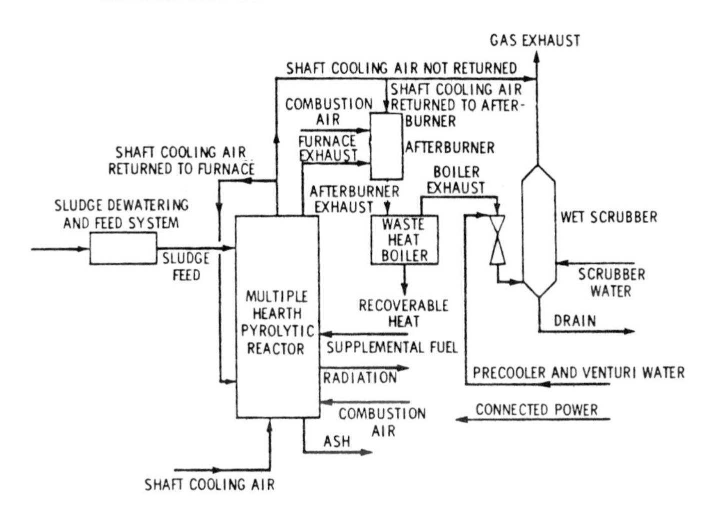

8.3.10 Performance

Unit can operate without auxiliary fuel, including afterburner, with sludge dewatered to the range of 29% to 39% solids. Based on a limited number of pilot-scale tests, the off-gas from an MHF unit operating in the SAC mode, with sludge alone, ranges from 18 to 73 Btu/standard cubic foot.

8.3.11 References

1. Innovative and Alternative Technology Assessment Manual, EPA-430/9-78-009 (draft), U.S. Environmental Protection Agency, Cincinnati, Ohio, 1978. 252 pp.

8.4 LANDFILLING (AREA FILL) [1]

8.4.1 Function

Landfilling is used for the final disposal of sludge by mixing and covering with soil.

8.4.2 Description

Landfilling is a sludge disposal operation in which sludge is placed above the original earth cover and subsequently covered with soil. To achieve stability and soil-bearing capacity, sludge is mixed with a bulking agent, usually soil. The soil absorbs excess moisture from the sludge and increases its workability. The large quantities of soil required may require hauling from elsewhere. Provisions must be made to keep the stockpiled soil dry. Installation of a liner is generally required for groundwater control and provisions must be made for surface drainage control, gas migration, dust, vectors and/or aesthetics. Area fills are more specifically categorized as follows:

Area Fill Mound. In an area fill mound, sludge is mixed with a bulking agent, usually soil, and the mixture is hauled to the filling area where it is stacked in mounds approximately 6 feet high. Cover material is then applied in a 3-foot thickness. This cover thickness may be increased to 5 feet if additional mounds are applied atop the first lift. The appropriate sludge/soil bulking ratio and soil cover thickness depend upon the solids content of the sludge as received, the need for mound stability, and the bearing capacity as dictated by the number of lifts and equipment weight. Lightweight equipment with swamp pad tracks is appropriate for area fill mound operations; heavier wheel equipment is appropriate in transporting bulking material to and from stockpiles. Construction of earthen containments is useful to minimize mound slumping and for sloping sites.

Area Fill Layer. In an area fill layer, sludge is mixed with soil on or off site and spread evenly in consecutive layers 0.5 to 3 feet thick. Interim cover between layers may be applied in 0.5 to 1 foot thick applications. Layering may continue to an indefinite height before final cover is applied. Lightweight equipment with swamp pad tracks is appropriate for area fill layer operations; heavier wheel equipment is appropriate for hauling soil. Slopes should be relatively flat to prevent sludge from flowing downhill. However, if sludge solids content is high and/or sufficient bulking soil is used, the effect can be prevented and layering performed on mildly sloping terrain.

Diked Containment. Dikes are constructed on level ground around all four sides of a containment area. Alternatively, the containment area may be placed at the toe of the hill so that the steep slope can be utilized as containment on one or two sides.

Dikes are then constructed around the remaining sides. Access is provided to the top of the dikes so that haul vehicles can dump sludge directly into the containment. A 1- to 3-foot interim cover may be applied at certain points during the filling; a 3- to 5-foot thick final cover should be applied when filling is discontinued. Cover material is applied either by a dragline based on solid ground atop the dikes or by track dozers directly on top of the sludge, depending upon sludge bearing capacity. Usually, operations are conducted without the addition of soil bulking agents, but occasionally soil bulking is added. Typical dimensions are 50 to 100 feet wide, 100 to 200 feet long, 10 to 30 feet deep.

8.4.3 Technology Status

Landfill technology is relatively new and not in widespread use.

8.4.4 Applications

Landfilling is suitable when subsurface placement is impossible due to shallow groundwater or bedrock. Area fill mounds are suitable for stabilized sludge; they offer good land utilization but have higher manpower and equipment requirements due to the constant need to push and stack slumping mounds. Area fill layers are suitable for stabilized sludge; they involve poor land utilization but offer lower manpower and equipment requirements. Diked containment offers efficient land use; suitable for stabilized or unstabilized sludge and have a low soil requirement.

8.4.5 Limitations

Rainfall causes mounds to slump; operating difficulties are encountered in wet and freezing weather.

8.4.6 Chemicals Required

Lime and masking agents are required to control odors.

8.4.7 Environmental Impact

Landfilling involves potential soil erosion, dust, vectors, noise, and odor problems. Leachate and gas continue to be produced for many years after the fill is completed. Leachate must be properly controlled to avoid groundwater and surface water contamination. Gas is explosive and can migrate to nearby structures, or can stunt or kill vegetation if not properly controlled. Mud can be transferred to local roads by transport vehicles; mud can be alleviated by a wash pad located near the exit gate. The area fill layer is relatively land intensive.

8.4.8 Reliability

Landfilling is a very reliable disposal method.

8.4.9 Design Criteria

Parameter	Area fill mound	Area fill layer	Diked containment
Sludge solids content	Greater than 20%	Greater than 15%	20 to 28% for land-based equipment; more than 28% for sludge-based equipment
Sludge characteristics	Stabilized	Stabilized	Stabilized or unstabilized
Ground slopes	No limitation if suitably prepared	Level ground preferred	Level ground or steep terrain if suitably prepared
Bulking required	Yes	Yes	Occasionally
Bulking ratio of soil to sludge	0.5 to 2 parts soil to 1 part sludge	0.25 to 1 part soil to 1 part sludge	0 to 0.5 part soil to 1 part sludge
Sludge application rate	3,000 to 14,000 yd^3/acre	2,000 to 9,000 yd^3/acre	4,800 to 15,000 yd^3/acre
Equipment	Track loader; backhoe with loader; track dozer	Track dozer; grader; track loader	Dragline; track dozer; scraper

8.4.10 Flow Diagram

8.4.11 References

1. Innovative and Alternative Technology Assessment Manual. EPA-430/9-78-009 (draft), U.S. Environmental Protection Agency, Cincinnati, Ohio, 1978. 252 pp.

8.5 LAND APPLICATION [1]

8.5.1 Function

Land application is used for the ultimate disposal of sludge.

8.5.2 Description

Land application techniques for applying liquid sludge, dried sludge, and sludge cake to the land include tank truck, injection, ridge and furrow spreading, and spray irrigation. Sludge can be incorporated into the soil by plowing, discing, or similar methods. Ridge and furrow methods involve spreading sludge in the furrows and planting crops on the ridges. Utilization of this technique is generally best suited to relatively flat land and is well suited to certain row crops. Spray irrigation systems are more flexible, require less soil preparation, and can be used with a wider variety of crops. High application rates are commonly used to reclaim strip mine spoils or other low quality land. Sludge spreading in forests has been limited; however, it offers opportunities for improved soil fertility and increased tree growth.

8.5.3 Applications

Land application is popular as a disposal method because it is simple. The technique also serves as a utilization measure because it is beneficial as a soil conditioner for agricultural, marginal, or drastically disturbed land. It contains considerable quantities of organic matter, all of the essential plant nutrients, and a capacity to produce water-retaining humus.

8.5.4 Limitations

Constituents of sludge may limit the acceptable rate of application, the crop that can be grown, or the management or location of the site. Trace elements added to soil may accumulate in a concentration that is toxic to plants or is taken up and concentrated in edible portions of plants in a concentration that is harmful to animals or man. The trace elements problem can be prevented by limiting the amount of sludge to be applied, industrial pretreatment, selection of tolerant or nonaccumulating crops, selection of crops not used in the human food chain, and adapting appropriate agronomic practices such as liming of the soil. Where population is concentrated and agricultural land is limited, sufficient land for sludge application may not be available. Terrain must be carefully selected; steep slopes and low lying fields are less suitable and require more careful management. Equipment with standard tires can cause ruts, compacted soil, and crop damage; vehicles with standard tires can get stuck in muddy terrain.

8.5.5 Toxics Management

Soil has a variable capacity to filter, buffer, absorb, and chemically and biologically react with sludge constituents. Toxics may pass through the soil unchanged, be degraded by microorganisms, react with organic or inorganic compounds to form soluble or insoluble compounds, be adsorbed on soil colloids, or be volatilized from the soil. Factors influencing these pathways include the physical and chemical state of the material and of soil constituents, microbial population, solubility, pH, cation exchange capacity, soil aeration, moisture, and temperature. Generally, most heavy metals applied to the surface are bound in the soil.

8.5.6 Environmental Impact

Land application has the potential for toxics and pathogens to contaminate soil, water, air, vegetation and animal life, and ultimately to be hazardous to humans. Accumulation of toxics in the soil may cause phytotoxic effects, the degree of which varies with the tolerance level of the particular plant specie and variety. Toxic substances such as cadmium that accumulate in plant tissues can subsequently enter the food chain and reach human beings directly by ingestion or indirectly through animals. If available nitrogen exceeds plant requirements, it can be expected to reach groundwater in the nitrate form. Toxic materials and pathogens can contaminate groundwater supplies or can be transported by runoff or erosion to surface waters if improper loading occurs. Aerosols that contain pathogenic organisms may be present in the air over a landspreading site, particularly where spray irrigation is the means of sludge application. Some pathogens remain viable in the soil and on plants for periods of several months; some parasitic ova can survive for a number of years. Other potential impacts include public acceptance and odor.

8.5.7 Reliability

As a disposal process, land application is very reliable; when employed as a utilization process, careful control should be exercised.

8.5.8 Design Criteria

Application rates depend on sludge composition, soil characteristics (usually 3% nitrogen; 2% phosphorus; 0.25% potassium), climate, vegetation, and cropping practices. Annual application rates have varied from 0.5 to more than 100 tons per acre. Applying sludge at a rate to support the nitrogen needs of a crop of about 5 to 10 tons of digested solids in the liquid form, avoids problems associated with overloading the soil. Rates based on phosphorus needs are lower. A pH of 6.5 or greater will minimize heavy metal uptake by most crops.

8.5.9 Performance

If the sludge contains all of the essential plant nutrients, it can be applied at rates that will supply all of the nitrogen and phosphorus needed by most crops. It may also increase the concentration in plants of certain elements that are at or near deficiency levels for animals. For instance, animal diets are often deficient in trace elements such as zinc, copper, nickel, chromium, and selenium. Thus, sludge application may improve the quality of feeds and forages used for animal consumption. Sludge as fertilizer can provide agricultural needs as shown below.

Fertilizer type	Material provided		
	Nitrogen	Phosphate	Potash
1 ton dry sludge	60 lb (50% available)	40 lb	5 lb
Typical corn fertilizer	180 lb/acre	50 lb/acre	60 lb/acre
6 tons dry sludge applied per acre	180 lb/acre (available)	240 lb/acre (available)	30 lb/acre (available)

8.5.10 References

1. Innovative and Alternative Technology Assessment Manual. EPA-430/9-78-009 (draft), U.S. Environmental Protection Agency, Cincinnati, Ohio, 1978. 252 pp.

8.6 COMPOSTING [1]

8.6.1 Function

Composting is the microbial degradation of sludge and other putrescible organic solid material by an aerobic metabolism in piles or windrows on a surfaced outdoor area.

8.6.2 Description

There are two types of composting: static pile composting and windrow composting.

Static Pile Composting. In static pile composting, sewage sludge is converted to compost in approximately eight weeks in a four-step process involving preparation, digestion, drying and screening, and curing.

In the preparation step, sludge is mixed with a bulking material such as wood chips or leaves to facilitate handling, provide the necessary structure and porosity for aeration, and lower the moisture content of the biomass to 60% or less. Following mixing, the aerated pile is constructed and positioned over porous pipe through which air is drawn. The pile is covered for insulation.

In the digestion step, the aerated pile undergoes decomposition by thermophilic organisms, whose activity generates a concomitant elevation in temperature to 60°C (140°F) or more. Aerobic composting conditions are maintained by drawing air through the pile at a predetermined rate. The effluent air stream is conducted into a small pile of screened, cured compost where odorous gases are effectively absorbed. After about 21 days, the composting rates and temperatures decline, and the pile is taken down, the plastic pipe is discarded, and the compost is either dried or cured depending upon weather conditions.

In the drying and screening step, drying to 40% to 45% moisture facilitates clean separation of compost from wood chips. The unscreened compost is spread out with a front-end loader to a depth of 12 inches. Periodically a tractor-drawn harrow is employed to facilitate drying. Screening is performed using a rotary screen. The chips are recycled.

In the curing step, the compost is stored in piles for about 30 days to assure that no offensive odors remain and to complete stabilization. The compost is then ready for utilization as a low-grade fertilizer, a soil amendment, or a material for land reclamation.

Windrow Composting. In windrow composting, the piles are turned periodically to provide oxygen for the microorganisms to carry out the stabilization and carry off the excess heat generated by the process. When masses of solids are assembled, and conditions of moisture, aeration, and nutrition are favorable for microbial activity and growth, the temperature rises spontaneously. As a result of biological self-heating, composting masses easily reach 60°C (140°F) and commonly exceed 70°C (150°F). Peak composting temperatures approaching 90°C (194°F) have been recorded. Temperatures of 140°F to 160°F serve to kill pathogens, insect larvae, and weed seeds. Nuisances such as odors, insect breeding, and vermin harborage are controlled through rapid destruction of putrescible materials. Sequential steps involved in composting are preparation, composting, curing, and finishing.

The preparation step involves adjusting the sludge properties in order to permit composting. To be compostable, a waste must have at least a minimally porous structure and a moisture content of 45% to 65%. Therefore, sewage sludge cake, which is usually about 20% solids, cannot be composted by itself but must be combined with a bulking agent, such as soil, sawdust, wood chips, refuse, or previously manufactured compost. Sludge and refuse make an ideal process combination. Refuse brings porosity to the mix, while sludge provides needed moisture and nitrogen; both are converted synergistically to an end product amenable to resource recovery. The sludge is suitably prepared and placed in piles or windrows.

In the composting step, the composting period is characterized by rapid decomposition. Air is supplied by periodic turnings. The reaction is exothermic, and wastes reach temperatures of 140°F to 160°F or higher. Pathogen kill and the inactivation of insect larvae and weed seeds are possible at these temperatures. The period of digestion is normally about six weeks.

Curing is characterized by a slowing of the decomposition rate. The temperature drops back to ambient, and the process is brought to completion. The period takes about two more windrow weeks.

In finishing, some sort of screening or other removal procedure is necessary if municipal solid waste fractions containing nondigestible debris have been included, or if the bulking agent such as wood chips is to be separated and recycled. The compost may be pulverized with a shredder, if desired.

8.6.3 Technology Status

Static Pile Composting. Static pile composting has been successfully demonstrated at four locations and projected to be capable of serving large areas. Experiments are ongoing on various operating parameters.

Windrow Composting. Windrow composting has been successfully demonstrated.

8.6.4 Applications

Static Pile Composting. This process is suitable for converting digested and undigested sludge cake to an end product of some economic value. Insulation of the pile and a controlled aeration rate enable better odor and quality control than the windrow process from which it evolved.

Windrow Composting. This is a sludge treatment method that successfully kills pathogens, larvae, and weed seeds. It is suitable for converting undigested primary and/or secondary sludge to an end product amenable to resource recovery with a minimum capital investment and relatively small operating commitment.

8.6.5 Limitations

Static Pile Composting. The drying process is weather-dependent and requires at least two rainless days. The use of compost on land is limited by the extent to which sludge is contaminated by heavy metals and industrial chemicals.

Windrow Composting. A small porous windrow may permit such rapid air movement that temperatures remain too low for effective composting. The outside of the pile may not reach temperatures sufficiently high for pathogen destruction. Pathogens may survive and regrow. Sale of product may be difficult.

8.6.6 Environmental Impact

Static Pile Composting. Potential odor problems can occur for a brief period between the time a malodorous sludge arrives at the site, is mixed, and is covered by the insulating layer. Human pathogen generation and aerosol distribution potential dictate careful attention to downwind land use.

Windrow Composting. This process is relatively land intensive; there is a potential for odors and the process may be aesthetically unacceptable. The compost product represents an environmental benefit when used as a soil amendment. Other uses include wallboard production, livestock feed, litter for the chicken industry, and adsorbent for oil spill cleanup.

8.6.7 Reliability

Static Pile Composting. This process offers a high degree of reliability through simplicity of operation. Thoroughness (percent stabilization) is a function of recycle scheme, porosity distribution in pile, and manifold design.

Windrow Composting. This process is highly reliable. Ambient temperatures and moderate rainfall do not affect the process.

8.6.8 Design Criteria

Static Pile Composting. Construction of the pile for a 10 dry ton/d (43 wet tons) operation involves the following steps: 1) a 6-inch layer of unscreened compost for base; 2) a 94-foot loop of 4-inch diameter perforated plastic pipe placed on top (0.25 inch hole diameter); 3) a 6-inch layer of unscreened compost or wood chips covering the pipe; 4) connection of the loop to a 1/3 hp blower using 14 feet of solid pipe fitted with a water trap to collect condensate; 5) a timer set for a cycle of 4 minutes on and 16 minutes off; 6) a blower connected to a covered scrubber pile (2 yd^3 wood chips covered with 10 yd^3 screened compost) using 16 feet of solid pipe; 7) placement of a wet sludge-wood chip mixture, in a volumetric ratio of 1:2.5, on a prepared base; and 8) placement of a 12-inch layer of screened compost on top for insulation. Air flow is 100 ft^3/h/ton of sludge. Approximately 3.5 acres of land area are required for processing 10 dry tons daily; the area includes a runoff collection pond; bituminous surface for roads, mixing, composting, drying, and storage; and an administration area. Pile dimensions are 53 ft x 12 ft x 8 ft high.

Windrow Composting. Approximate land requirement is 1/3 acre/dry ton sludge daily production. Windrows can be 4- to 8-feet high, 12- to 25-feet wide at the base, and of variable length. Sludge cannot be composted by itself but must be combined with a bulking agent to provide the biomass with the necessary porosity and moisture content. Biomass criteria include the following: 1) moisture content of 45% to 65%; 2) carbon/nitrogen ratio between 30% and 35%; 3) carbon/phosphorus ratio between 75/1 and 150/1; 4) air flow of 10 to 30 ft^3 air/d/lb vs; 5) detention time of 6 weeks to 1 year.

8.6.9 Flow Diagram

Static Pile Composting.

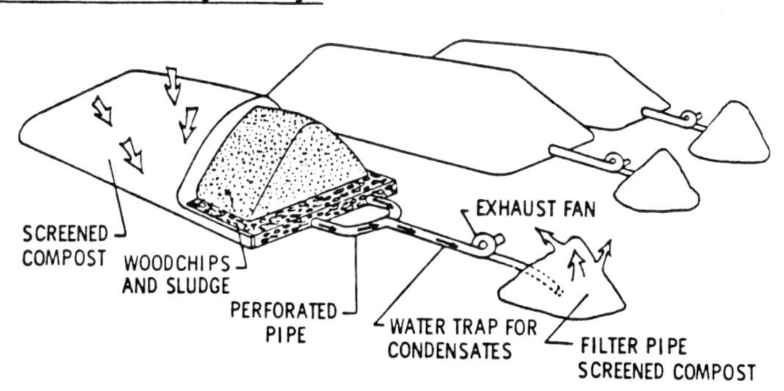

Windrow Composting.

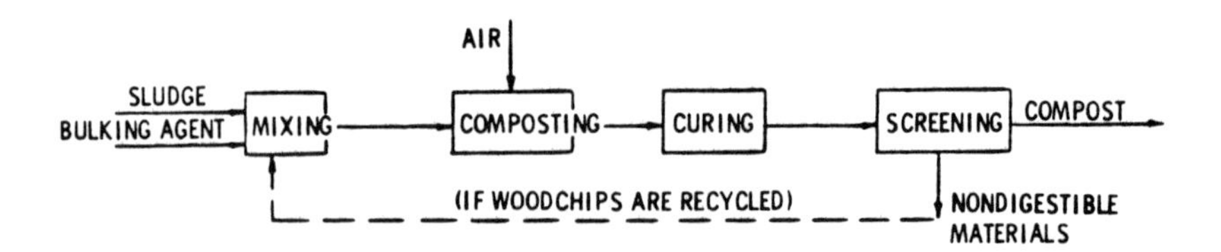

8.6.10 Performance

Static Pile Composting. Sludge is generally stabilized after 21 days at elevated temperatures. Maximum temperatures between 60 and 80°C are produced during the first three-to-five days, during which time odors, pathogens, and weed seeds are destroyed. Temperatures above 55°C (131°F) for sufficient periods can effectively destroy most human pathogens. The finished compost is a humus-like material, free of malodors, and useful as a soil conditioner containing low levels of essential plant macronutrients such as nitrogen and phosphorus and often adequate levels of micronutrients such as copper and zinc.

Windrow Composting. Sludge is converted to a relatively stable organic residue, reduced in volume by 20% to 50%. The residue loses its original identity with respect to appearance, odor, and structure. The end product is humified and has earthy characteristics; pathogens, weed seeds and insect larvae are destroyed.

8.6.11 References

1. Innovative and Alternative Technology Assessment Manual. EPA-430/9-78-009 (draft), U.S. Environmental Protection Agency, Cincinnati, Ohio, 1978. 252 pp.

8.7 LANDFILLING (TRENCHING) [1]

8.7.1 Function

Landfilling is the final disposal of sludge in excavated trenches.

8.7.2 Description

In landfilling, stabilized or unstabilized sludge is placed within a subsurface excavation and covered with soil. Trench operations are more specifically categorized as narrow trench and wide trench. Narrow trenches are defined as having widths less than 10 feet, while wide trenches are defined as having widths greater than that value. The width of the trench is determined by the solids content of the receiving sludge and its capability of supporting cover material and equipment. Distances between trenches should be large enough to provide sidewall stability, as well as space for soil stockpiles, operating equipment, and haul vehicles. Design considerations should include provisions to control leachate and gas migration, dust, vectors, and/or aesthetics. Leachate control measures include the maintenance of 2 to 5 feet of soil thickness between the trench bottom and the highest groundwater level or bedrock (2 feet for clay to 5 feet for sand), or membrane liners and leachate collection and treatment system. Installation of gas control facilities may be necessary if inhabited structures are nearby.

In narrow trench operations, sludge is disposed in a single application, and a single layer of cover soil is applied atop this sludge. Trenches are usually excavated by equipment based on solid ground adjacent to the trench; equipment does not enter the excavation. Backhoes, excavators, and trenching machines are particularly useful. Excavated material is usually applied immediately as cover over an adjacent sludge-filled trench. Sludge is placed in trenches either directly from haul vehicles, through a chute extension, or by pumping. The main advantage of a 2- to 3-foot narrow trench is its ability to handle sludge with a relatively low solids content (15% to 20%). Instead of sinking to the bottom of the sludge, the cover soil bridges over the trench and receives support from undisturbed soils along each side of the trench. A 3- to 10-foot width is more appropriate for sludge with solids content of 20% to 28%, which is high enough to support cover soil.

Wide trench operations are usually excavated by equipment operating inside the trench. Track loaders, draglines, scrapers, and track dozers are suitable. Excavated material is stockpiled on solid ground adjacent to the trench for subsequent application as cover material. If sludge is incapable of supporting equipment, cover is applied by equipment based on solid undisturbed ground adjacent to the trench. A front-end loader is suitable for trenches up to 10 feet wide; a dragline is suitable for trench widths up to 50 feet. If sludge can support equipment, a track dozer applies cover from within the trench. Sludge is placed in trenches from haul vehicles directly entering the trench or from haul vehicles dumping from the top of the trench. Dikes can be used to confine sludge to a specific area in a continuous trench.

After maximum settlement has occurred, in approximately one year, the area should be regraded to ensure proper drainage.

8.7.3 Technology Status

Landfilling has been fully demonstrated.

8.7.4 Applications

Landfilling is a relatively simple sludge disposal method suitable for stabilized or unstabilized sludge. The method does not require special expertise beyond the skills necessary to operate the above-mentioned equipment, plus administrative skills. Narrow trench systems are particularly well suited for smaller plants.

8.7.5 Limitations

Frozen soil conditions and precipitation cause operating difficulties in landfilling.

8.7.6 Chemicals Required

Lime and masking agents are used to control odors.

8.7.7 Environmental Impact

Landfilling involves potential soil erosion and odor problems. Leachate and gas continue to be produced for many years after the fill is completed. Leachate must be properly controlled to avoid groundwater and surface water contamination. Gas is explosive or can stunt or kill vegetation if not properly controlled. Narrow trenches are relatively intensive.

8.7.8 Reliability

Landfilling is a very reliable disposal method.

8.7.9 Design Criteria

Parameter	Value	
	Narrow trench	Wide trench
Sludge solids content	15% to 20% for 2- to 3-ft widths; 20% to 28% for 3- to 10-ft widths	20% to 28% for land-based equipment; more than 28% for sludge-based equipment
Ground slopes	Less than 20%	Less than 10%
Cover soil thickness	2 to 3 ft for 2- to 3-ft widths; 3 to 4 ft for 3- to 10-ft widths	3 to 4 ft for land-based equipment; 4 to 5 ft for sludge-based equipment
Sludge application rate	1,200 to 5,600 yd^3/acre	3,200 to 14,500 yd^3/acre
Equipment	Backhoe with loader, excavator, trenching machine	Track loader, dragline, scraper, track dozer

8.7.10 Flow Diagram

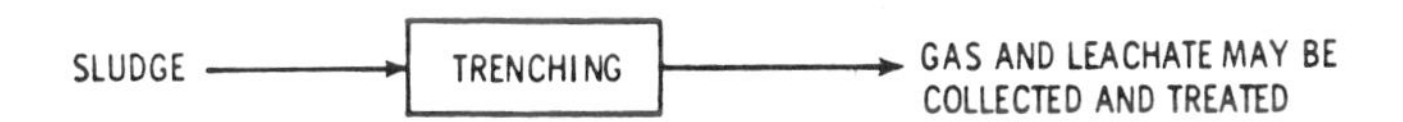

8.7.11 References

1. Innovative and Alternative Technology Assessment Manual. EPA-430/9-78-009 (draft), U.S. Environmental Protection Agency, Cincinnati, Ohio, 1978. 252 pp.

8.8 DEEP-WELL INJECTION [1]

8.8.1 Function

Deep-well injection is used for the ultimate disposal of waste via injection deep underground.

8.8.2 Description

The prefix deep when applied to the phrase "deep-well injection" does not refer to any specific depth nor to the soil cover, but it refers to rock formations that are below and isolated from freshwater aquifers. The concept of underground injection of liquid wastes was developed by the oil industry early in its history, when surface drainage was becoming contaminated by oil-field brines. The operators developed the now-common practice of injecting salt water wastes back into the subsurface reservoirs from which the oil and salt water originally were produced, and also into shallower or deeper reservoirs which had the storage capacity to accept the injected brine. This practice was not long established when it became obvious that the injected brines were contaminating fresh water sands that were being used for domestic supplies. This disposal practice for oilfield brines matured, and under the control of state laws has become a standard and accepted practice. More recently, numerous other industries have turned to deep-well injections as a possible solution for their chronic waste disposal problem. In about 1958, the AEC began serious consideration of this method as a possible means for disposing of certain types of radioactive waste liquids.

The concept was developed to the degree that in 1974, in addition to over 40,000 oilfield brine disposal wells, there were 209 operating industrial waste disposal wells in the United States, as well as approximately five to eight thousand sewage and storm-water disposal wells in the states of Oregon and Idaho. In addition to these wells in the United States, there are an estimated 20 wells operating in Canada and two or three wells operating in Mexico.

8.8.3 Technology Status

Deep-well injection is a well-developed technology and widely used in acceptable geographical areas.

8.8.4 Applications

The present inventory shows at least 383 deep-well injection sites for which permits had been granted, located in 25 states. Wells were drilled on 322 of the sites, and 209 were being used for waste injection in 1974. The following figure shows the growth of operating injection wells in the United States by year. The rapid development of industrial waste injection wells indicated is of concern to federal, state, and local governments because of the many uncertainties related to deep-well injection that relate directly to the degradation of the environment.

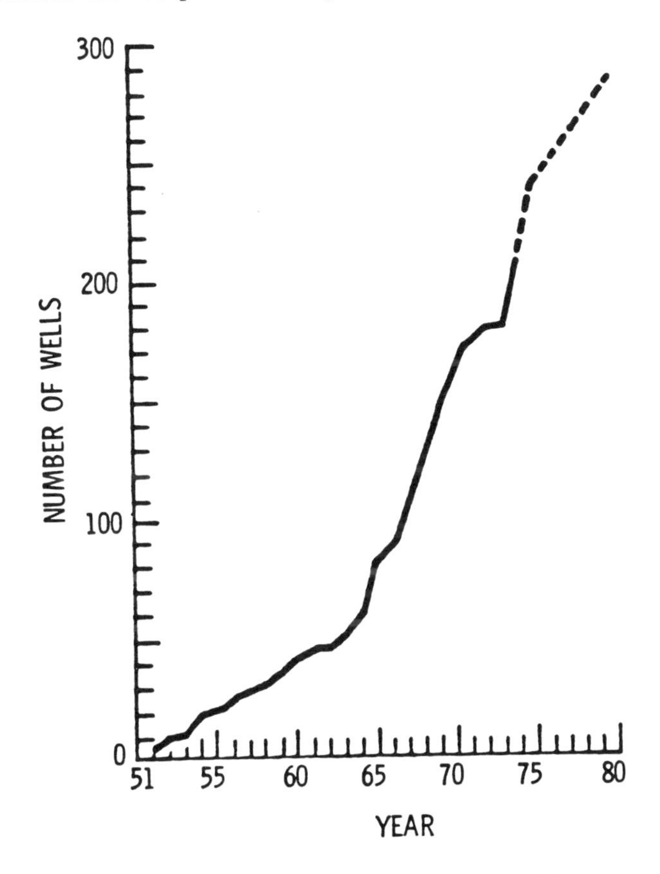

Growth of operating deep-well injection systems in the United States

The following tables indicate the standard industrial classification of injection wells and well completion depths, respectively.

STANDARD INDUSTRIAL CLASSIFICATION OF INJECTION WELLS
(268 wells)

Industry	No. of wells	Percentage
MINING (9.3%)		
10 Metal mining	2	0.7
12 Coal	1	0.4
13 Oil and gas extraction	17	6.4
14 Non-metallic mining	5	1.9
MANUFACTURING (80.6%)		
20 Food	6	2.2
26 Paper	3	1.1
28 Chemical and allied products	131	48.9
29 Petroleum refining	51	19
32 Stone and concrete	1	0.4
33 Primary metals	16	5.9
34 Fabricated metals	3	1.1
35 Machinery - except electronics	1	0.4
38 Photographics	3	1.1
TRANSPORTATION, GAS, AND SANITARY SERVICES (9.8%)		
47 Transportation service	1	0.4
49 Sanitary service	23	8.6
50 Wholesale trade - durable	1	0.4
55 Auto dealers and service	1	0.4
OTHER (0.4%)		
72 Personal service	1	0.4

WELL COMPLETION DEPTHS
(262 wells)

Depth	No. of wells	Percentage
0 - 1,000	20	7.6
1,001 - 2,000	56	21.4
2,001 - 3,000	33	12.6
3,001 - 4,000	34	12.8
4,001 - 5,000	39	14.8
5,001 - 6,000	44	16.7
6,001 - 7,000	18	7.2
7,001 - 8,000	12	4.8
8,001 +	3	1.2

8.8.5 Limitations/Environmental Impact

Overall, the injection of industrial liquid waste into subsurface reservoirs has been acceptable. To this overall smoothly operating scheme, there have been notable exceptions among which can be included the Rocky Mountain Arsenal well at Denver, Colorado, the Hammermill Paper Company well at Erie, Pennsylvania, and the Hercules Company well at Wilmington, North Carolina. It has been determined that major failures of injection systems can be directly related to two basic factors. The first is that the geology of the area was not studied throughly prior to well drilling, or, if studied, the results of the studies were ignored or slighted in the planning. The second factor is that poor engineering design was employed, or poor well completion practices were followed. Chemical and biological factors also have contributed to well failures; however, unlike the geologic and engineering factors, the chemical and biologic factors are often correctable if they are not allowed to progress too far.

8.8.6 References

1. Review and Assessment of Deep-Well Injection of Hazardous Waste, Volume I. EPA-600/2-27-029a, U.S. Enviromental Protection Agency, Cincinnati, Ohio, 1977. pp. 39-44.

References

A1 Revised Technical Review of the Best Available Technology, Best Demonstrated Technology, and Pretreatment Technology for the Timber Products Processing Point Source Category (draft contractors report). Contract 68 01 4827, U.S. Environmental Protection Agency, Washington, D.C., October 1978.

A2 Development Document for BAT Effluent Limitations Guidelines and New Source Performance Standards for Ore Mining and Dressing Industry. No. 6332-M.1, A Division of Calspan Corporation, Buffalo, New York, 1979.

A3 Effluent Limitations Guidelines (BATEA), New Source Performance Standards and Pretreatment Standards for the Petroleum Refining Point Source Category, March 1978, U.S. Environmental Protection Agency, Washington, D.C.

A4 Effluent Limitations Guidelines for the Paint Manufacturing Industry, January 1979, U.S. Environmental Protection Agency, Washington, D.C.

A5 Development Document for Effluent Limitations Guidelines and New Source Performance Standards for the Tire and Synthetic Segment of the Rubber Processing Point Source Category. EPA-440/1-74-013a, U.S. Environmental Protection Agency, Washington, D.C., February 1974. 193 pp.

A6 Technical Study Report BATEA-NSPS-PSES-PSNS: Textile Mills Point Source Category, November 1978. U.S. Environmental Protection Agency.

A7 Technical Review of the Best Available Technology, Best Demonstrated Technology, and Pretreatment Technology for the Gum and Wood Chemicals Point Source Category. No. 77-094, Environmental Science and Engineering Incorporation, Gainesville, Florida, 1978.

A8 Development Document for Interim Final Effluent Limitations Guidelines and Proposed New Source Performance Standards for the Gum and Wood Chemicals Manufacturing. EPA 440/1-76, U.S. Environmental Protection Agency, Washington, D.C., April 1976.

A9 Interim Final Supplement for Pretreatment to the Development Document for the Petroleum Refining Industry Existing Point Source Category. EPA-440/1-76, U.S. Environmental Protection Agency, March 1977.

A10 Effluent Limitations Guidelines for the Ink Manufacturing Industry (BATEA, NSPS, Pretreatment), January 1979, U.S. Environmental Protection Agency, Effluent Guidelines Division, Washington, D.C.

A11 Technical Assistance in the Implementation of the BAT Review of the Coal Mining Industry Point Source Category, March 9, 1979, Environmental Protection Agency, Washington, D.C.

A12 Development Document for Interim Final Effluent Limitations Guidelines and Proposed New Source Performance Standards for the Pharmaceutical Manufacturing, Point Source Category. EPA-440/1-75-060, U.S. Environmental Protection Agency, Washington, D.C., December 1976. 331 pp.

A13 Development Document for Effluent Limitations Guidelines and New Source Performance Standards for the Fish Meal, Salmon, Bottom Fish, Clam, Oyster, Sardine, Scallop, Herring, and Abalone, Segment of the Canned and Perserved Fish and Seafood Processing Industry, Point Source Category. EPA-440/1-75-041a, U.S. Environmental Protection Agency, Washington, D.C., September 1975. 485 pp.

A14 Development Document for Proposed Existing Source Pretreatment Standards for the Electroplating, Point Source Category. EPA-440/1-78-085, U.S. Environmental Protection Agency, Washington, D.C., February 1978. 532 pp.

A15 Development Document for Effluent Limitations Guidelines and New Source Performance Standards for the Leather Tanning and Finishing Point Source Category. EPA-440/1-74-016-a, U.S. Environmental Protection Agency, Washington, D.C., March 1974. 157 pp.

A16 Development Document for Effluent Limitations Guidelines for the Pesticide Chemicals Manufacturing Point Source Category. EPA-440/1-78-060-e, U.S. Environmental Protection Agency, Washington, D.C., April 1978. 316 pp.

A17 Development Document for Effluent Limitations Guidelines and New Source Performance Standards for the Dairy Products Processing Point Source Category. EPA-440/1-74-021a, U.S. Environmental Protection Agency, Washington, D.C., May 1974. 167 pp.

A18 Development Document for Interim Final Effluent Limitations Guidelines New Source Performance Standards for the Mineral Mining and Processing Industry Point Source Category. EPA-440/1-76-059-a, U.S. Environmental Protection Agency, Washington, D.C., June 1976. 432 pp.

A19 Development Document for Interim Final Effluent Limitations Guidelines and New Source Performance Standards for the Primary Copper Smelting Subcategory of the Copper Segment of the Nonferrous Metals Manufacturing Point Source Category. EPA-440/1-75-032b, U.S. Environmental Protection Agency, Washington, D.C., February 1975. 213 pp.

A20 Development Document for Interim Final Effluent Limitations Guidelines and Proposed New Source Performance Standards for the Raw Cane Sugar Processing Segment of the Sugar Processing Point Source Category. EPA-440/1-75-044, U.S. Environmental Protection Agency, Washington, D.C., February 1975. 291 pp.

A21 Development Document for Interim Final and Proposed Effluent Limitations Guidlines and New Source Performance Standards for the Fruits, Vegetables, and Specialties Segment of the Canned and Preserved Fruits and Vegetables Point Source Category. EPA-440/1-75-046, U.S. Environmental Protection Agency, Washington, D.C., October 1975. 520 pp.

A22 Development Document for Interim Final Effluent Limitations, Guidelines and Proposed New Source Performance Standards for the Hospital Point Source Category. EPA-440/1-76--060n, U.S. Environmental Protection Agency, Washington, D.C., April 1976. 131 pp.

A23 Development Document for Effluent Limitations Guidelines and New Source Performance Standards for the Synthetic Resins, Segment of the Plastics and Synthetic Materials Manufacturing Point Source Category. EPA-440/1-74-010-a, U.S. Environmental Protection Agency, Washington, D.C., March 1974. 238 pp.

A24 Development Document for Effluent Limitations Guidelines and New Source Performance Standards for the Plywood, Hardboard and Wood Preserving Segment of the Timber Products Processing Point Source Category. EPA-440/1-74-023-a, U.S. Environmental Protection Agency, Washington, D.C., April 1974.

A25 Development Document for Proposed Effluent Limitations Guidelines and New Source Performance Standards for the Major Organic Products Segment of the Organic Chemicals Manufacturing Point Source Category. EPA-440/1-73-009, U.S. Environmental Protection Agency, Washington, D.C., December 1973. 369 pp.

A26 Preliminary Data Base for Review of BATEA Effluent Limitations Guidelines, NSPS, and Pretratment Standards for the Pulp, Paper, and Paperboard Point Source Category, June 1979, U.S. Environmental Protection Agency, Washington, D.C.

A27 Foundry Industry (contractor's Draft Report). Contract No. 68-01-4379, U.S. Environmental Protection Agency, Washington, D.C., May 1979.

A28 Technical Support Document for Auto and Other Laundries Industry (draft contractor's report). Contract 68-03-2550, U.S. Environmental Protection Agency, Washington, D.C., August 1979.

A29 Draft Development Document for Inorganic Chemicals Manufacturing Point Source Category - BATEA, NSPS, and Pretreatment Standards (contractor's draft report). Contract 68-01-4492, U.S. Environmental Protection Agency, Effluent Guidelines Division, Washington, D.C., April 1979.

A30 Review of the Best Available Technology for the Rubber Processing Point Source Category, July 1978, U.S. Environmental Protection Agency, Washington, D.C.

A31 Draft Technical Report for Revision of Steam Electric Effluent Limitations Guidelines, September 1978, U.S. Environmental Protection Agency, Washington, D.C.

A32 Draft Contractor's Engineering Report for Development of Effluent Limitations Guidelines for the Pharmaceutical Manufacturing Industry (BATEA, NSPS, BCT, BMP, Pretreatment), July 1979, U.S. Environmental Protection Agency, Washington, D.C.

A33 Draft Development Document for Proposed Effluent Limitations Guidelines and Standards for the Iron and Steel Manufacturing Point Source Category, Volume V. EPA-440/1-79/024a, U.S. Environmental Protection Agency, Washington, D.C., October 1979. 435 pp.

A34 Development Document for Effluent Limitations Guidelines and Standards Leather Tanning and Finishing Point Source Category. EPA-440/1-79/016. U.S. Environmental Protection Agency, Washington, D.C., July 1979. 381 pp.

A35 Draft Development Document for Proposed Effluent Limitations Guidelines and Standards for the Iron and Steel Manufacturing Point Source Category, Volume VI. EPA-440/1-79/024a, U.S. Environmental Protection Agency, Washington, D.C., October 1979. 389 pp.

A36 Development Document for Effluent Limitations Guidelines and Standards for the Nonferrous Metals Manufacturing Point Source Category. EPA-440/1-79-019a, U.S. Environmental Protection Agency, Washington, D.C., September 1979. 622 pp.

A37 Draft Development Document for Proposed Effluent Limitations Guidelines and Standards for the Iron and Steel Manufacturing Point Source Category, Volume I. EPA-440/1-79/024a, U.S. Environmental Protection Agency, Washington, D.C., October 1979. 220 pp.

A38 Draft Development Document for Proposed Effluent Limitations Guidelines and Standards for the Iron and Steel Manufacturing Point Source Category, Volume II. EPA-440/1-79/024a, U.S. Environmental Protection Agency, Washington, D.C., October 1979. 272 pp.

A39 Draft Development Document for Proposed Effluent Limitation Guidelines and Standards for the Iron and Steel Manufacturing Point Source Category, Volume III. EPA-440/1-79/024a, U.S. Environmental Protection Agency, Washington, D.C., October 1979. 308 pp.

A40 Draft Development Document for Proposed Effluent Limitations Guidelines and Standards for the Iron and Steel Manufacturing Point Source Category, Volume IV. EPA-440/1-79/024a, U.S. Environmental Protection Agency, Washington, D.C., October 1979. 446 pp.

A41 Draft Development Document for Proposed Effluent Limitations Guidelines and Standards for the Iron and Steel Manufacturing Point Source Category, Volume VII. EPA-440/1-79/024a, U.S. Environmental Protection Agency, Washington, D.C., October 1979. 419 pp.

A42 Draft Development Document for Proposed Effluent Limitations Guidelines and Standards for the Iron and Steel Manufacturing Point Source Category, Volume VIII. EPA-440/1-79/024a, U.S. Environmental Protection Agency, Washington, D.C., October 1979. 768 pp.

A43 Draft Development Document for Proposed Effluent Limitations Guidelines and Standards for the Iron and Steel Manufacturing Point Source Category, Volume IX. EPA-440/1-79/024a, U.S. Environmental Protection Agency, Washington, D.C., October 1979. 758 pp.

B1 Kleper, M. H., A. Z. Gollan, R. L. Goldsmith and K. J. McNulty. Assessment of Best Available Technology Economically Achievable for Synthetic Rubber Manufacturing Wastewater. EPA-600/2-78-192, U.S. Environmental Protection Agency, Cincinnati, Ohio, August 1978. 182 pp.

B2 CoCo, J. H., E. Klein, D. Howland, J. H. Mayes, W. A. Myers, E. Pratz, C. J. Romero, and F. H. Yocum. Development of Treatment and Control Technology for Refractory Petrochemical Wastes (draft report). Project No. S80073, U.S. Environmental Protection Agency, Ada, Oklahoma. 220 pp.

B3 Klieve, J. R., and G. D. Rawlings. Source Assessment: Textile Plant Wastewater Toxics Study Phase II. Contract No. 68-02-1874, U.S. Environmental Protection Agency, Washington, D.C., April 1979. 127 pp.

B4 Schimmel, C., and D. B. Griffin. Treatment and Disposal of Complex Industrial Wastes. EPA-600/2-76-123. U.S. Environmental Protection Agency, Cincinnati, Ohio, November 1976.

B5 Rawlings, G. D. Source Assessment: Textile Plant Wastewater Toxics Study Phase I. EPA-600/2-78-004h, U.S. Environmental Protection Agency, Triangle Park, North Carolina, March 1979. 153 pp.

B6 Davis, H. J., F. S. Model, and J. R. Leal. PBI Reverse Osmosis Membrane for Chromium Plating Rinse Water. EPA-600/2-78-040. U.S. Environmental Protection Agency, Cincinnati, Ohio, March 1978. 28 pp.

B7 Chian, E. S. K., M. N. Aschauer, and H. H. P. Fang. Evaluation of New Reverse Osmosis Membranes for the Separation of Toxic Compounds from Wastewater. Contract No. DADA 17-73-C-3025, U.S. Army Medical Research and Development Command, Washington, D.C., October 1975. 309 pp.

B8 Bollyky, L. J. Ozone Treatment of Cyanide-Bearing Plating Waste. EPA-600/2-77-104, U.S. Environmental Protection Agency, Cincinnati, Ohio, June 1977. 43 pp.

B9 Kleper, M. H., R. L. Goldsmith, and A. Z. Gollan. Demonstration of Ultrafiltration and Carbon Adsorption for Treatment of Industrial Laundering Wastewater. EPA/2-78-177, U.S. Environmental Protection Agency, Cincinnati, Ohio, August 1978. 109 pp.

B10 Kleper, M. H., R. L. Goldsmith, T. V. Tran, D. H. Steiner, J. Pecevich, and M. A. Sakillaris. Treatment of Wastewaters from Adhesives and Sealants Manufacturing by Ultrafiltration. EPA-600/2-78-176, U.S. Environmental Protection Agency, Cincinnati, Ohio, August 1978.

B11 McNulty, K. J., R. L. Goldsmith, A. Gollan, S. Hossain, and D. Grant. Reverse Osmosis Field Test: Treatment of Copper Cyanide Rinse Waters, EPA-600/2-77-170, U.S. Environmental Protection Agency, Cincinnati, Ohio, August 1977. 89 pp.

B12 Brandon, C. A., and J. J. Porter. Hyperfiltration for Renovation of Textile Finishing Plant Wastewater. EPA-600/2-76-060, U.S. Environmental Protection Agency, Triangle Park, North Carolina, March 1976. 147 pp.

B13 Petersen, R. J., and K. E. Cobian. New Membranes for Treating Metal Finishing Effluents by Reverse Osmosis. EPA-600/2-76-197, U.S. Environmental Protection Agency, Cincinnati, Ohio, October 1976. 59 pp.

B14 Lang, W. C., J. H. Crozier, F. P. Drace, and K. H. Pearson. Industrial Wastewater Reclamation with a 400,000-gallon-per-day vertical tube evaporator. EPA-600/2-76-260, U.S. Environmental Protection Agency, Cincinnati, Ohio, October 1976. 90 pp.

B15 Study of Effectiveness of Activated Carbon Technology for the Removal of Specific Materials from Organic Chemical Processes. EPA Contract No. 68-03-2610. Final report on Pilot Operations at USS Chemical, Nevella.

B16 Selected Biodegradation Techniques for Treatment and/or Ultimate Disposal of Organic Materials. EPA-600/2-79-006, U.S. Environmental Protection Agency, Cincinnati, Ohio, March 1973. 377 pp.

B17 Rawlings, G. D. Evaluation of Hyperfiltration Treated Textile Wastewaters. Contract 68-02-1874, U.S. Environmental Protection Agency, Washington, D.C., November 1978.

B18 Extraction of Chemical Pollutants from Industrial Wastewaters with Volatile Solvents. EPA-600/2-76-220, U.S. Environmental Protection Agency, Ada, Oklahoma, December 1976. 510 pp.

B19 Treatment and Recovery of Fluoride Industrial Wastes. No. PB 234 447, Grumman Aerospace Corporation. Bethpage, N.Y., March 1974.

B20 Priority Pollutant Treatibility Review, Industrial Sampling and Assessment. Contract 68-03-2579, U.S. Environmental Protection Agency, Cincinnati, Ohio, July 1978. 47 pp.

B21 Effects of Liquid Detergent Plant Effluent on the Rotating Biological Contactor. EPA-600/2-78-129, U.S. Environmental Protection Agency, Cincinnati, Ohio, June 1978. 58 pp.

B22 Olem, H. The Rotating Biological Contactor for Biochemical Ferrous Iron Oxidation in the Treatment of Coal Mine Drainage. No. W77-05337, Penn State University, Pennsylvania, November 1975.

C1 Brunotts, V. A., R. S. Lynch, G. R. Van Stone. Granular Carbon Handles Concentrated Waste. Chemical Engineering Progress, 6(8):81-84, 1973.

C2 Putting Powdered Carbon in Wastewater Treatment. Environmental Science and Technology, Volume II, No. 9, September 1977.

D1 De, J. and B. Paschal. The Effectiveness of Granular Activated Carbon in Treatability Municipal and Industrial Wastewaters. In: Third National Conference on Complete Water Reuse, AIChE and EPA Technology Transfer, June 1976. pp. 204-211.

D2 De, J., B. Paschal, and A. D. Adams. Treatment of Oil Refinery Wastewaters with Granular and Powdered Activated Carbon. In: Thirtieth Industrial Waste Conference, Purdue University, Indiana, May 1975. pp. 216-232.

D3 Argaman, Yerachmiel, and C. L. Weddle. Fate of Heavy Metals Physical Treatment Processes. In: AIChE Symposium Series, Volume 70, No. 136.

Glossary

AAP: Army Ammunitions Plant.

AN: Ammonium Nitrate.

ANFO: Ammonium Nitrate/Fuel Oil.

BATEA: Best Available Technology Economically Achievable.

BAT: Best Available Technology.

BDL: Below Detection Limit.

BEJ: Best Engineering Judgement.

BOD: Biochemical Oxygen Demand.

clarification: Process by which a suspension is clarified to give a "clear" supernatant.

cryolite: A mineral consisting of sodium-aluminum fluoride.

CWA: Clean Water Act.

cyanidation process: Gold and/or silver are extracted from finely crushed ores, concentrates, tailings, and low-grade mine-run rock in dilute, weakly alkaline solutions of potassium or sodium cyanide.

comminutor: Mechanical devices that cut up material normally removed in the screening process.

effluent: A waste product discharged from a process.

EGD: Effluent Guidelines Division.

elutriation: The process of washing and separating suspended particles by decantation.

extraction: The process of separating the active constituents of drugs by suitable methods.

fermentation: A chemical change of organic matter brought about by the action of an enzyme or ferment.

flocculation: The coalescence of a finely-divided precipitate.

fumigant: A gaseous or readily volatilizable chemical used as a disinfectant or pesticide.

GAC: Granular Activated Carbon.

gravity concentration: A process which uses the differences in density to separate ore minerals from gangue.

gravity separation/settling: A process which removes suspended solids by natural gravitational forces.

grit removal: Preliminary treatment that removes large objects, in order to prevent damage to subsequent treatment and process equipment.

influent: A process stream entering the treatment system.

intake: Water, such as tap or well water, that is used as makeup water in the process.

lagoon: A shallow artificial pond for the natural oxidation of sewage or ultimate drying of the sludge.

LAP: Loading Assembly and Packing operations.

MGD: Million Gallons per Day.

MHF: Multiple Hearth Furnace.

NA: Not Analyzed.

ND: Not Detected.

neutralization: The process of adjusting either an acidic or a basic wastestream to a pH near seven.

NPDES: National Pollutant Discharge Elimination System.

NRDC: National Resources Defense Council.

NSPS: New Source Performance Standards.

photolysis: Chemical decomposition or dissociation by the action of radiant energy.

PCB: PolyChlorinated Biphenyl.

POTW: Publicly Owned Treatment Works.

PSES: Pretreatment Standards for Existing Sources.

purged: Removed by a process of cleaning; take off or out.

screening process: A process used to remove coarse and/or gross solids from untreated wastewater before subsequent treatment.

SIC: Standard Industrial Classification.

SS:. Suspended Solids.

SRT: Solids Retention Time.

starved air combustion: Used for the volumetric and organic reduction of sludge solids.

terpene: Any of a class of isomeric hydrocarbons.

thermal drying: Process in which the moisture in sludge is reduced by evaporation using hot air, without the solids being combusted.

TKN: Total Kjeldahl Nitrogen.

TOC: Total Organic Carbon.

trickling filter: Process in which wastes are sprayed through the air to absorb oxygen and allowed to trickle through a bed of rock or synthetic media coated with a slime of microbial growth to removed dissolved and collodial biodegradable organics.

TSS: Total Suspended Solids.

vacuum filtration: Process employed to dewater sludges so that a is produced having the physical handling characteristics and contents required for processing.

VSS: Volatile Suspended Solids.

WQC: Water Quality Criterion.